Recent Developments in Environmentally Sustainable and Cost-Effective Construction Materials

Recent Developments in Environmentally Sustainable and Cost-Effective Construction Materials

Editor

Ahmed Salih Mohammed

MDPI

Basel • Beijing • Wuhan • Barcelona • Belgrade • Novi Sad • Cluj • Manchester

Editor
Ahmed Salih Mohammed
University of Sulaimani
Sulaimaniyah, Iraq

Editorial Office
MDPI
St. Alban-Anlage 66
4052 Basel, Switzerland

This is a reprint of articles from the Special Issue published online in the open access journal *Sustainability* (ISSN 2071-1050) (available at: https://www.mdpi.com/journal/sustainability/special_issues/costeffective_construction_materials).

For citation purposes, cite each article independently as indicated on the article page online and as indicated below:

Lastname, A.A.; Lastname, B.B. Article Title. *Journal Name* **Year**, *Volume Number*, Page Range.

ISBN 978-3-0365-9403-3 (Hbk)
ISBN 978-3-0365-9402-6 (PDF)
doi.org/10.3390/books978-3-0365-9402-6

Contents

sustainability

MDPI

Article

Soft Computing and Machine Learning-Based Models to Predict the Slump and Compressive Strength of Self-Compacted Concrete Modified with Fly Ash

Dilshad Kakasor Ismael Jaf

Engineering Department, College of Engineering, Salahaddin University, Erbil 44002, Iraq; dilshad.jaf@su.edu.krd

Abstract: Self-compacted concrete (SCC) is a special type of concrete; it is a liquid mixture appropriate for structural elements with excessive reinforcement without vibration. SCC is commonly produced by increasing the paste volume and cement content. As cement production is one of the huge factors in releasing CO_2 gas into the atmosphere, by-product materials such as fly ash are utilized as a cement replacement in concrete. In addition to the positive environmental impact, fly ash can maintain an excellent fresh and mechanical property. Incorporating fly ash into self-compacted concrete is widely applied in practice. However, its application is frequently limited by a lack of knowledge about the mixed material gained from laboratory tests. The most significant mechanical property for all concrete types is compressive strength (CS); also, the slump flow diameter (SL) in the fresh state is a crucial property for SCC. Hence, developing an accurate and reliable model for predicting the CS and SL is very important for saving time and energy, as well as lowering the cost. This research study proposed a projection of both the CS and SL of SCC modified with fly ash by three different model approaches: Nonlinear regression (NLR), Multi-Linear regression (MLR), and Artificial Neural Networks (ANN). In this regard, two different datasets were collected and analyzed for developing models: 308 data samples were used for predicting the CS, and 86 data samples for the SL. Each database included the same five independent parameters. The ranges for CS prediction were: cement (134.7–583 kg/m^3), water-to-binder ratio (0.27–0.9), fly ash (0–525 kg/m^3), sand (478–1180 kg/m^3), coarse aggregate (578–1125 kg/m^3), and superplasticizer (0–1.4%). The dependent parameter (CS) ranged from 9.7 to 81.3 MPa. On the other hand, the data ranges for the SL prediction included independent parameters such as cement (83–733 kg/m^3), water-to-binder ratio (0.26–0.58), fly ash (0–468 kg/m^3), sand (624–1038 kg/m^3), coarse aggregate (590–966 kg/m^3), and superplasticizer (0.087–21.84%). Also, the dependent parameter (SL) ranged from 615 to 800 m. Various statistical assessment tools, such as the coefficient of determination (R^2), Root Mean Squared Error (RMSE), Mean Absolute Error (MAE), Objective value (OBJ), and Scatter Index (SI), were used to evaluate the performance of the developed models. The results showed that the ANN model best predicted the CS and SL of SCC mixtures modified with fly ash. Furthermore, the sensitivity analysis demonstrated that the cement content is the most effective factor in predicting the CS and SL of SCC mixtures.

Keywords: self-compacting concrete; mix proportion; slump flow; compressive strength; modeling

Citation: Ismael Jaf, D.K. Soft Computing and Machine Learning-Based Models to Predict the Slump and Compressive Strength of Self-Compacted Concrete Modified with Fly Ash. *Sustainability* **2023**, *15*, 11554. https://doi.org/10.3390/su151511554

Academic Editor: Ahmed Salih Mohammed

Received: 2 July 2023
Revised: 14 July 2023
Accepted: 17 July 2023
Published: 26 July 2023

1. Introduction

Self-compacted concrete (SCC) is one of the special types of concrete with a high viscosity without any requirement for compaction. SCC is vital in the concrete industry due to its numerous benefits. The main idea of SCC is to increase the amount of paste volume to increase the rheological property of the mixture. Thus, the SCC can spread completely inside the formwork by its weight without any bleeding or segregation. Thus, the labor cost will be eliminated. However, utilizing excessive cement leads to a higher cost of SCC production [1–3]. SCC is an excellent choice for congestion structural elements due to its high flowability. The high flowability and stability of SCC can be achieved by utilizing different mineral and chemical admixtures; fly ash and superplasticizer, respectively, are commonly used [4]. Fly ash is one of the common replacements for cement in concrete. It can provide lower costs by reducing the utilization of cement and improving the mixture's flowability due to its rounded shape. One of the advantages of SCC over the other special concretes is that SCC does not require any special curing method; the same methods and procedures for normal concrete can also be used for SCC [5].

In the fresh state, the slump flow diameter (SL) of SCC is an important property that should be checked. Also, among the mechanical properties in the hardened state, the compressive strength (CS) of SCC is one of the significant parameters in the design of engineering structures. Other mechanical properties and the durability of SCC have a direct or indirect relation to compressive strength, and they can be derivate from the CS [6,7].

The term 'water-to-binder ratio' (w/b) refers to the ratio of water to cement plus pozzolanic material. One of the most common and most-well known pozzolanic materials is fly ash, which has been used to improve some significant properties of self-compacted concrete, such as workability. The compressive strength of SCC is improved by increasing the water-to-binder ratio [8]. Karamoozian et al. [9] noticed that, for the same water content, increasing the cement content from 360 to 450 kg/m^3 increased the compressive strength of SCC from 48 to 52 MPa at 28 days of curing. Similar findings have been reported in other studies [10,11]. The w/b is one of the significant factors influencing the compressive strength of SCC, similar to normal concrete. For SCC, the ratio is normally lower than normal concrete. A lower w/b refers to lower water content and a higher amount of binder materials, leading to a higher CS and a more homogeneous matrix. Ahmadi et al. [12] stated that the w/b ratio has a greater impact on the compressive strength of normal concrete than on the self-compacting compressive strength. Naderpour and Abbasi [13] investigated the effect of different w/b ratios on the compressive strength of SCC. It was noticed that lowering the w/b ratio from 0.45 to 0.35 improved the CS of SCC at all curing ages (from 3 to 56 days).

Since the volume of aggregates in the SCC is about 60–70% of its total volume, aggregate greatly influences the rheological and mechanical properties of SCC. Compared to conventional concrete, lower coarse aggregate content is utilized in the production of SCC [1,3,14]. The study by [15] showed that the coarse aggregate had a greater impact on the CS of SCC than the fine aggregate content. However, the fine aggregate has more influence on the homogeneity and viscosity of the SCC mixture [16].

In addition to adding pozzolanic materials, SCC production is impossible without superplasticizers. This is due to the utilization of high cement and binder content in SCC. Therefore, high workability (slump flow diameter) can be achieved in SCC only when the SP content is increased. Adding SP to the SCC mixture improves the rheological performance but has different effects on the compressive strength based on the SP dosage. A previous study by Sor N. A. [17] concluded that increasing the SP dosage from 5.5 kg/m^3 to 8.25 kg/m^3 increased the compressive strength of SCC from 35 MPa to 45.5 MPa at 28 days.

As illustrated above, the compressive strength of SCC is sensitive and affected by several parameters. Therefore, apart from experimental work in the laboratory, which is quite costly, it is important to utilize more advanced techniques to predict the CS value through various statistical tools and numerical equations. Soft computing techniques [18–22] are currently one of the most common and ideal approaches for predicting measured values; they may be a good alternative for solving linear or nonlinear problems where mathematical models cannot simply indicate the relationship between the involved parameters in the problem [23]. Mohammed et al. [24] created a systematic multiscale model to predict the CS of fly ash-based concrete. A total of 450 experimental data were used for modeling in their study. For the qualifications, the Linear regression (LR), Nonlinear regression (NLR), Multi-Linear regression (MLR), M5P-tree, and Artificial Neural Network (ANN) models were all developed. It was concluded that the MLR, M5P-tree, and ANN models were the most accurate and reliable in predicting the CS of high-volume fly ash concrete, with higher R^2 values and lower RMSE and MAE values.

In this study, two different databases of fly ash-based self-compacted concrete mixtures with the same parameters were prepared. The first database consisted of 308 data samples of SCC mixtures, and it was used to predict the compressive strength; the second one had 86 data points used to predict the slump flow diameter of SCC. Thus, the CS and SL are were dependent parameters predicted independently using collected databases. The independent parameters of SCC were in various ranges, which included cement (C), water-to-binder ratio (w/b), fly ash (FA), sand (S), coarse aggregate (CA), and superplasticizer (SP).

As fly ash has a wide range of applications for producing SCC mixtures, a reliable and accurate model cannot be found in the literature regarding the efficient use of FA in SCC mixtures for the construction industry. As a result, this study attempted to evaluate and quantify the effect of a wide range of mixture proportions on the CS and SL of SCC, including the cement content (kg/m^3), water-to-binder ratio, fly ash content (kg/m^3), sand content (kg/m^3), coarse aggregate content (kg/m^3), and superplasticizer percentage amount.

Three different model techniques, namely NLR, MLR, and ANN, were performed to predict the CS [25–28] and SL of SCC modified with FA using the databases collected from the literature. Furthermore, the most accurately developed model was applied to different CS and SL ranges. The compressive strength ranges were as follows: low-strength concrete (LSC), lower than 20 MPa; Normal-strength concrete (NSC), between 20 and 55 MPa; and High-strength concrete (HSC), CS greater than 50 MPa [29]. The slump flow diameter was divided into three different classes: the SL from 550 to 650 mm (Class 1), SL between 650 and 750 mm (Class 2), and SL of greater than 750 mm (Class 3) [5]. In addition, different assessment criteria, such as the Correlation Coefficient (R^2), Mean Absolute Error (MAE), Root Mean Squared Error (RMSE), Objective (OBJ), and Scatter Index (SI), were used to evaluate the performance of the developed models.

Research Objectives

This study aims to evaluate the effect of fly ash on the compressive strength and slump flow diameter of self-compacted concrete mixtures collected from literature; the following are the main objectives:

I. Perform statistical analysis to determine the influence of concrete ingredients, such as the cement, water-to-binder ratio, fly ash, sand, coarse aggregate, and superplasticizer, on self-compacted concrete's compressive strength and slump flow diameter.
II. Provide a systematic multiscale model and propose to predict the compressive strength and slump flow diameter of self-compacted concrete containing up to 70% of fly ash, with a variety of cement, sand, and coarse aggregate content, as well as different water-to-binder ratios and superplasticizer percentages.
III. Apply the most accurately developed model on different compressive strength ranges and slump flow diameter classes.
IV. As an alternative to the developed model techniques (NLR, MLR, and ANN), determine the most reliable and accurate model based on different statistical assessment criteria to predict the CS and SL of fly ash-based self-compacted concrete.
V. The overall and main objective of the current study is to model compressive strength as one of the significant mechanical properties of concrete and slump flow diameter as a fresh state property of SCC modified with different FA content.

2. Methodology

2.1. Data Collection

Two databases were prepared for each targeted parameter to develop a reliable and applicable model to predict both the compressive strength and slump flow diameter of self-compacted concrete modified with fly ash. The independent parameters included six main parameters: cement content (kg/m^3), water-to-binder ratio, fly ash content (kg/m^3), sand (kg/m^3), coarse aggregate (kg/m^3), and superplasticizer (%). For the compressive strength and slump flow, 308 data samples [30–51] and 86 data samples [31,37,38,41,42,50–63] were selected, respectively, as shown in Table 1. According to the literature [29,64–66], the datasets were divided into two groups, training (by 70%) and testing (by 30%), using the Rand Function. The training dataset included 216 samples for CS data, while the testing data included 92 samples to check the validity of the developed models based on the training datasets. The training and testing datasets for the SL database were determined as 60 samples and 26 samples, respectively. Figure 1 shows the methodology of the current study.

Table 1. Summary of collected data for CS and SL prediction in FA-modified SCC.

References	Cement, C (kg/m³)	Water-to-Binder Ratio (w/b)	Fly Ash, FA (kg/m³)	Sand, S (kg/m³)	Coarse Aggregate, CA (kg/m³)	Superplasticize SP (%)	Compressive Strength, CS (MPa)
[30]	134.7–540	0.27–0.9	0–525	487–1135	600–1125	0–1.36	9.74–79.19
[31]	160–280	0.34–0.45	120–240	808–1034	900	0.1–0.6	31–52
[32]	280–400	0.55–0.87	0–120	718–1042	850	0.12–0.75	13.3–41.2
[33,34]	183–317	0.38–0.65	100–261	478–919	837	0–1	Oct-43
[35]	533–583	0.31–0.33	50–215	813–835	745–766	0.24–0.46	50–81
[36]	161–247	0.35–0.45	159–254	842–866	843–864	0–0.4	26.2–38.0
[37]	250–427	0.31–0.59	90–257	768–988	659–923	0.09–0.9	47–66
[38]	220–440	0.32	110–330	686–714	881–917	0.62–0.69	48–70
[39]	300–350	0.38–0.4	150–200	830–845	860–876	0.818–0.827	21.6–26.5
[40]	380	0.38	20	1180	578	0.398	40.4
[41]	275–350	0.34–0.36	150–325	611–707	777–901	0.795–1.25	50–72
[42]	165–275	0.37–0.58	275–385	735–796	865–937	0.836–0.74	37.92–63.32
[43]	215	0.38	215	925	905	0.15	20.4
[44]	290	0.38	290	975	650	0.45	37.97
[45]	300	0.28	300	787	720	0.33	52.7
[46]	420	0.33	80	785	860	0.3	56
[47]	350	0.35	150	900	600	1.0	37.18
[48]	360	0.28	240	853	698	0.3	63.5
[49]	344–399	0.35	100–147	814	881–882	0.116–0.146	48.75–55
[50]	225	0.35	275	908	652	0.70	41.42
[51]	480	0.38	96	819	699	0.94	53
References	**Cement, C (kg/m³)**	**Water-to-binder ratio (w/b)**	**Fly ash, FA (kg/m³)**	**Sand, S (kg/m³)**	**Coarse aggregate, CA (kg/m³)**	**Superplasticize SP (%)**	**Slump flow diameter, SL (mm)**
[52]	450–480	0.40–0.45	0–144	890	810	4.8–13.3	650–695
[50]	500	0.35	0–275	908–967	652–694	0.7–8	630–700
[37]	220–427	0.31–0.41	90–330	686–988	659–923	0.18–0.9	670–749
[38]	550	0.32–0.44	0–110	728–826	855–935	3.2–8.43	670–675
[53]	530	0.45	0–265	768	668	0.09–4.55	660–690
[41]	83–385	0.31–0.41	165–468	624–732	794–931	1–1.25	680–800
[31]	430–450	0.36–0.39	202.5–232.2	872–808	900	1.58–2.15	680–710
[54]	465–550	0.41–0.44	83–193	910	590	0.97–11	635–690
[55]	450–500	0.39–0.43	135–225	724–789	850–926	2.5–6.15	640–680
[56]	500	0.35	150–250	900	600	10.5–11	660–680
[57]	550	0.41–0.44	83–193	910	590	9.91–11.01	633–690
[58]	180–270	0.44	180–270	788–801	829–842	0.27–0.28	720–730

Table 1. *Cont.*

References	Cement, C (kg/m³)	Water-to-binder ratio (w/b)	Fly ash, FA (kg/m³)	Sand, S (kg/m³)	Coarse aggregate, CA (kg/m³)	Superplasticizer SP (%)	Slump flow diameter, SL (mm)
[42]	165–385	0.29–0.58	165–385	735–821	865–966	0.74–0.84	670–730
[59]	567–670	0.26–0.31	0–156	656–846	729–875	12.39–21.84	615–655
[60]	733	0.26	271.21	748	698	8.40	660
[49]	399	0.35	100	814	882	0.146	690
[61]	500	0.35	0	1038	639	6.75	665
[51]	480	0.38	96	819	699	0.94	680
[62]	437	0.34	80	743	924	0.43	700
[63]	321.75	0.36	173.25	862.45	729.18	0.545	696

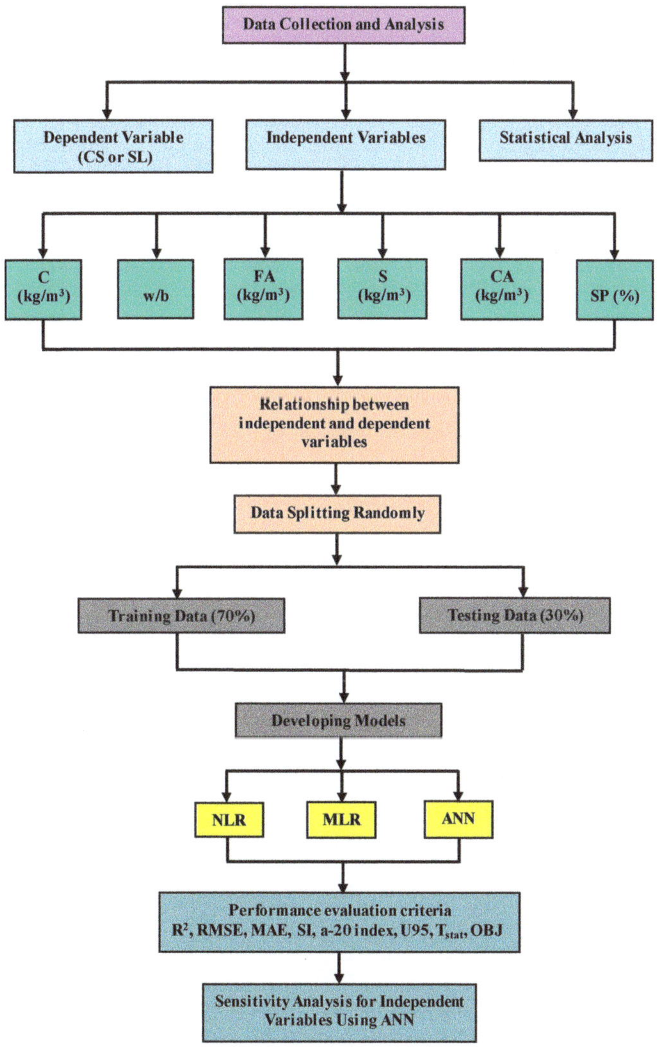

Figure 1. Flowchart of the study.

2.2. Pre-Processing

The pre-processing technique is a vital step before using a dataset. The process can improve the performance of created models. In the current study, each independent variable was converted to a value between zero and one utilizing Equation (1) [29,67]. The converted values in both the CS and SL databases are demonstrated in the box plot (Figure 2).

$$N_f = \frac{(N_i - N_i min)\left(N_f max - N_f min\right)}{(N_i max - N_i min)} + N_f min \quad (1)$$

where N_i represents the old value and N_f represents the new value. The $N_f min$ is zero and $N_f max$ is one.

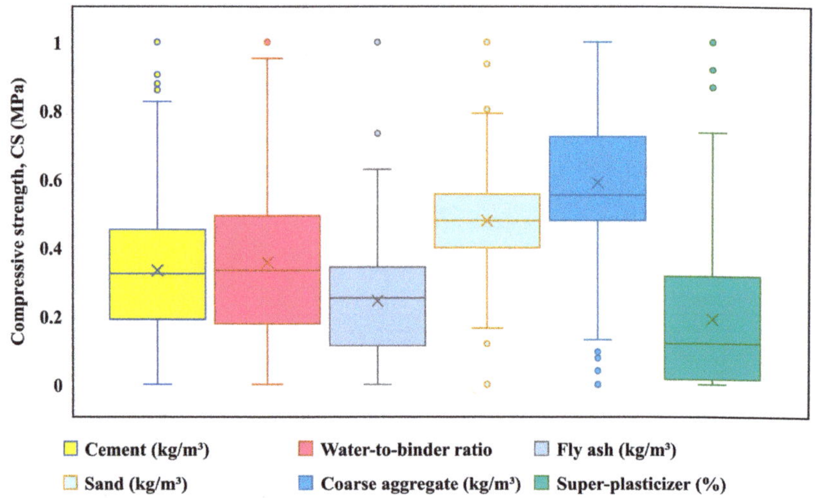

(a)

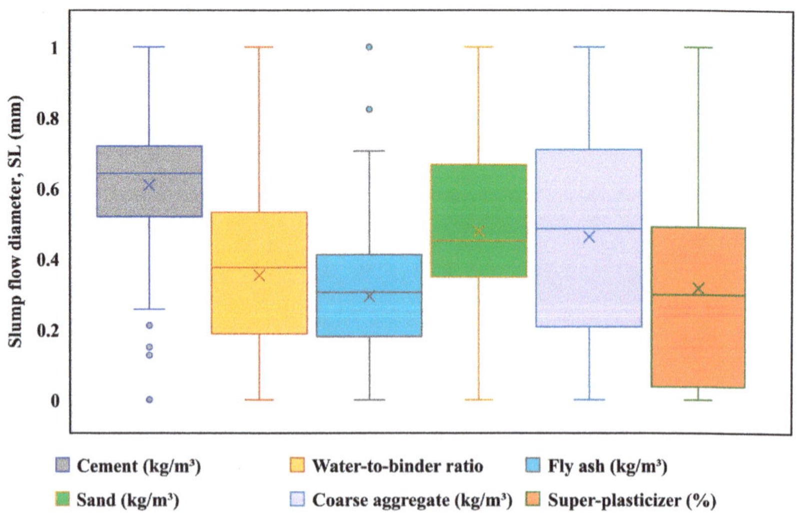

(b)

Figure 2. Box plot of the independent parameters in (**a**) the CS dataset and (**b**) the SL dataset.

2.3. Statistical Evaluation

The data collected to predict the compressive strength and slump flow diameter of SCC mixtures were statistically analyzed to determine the relationship between each independent variable (C, w/b, FA, S, CA, and SP) with the dependent variables (CS and SL). In this regard, statistical functions such as Mean, Median, Mode, Standard Deviation (SD), Skewness (Skew), Kurtosis (Kur), Variance (Var), Maximum (Max), and Minimum (Min) were calculated. Skewness depicts the distribution of variables, whether positive or negative, on the right or left, with positive values on the right and negative values on the left. A negative value for kurtosis indicates a short distribution tail, whereas a positive value indicates a longer tail. The statistical analysis is summarized in Table 2 for both the CS and SL datasets. The histogram of each independent variable and its relationship with compressive strength and the slump flow are plotted in Figures 3 and 4, respectively.

Table 2. Summary of the statistical analysis of the FA-modified SCC parameters for the CS and SL prediction.

	Variables	C (kg/m^3)	w/b	FA (kg/m^3)	S (kg/m^3)	CA (kg/m^3)	SP (%)	CS (MPa)
Compressive strength database	Mean	283.9	0.5	128.3	813.5	900.7	0.3	36.6
	Median	279.8	0.5	133	813.5	881	0.2	34.5
	Mode	250	0.55	0	916	837	0	49
	SD	87.78	0.13	86.4	95.24	109.26	0.28	15.08
	Var	7705.83	0.02	7465.52	9070.26	11,937.73	0.08	227.5
	Kurt	0.2227	−0.1085	1.0307	2.2695	0.2755	0.9795	−0.2692
	Skew	0.5491	0.5752	0.4626	0.2461	−0.0674	1.1913	0.4736
	Min	134.7	0.27	0	478	578	0	9.7
	Max	583	0.9	525	1180	1125	1.4	81.3
Slump flow database	Variables	C (kg/m^3)	w/b	FA (kg/m^3)	S (kg/m^3)	CA (kg/m^3)	SP (%)	SL (mm)
	Mean	478.3	0.37	137.7	821.5	763.5	6.97	674.9
	Median	500	0.38	142.9	810.5	772	6.58	675
	Mode	550	0.35	0	910	590	4.55	680
	SD	122.18	0.065	91.65	82.42	113.97	5.94	31.52
	Var	14,928.1	0.00428	8399.5	6792.83	12,989.31	35.33	993.62
	Kurt	0.8148	−0.2208	1.2492	−0.5606	−1.2545	−0.5945	1.9597
	Skew	−0.8308	−0.0041	0.6296	0.0668	−0.1665	0.569	0.8086
	Min	83	0.26	0	624	590	0.087	615
	Max	733	0.58	468	1038	966	21.84	800

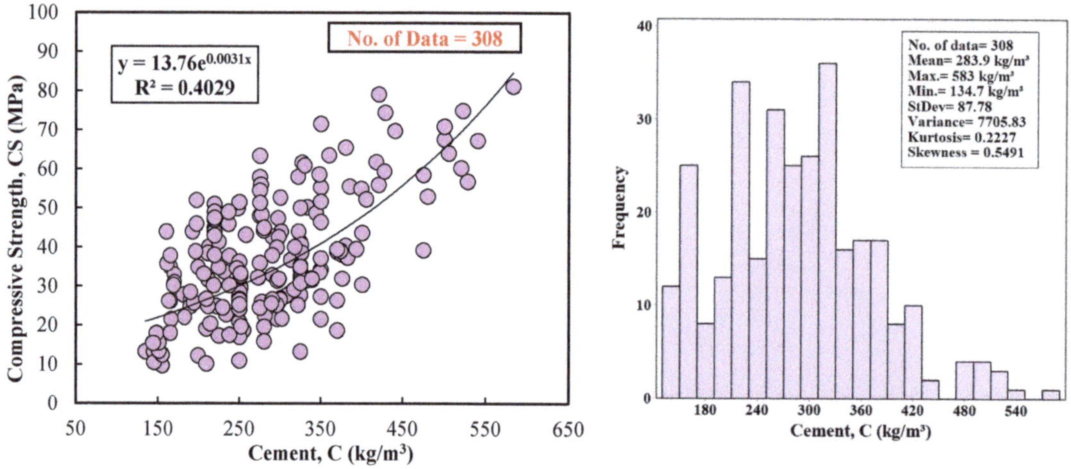

(**a**)

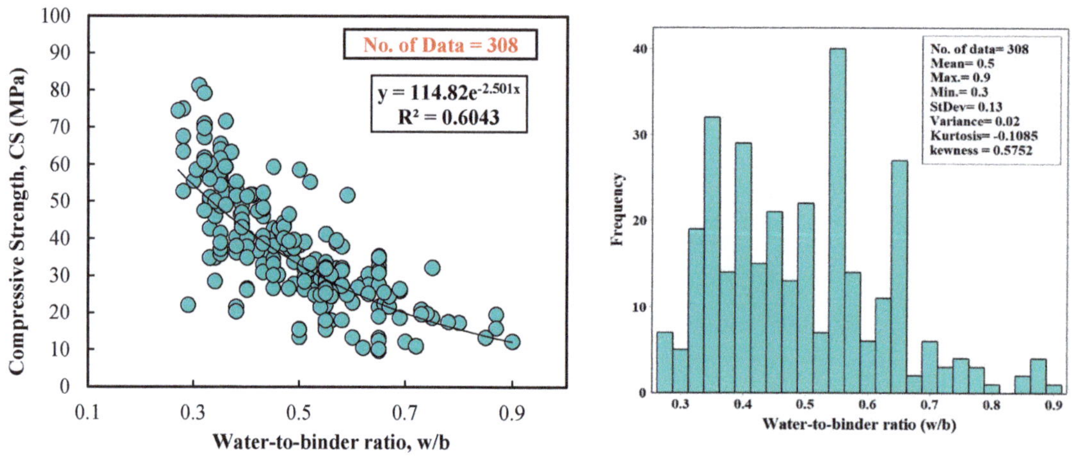

(**b**)

Figure 3. *Cont.*

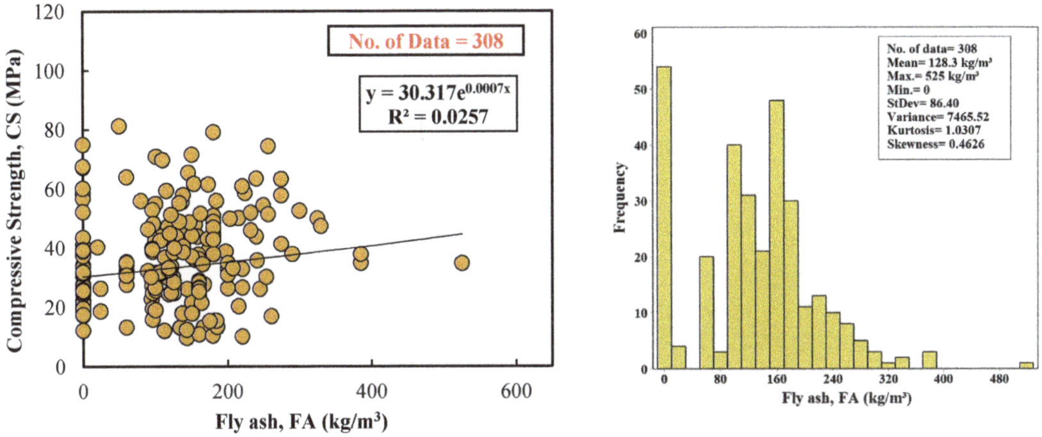

(c)

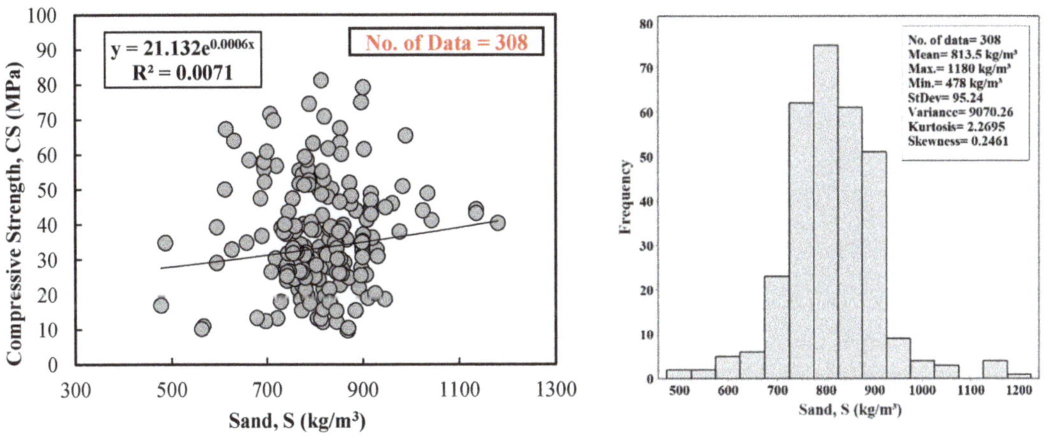

(d)

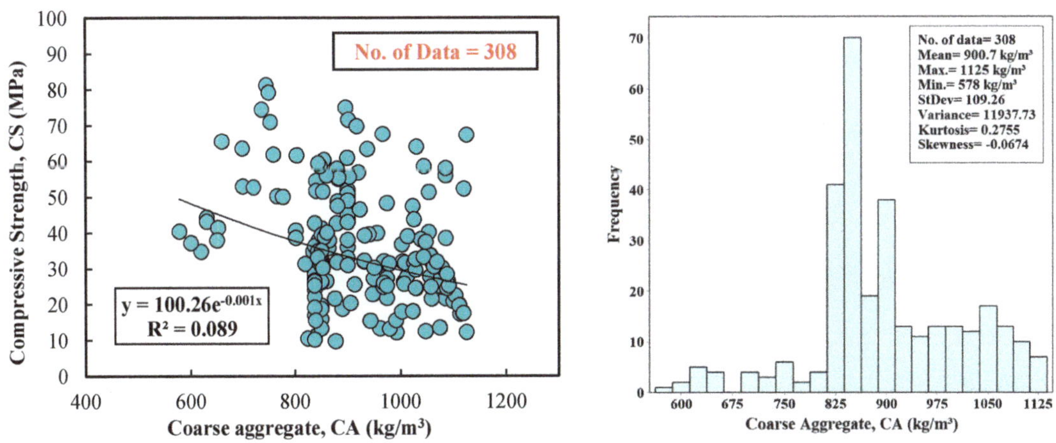

(e)

Figure 3. *Cont.*

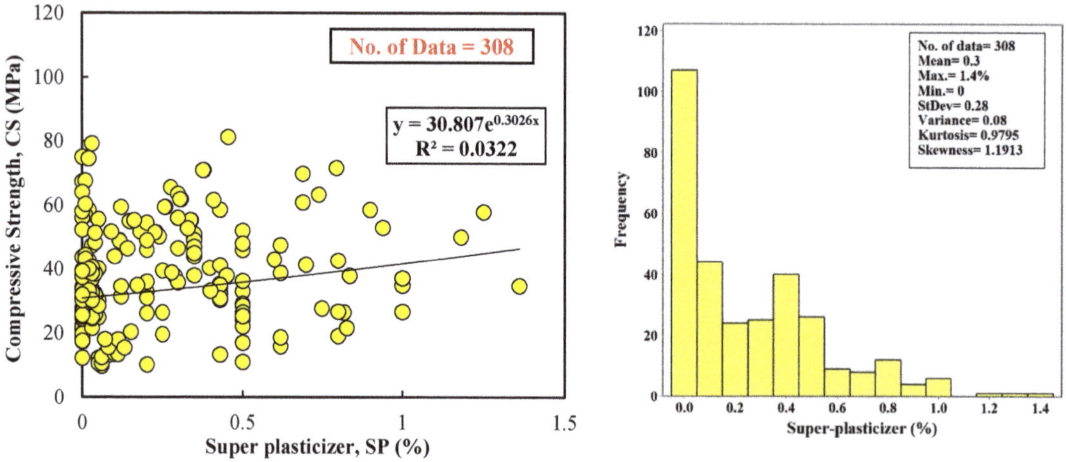

(f)

Figure 3. Histogram and Marginal plots for the compressive strength of FA-modified SCC with (**a**) cement (kg/m^3), (**b**) water-to-binder ratio, (**c**) fly ash (kg/m^3), (**d**) fine aggregate (kg/m^3), (**e**) coarse aggregate (kg/m^3), and (**f**) superplasticizer (%).

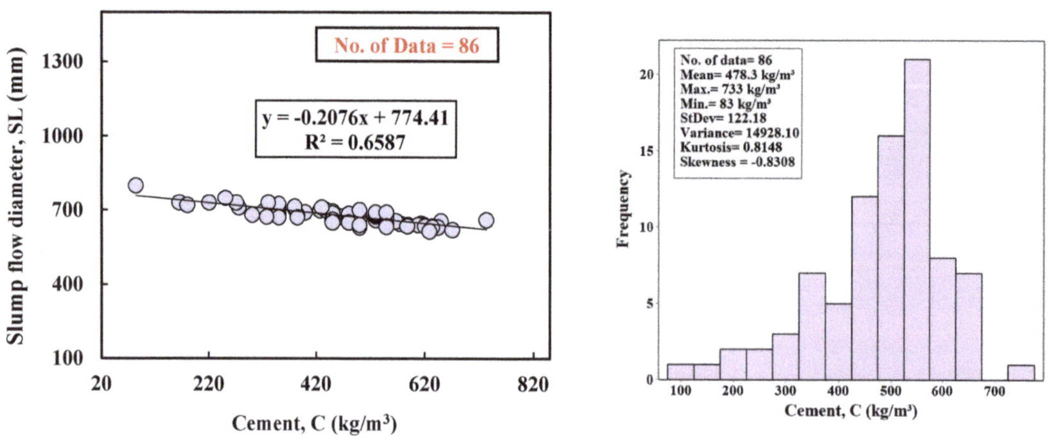

(a)

Figure 4. *Cont.*

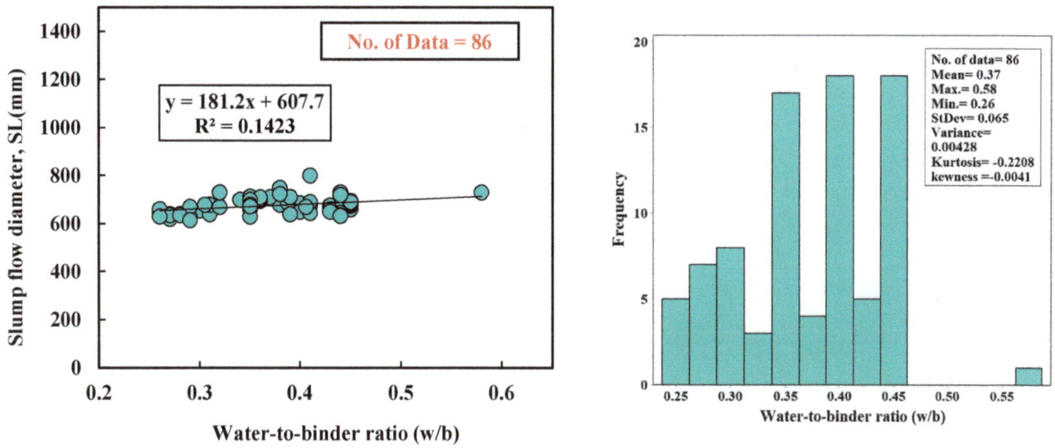

(b)

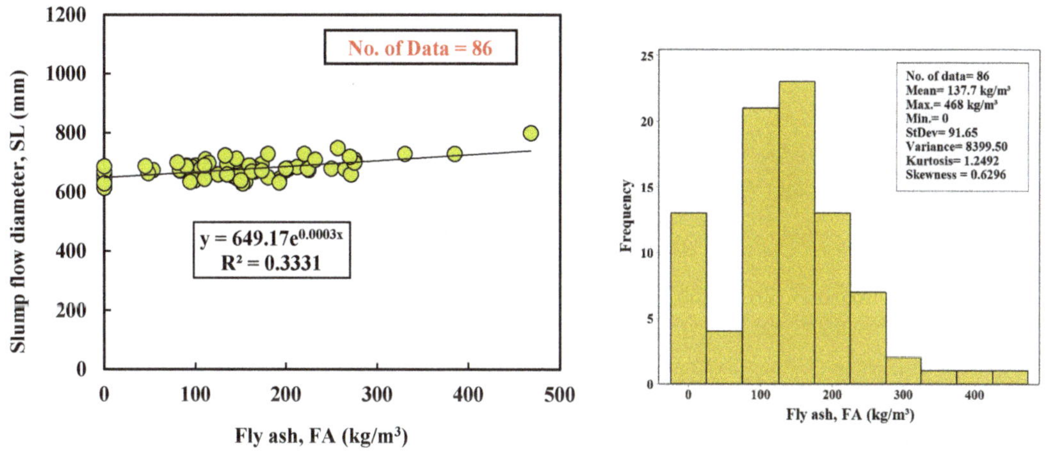

(c)

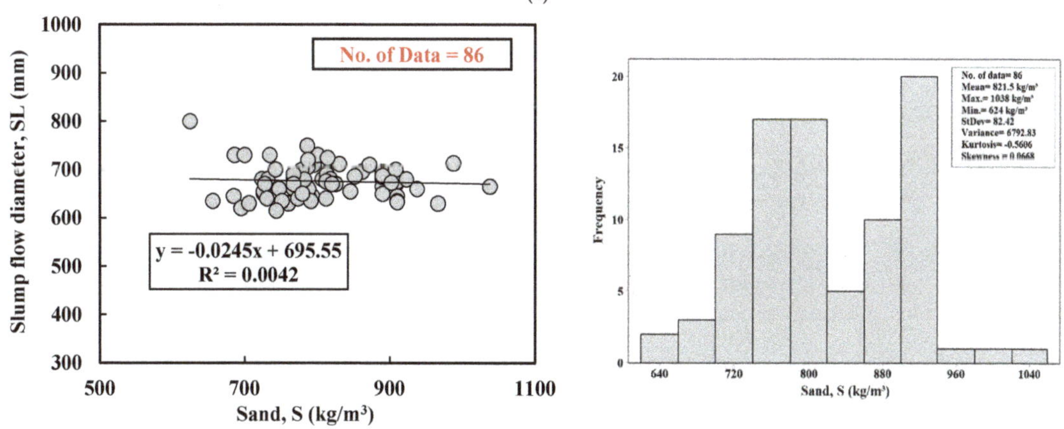

(d)

Figure 4. *Cont.*

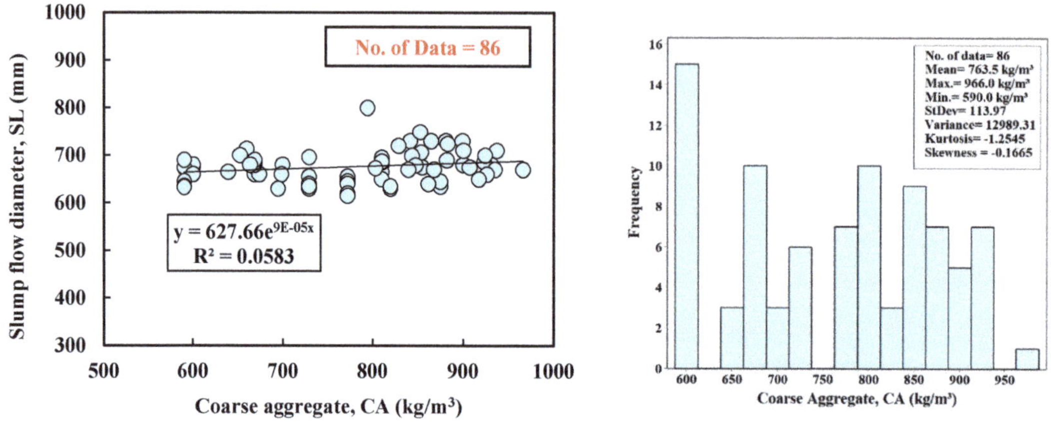

(e)

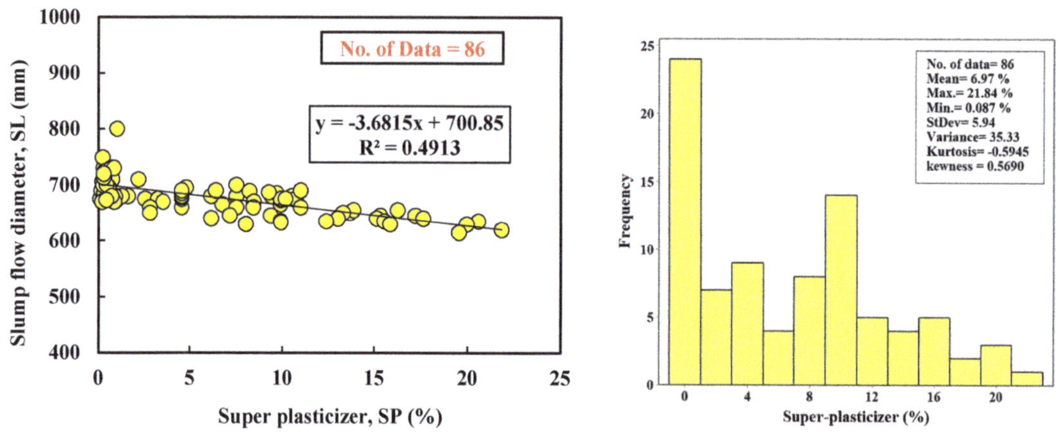

(f)

Figure 4. Histogram and Marginal plots for the slump flow diameter of FA-modified SCC with (**a**) cement (kg/m^3), (**b**) water-to-binder ratio, (**c**) fly ash (kg/m^3), (**d**) fine aggregate (kg/m^3), (**e**) coarse aggregate (kg/m^3), and (**f**) superplasticizer (%).

2.4. Modeling

As illustrated in Figure 5, the relationships between compressive strength or slump flow diameter and other compositions of FA-modified SCC mixtures, such as cement, water-to-binder ratio, fly ash, sand, coarse aggregate, and superplasticizer, were obtained. Based on the correlation matrix, a good correlation between cement and compressive strength was observed, which was 0.632. However, a poor correlation was noted for the other variables. The correlations were −0.748, 0.161, 0.082, −0.301, and 0.185, respectively. On the other hand, the relationships between independent parameters and the slump flow diameter were determined. The highest correlation between the FA and the SL was 0.572. However, poor relationships were found between C, w/b, S, CA, SP, and SL. The relations were −0.814, 0.397, −0.052, 0.236, and −0.705, respectively.

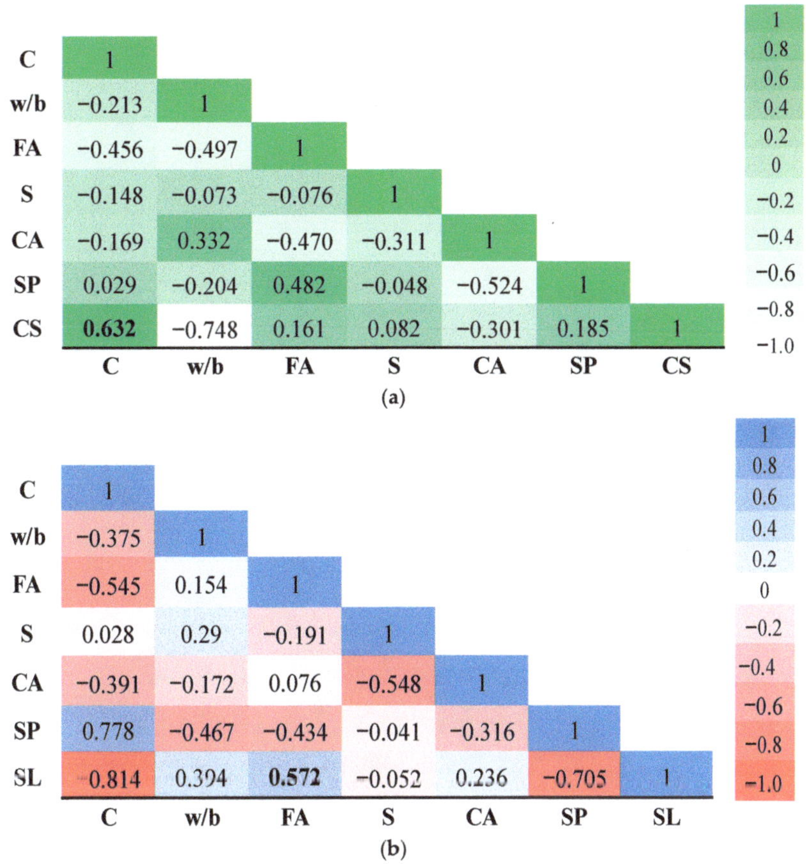

Figure 5. Correlation matrix plot between the dependent and independent variables of FA-modified SCC based on (**a**) CS and (**b**) SL.

Therefore, three models were proposed below to evaluate the effect of the various mixture proportions mentioned above on the CS and SL of SCC modified with FA. This study used NLR, MLR, and ANN models to predict the CS and SL of SCC mixtures. The most accurate and reliable model was proposed based on the following criteria: the model must be scientifically valid, with a lower percentage error between the measured and predicted data and a lower RMSE, MAE, OBJ, SI, and a higher R^2 value.

2.4.1. Nonlinear Regression (NLR) Model

Equation (2) can be used as a general form to develop a nonlinear regression model to determine [24,68] both the compressive strength and slump flow diameter, including the fly ash content and the self-compacted concrete components. The model was developed for each dependent parameter independently using the mentioned collected database. NLR is an advanced representation of the MLR model that is accurate and reliable. However, the model has a disadvantage in the form of mathematical complications [69].

$$CS, SL = \alpha_1(C)^{\alpha_2} + \alpha_3(w/b)^{\alpha_4} + \alpha_5(FA)^{\alpha_6} + \alpha_7(S)^{\alpha_8} + \alpha_9(CA)^{\alpha_{10}} + \alpha_{11}(SP)^{\alpha_{12}} \quad (2)$$

where $\alpha_1, \alpha_2, \alpha_3 \ldots \alpha_{11}$ and α_{12} are the model parameters. CS, SL, w/b, FA, S, CA, and SP are the compressive strength (MPa), slump flow diameter (mm), cement (kg/m^3), water-to-binder ratio, fly ash (kg/m^3), sand (kg/m^3), coarse aggregate (kg/m^3), and superplasticizer (%), respectively.

2.4.2. Multi-Linear Regression (MLR) Model

The Multi-Linear Regression model can predict the compressive strength and slump flow diameter of the fly ash-modified self-compacted concrete with different mix design components. The equation of MLR includes the product of significant parameters affecting the CS and SL of the self-compacted concrete in exponential and constant terms (Equation (3)). The MLR model has several advantages, including simple mathematical operation and ease of implementation. However, this model is of poor quality because it is highly dependent on the number of data used; fewer data points provide less accuracy [70].

$$CS, SL = \alpha_1(C)^{\alpha_2} * (w/b)^{\alpha_3} * (FA)^{\alpha_4} * (S)^{\alpha_5} * (CA)^{\alpha_6} * (SP)^{\alpha_7} \quad (3)$$

where the CS, SL, C, w/b, FA, S, CA, and SP are the compressive strength, slump flow diameter, cement (kg/m^3), water-to-binder ratio, fly ash (kg/m^3), sand (kg/m^3), coarse aggregate (kg/m^3), and superplasticizer (%), respectively. In addition, $\alpha_1, \alpha_2, \alpha_3, \alpha_4, \alpha_5, \alpha_6$, and α_7 are the model parameters.

2.4.3. Artificial Neural Network (ANN) Model

The ANN model [71–75] is a computer system of artificial neurons that function as fundamental units and mimic the parallel processes to analyze data like the human brain. The pattern of neuron connections influences the behavior of ANN networks, which also determines the network class. The ANN model can handle a mapping problem by estimating the relationship between input and output variables and distinguishing it from other expert systems by learning automatically from the obtained training patterns [76]. The ANN model is a machine learning system used in construction engineering for various numerical predictions and challenges [76–79]. The model is constructed based on three layers, input, hidden, and output, linked by biases and weights [80,81]. Several parameters affect the final model result, such as the number of hidden layers and neurons, the training algorithm, and the transfer function [82]. The ANN structure can be discovered by tuning the required parameters through trial and error.

The current study designed a multi-layer feed-forward network with SCC components (C, w/b, FA, S, CA, and SP) as input and the CS or SL as output. In the output layer, a sigmoid activation function was utilized. The typical process of the ANN result is shown in Figure 6. Equation (4) can be considered as a general formula for the calculation of an ANN output with only one node:

$$Output = bias + \sum_{j=1}^{n} (x_j \times w_j) \quad (4)$$

$$\beta_n = a_n(C) + b_n(w/b) + c_n(FA) + d_n(S) + e_n(CA) + f_n(SP) \quad (4a)$$

$$CS, SL = \frac{Node1}{1 + e^{-\beta_1}} + \frac{Node2}{1 + e^{-\beta_2}} + \cdots + \frac{Node_n}{1 + e^{-\beta_n}} + Threshold \quad (4b)$$

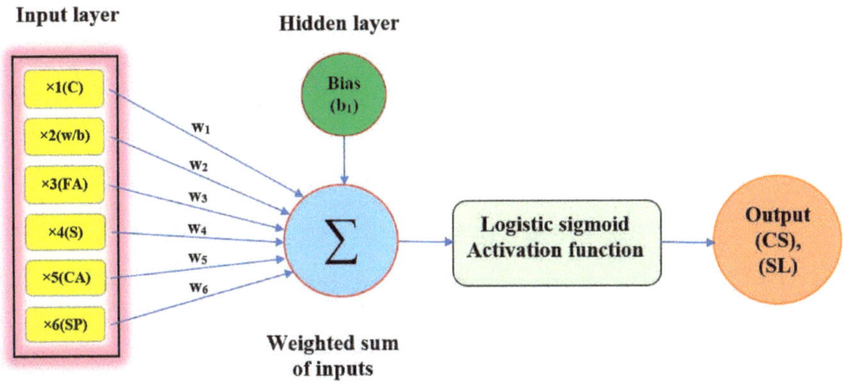

Figure 6. Typical ANN output procedure in a single node.

2.5. Metrics for Assessing Developed Models

The performance of the developed models was evaluated and characterized based on various assessment tools such as the Correlation Coefficient (R^2), Mean Absolute Error (MAE), Root Mean Squared Error (RMSE), Objective (OBJ), and Scatter Index (SI). The equations of these parameters are well defined in Table 3. R^2 is the squared correlation between the measured and predicted values. The greater the R-squared value, the more accurate the model. The average error made by models in predicting the outcome of the observation was measured by the RMSE. The OBJ function identified evaluation objectives based on the training and testing datasets and a variety of data.

Table 3. The performance evaluation criteria for the developed models.

Parameter	Equation	Range	Best Value
R^2 [58,80]	$R^2 = \left(\frac{\Sigma_i \left(vp-\bar{u}\right) \times \left(vi-\bar{v}\right)}{\sqrt{\Sigma_i \left(vp-\bar{u}\right)^2} \times \sqrt{\Sigma_i \left(vi-\bar{v}\right)^2}} \right)^2$	$0-1$	1
MAE [24,80]	$MAE = \frac{\sum_{i=1}^{n}(\lvert vi-vp \rvert)}{n}$	$0-\infty$	0
$RMSE$ [24,69]	$RMSE = \sqrt{\frac{\sum_{i=1}^{n}(vi-vp)^2}{N}}$	$0-\infty$	0
OBJ [58,80]	$OBJ = \left(\frac{n_{tr}}{n_{all}} \times \frac{MAE_{tr}+RMSE_{tr}}{R_{tr}^2+1}\right) + \left(\frac{n_{tst}}{n_{all}} \times \frac{MAE_{tst}+RMSE_{tst}}{R_{tst}^2+1}\right)$	$0-\infty$	0
SI [69,80]	$SI = \frac{RMSE}{vi}$	<0.1	Excellent
		0.1 to 0.2	Good
		0.2 to 0.3	fair
		>0.3	Poor

Notes: vp = predicted value of compressive strength or slump flow; vi = experimental value of compressive strength or slump flow; $\bar{u}$ = average of predicted compressive strength or slump flow; $\bar{v}$ = average of experimental compressive strength or slump flow data; n = number of the dataset (training or testing); n_{tst} = number of the testing dataset; n_{tr} = number of the training dataset; n_{all} = total number of datasets (including training and testing).

3. Results and Discussion

3.1. Relation between Predicted and Experimental Values

3.1.1. Nonlinear Regression (NLR) Model

The Nonlinear Regression model was utilized to predict the compressive strength and slump flow diameter of FA-modified SCC. The NLR model results for CS and SL prediction are shown in Equations (5) and (6), respectively. The relationship between the measured and predicted CS and SL values is displayed in Figure 7. When predicting the compressive strength, the training dataset was observed to have an R^2 of 0.81 and an RMSE value of 5.82 MPa. The testing dataset had an R^2 of 0.84 and RMSE of 7.67 MPa. In the testing dataset, the error line was from +50 to −25%, indicating that 25% of the data fell between 0.75 and 1.5 for the predicted to measured compressive strength ratio.

Regarding predicting the SL of FA-modified SCC, the training dataset had an R^2 of 0.82 and an RMSE of 11.6 mm. Also, the testing dataset had an R^2 of 0.57 and an RMSE of 27.4 mm. The error line was from +20 to −6% for the training data, indicating that 74% of the data fell between 0.94 and 1.2 for the predicted to experimental slump flow diameter ratio.

The NLR model provided nearly the same relationship value regarding R^2 based on the training dataset for both the CS and SL predictions.

$$CS = 14.1(C)^{0.32} + 1151.8(w/b)^{-0.03} + 0.42(FA)^{0.52} - 4.0(S)^{-14.45} - 5.1(CA)^{-13.96} - 1226(SP)^{0.00006} \quad (5)$$

No. of training dataset = 216, R^2 = 0.81, RMSE = 5.82 MPa

$$SL = 1242(C)^{-0.107} + 0.0001(w/b)^{0.0001} + 31.4(FA)^{0.026} - 0.0004(S)^{-0.0004} - 0.0002(CA)^{-0.0002} + 0.00001(SP)^{0.0002} \quad (6)$$

No. of training dataset = 60, R^2 = 0.82, RMSE = 11.6 mm

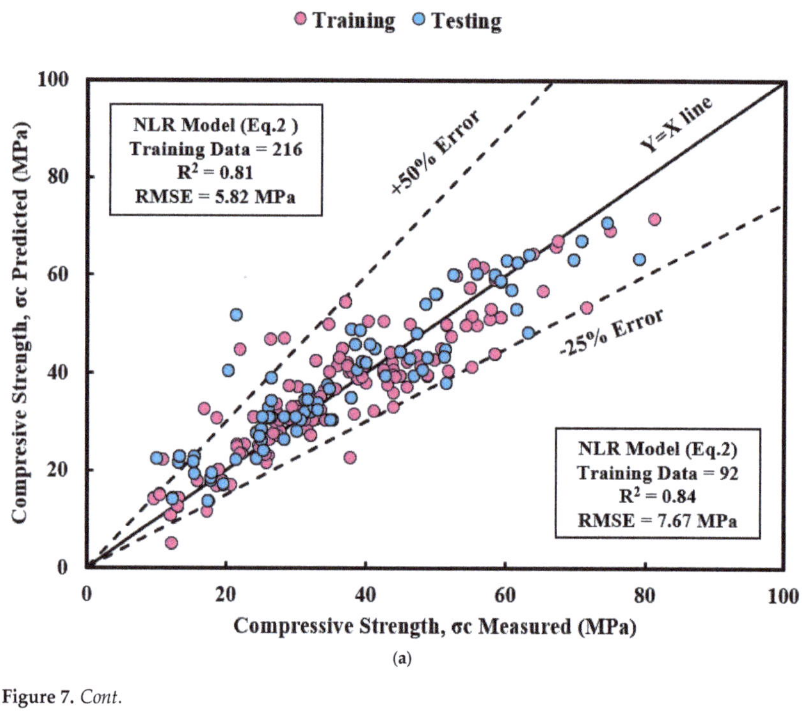

(a)

Figure 7. *Cont.*

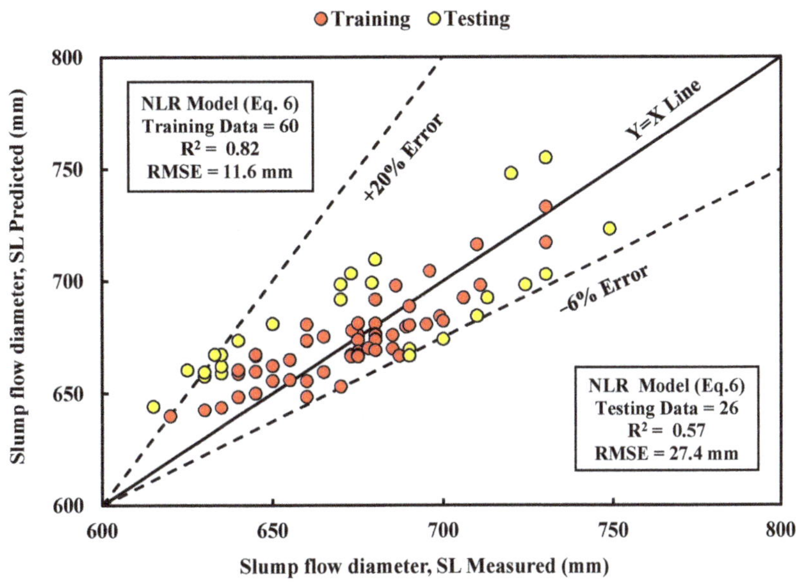

(**b**)

Figure 7. Comparison between the measured and predicted (**a**) CS and (**b**) SL using the NLR model for the training and testing datasets.

3.1.2. Multi-Linear Regression (MLR) Model

The Multi-Linear Regression model is another model used to predict the compressive strength and slump flow diameter of self compacted concrete modified with fly ash. The MLR model has a simple mathematical expression. Therefore, it is considered one of the least effective models. The MLR model formula comprises constant terms and terms raised to the power of constant variables. In predicting compressive strength, the variables and their relationships are detailed in Equation (7). The relationship between variables in predicting the slump flow diameter is presented in Equation (8). The relationships between the predicted CS and SL with the experimental values are illustrated in Figure 8.

As displayed in Figure 8a, while predicting the CS of FA-modified SCC, the R^2 and RMSE were 0.81 and 6.04 MPa for training and 0.82 and 7.92 MPa for testing data, respectively. The error line was from +30 to −25% for training, meaning 45% of the data fell between 0.75 and 1.3 for the predicted to measured compressive strength ratio. Figure 8b presents the MLR model results when predicting the SL. The training dataset has an R^2 of 0.86 and an RMSE of 10.3 mm. However, the testing dataset has an R^2 of 0.57 and an RMSE of 26.8 mm. The error line ranges for training from +12 to −15%, implying that 73% of the data falls between 0.85 and 1.12 for the predicted to measured slump flow ratio.

$$CS = 0.0000003(C)^{0.8}(w/b)^{-0.88}(FA)^{0.015}(S)^{0.89}(CA)^{1.1}(SP)^{0.0063} \tag{7}$$

No. of training dataset = 216, R^2 = 0.81, RMSE = 6.04 MPa

$$SL = 1200(C)^{-0.09}(w/b)^{0.04}(FA)^{0.0009}(S)^{-0.0054}(CA)^{0.0086}(SP)^{-0.0028} \tag{8}$$

No. of training dataset = 60, R^2 = 0.86, RMSE = 10.3 mm.

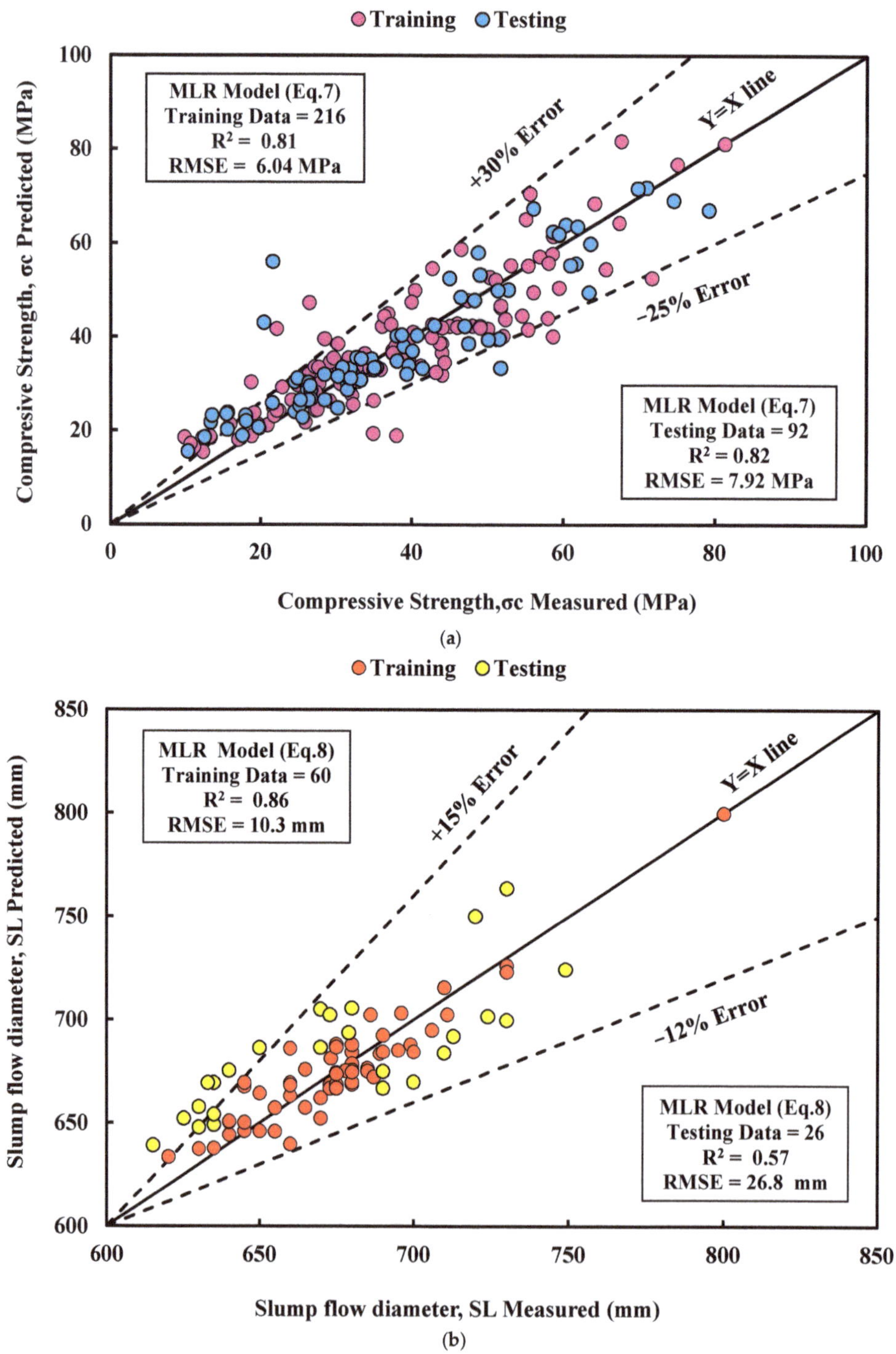

Figure 8. Comparison between the measured and predicted (**a**) CS and (**b**) SL using thr MLR model for the training and testing datasets.

3.1.3. Artificial Neural Network (ANN) Model

An Artificial Neural Network was the last model used to predict the compressive strength and slump flow diameter of FA-modified SCC. The ANN network structure is known to be an excellent model for prediction. The ANN model was developed for various trials. As shown in Figure 9, the five trials (4, 6, 8, 10, and 12) were chosen based on their RMSE and MAE values for both the CS and SL training datasets independently. Then, the network with one hidden layer and six neurons had the lowest RMSE and MAE values in both predictions. Therefore, the ANN network having one hidden layer with six neurons with a Learning rate of 0.2, a Learning time of 2000, and 0.1 of momentum was selected (Figure 10).

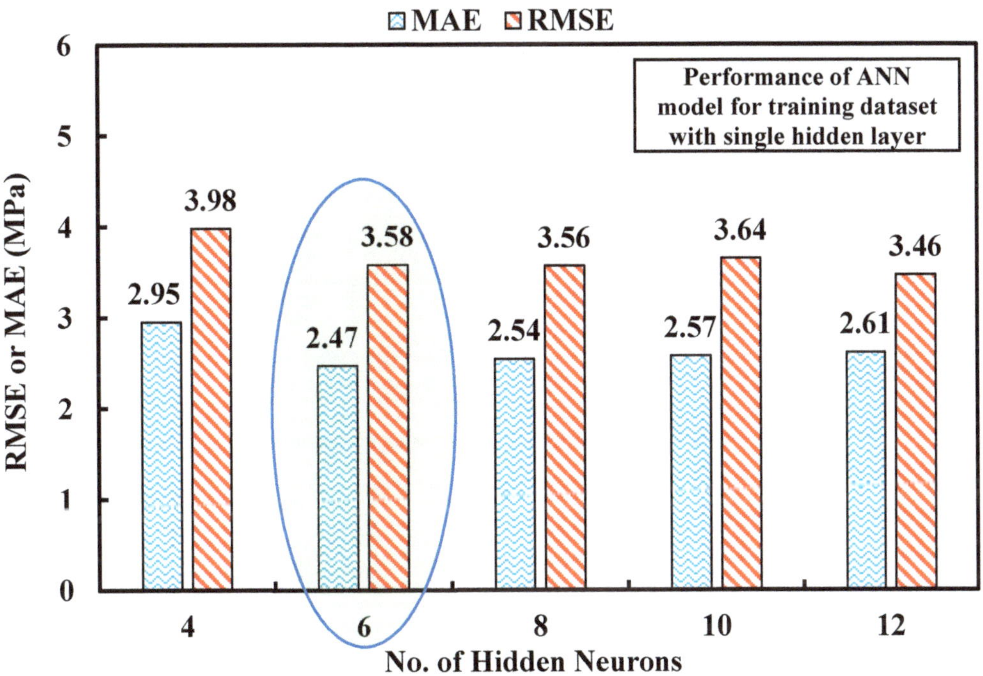

(a)

Figure 9. *Cont.*

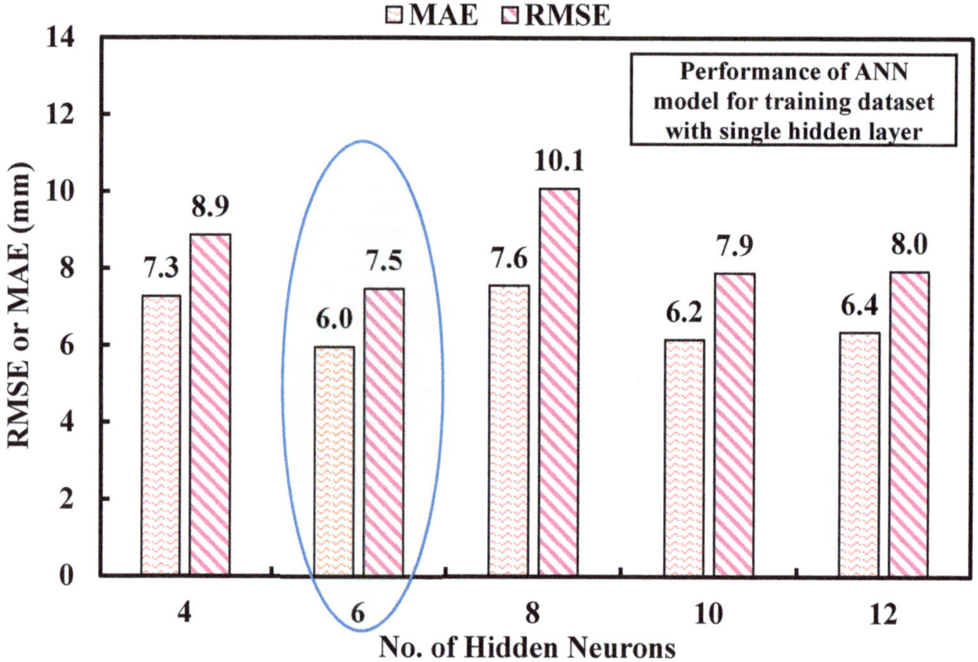

(b)

Figure 9. The RMSE and MAE values to select the optimum ANN. (**a**) CS and (**b**) SL.

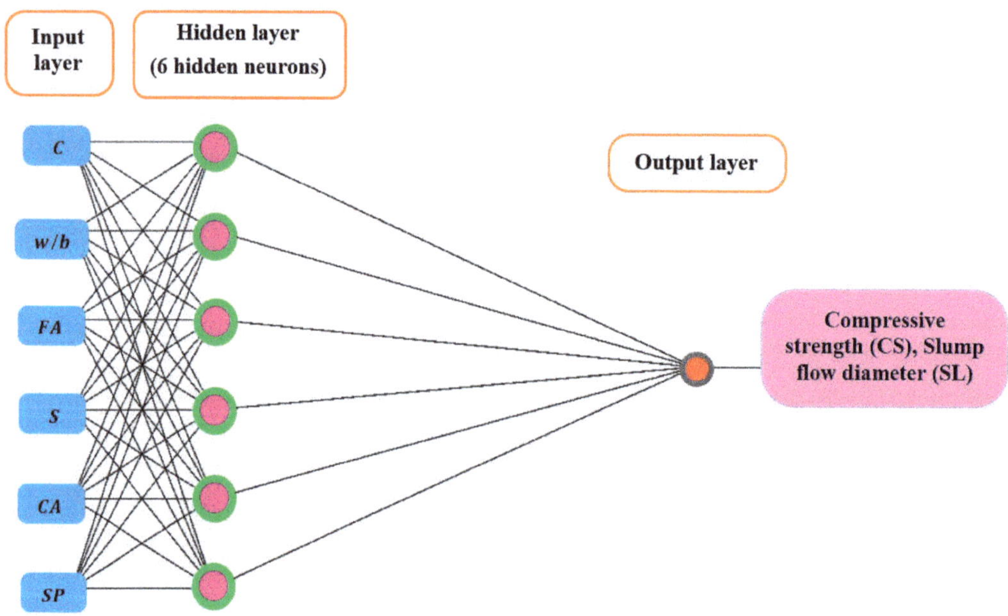

Figure 10. Optimal ANN network structures containing one hidden layer and six hidden neurons for predicting the CS and SL.

Equations (9) and (10) show the ANN formula for the CS and SL prediction, including weights and biases. Based on the training dataset, the ANN network analysis predicted a CS with an R^2 of 0.94 and an RMSE of 3.56 MPa. When the testing data were used, the model had an R^2 of 0.95 and an RMSE of 3.49 MPa. The training dataset had an error line from +20 to −20%, indicating that 80% of the data fell between 0.8 and 1.2 for the predicted to measured compressive strength ratio (Figure 11a). Concerning the SL prediction, the ANN network analysis predicted the SL with an R^2 of 0.93 and an RMSE of 7.5 mm when using the training data. However, when the tested data were used, the model had an R^2 of 0.997 and an RMSE of 2.2 mm. The error line for the training dataset was from +6 to −6%, indicating that 94% of the data fell between 0.94 and 1.06 for the predicted to measured SL ratio (Figure 11b).

$$\begin{bmatrix} 3.359 & 2.124 & 2.148 & 3.227 & -3.055 & -1.632 & -0.721 \\ -1.44 & -0.375 & 1.352 & 0.85 & -4.672 & 0.748 & -1.398 \\ -2.898 & 2.86 & -3.685 & -1.624 & -1.919 & -1.0 & -1.278 \\ 1.676 & 2.417 & -0.28 & 0.863 & -2.15 & -0.516 & -2.244 \\ 0.566 & 0.664 & 2.387 & -0.666 & 2.182 & -0.175 & -1.704 \\ -5.371 & 0.485 & -0.725 & -4.055 & 1.378 & -0.933 & -5.319 \end{bmatrix} \times \begin{bmatrix} c \\ w/b \\ FA \\ S \\ CA \\ SP \\ 1 \end{bmatrix} = \begin{bmatrix} \beta_1 \\ \beta_2 \\ \beta_3 \\ \beta_4 \\ \beta_5 \\ \beta_6 \end{bmatrix}$$

$$CS = \frac{1.567}{1+e^{-\beta_1}} - \frac{0.795}{1+e^{-\beta_2}} - \frac{0.848}{1+e^{-\beta_3}} - \frac{1.771}{1+e^{-\beta_4}} + \frac{1.392}{1+e^{-\beta_5}} - \frac{0.681}{1+e^{-\beta_6}} + 0.406 \quad (9)$$

No. of training dataset = 216, R^2 = 0.94, RMSE = 3.65 MPa.

$$\begin{bmatrix} -0.727 & -1.185 & -0.596 & 0.933 & -1.821 & -0.152 & -0.964 \\ 0.994 & 2.508 & 0.733 & 0.542 & 1.019 & 0.152 & -1.545 \\ 2.787 & -2.651 & -2.147 & -2.421 & 0.832 & -1.129 & -0.083 \\ -0.933 & -0.718 & 0.339 & 1.166 & 3.153 & 2.376 & -1.338 \\ 0.506 & 0.705 & 1.46 & 0.25 & -0.421 & -0.534 & -1.808 \\ -1.071 & 0.003 & 1.204 & -0.509 & -0.073 & 0.813 & -1.665 \end{bmatrix} \times \begin{bmatrix} c \\ w/b \\ FA \\ S \\ CA \\ SP \\ 1 \end{bmatrix} = \begin{bmatrix} \beta_1 \\ \beta_2 \\ \beta_3 \\ \beta_4 \\ \beta_5 \\ \beta_6 \end{bmatrix}$$

$$SL = -\frac{1.508}{1+e^{-\beta_1}} - \frac{1.372}{1+e^{-\beta_2}} - \frac{1.025}{1+e^{-\beta_3}} - \frac{1.441}{1+e^{-\beta_4}} - \frac{1.326}{1+e^{-\beta_5}} + \frac{0.933}{1+e^{-\beta_6}} + 1.714 \quad (10)$$

No. of training dataset = 60, R^2 = 0.93, RMSE = 7.5 mm.

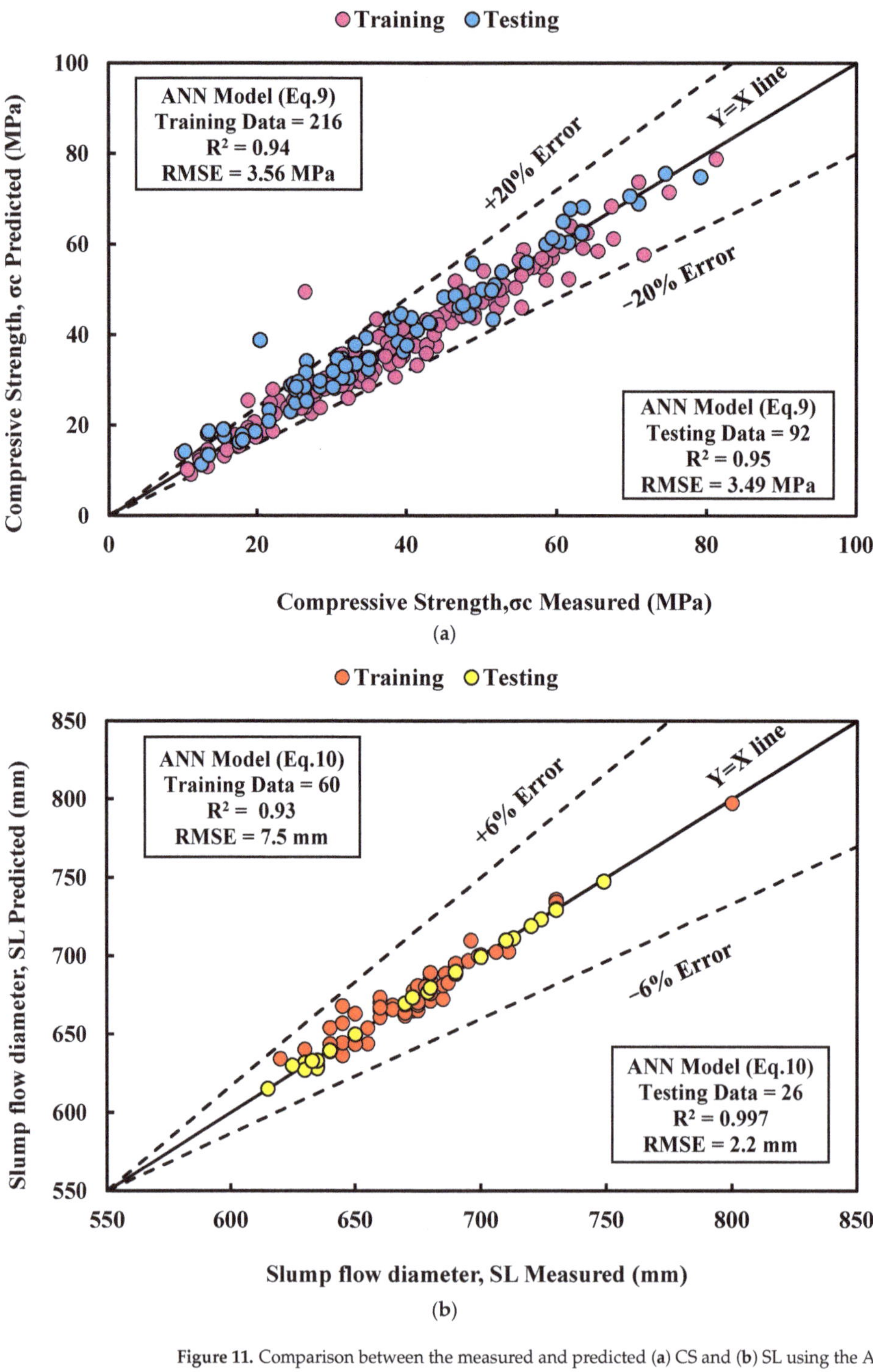

Figure 11. Comparison between the measured and predicted (**a**) CS and (**b**) SL using the ANN model for the training and testing dataset.

Since the measured compressive strength values in the training dataset were divided into three stages—LSC (less than 20 MPa), NSC (20 to 50 MPa), and HSC (greater than 50 MPa)—the ANN model was applied to all three stages, and the resulting R^2, RMSE, and MAE values were all reported. The samples were 29, 143, and 44 for LSC, NSC, and HSC, respectively. The ANN model for LSC had an R^2 of 0.76, RMSE of 1.84 MPa, and MAE of 1.34 MPa. The model result differed for the middle stage (NSC); the R^2 was 0.77, the RMSE was 3.77 MPa, and the MAE had a value of 2.624 MPa. The HSC maintained greater result values, with an R^2 of 0.79, RMSE of 4.13 MPa, and MAE of 2.98 MPa (Figure 12).

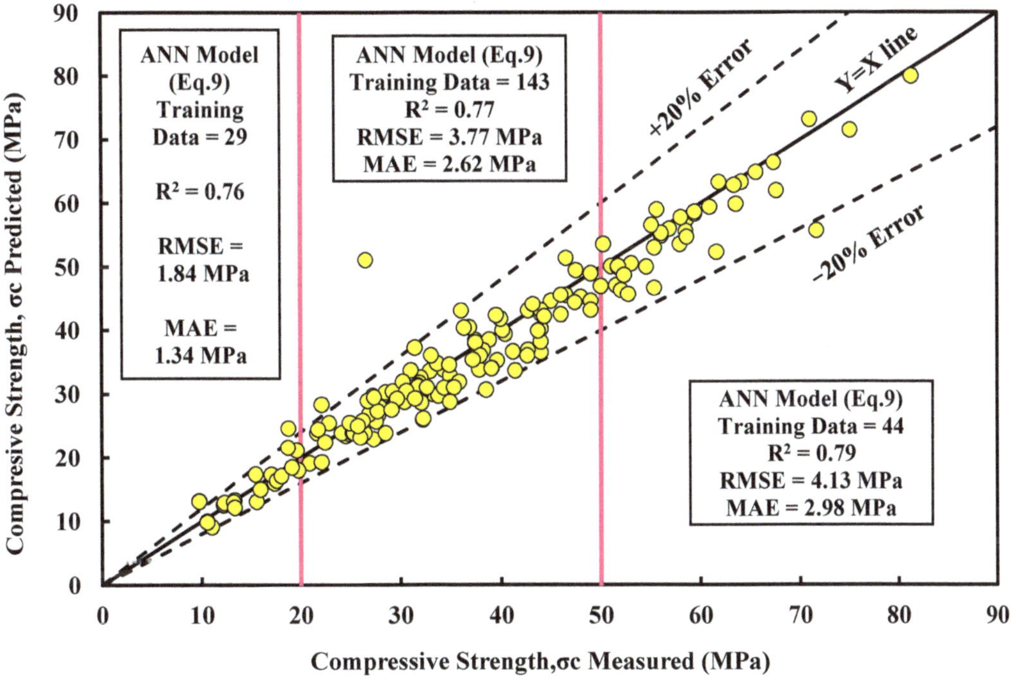

Figure 12. Relationship between the measured and predicted CS of FA-modified SCC for different ranges.

Furthermore, based on the training dataset, the slump flow diameter values were divided into classes 1, 2, and 3. The number of samples was 12 in the first class, 47 in the second class, and only 1 in the last. Due to the high sample number of class 2, the ANN model was applied to the class. The model provided an R^2 of 0.88, RMSE of 6.2 mm, and MAE of 5.2 mm (Figure 13).

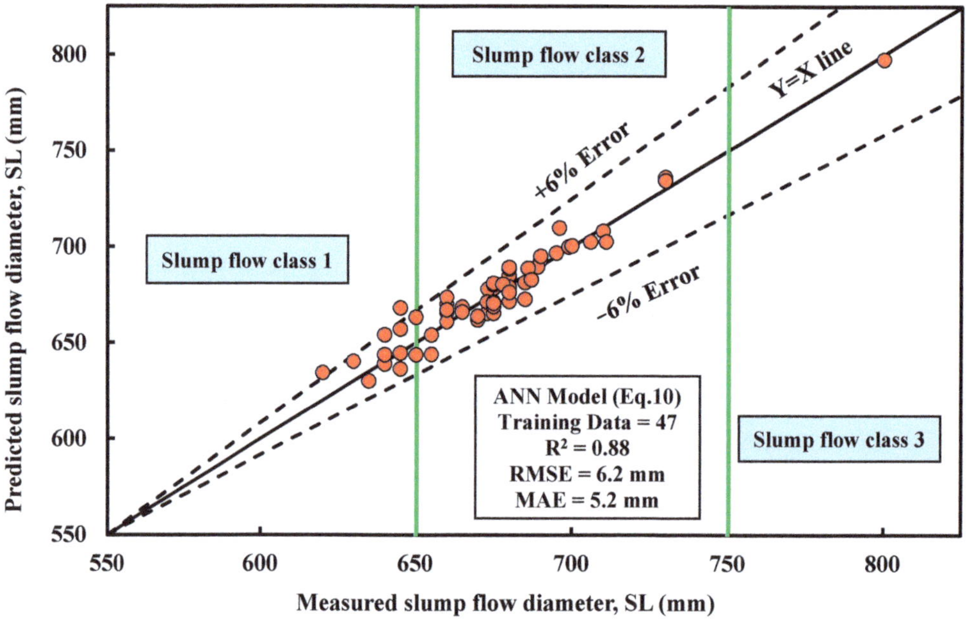

Figure 13. Relationship between the measured and predicted SL of FA-modified SCC for different classes.

3.2. *Effective Factors*

The effect of independent parameters such as cement, water-to-binder ratio, fly ash, sand, coarse aggregate, and superplasticizer on the compressive strength and slump flow diameter of self-compacted concrete was evaluated using the MLR model. The effect of a single parameter was found by changing its value from the minimum to the maximum by fixing other parameters on either the minimum value or the maximum value. The predicted CS and SL results were recorded in both cases, fixing independent variables at the minimum and the maximum. Figure 14 shows all the effects of both the CS and SL predictions.

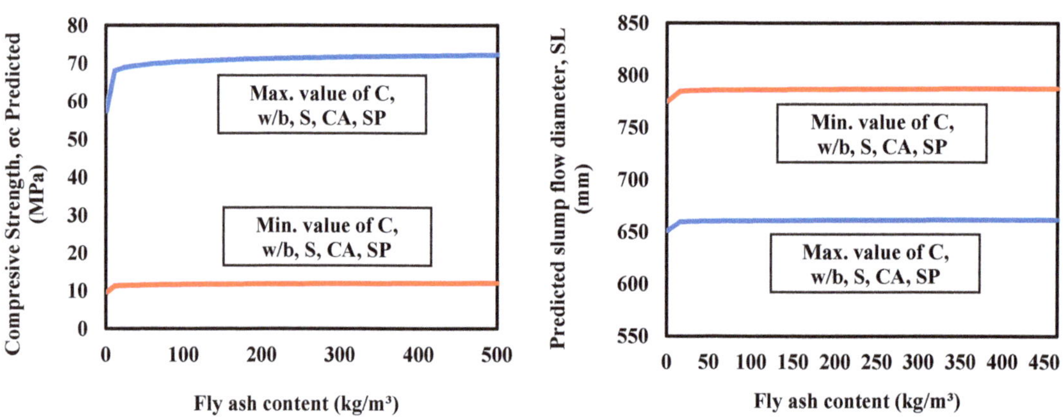

(a)

Figure 14. *Cont.*

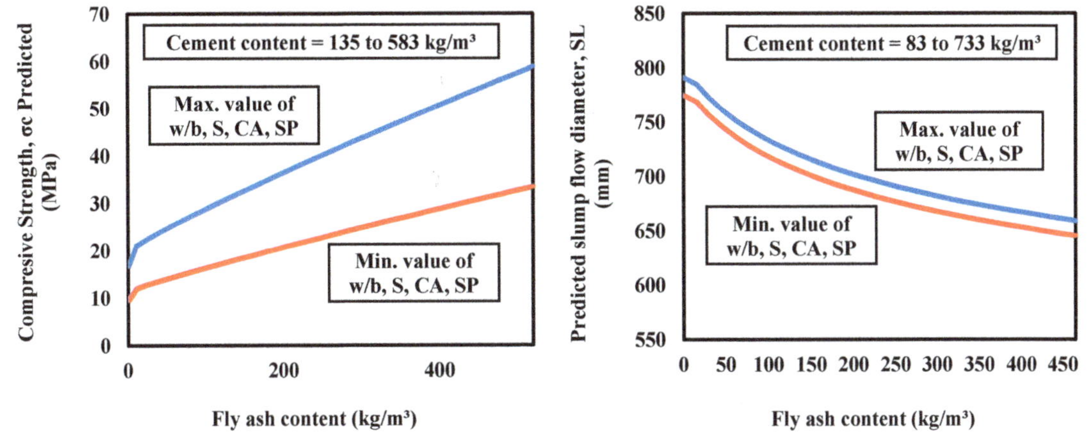

(b)

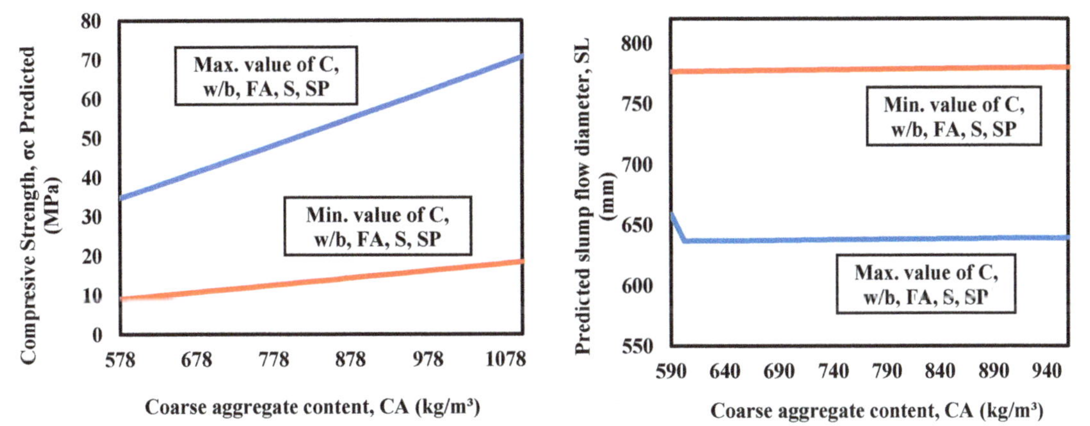

(c)

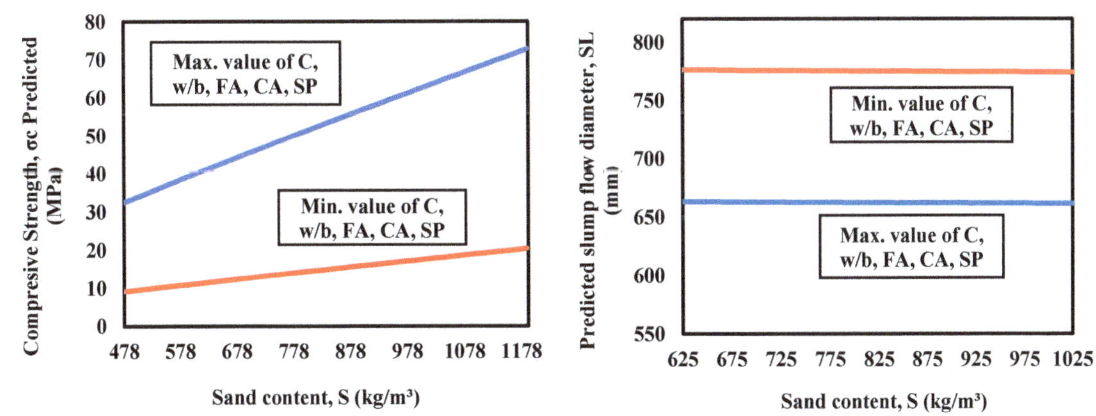

(d)

Figure 14. *Cont.*

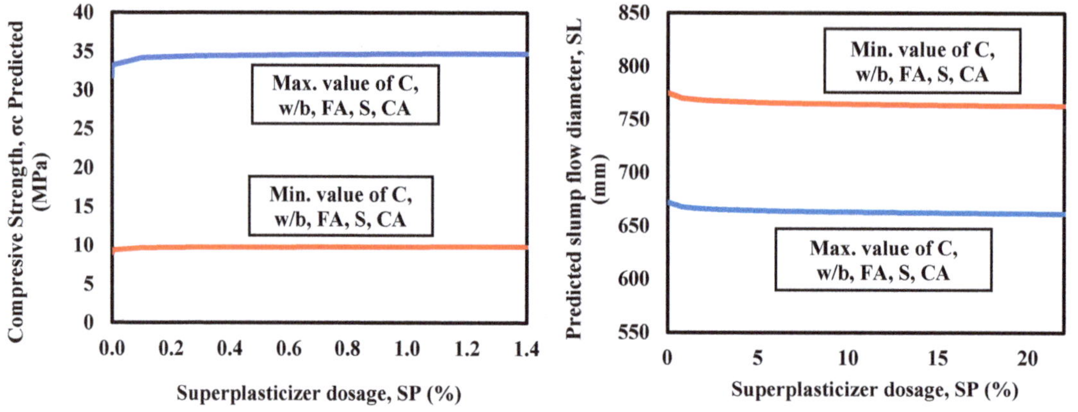

(e)

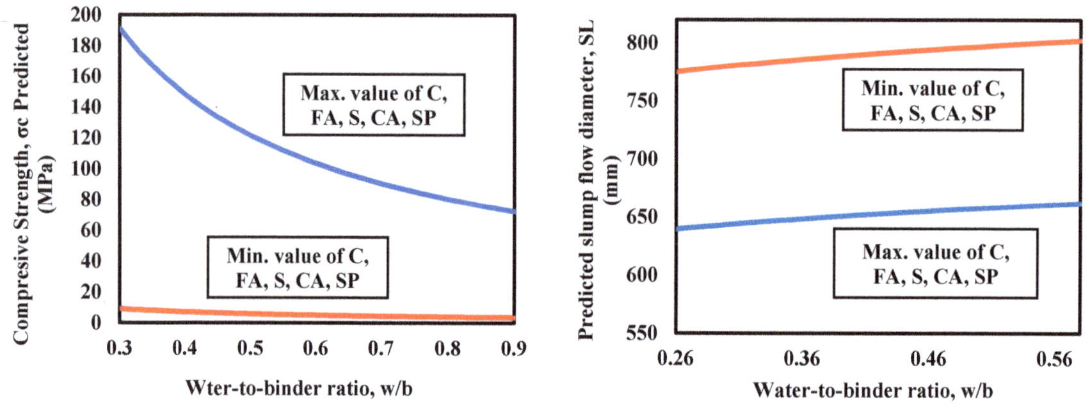

(f)

Figure 14. Effect of (**a**) fly ash content, (**b**) fly ash with cement content, (**c**) coarse aggregate content, (**d**) sand content, (**e**) superplasticizer dosage, and (**f**) water-to-binder ratio on the CS and SL of SCC.

The MLR model result noted that increasing the fly ash content caused an increased CS but had very little effect on the SL of SCC. Meanwhile, increasing the cement and fly ash content at the same time decreased the SL but increased the CS. The coarse aggregate and sand content greatly affected the CS. Increasing the CA and S content increased the CS. However, the effect of aggregates was less on the SL. Increasing the CA content increased the SL but decreased with the S content. On the other hand, superplasticizer was observed to have a small effect on the CS and SL.

From the CS prediction, it was noted that all the independent variables in their maximum value provided a greater value of CS when the value of any single parameter was changed. In contrast to CS, the greater value of SL was achieved while applying the minimum value of the independent variables. The summary of the most affected factors for both CS and SL is shown in Figure 15.

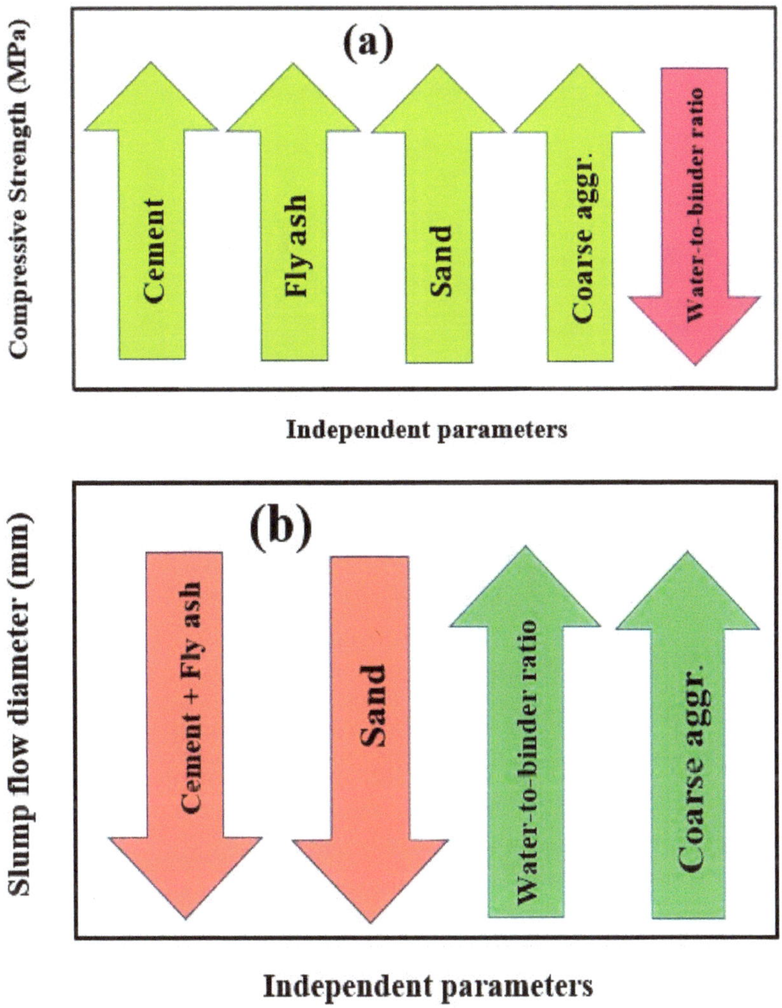

Figure 15. Effect of independent parameters on the (**a**) CS and (**b**) SL of SCC.

4. Evaluation of Developed Models

The study was conducted to determine the effect of different fly ash content on the compressive strength and slump flow diameter of self-compacted concrete. The experiment included predicting the CS and SL using three alternative models; NLR, MLR, and ANN. Each model provided a formula based on several mathematical parameters, and various assessment criteria were used to assess the performance of each constructed model.

Based on the R^2, RMSE, and MAE values, the ANN model provided the highest accuracy and reliability for predicting compressive strength and slump flow diameter using the training dataset. For the CS prediction, the ANN model had an R^2 of 0.94, RMSE of 3.56 MPa, and MAE of 2.54 MPa based on the training dataset, as well as an R^2 of 0.95, RMSE of 3.49 MPa, and MAE of 2.45 MPa for the testing dataset. In terms of the SL prediction, the ANN model had an R^2 of 0.93, RMSE of 7.5 mm, and MAE of 5.97 mm based on the training dataset, and an R^2 of 0.997, RMSE of 2.2 mm, and MAE of 1.39 mm based on the testing dataset. All statistical results for all models are summarized in Table 4. Considering the error lines, the ANN was noted to have more data along the Y=X line. For

CS, the model had an error line from +20 to −20% for the training dataset, indicating that 80% of the data were between 0.8 and 1.2 (predicted CS/experimental CS). However, in predicting the SL, the ANN model had an error line from +6 to −6% for the training dataset, indicating that 94% of the data were between 0.94 and 1.06 (predicted SL/experimental SL).

Table 4. Summary of the evaluation assessment criteria for the developed models in FA-modified SCC for the CS prediction.

	Model	Figure (No)	Equation (No.)	Training			Testing			Ranking
				R^2	RMSE (MPa)	MAE (MPa)	R^2	RMSE (MPa)	MAE (MPa)	
Compressive strength	**NLR**	7a	5	0.81	5.82	4.67	0.84	7.67	4.72	2
	MLR	8a	7	0.81	6.04	4.69	0.82	7.92	4.65	3
	ANN	11a	9	0.94	3.56	2.54	0.95	3.49	2.45	1
	Model	**Figure (No)**	**Equation (No.)**	**Training**			**Testing**			**Ranking**
				R^2	RMSE (mm)	MAE (mm)	R^2	RMSE (mm)	MAE (mm)	
Slump flow diameter	**NLR**	7b	6	0.82	11.6	10.12	0.57	27.4	27.1	3
	MLR	8b	8	0.86	10.3	8.54	0.57	26.8	25.93	2
	ANN	11b	10	0.93	7.5	5.97	0.997	2.2	1.39	1

Furthermore, the second-ranked model was the NLR for CS prediction; it had an R^2 of 0.81, RMSE of 5.82 MPa, and MAE of 4.67 MPa for the training dataset, and R^2 of 0.84, RMSE of 7.67 MPa, and MAE of 4.72 MPa for the testing dataset. The model ranged between the +30 and −25% error lines for the training dataset. However, the MLR was second-ranked for the SL prediction; it had an R^2 of 0.86, RMSE of 10.3 mm, and MAE of 8.54 mm for the training dataset, and an R^2 of 0.57, RMSE of 26.8 mm, and MAE of 25.93 mm for the testing dataset. The error line was between +12 and −15% for the MLR model based on the training dataset.

Moreover, the training dataset from the collected data for CS prediction was divided into three different ranges, and then the ANN model was applied. All the R^2, RMSE, and MAE values were calculated. The ANN model for the low CS strength range had an R^2 of 0.76, RMSE of 1.84 MPa, and MAE of 1.34 MPa. The model result differed for the middle stage; the R^2 was 0.77, the RMSE was 3.77 MPa, and the MAE was 2.62 MPa. The high CS strength range maintained higher results, with an R^2 of 0.79, RMSE of 4.13 MPa, and MAE of 2.98 MPa. On the other hand, the training dataset for predicting the SL was divided into three classes, and the ANN model was used to calculate each class's R^2, RMSE, and MAE values. The model had an R^2 of 0.88, an RMSE of 6.2 mm, and an MAE of 5.2 mm for the mid-class (from 650 to 750 mm).

The testing dataset had lower R^2, RMSE, and MAE values than the training dataset for all the developed models predicting CS and SL. The variations to the measured values were plotted as shown in Figure 16 for the compressive strength and slump flow diameter of self-compacted concrete. Model values fell between the +30 and −20% error lines for CS prediction, and between −20% and +30% for the SL prediction, indicating poor performance.

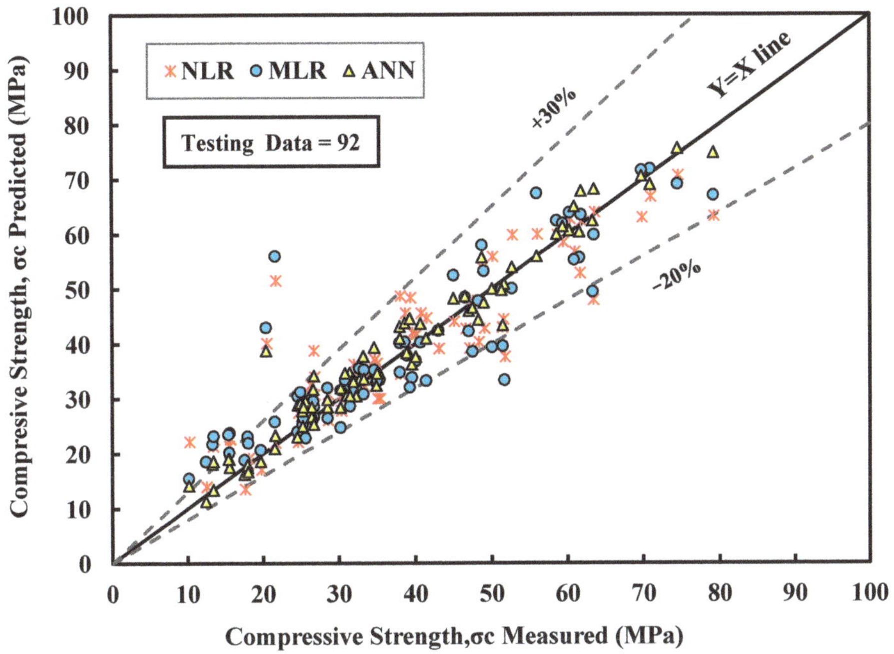

(a)

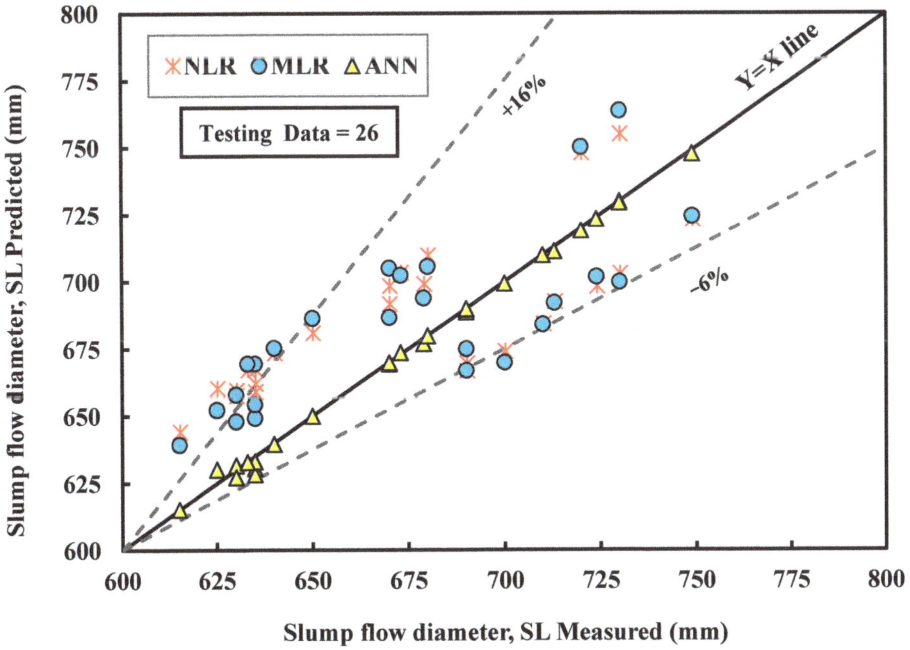

(b)

Figure 16. Relationship between the measured and predicted (**a**) CS and (**b**) SL for the developed models using the testing dataset.

The performance of the developed models was also evaluated using the training and testing datasets through the OBJ function and SI value. According to the objective function, the ANN model maintained the lowest value using the training dataset (Figure 17). The OBJ value was 3.12 and 5.5 for the CS and SL, respectively. Based on the SI value, the NLR model showed an excellent performance in predicting the CS. The SI value was 0.10 for both the training and testing datasets. In predicting the SL, the NLR, MLR, and ANN models showed excellent performances for both the training and testing datasets. The SI value was 0.02 for both the NLR and MLR models and 0.01 for the ANN model (Figure 18).

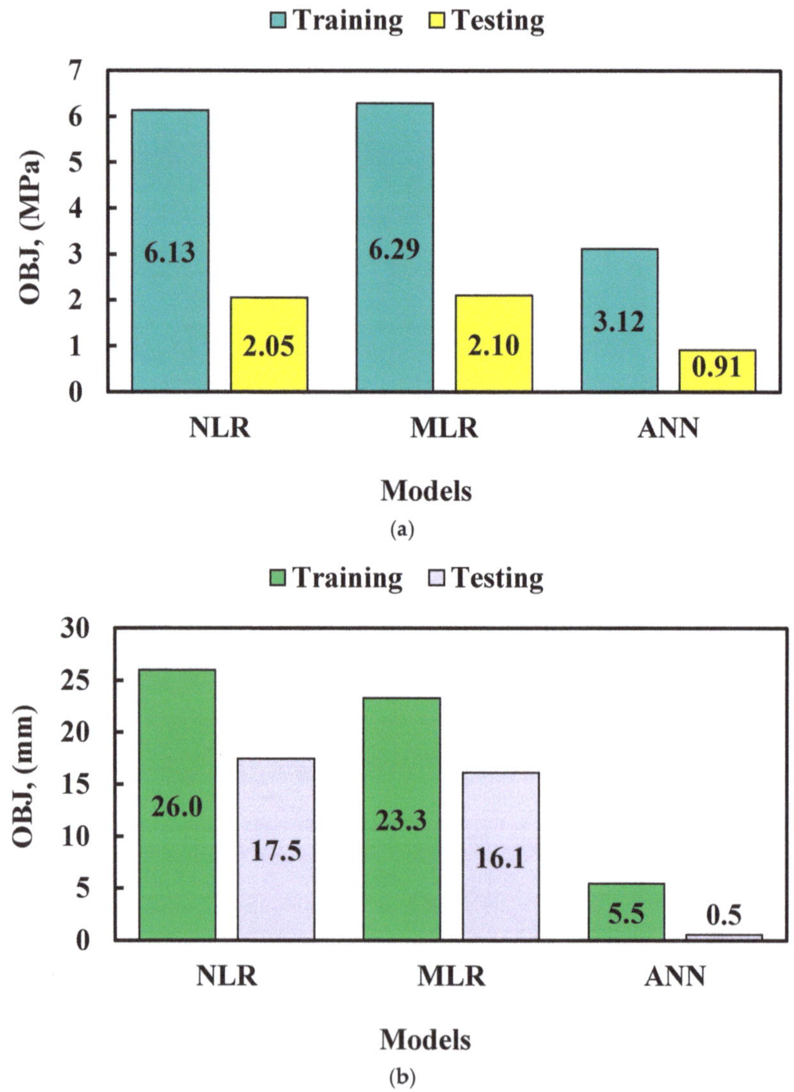

Figure 17. Comparison of the developed models using OBJ: (**a**) CS and (**b**) SL.

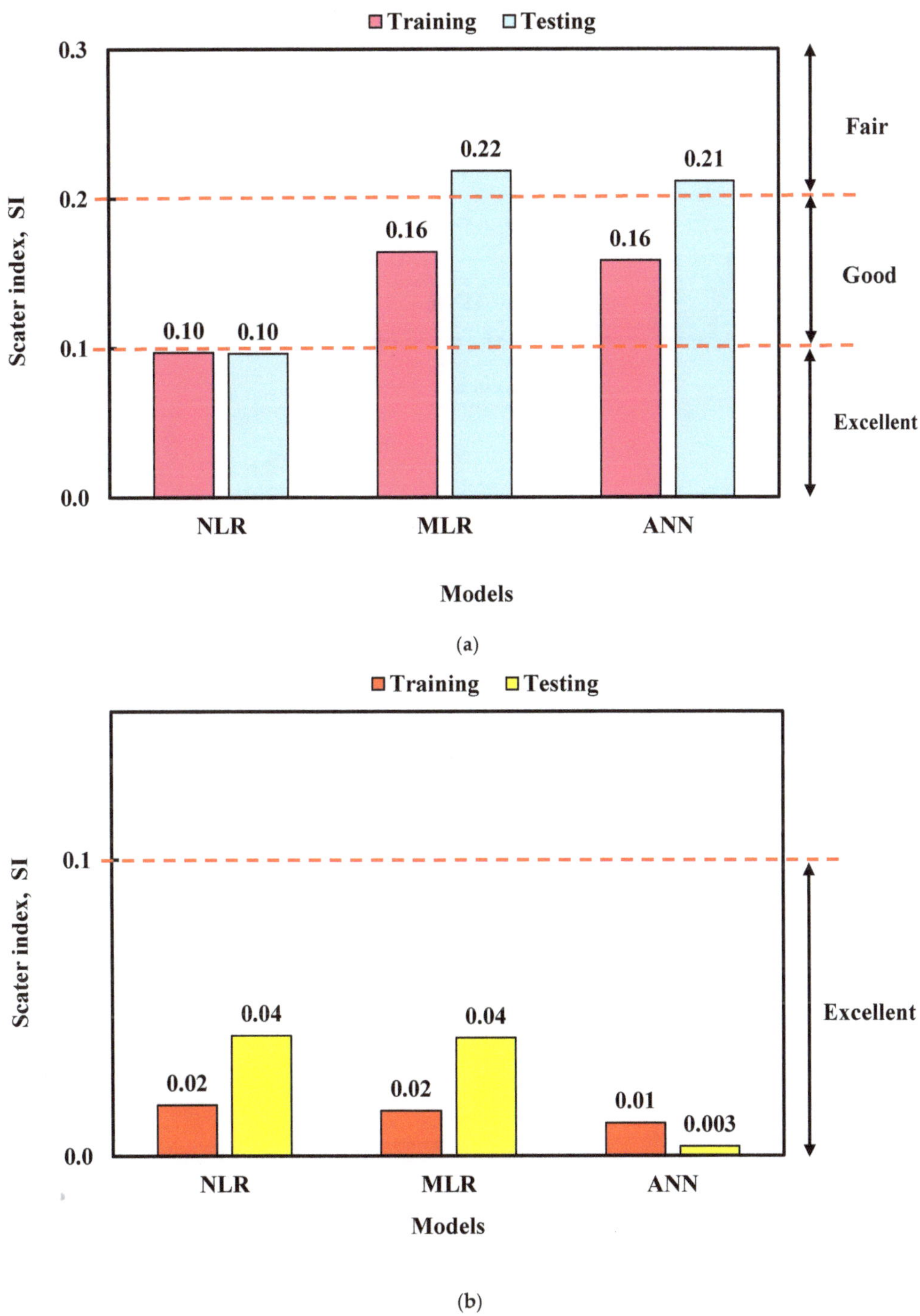

Figure 18. Comparison of the developed models using SI: (**a**) CS and (**b**) SL.

In addition, the created models were compared using residual error, as shown in Figure 19, for both the CS and SL predictions. The residual error value was obtained by subtracting the expected value from the measured value. The results of the CS prediction showed that the ANN model had the lowest error value, ranging from −23.0 to +13.91 MPa. The residual error for the NLR and MLR models ranged from −18.27 to 29.98 MPa and from −19.06 to 34.41 MPa, respectively. In contrast to the CS, the ANN model provided the lowest error value in the SL prediction. The error value was from −13.91 to 23.0 mm for the MLR model. The results of the NLR and ANN models were from −27.02 to 35.35 mm and from −30.13 to 36.46 mm, respectively.

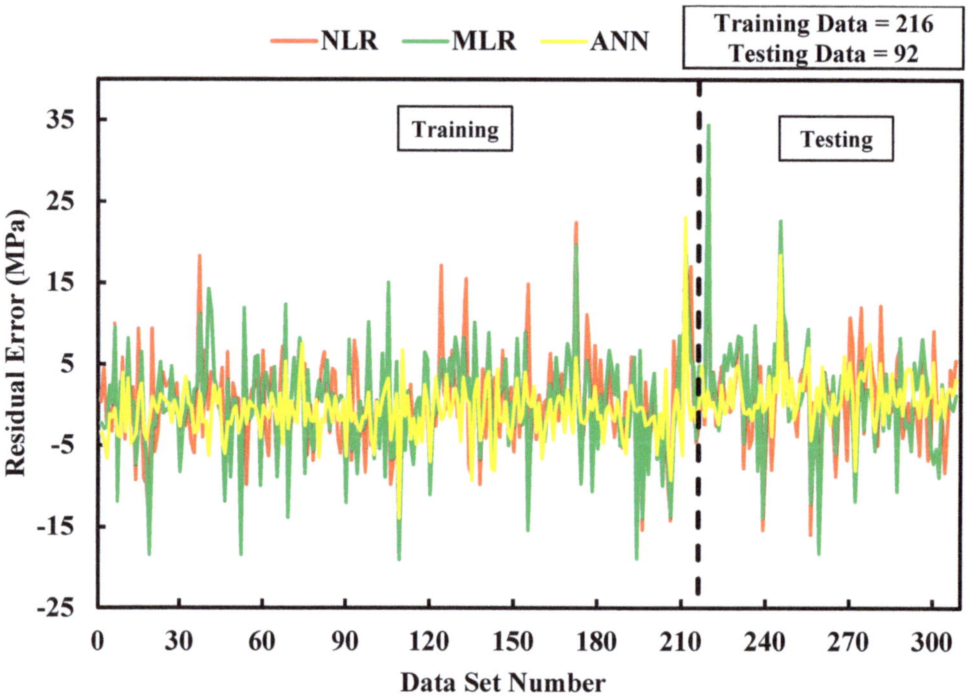

(a)

Figure 19. *Cont.*

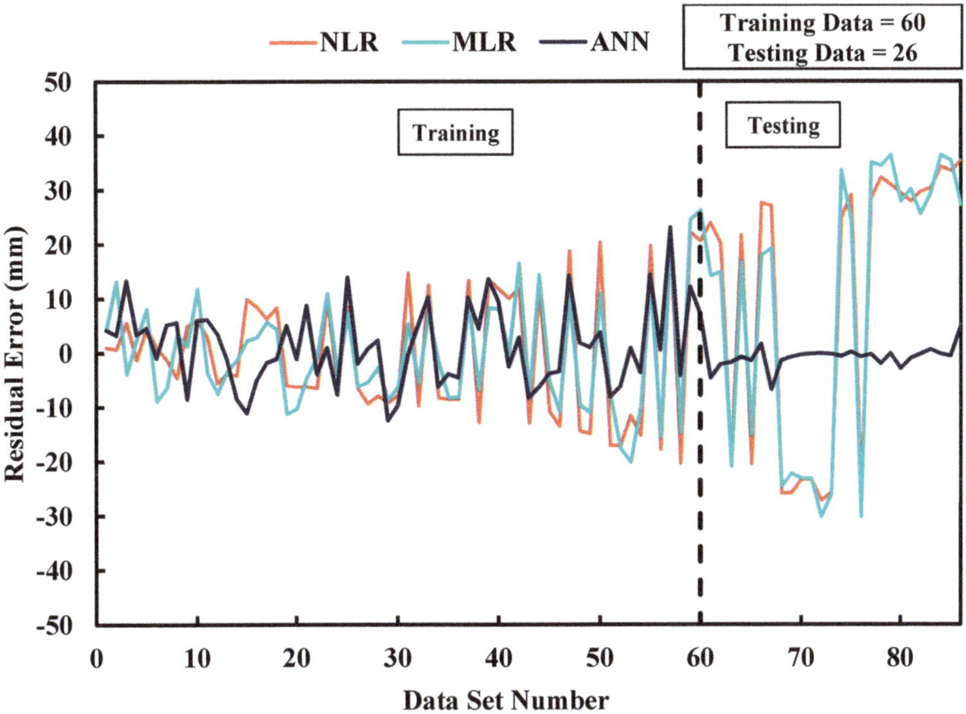

(b)

Figure 19. Variation between the measured and predicted (**a**) CS and (**b**) SL for the developed models based on the residual errors of NLR, MLR, and the ANN.

5. Sensitivity Investigation

Sensitivity analysis is an effective way to find and evaluate the effect of each independent variable on the modeled dependent variable, CS, and SL of fly ash-modified SCC [20]. For this purpose, the most accurate and efficient model was determined and selected for the analysis; in this study, the ANN was used for both CS and SL training datasets, as the analysis was performed for both collected data independently. During the sensitivity analysis, one parameter was excluded each time, and the assessment tools' results, such as R^2, RMSE, and MAE, were independently calculated for each trial. The sensitivity results for CS and SL are shown in Table 5. The findings show that the cement content was the most effective parameter in predicting both CS and SL of SCC, followed by coarse aggregate and fly ash content (Figure 20).

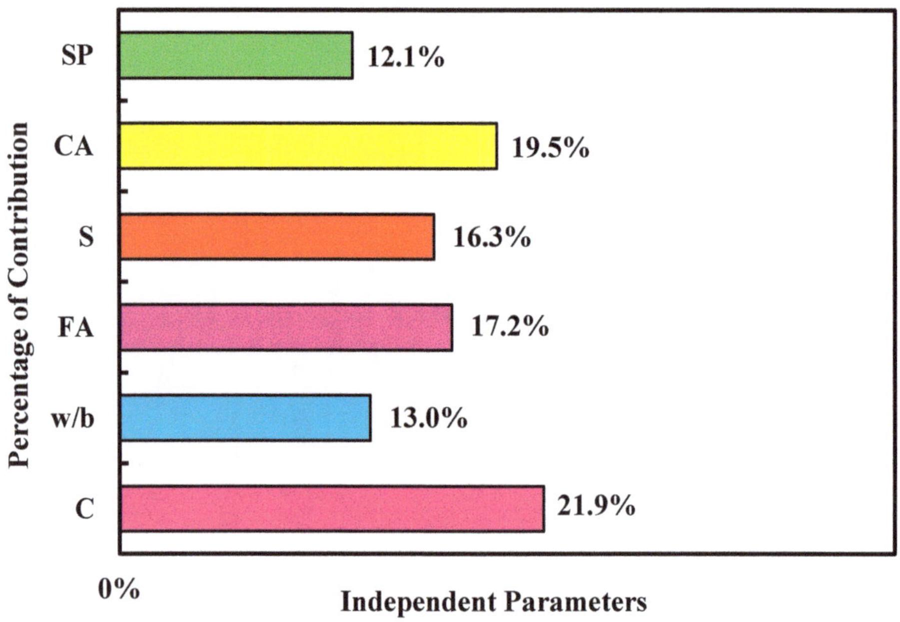

(a)

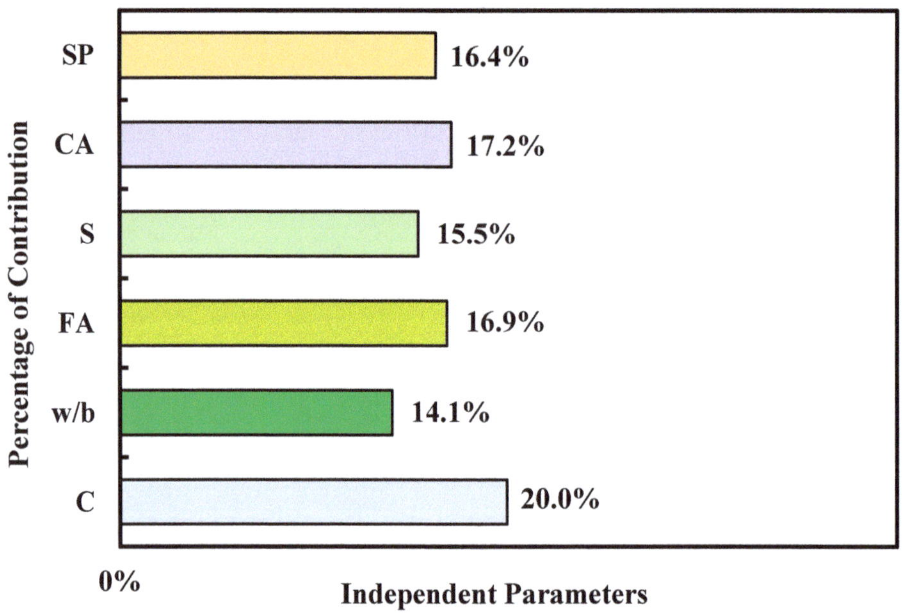

(b)

Figure 20. The percentage contribution of input variables in predicting; (**a**) CS and (**b**) SL of FA-modified SCC using the ANN model.

Table 5. Sensitivity analysis results of the ANN model applied to the training dataset of FA-modified SCC for CS prediction.

	No.	Combination	Removed Parameter	R^2	RMSE (MPa)	MAE (MPa)	Ranking Based on RMSE and MAE
Compressive strength	1	C, w/b, FA, S, CA, SP	-	0.94	3.65	2.52	-
	2	w/b, FA, S, CA, SP	C	**0.82**	**6.19**	**4.7**	1
	3	C, FA, S, CA, SP	w/b	0.93	3.85	2.79	5
	4	C, w/b, S, CA, SP	FA	0.9	4.92	3.7	3
	5	C, w/b, FA, CA, SP	S	0.91	4.75	3.5	4
	6	C, w/b, FA, S, SP	CA	0.89	5.46	4.19	2
	7	C, w/b, FA, S, CA	SP	0.94	3.83	2.6	6
	No.	**Combination**	**Removed Parameter**	**R^2**	**RMSE (mm)**	**MAE (mm)**	**Ranking based on RMSE and MAE**
Slump flow diameter	1	C, w/b, FA, S, CA, SP	-	0.93	7.5	6	-
	2	w/b, FA, S, CA, SP	C	**0.87**	**10.9**	**9.2**	1
	3	C, FA, S, CA, SP	w/b	0.91	8.7	6.5	6
	4	C, w/b, S, CA, SP	FA	0.88	9.5	7.8	3
	5	C, w/b, FA, CA, SP	S	0.9	9	7.2	5
	6	C, w/b, FA, S, SP	CA	0.88	9.5	7.9	2
	7	C, w/b, FA, S, CA	SP	0.9	9.4	7.6	4

6. Conclusions

The current study aimed to find and propose an accurate and reliable model to predict self-compacted concrete's compressive strength and slump flow diameter modified with different fly ash types and quantities. Overall, 216 and 86 data samples for fly-ash-modified self-compacted concrete with different mixture proportions, cement content, water-to-binder ratio, sand content, coarse aggregate content, and superplasticizer dosage were collected from the literature. Based on the collected data and the results of three different model approaches, the following conclusions can be drawn:

1. The database for predicting CS included fly ash content ranging between 0 and 525 kg/m^3, while that for predicting SL ranged between 0 and 468 kg/m^3.
2. Increasing fly ash content caused an increase in the CS, but a lower impact was found on the SL. However, the impact of fly ash was found when the cement content was increased with an increase in the fly ash content simultaneously. It decreased the SL but increased the CS.
3. The compressive strength was more affected by aggregates rather than the slump flow. Increasing the CA and S content increased the CS but led to small changes in the SL. The influence of CA and S was noted to be higher at the maximum values of the variables. These findings highlight the importance of aggregates, specifically coarse and fine aggregates, in determining the compressive strength of concrete. Whereas

the slump flow, which measures the workability and fluidity of the mixture, did not substantially impact the CS, the composition and content of aggregates played a crucial role in enhancing the concrete's overall strength.

4. According to the various assessment criteria, such as R^2, RMSE, and MAE, the ANN model was noted to have the highest accuracy and reliability for predicting both compressive strength and slump flow diameter of self-compacted concrete.
5. When predicting the CS, the ANN model had the highest R^2 of 0.94 for training and 0.95 for testing datasets. The lowest RMSE and MAE values were found to be 3.56 MPa and 2.54 MPa for training and 3.49 MPa and 2.45 MPa for testing datasets, respectively. However, in predicting the SL, the ANN model had an R^2 value of 0.93, RMSE of 7.5 mm, and MAE of 5.97 mm for the training dataset. The testing dataset's R^2, RMSE, and MAE values were 0.997, 2.2 mm, and 1.39 mm, respectively.
6. Other statistical assessment tools, such as the OBJ function and SI value, were used. The ANN model maintained the lowest OBJ value of 3.12 and 5.5 for the CS and SL, respectively. Regarding the SI value, excellent performance was observed from the NLR model when predicting the CS, which was 0.10 for both the training and testing datasets. However, all models were observed to predict the SL. The SI value was 0.02 for both the NLR and MLR models and 0.01 for the ANN model.
7. The application of the Artificial Neural Network (ANN) model to different ranges of concrete strength (CS) and different classes of specimen length (SL) demonstrates its versatility and effectiveness. The higher CS strength range yielded more favorable outcomes, as indicated by an R2 (coefficient of determination) value of 0.79, an RMSE (Root Mean Square Error) of 4.13 MPa, and an MAE (Mean Absolute Error) of 2.98 MPa. These metrics signify a strong correlation and relatively low prediction errors, suggesting the model performed well in estimating axial strength for high CS levels. Overall, the reported R2 values demonstrated a good fit between the predicted and actual values, while the RMSE and MAE values indicated relatively small errors in the model's predictions. These findings suggest that the ANN model can effectively capture the relationships between the CS, SL, and axial strength, highlighting its potential as a reliable tool for estimating concrete strength in various scenarios and ranges.
8. Sensitivity analysis illustrates the cement content as the most effective parameter for both the CS and SL prediction of SCC.

7. Limitations and Future Work

1. Other soft computing models should be used to predict the slump flow diameter and compressive strength of the fly ash-based self-compacted concrete.
2. It is possible to assess other fly ash types and sources.
3. The prediction of other types of workability tests can be investigated.
4. Experiments need to be carried out to verify the produced models.
5. It is also important to determine the effect of fly ash content on flexural and tensile strength.

Funding: This research received no external funding.

Institutional Review Board Statement: Not applicable.

Informed Consent Statement: Not applicable.

Data Availability Statement: The data supporting the conclusions of this article are included with the article.

Conflicts of Interest: The author declares no conflict of interest.

References

1. Faraj, R.H.; Sherwani, A.F.H.; Daraei, A. Mechanical, fracture and durability properties of self-compacting high strength concrete containing recycled polypropylene plastic particles. *J. Build. Eng.* **2019**, *25*, 100808. [CrossRef]
2. Faraj, R.H.; Sherwani, A.F.H.; Jafer, L.H.; Ibrahim, D.F. Rheological behavior and fresh properties of self-compacting high strength concrete containing recycled PP particles with fly ash and silica fume blended. *J. Build. Eng.* **2020**, *34*, 101667. [CrossRef]
3. Faraj, R.H.; Ali, H.F.H.; Sherwani, A.F.H.; Hassan, B.R.; Karim, H. Use of recycled plastic in self-compacting concrete: A comprehensive review on fresh and mechanical properties. *J. Build. Eng.* **2020**, *30*, 101283. [CrossRef]
4. Deilami, S.; Aslani, F.; Elchalakani, M. An experimental study on the durability and strength of SCC incorporating FA, GGBS and MS. *Proc. Inst. Civ. Eng. Struct. Build.* **2019**, *172*, 327–339. [CrossRef]
5. Faraj, R.H.; Mohammed, A.A.; Mohammed, A.; Omer, K.M.; Ahmed, H.U. Systematic multiscale models to predict the compressive strength of self-compacting concretes modified with nanosilica at different curing ages. *Eng. Comput.* **2022**, *38* (Suppl. 3), 2365–2388. [CrossRef]
6. Neville, A.M.; Brooks, J.J. *Concrete Technology*; Longman Scientific & Technical: London, UK, 1987; pp. 242–246.
7. Neville, A.M. *Properties of Concrete*; Longman Scientific & Technical: London, UK, 1995; Volume 4.
8. Silvestre, J.; Silvestre, N.; De Brito, J. Review on concrete nanotechnology. *Eur. J. Environ. Civil. Eng.* **2016**, *20*, 455–485. [CrossRef]
9. Karamoozian, A.; Karamoozian, M.; Ashrafi, H. Effect of Nano Particles on Self Compacting Concrete: An Experimental Study. *Life Sci. J.* **2013**, *10*, 95–101.
10. Larsen, O.; Naruts, V.; Aleksandrova, O. Self-compacting concrete with recycled aggregates. *Mater. Today Proc.* **2019**, *19*, 2023–2026. [CrossRef]
11. Ghasemi, M.; Ghasemi, M.R.; Mousavi, S.R. Studying the fracture parameters and size effect of steel fiber-reinforced self-compacting concrete. *Constr. Build. Mater.* **2019**, *201*, 447–460. [CrossRef]
12. Ahmadi, M.A.; Alidoust, O.; Sadrinejad, I.; Nayeri, M. Development of mechanical properties of self compacting concrete contain rice husk ash. *Int. J. Comput. Inf. Syst. Sci. Eng.* **2007**, *1*, 259–262.
13. Dinakar, P.; Manu, S.N. Concrete mix design for high strength self-compacting concrete using metakaolin. *Mater. Des.* **2014**, *60*, 661–668. [CrossRef]
14. Güneyisi, E.; Gesoglu, M.; Al-Goody, A.; İpek, S. Fresh and rheological behavior of nano-silica and fly ash blended self-compacting concrete. *Constr. Build. Mater.* **2015**, *95*, 29–44. [CrossRef]
15. Madandoust, R.; Ranjbar, M.M.; Ghavidel, R.; Shahabi, S.F. Assessment of factors influencing mechanical properties of steel fiber reinforced self-compacting concrete. *Mater. Des.* **2015**, *83*, 284–294. [CrossRef]
16. Nikbin, I.M.; Beygi, M.H.A.; Kazemi, M.T.; Amiri, J.V.; Rahmani, E.; Rabbanifar, S.; Eslami, M. A comprehensive investigation into the effect of aging and coarse aggregate size and volume on mechanical properties of self-compacting concrete. *Mater. Des.* **2014**, *59*, 199–210. [CrossRef]
17. Mahmood, W.; Mohammed, A. Performance of ANN and M5P-tree to forecast the compressive strength of hand-mix cement-grouted sands modified with polymer using ASTM and BS standards and evaluate the outcomes using SI with OBJ assessments. *Neural Comput. Appl.* **2022**, *34*, 15031–15051. [CrossRef]
18. Asteris, P.G.; Lourenço, P.B.; Hajihassani, M.; Adami, C.-E.N.; Lemonis, M.E.; Skentou, A.D.; Marques, R.; Nguyen, H.; Rodrigues, H.; Varum, H. Soft computing based models for the prediction of masonry compressive strength. *Eng. Struct.* **2021**, *248*, 113276. [CrossRef]
19. Rahimi, I.; Gandomi, A.H.; Asteris, P.G.; Chen, F. Analysis and prediction of COVID-19 using sir, seiqr and machine learning models: Australia, italy and uk cases. *Information* **2021**, *12*, 109. [CrossRef]
20. Bardhan, A.; Gokceoglu, C.; Burman, A.; Samui, P.; Asteris, P.G. Efficient computational techniques for predicting the California bearing ratio of soil in soaked conditions. *Eng. Geol.* **2021**, *291*, 106239. [CrossRef]
21. Ahmed, H.U.; Mohammed, A.S.; Mohammed, A.A.; Faraj, R.H. Systematic multiscale models to predict the compressive strength of fly ash-based geopolymer concrete at various mixture proportions and curing regimes. *PLoS ONE* **2021**, *16*, e0253006. [CrossRef]
22. Koopialipoor, M.; Asteris, P.G.; Mohammed, A.S.; Alexakis, D.E.; Mamou, A.; Armaghani, D.J. Introducing stacking machine learning approaches for the prediction of rock deformation. *Transp. Geotech.* **2022**, *34*, 100756. [CrossRef]
23. Ahmed, H.U.; Mohammed, A.S.; Faraj, R.H.; Abdalla, A.A.; Qaidi, S.M.; Sor, N.H.; Mohammed, A.A. Innovative modeling techniques including MEP, ANN and FQ to forecast the compressive strength of geopolymer concrete modified with nanoparticles. *Neural Comput. Appl.* **2023**, *35*, 12453–12479. [CrossRef]
24. Asteris, P.G.; Ashrafian, A.; Rezaie-Balf, M. Prediction of the Compressive Strength of Self-Compacting Concrete using Surrogate Models. *Comput. Concr.* **2019**, *24*, 137–150.
25. Emad, W.; Mohammed, A.S.; Bras, A.; Asteris, P.G.; Kurda, R.; Muhammed, Z.; Hassan, A.M.; Qaidi, S.M.A.; Sihag, P. Meta-model techniques to estimate the compressive strength of UHPFRC using various mix proportions and a high range of curing temperatures. *Constr. Build. Mater.* **2022**, *349*, 128737. [CrossRef]
26. Mahmood, W.; Mohammed, A.S.; Asteris, P.G.; Kurda, R.; Armaghani, D.J. Modeling flexural and compressive strengths behaviour of cement-grouted sands modified with water reducer polymer. *Appl. Sci.* **2022**, *12*, 1016. [CrossRef]

27. Asteris, P.G.; Koopialipoor, M.; Armaghani, D.J.; Kotsonis, E.A.; Lourenço, P.B. Prediction of Cement-based Mortars Compressive Strength using Machine Learning Techniques. *Neural Comput. Appl.* **2021**, *33*, 13089–13121. [CrossRef]
28. Jaf, D.K.I.; Abdulrahman, P.I.; Abdulrahman, A.S.; Mohammed, A.S. Effitioned soft computing models to evaluate the impact of silicon dioxide (SiO_2) to calcium oxide (CaO) ratio in fly ash on the compressive strength of concrete. *J. Build. Eng.* **2023**, *74*, 106820.
29. Farooq, F.; Czarnecki, S.; Niewiadomski, P.; Aslam, F.; Alabduljabbar, H.; Ostrowski, K.A.; Śliwa-Wieczorek, K.; Nowobilski, T.; Malazdrewicz, S. A comparative study for the prediction of the compressive strength of self-compacting concrete modified with fly. *Materials* **2021**, *14*, 4934. [CrossRef]
30. Patel, R.; Hossain, K.M.A.; Shehata, M.; Bouzoubaa, N.; Lachemi, M. Development of statistical models for mixture design of high-volume fly ash self-consolidating concrete. *Mater. J.* **2004**, *101*, 294–302.
31. Ghezal, A.; Khayat, K.H. Optimizing self-consolidating concrete with limestone filler by using statistical factorial design methods. *Mater. J.* **2002**, *99*, 264–272.
32. Sonebi, M. Applications of statistical models in proportioning medium-strength self-consolidating concrete. *Mater. J.* **2004**, *101*, 339–346.
33. Sonebi, M. Medium strength self-compacting concrete containing fly ash: Modelling using factorial experimental plans. *Cem. Concr. Res.* **2004**, *34*, 1199–1208. [CrossRef]
34. Sukumar, B.; Nagamani, K.; Raghavan, R.S. Evaluation of strength at early ages of self-compacting concrete with high volume fly ash. *Constr. Build. Mater.* **2008**, *22*, 1394–1401. [CrossRef]
35. Bouzoubaâ, N.; Lachemi, M. Self-compacting concrete incorporating high volumes of class F fly ash: Preliminary results. *Cem. Concr. Res.* **2001**, *31*, 413–420. [CrossRef]
36. Bui, V.K.; Akkaya, Y.; Shah, S.P. Rheological model for self-consolidating concrete. *Mater. J.* **2002**, *99*, 549–559.
37. Güneyisi, E.; Gesoğlu, M.; Özbay, E. Strength and drying shrinkage properties of self-compacting concretes incorporating multi-system blended mineral admixtures. *Constr. Build. Mater.* **2010**, *24*, 1878–1887. [CrossRef]
38. Pathak, N.; Siddique, R. Properties of self-compacting-concrete containing fly ash subjected to elevated temperatures. *Constr. Build. Mater.* **2012**, *30*, 274–280. [CrossRef]
39. Jalal, M.; Mansouri, E. Effects of fly ash and cement content on rheological, mechanical, and transport properties of high-performance self-compacting concrete. *Sci. Eng. Compos. Mater.* **2012**, *19*, 393–405. [CrossRef]
40. Dinakar, P.; Babu, K.G.; Santhanam, M. Mechanical properties of high-volume fly ash self-compacting concrete mixtures. *Struct. Concr.* **2008**, *9*, 109–116. [CrossRef]
41. Dinakar, P. Design of self-compacting concrete with fly ash. *Mag. Concr. Res.* **2012**, *64*, 401–409. [CrossRef]
42. Nehdi, M.; Pardhan, M.; Koshowski, S. Durability of self-consolidating concrete incorporating high-volume replacement composite cements. *Cem. Concr. Res.* **2004**, *34*, 2103–2112. [CrossRef]
43. Barbhuiya, S. Effects of fly ash and dolomite powder on the properties of self-compacting concrete. *Constr. Build. Mater.* **2011**, *25*, 3301–3305. [CrossRef]
44. Venkatakrishnaiah, R.; Sakthivel, G. Bulk utilization of flyash in self compacting concrete. *KSCE J. Civil. Eng.* **2015**, *19*, 2116–2120. [CrossRef]
45. Sun, Z.J.; Duan, W.W.; Tian, M.L.; Fang, Y.F. Experimental research on self-compacting concrete with different mixture ratio of fly ash. In *Advanced Materials Research*; Trans Tech Publications Ltd.: Bäch, Switzerland, 2011; Volume 236, pp. 490–495.
46. Ramanathan, P.; Baskar, I.; Muthupriya, P.; Venkatasubramani, R. Performance of self-compacting concrete containing different mineral admixtures. *KSCE J. Civil. Eng.* **2013**, *17*, 465–472. [CrossRef]
47. Boel, V.; Audenaert, K.; De Schutter, G.; Heirman, G.; Vandewalle, L.; Desmet, B.; Vantomme, J. Transport properties of self compacting concrete with limestone filler or fly ash. *Mater. Struct.* **2007**, *40*, 507–516. [CrossRef]
48. Douglas, R.P.; Bui, V.K.; Akkaya, Y.; Shah, S.P. Properties of Self-consolidating concrete containing class f fly ash: With a Verification of the minimum paste volume method. *Spec. Publ.* **2006**, *233*, 45–64.
49. Bingöl, A.F.; Tohumcu, İ. Effects of different curing regimes on the compressive strength properties of self compacting concrete incorporating fly ash and silica fume. *Mater. Des.* **2013**, *51*, 12–18. [CrossRef]
50. Hemalatha, T.; Ramaswamy, A.; Chandra Kishen, J.M. Micromechanical analysis of self compacting concrete. *Mater. Struct.* **2015**, *48*, 3719–3734. [CrossRef]
51. Krishnapal, P.; Yadav, R.K.; Rajeev, C. Strength characteristics of self compacting concrete containing fly ash. *Res. J. Eng. Sci. ISSN* **2013**, *2278*, 9472.
52. Dhiyaneshwaran, S.; Ramanathan, P.; Baskar, I.; Venkatasubramani, R. Study on durability characteristics of self-compacting concrete with fly ash. *Jordan J. Civ. Eng.* **2013**, *7*, 342–353.
53. Siddique, R. Properties of self-compacting concrete containing class F fly ash. *Mater. Des.* **2011**, *32*, 1501–1507. [CrossRef]
54. Mahalingam, B.; Nagamani, K. Effect of processed fly ash on fresh and hardened properties of self compacting concrete. *Int. J. Earth Sci. Eng.* **2011**, *4*, 930–940.
55. Muthupriya, P.; Sri, P.N.; Ramanathan, M.P.; Venkatasubramani, R. Strength and workability character of self compacting concrete with GGBFS, FA and SF. *Int. J. Emerg. Trends Eng. Dev.* **2012**, *2*, 424–434.

56. Siddique, R.; Aggarwal, P.; Aggarwal, Y. Influence of water/powder ratio on strength properties of self-compacting concrete containing coal fly ash and bottom ash. *Constr. Build. Mater.* **2012**, *29*, 73–81. [CrossRef]
57. Gesoğlu, M.; Güneyisi, E.; Özbay, E. Properties of self-compacting concretes made with binary, ternary, and quaternary cementitious blends of fly ash, blast furnace slag, and silica fume. *Constr. Build. Mater.* **2009**, *23*, 1847–1854. [CrossRef]
58. Nepomuceno, M.C.; Pereira-de-Oliveira, L.A.; Lopes, S.M.R. Methodology for the mix design of self-compacting concrete using different mineral additions in binary blends of powders. *Constr. Build. Mater.* **2014**, *64*, 82–94. [CrossRef]
59. Gettu, R.; Izquierdo, J.; Gomes, P.C.C.; Josa, A. Development of high-strength self-compacting concrete with fly ash: A four-step experimental methodology. In Proceedings of the 27th Conference on Our World in Concrete & Structures, Singapore, 29–30 August 2002; Tam, C.T., Ho, D.W., Paramasivam, P., Tan, T.H., Eds.; Premier Pte. Ltd.: Singapore, 2002; pp. 217–224.
60. Şahmaran, M.; Yaman, İ.Ö.; Tokyay, M. Transport and mechanical properties of self consolidating concrete with high volume fly ash. *Cem. Concr. Compos.* **2009**, *31*, 99–106. [CrossRef]
61. Liu, M. Self-compacting concrete with different levels of pulverized fuel ash. *Constr. Build. Mater.* **2010**, *24*, 1245–1252. [CrossRef]
62. Piro, N.S.; Mohammed, A.S.; Hamad, S.M.; Kurda, R.; Qader, B.S. Multifunctional computational models to predict the long-term compressive strength of concrete incorporated with waste steel slag. *Struct. Concr.* **2023**, *24*, 2093–2112. [CrossRef]
63. Piro, N.S.; Mohammed, A.; Hamad, S.M.; Kurda, R. Artificial neural networks (ANN), MARS, and adaptive network-based fuzzy inference system (ANFIS) to predict the stress at the failure of concrete with waste steel slag coarse aggregate replacement. *Neural Comput. Appl.* **2023**, *35*, 13293–13319. [CrossRef]
64. Ibrahim, A.K.; Dhahir, H.Y.; Mohammed, A.S.; Omar, H.A.; Sedo, A.H. The effectiveness of surrogate models in predicting the long-term behavior of varying compressive strength ranges of recycled concrete aggregate for a variety of shapes and sizes of specimens. *Arch. Civil. Mech. Eng.* **2023**, *23*, 61. [CrossRef]
65. Ahmed, A.; Abubakr, P.; Mohammed, A.S. Efficient models to evaluate the effect of C3S, C2S, C3A, and C4AF contents on the long-term compressive strength of cement paste. *Structures* **2023**, *47*, 1459–1475. [CrossRef]
66. Mohammed, A.; Rafiq, S.; Sihag, P.; Mahmood, W.; Ghafor, K.; Sarwar, W. ANN, M5P-tree model, and nonlinear regression approaches to predict the compression strength of cement-based mortar modified by quicklime at various water/cement ratios and curing times. *Arab. J. Geosci.* **2020**, *13*, 1216. [CrossRef]
67. Mohammed, A.; Rafiq, S.; Sihag, P.; Kurda, R.; Mahmood, W. Soft computing techniques: Systematic multiscale models to predict the compressive strength of HVFA concrete based on mix proportions and curing times. *J. Build. Eng.* **2021**, *33*, 101851. [CrossRef]
68. Mohammed, A.S.; Emad, W.; Sarwar Qadir, W.; Kurda, R.; Ghafor, K.; Kadhim Faris, R. Modeling the Impact of Liquid Polymers on Concrete Stability in Terms of a Slump and Compressive Strength. *Appl. Sci.* **2023**, *13*, 1208. [CrossRef]
69. Ali, R.; Muayad, M.; Mohammed, A.S.; Asteris, P.G. Analysis and prediction of the effect of Nanosilica on the compressive strength of concrete with different mix proportions and specimen sizes using various numerical approaches. *Struct. Concr.* **2022**, *24*, 4161–4184. [CrossRef]
70. Abdalla, A.; Mohammed, A.S. Hybrid MARS-, MEP-, and ANN-based prediction for modeling the compressive strength of cement mortar with various sand size and clay mineral metakaolin content. *Arch. Civil. Mech. Eng.* **2022**, *22*, 194. [CrossRef]
71. Asteris, P.G.; Nozhati, S.; Nikoo, M.; Cavaleri, L.; Nikoo, M. Krill herd algorithm-based neural network in structural seismic reliability evaluation. *Mech. Adv. Mater. Struct.* **2019**, *26*, 1146–1153. [CrossRef]
72. Asteris, P.G.; Lemonis, M.E.; Le, T.-T.; Tsavdaridis, K.D. Evaluation of the ultimate eccentric load of rectangular CFSTs using advanced neural network modeling. *Eng. Struct.* **2021**, *248*, 113297. [CrossRef]
73. Asteris, P.G.; Lourenço, P.B.; Adami, C.A.; Roussis, P.C.; Armaghani, D.J.; Cavaleri, L.; Chalioris, C.E.; Hajihassani, M.; Lemonis, M.E.; Mohammed, A.S.; et al. Revealing the nature of metakaolin-based concrete materials using Artificial Intelligence Techniques. *Constr. Build. Mater.* **2022**, *322*, 126500. [CrossRef]
74. Asteris, P.G.; Armaghani, D.J.; Hatzigeorgiou, G.D.; Karayannis, C.G.; Pilakoutas, K. Predicting the shear strength of reinforced concrete beams using Artificial Neural Networks. Computers and Concrete. *An. Int. J.* **2019**, *24*, 469–488.
75. Armaghani, D.J.; Asteris, P.G. A comparative study of ANN and ANFIS models for the prediction of cement-based mortar materials compressive strength. *Neural Comput. Appl.* **2021**, *33*, 4501–4532. [CrossRef]
76. Mohammed, A.; Kurda, R.; Armaghani, D.J.; Hasanipanah, M. Prediction of compressive strength of concrete modified with fly ash: Applications of neuro-swarm and neuro-imperialism models. *Comput. Concr.* **2021**, *27*, 489.
77. Qadir, W.; Ghafor, K.; Mohammed, A. Evaluation the effect of lime on the plastic and hardened properties of cement mortar and quantified using Vipulanandan model. *Open Eng.* **2019**, *9*, 468–480. [CrossRef]
78. Sarwar, W.; Ghafor, K.; Mohammed, A. Regression analysis and Vipulanandan model to quantify the effect of polymers on the plastic and hardened properties with the tensile bonding strength of the cement mortar. *Results Mater.* **2019**, *1*, 100011. [CrossRef]
79. Qaidi, S.; Al-Kamaki, Y.S.; Al-Mahaidi, R.; Mohammed, A.S.; Ahmed, H.U.; Zaid, O.; Althoey, F.; Ahmad, J.; Isleem, H.F.; Bennetts, I. Investigation of the effectiveness of CFRP strengthening of concrete made with recycled waste PET fine plastic aggregate. *PLoS ONE* **2022**, *17*, e0269664. [CrossRef]
80. Abdalla, A.A.; Salih Mohammed, A. Theoretical models to evaluate the effect of SiO2 and CaO contents on the long-term compressive strength of cement mortar modified with cement kiln dust (CKD). *Arch. Civ. Mech. Eng.* **2022**, *22*, 105. [CrossRef]

81. Barkhordari, M.S.; Armaghani, D.J.; Mohammed, A.S.; Ulrikh, D.V. Data-driven compressive strength prediction of fly ash concrete using ensemble learner algorithms. *Buildings* **2022**, *12*, 132. [CrossRef]
82. Piro, N.S.; Salih, A.; Hamad, S.M.; Kurda, R. Comprehensive multiscale techniques to estimate the compressive strength of concrete incorporated with carbon nanotubes at various curing times and mix proportions. *J. Mater. Res. Technol.* **2021**, *15*, 6506–6527. [CrossRef]

sustainability

MDPI

Article

Mechanical Properties of Polyamide Fiber-Reinforced Lime–Cement Concrete

Mohammad Mostafa Jafari [1], Soheil Jahandari [2,3], Togay Ozbakkaloglu [4], Haleh Rasekh [5], Danial Jahed Armaghani [5,*] and Aida Rahmani [2,3]

1 Department of Civil Engineering, Technical and Vocational University (TVU), Tehran 1435761137, Iran; jafari-moh@tvu.ac.ir
2 Chem Concrete Pty Ltd., Seven Hills, NSW 2147, Australia; soheil@chemconcrete.com.au (S.J.); aida@chemconcrete.com.au (A.R.)
3 Centre for Infrastructure Engineering, Western Sydney University, Penrith, NSW 2751, Australia
4 Department of Civil Engineering, Texas State University, San Marcos, TX 78666, USA; togay.oz@txstate.edu
5 School of Civil and Environmental Engineering, University of Technology Sydney, Sydney, NSW 2007, Australia; haleh.rasekh@uts.edu.au
* Correspondence: danial.jahedarmaghani@uts.edu.au

Abstract: Lime–cement concrete (LCC) is a type of lime-based concrete in which lime and cement are utilized as the main binding agents. This type of concrete has been extensively used to construct support layers for shallow footings and road backfills in some warm regions. So far, there has been no systematic research conducted to investigate the mechanical characteristics of polyamide fiber-reinforced LCC. To address this gap, LCC specimens were prepared with 0%, 0.5%, 1%, and 2% of polyamide fibers (a synthetic textile made of petroleum-based plastic polymers). Specimens were then cured for 3, 7, and 28 days at room and oven temperatures. Then, the effects of the fibers' contents, curing conditions, and curing periods on the mechanical characteristics of LCC, such as secant modulus, deformability index, bulk modulus, shear modulus, stiffness ratio, strain energy, failure strain, strength ratio, and failure patterns, was investigated. The results of the unconfined compressive strength (UCS) tests showed that specimens with 1% fiber had the highest UCS values. The curing condition and curing period had significant effects on the strength of the LCC specimens, and oven-cured specimens developed higher UCS values. The aforementioned mechanical properties of the LCC specimens and the ability of the material to absorb energy significantly improved when the curing period under the oven-curing condition was increased, as well as through the application of fibers in the mix design. Based on the test results, a simple mathematical model was also established to forecast the mechanical properties of fiber-reinforced LCC. It is concluded that the use of polyamide fibers in the mix design of LCC can both improve mechanical properties and perhaps address the environmental issues associated with waste polyamide fibers.

Keywords: lime-based concrete; mechanical characteristics; fiber reinforcement; curing conditions; curing periods

Citation: Jafari, M.M.; Jahandari, S.; Ozbakkaloglu, T.; Rasekh, H.; Jahed Armaghani, D.; Rahmani, A. Mechanical Properties of Polyamide Fiber-Reinforced Lime–Cement Concrete. *Sustainability* **2023**, *15*, 11484. https://doi.org/10.3390/su151511484

Academic Editor: Mahdi Kioumarsi

Received: 14 April 2023
Revised: 4 July 2023
Accepted: 20 July 2023
Published: 25 July 2023

1. Introduction

Soil stabilization is a common method for improving mechanical properties, enhancing stability, and reducing the lateral deformation and settlement of soil [1–4]. Depending on the thickness of a soil layer, different stabilization methods can be adopted, including pre-loading, the use of chemical stabilizer additives, soft deposit replacement and removal, stone columns, surface mattresses, lightweight fills, surcharge loading, and control of compaction. However, some of the materials used in the treatment are costly and ineffective [4,5]. For example, replacement and removal, control of compaction, surcharge loading, and column of stone are somewhat costly. Meanwhile, the addition of gypsum is often found to be ineffective [6,7]. Consequently, new methods have emerged to deal with soil problems, aiming to enhance strength and decrease swelling behavior.

Lime stabilization is an attractive method for improving the properties of clayey soils. The main application of lime, as a well-established method in the construction industry, is the stabilization of problematic soils such as expansive soils, collapsible soils, and soft clays [8,9]. Lime concrete (LC), lime-stabilized soil, lime–hemp concrete, lime mortar, lime–cement concrete (LCC) have been used in various construction sites; however, the sample preparation process and their applications make them very different from each other [10–15]. In the lime application for construction projects, the clay soil, lime, and water are compacted using a roller after being mixed well. Improving the rail tracks and pavement layers are among the main applications of this method [7,16]. Another lime application is lime mortar, which is used to bond masonry blocks [17,18]. To make lime blocks, lime–hemp concrete is manufactured using water, clay, lime, hemp, and coarse aggregates [19,20].

Various studies have reported on the feasibility of using lime as a clay soil stabilizer. Because of the strong pozzolanic reaction between lime and soil, the treatment can reduce the plasticity index, increase the unconfined compressive strength (UCS), and control the swelling potential [21]. For example, Bartlett and Farnsworth [22] concluded that soil stabilized with LCC has a lower initial settlement by about 0.8 m in the stabilized area in comparison with the untreated area. Rogers et al. [8] concluded that the use of lime improves the resilient modulus and strength of the soil and that the addition of lime and cement can significantly reduce the plastic strain. Chand and Subbarao [23] reported that specimens stabilized with 14% lime cured for 180 days had higher UCS, rebound hammer numbers, slake durability, and point load strength index scores compared to specimens treated with 10% lime at different curing times. The study by Malekpoor and Toufigh [24] showed substantial growth in the load-bearing capacity of soft soil improved with LC columns containing 20% lime and 22% clay and a substantial strength reduction while in a saturated condition. Toohey et al. [25] reported that the UCS results for the specimens cured at 23 °C for 28 days were lower compared with those cured at 41 °C for 7 days.

The use of lime in soil stabilization and the production of LCC results in improved mechanical properties and reduced swelling potential and ductility. Adding fibers to the mix design of LCC and lime-stabilized soils is considered to be a cost-effective and environmentally friendly method to improve ductility performance [26]. Fibers have a wide application in various fields of engineering [27–29]. The application of fibers as soil reinforcement with or without cement has been reported previously [30–36]. Miraftab and Lickfold [35] reported that, in recent years, many waste producers have sought to utilize waste fiber products to improve the performance of various materials. Improving the mechanical properties of soils with the addition of fibers can reduce the required cement content. Mandal and Murti [37] found that a mixture of soil and high-tensile-strength fibers can improve the engineering characteristics of soil.

In recent decades, various natural fibers such as coconut fibers [38], sugarcane [39], sisal [40], hemp [41], and palm fiber [42] have been used for soil reinforcement. In many cases, synthetic fibers have also been used for this purpose. Among the different types of synthetic fibers used for soil reinforcement, polyethylene fiber [42,43], polyester [44,45], glass [44,46–48], carbon [49,50], and iron [51] can be mentioned. But, it is important to pay attention to the fact that, despite the cheapness and better ductility of natural fibers compared to synthetic fibers, they have much less resistance to atmospheric conditions and environmentally corrosive factors, especially compared to polymer synthetic fibers [52]. The use of polyamide fibers, as a synthetic textile made of petroleum-based plastic polymers, can both reinforce soils and concretes and also may contribute to the protection of the environment [53].

Synthetic fibers are increasingly being used to reinforce soil. Jiang et al. [54] found that the 0.3% (by weight) polypropylene fiber is the optimal fiber percentage to increase the internal friction angle, cohesion, and UCS of soil. However, by increasing the fiber content to above 0.3%, the abovementioned parameters were reduced. Consoli et al. [55] reported that an increase in fiber content, fiber aspect ratio, and fiber length improved the strength of the soil. However, by increasing the fiber diameter, the deviatoric stress decreased. Zaimoglu [56] concluded that, by increasing the polypropylene fiber content, the ductility of fiber-reinforced, fine-grained soils decreased significantly. Miller and Rifai [57] reported a reduction in the formation of cracks and hydraulic conductivity of compacted soil after using polypropylene fibers.

Stefanidou et al. [58] studied the effects of wood and cannabis, polypropylene, cellulose, and carbon fibers in pure lime mortars. Lime mortars reinforced with polypropylene and cannabis had a significantly higher amount of fracture energy compared to other mortars. The mechanical properties of wood-modified lime mortars improved by almost 25% in comparison to the reference sample; however, the cohesion of the matrix was problematic. Almerich et al. [59] used Glass Fiber-Reinforced Polymer (GFRP) bars as a replacement for steel bars in reinforced lime concrete. It was observed that the lime concrete modified with GFRP had high UCS, acceptable tensile strength, a higher elastic modulus and adhesion, good compatibility, and reduced cement alkalinity.

A review of the previous research studies indicated that there are limited studies available on the mechanical characteristics of lime-based concretes such as LCC, even though there are many studies available on lime–cement-stabilized soils, lime–hemp concrete and lime–cement mortars. Moreover, a comprehensive literature review by the authors demonstrated that, at present, there has been no research performed to evaluate the effect of polyamide fibers on the mechanical properties of LCC as a widely used, lime-based concrete in some warm regions. To address this research gap, LCC samples were prepared with different percentages of polyamide fibers and cured at different times at two different temperatures. Then, the effects of the fibers' contents, curing conditions, and curing periods on the mechanical properties of LCC, such as the deformability index, secant modulus, bulk modulus, shear modulus, stiffness ratio, strain energy, failure strain, strength ratio, and failure patterns, were investigated. Based on the test results, a simple mathematical model was also established to forecast the mechanical properties of polyamide fiber-reinforced LCC.

2. Raw Materials

The coarse aggregates utilized in the current study were well-graded sand with silt (SW-SM) that were sourced from a quarry in Iran, named Ekhtiarabad, where most of the coarse aggregates utilized in concreting projects in Kerman, a city in Iran, are sourced from. In most construction sites in Kerman, layers of fine-grained soils are found about 30 m beneath the ground surface. Therefore, the fine-grained soil utilized in the current work was sourced from a building site in Kerman, which can reasonably represent the soil properties that are available in most construction sites in Kerman. The sieve and hydrometer analysis test results and the engineering properties of the used soils are presented in Figure 1 and Table 1. In Table 1, D10 represents the diameter where 10% of the sample's mass consists of smaller particles. The uniformity coefficient (Cu) is determined by the ratio of D60 to D10. The coefficient of curvature (C_c) is given by the formula: $C_c = \frac{D30^2}{D60 \cdot D10}$. Additionally, the abbreviations PL, LL, and PI represent plastic limit, liquid limit, and plastic index, respectively. The chemical compositions of the materials were determined using X-ray Fluorescence (XRF) analysis using Bruker S_4-Explorer. As can be seen in Table 2, the clayey soil is mostly characterized by 41.75% SiO_2, 15.15% Al_2O_3, and 5.2% Fe_2O_3, which has a total share of 62.1% (lower than 70%, which is specified as the minimum requirement based on ASTM C618-15 [60]). Therefore, increasing the compressive strength and reducing the shrinkage and swelling properties of the soil could be achieved through the application of pozzolanic materials such as cement, lime, or a blend of cement and lime [5,61]. Table 2 shows that calcium oxide (CaO) is the main component of lime and that CaO and silica (SiO_2) are key components of cement. Similar types of materials were used in the authors' previous studies [13,15].

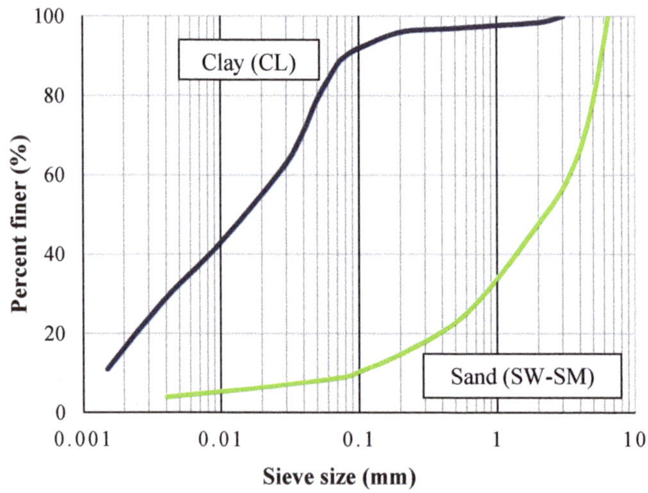

Figure 1. Sieve analysis of tested soils [15].

Table 1. Properties of tested soils.

Characteristics	Results	Used Standards
Coarse aggregates type	SW-SM	ASTM 2487-11 [62]
D_{10}	0.11	ASTM D422-63 [63]
C_u	33	ASTM D422-63 [63]
C^c	1.86	ASTM D422-63 [63]
PL	21%	ASTM D424-54 [64]
LL	26%	ASTM D423-66 [65]
PI	5%	Das 2019 [66]
Fine aggregates type	Cay (CL)	ASTM 2487-11 [67]
Mineral	Kaolinite	Das 2019 [66]
Activity degree	0.48	Das 2019 [66]
D_{10}	0.0016	ASTM D422-63 [63]
C_u	18	ASTM D422-63 [63]
C_C	0.40	ASTM D422-63 [63]
PL	23%	ASTM D424-54 [64]
LL	33%	ASTM D423-66 [65]
PI	10%	Das 2019 [66]
Optimum moisture	15%	AASHTO T180 [68]
Maximum specific weight	18.76 kN/m^3	AASHTO T180 [68]
Specific gravity (G_s)	2.46	ASTM D854-10 [69]
UCS	111.33 kPa	ASTM D2166 [70]

Table 2. Oxide compounds of cement, lime, and clay.

Component Oxides	Clay Composition (%)	Lime Composition (%)	Cement Composition (%)
Calcium oxide (CaO)	13.20	73.70	63.41
Silica (SiO_2)	41.75	1.15	21.66
Alumina (Al_2O_3)	15.15	0.11	4.21
Iron oxide (Fe_2O_3)	5.20	0.24	3.10
Magnesium oxide (MgO)	5.13	1.619	2.82
Sulfur trioxide (SO_3)	3.48	0.015	2.61
Chloride as NaCl	0.08	0.011	-
Manganese (Mn)	-	0.005	-
Loss on ignition	12.58	23.15	0.81

3. Methodology, Sample Preparation, and Testing

A total of 48 cylindrical LCC specimens were prepared with different fiber contents for different curing days and curing conditions. To increase the accuracy of the results, three (3) similar specimens were cast for each test, and the average testing results were calculated and reported.

Polyamide fibers are known to have ultrahigh tensile strength at low weight, a high melting point, and excellent heat and flame resistance as well as appropriate solvent resistance and dimensional stability at elevated and room temperatures.

Table 3 shows the mixture proportion, curing time, and curing condition for the different mixes. The percentages of cement and lime in all specimens were kept constant at 4% and 3% of the dry weight of the clay and coarse-grained soil mixtures, respectively. The water content was 24.04% of the total dry weight of the clay, coarse-grained soil, cement, and lime mixture; moreover, in all the specimens, the clay content was equal to 23% of the dry weight of coarse-grained soil. The fibers were added to the mixes at 0.5%, 1%, and 2% of the total dry weight of the coarse-grained soil, cement, lime, and clay mixture.

Table 3. Composition of LCC specimens.

Specimen	Clay Content (%)	Lime Content (%)	Cement Content (%)	Water Content (%)	Fiber Content (%)	Curing Temp. (°C)	Curing Time (Days)	Number of Samples	Curing Condition
* F0-7-L	23	3	4	24.04	0	20	7	3	Lab
F0.5-7-L	23	3	4	24.04	0.5	20	7	3	Lab
F1-7-L	23	3	4	24.04	1	20	7	3	Lab
F2-7-L	23	3	4	24.04	2	20	7	3	Lab
F0-28-L	23	3	4	24.04	0	20	28	3	Lab
F0.5-28-L	23	3	4	24.04	0.5	20	28	3	Lab
F1-28-L	23	3	4	24.04	1	20	28	3	Lab
F2-28-L	23	3	4	24.04	2	20	28	3	Lab
F0-3-O	23	3	4	24.04	0	50	3	3	Oven
F0.5-3-O	23	3	4	24.04	0.5	50	3	3	Oven
F1-3-O	23	3	4	24.04	1	50	3	3	Oven
F2-3-O	23	3	4	24.04	2	50	3	3	Oven
F0-7-O	23	3	4	24.04	0	50	7	3	Oven
F0.5-7-O	23	3	4	24.04	0.5	50	7	3	Oven
F1-7-O	23	3	4	24.04	1	50	7	3	Oven
F2-7-O	23	3	4	24.04	2	50	7	3	Oven

* Explanation of abbreviations: F(fiber content)-(curing days)-(curing condition); for example: F0-7-L = sample with 0% fiber and cured for 7 days in laboratory condition.

In order to prepare the LCC samples, coarse and fine aggregates were properly blended together. Then, a lime–cement slurry was made by blending lime and cement with optimum water content. Then, the mixture of coarse-grained and clay soils was added to the lime–cement slurry. The process of mixing continued until a homogeneous paste was attained. The fresh mixture was cast into several cylinders that were 60 mm in diameter and 120 mm in height. After 72 h of concrete casting, the samples were demolded and then placed in a plastic bag to avoid moisture loss. Half of the samples were oven-cured for 3 or 7 days at 50 °C, and the rest were cured for 7 or 28 days at an ambient temperature of 20 °C. The samples were then tested under uniaxial compression after curing because compressive strength is the most important factor in concrete design [71].

4. Results and Discussion

4.1. Failure Pattern

Figures 2 and 3 show the failure patterns of the control and LCC samples. The LCC without fibers had diagonal cracks and gradually expanded toward the middle of the sample. Therefore, the LCC surface clod spalled off, leading to a brittle failure. For the

LCC samples with 0.5% or 1% fibers, the crack width became smaller with increasing fiber content. The uniformly dispersed fibers in the LCC created a mesh constraint on the soil particles. Thus, crack initiation and development were inhibited. By adding extra fibers (2%) to the LCC, the fibers agglomerated in it. This was due to the development of cracks in the LCC, destroying it gradually due to fiber agglomeration, i.e., when the frictional resistance between fibers is smaller compared with that between the fiber and soil particles. Duan and Zhang [72] have also reported a similar failure pattern.

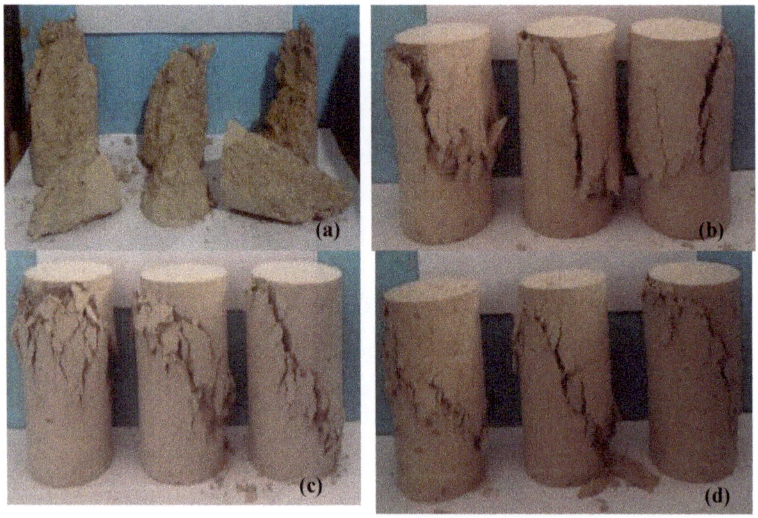

Figure 2. A view of oven-cured specimens after failure: (**a**) 0%, (**b**) 0.5%, (**c**) 1%, and (**d**) 2% fiber content.

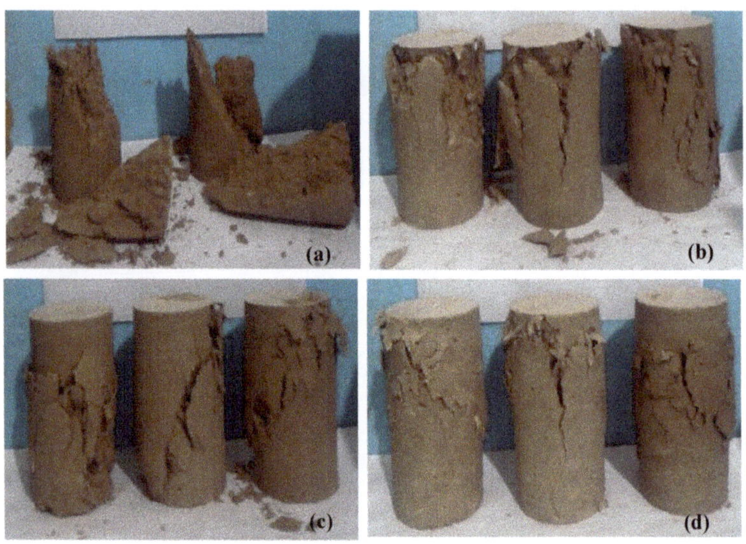

Figure 3. A view of 28-day, ambient-cured specimens after failure: (**a**) 0%, (**b**) 0.5%, (**c**) 1%, and (**d**) 2% fiber content.

4.2. Unconfined Compressive Strength and Stress–Strain Properties of Lime–Cement Concrete

The results of the UCS tests on the LCC specimens with fiber contents of 0%, 0.5%, 1%, and 2% cured at various times and conditions are shown in Figures 4 and 5. The results show that, by adding any percentage of fiber content, the UCS increased compared to the specimens without fibers. The tensile strength of fiber is high compared to concrete with a low tensile strength. Therefore, by adding fibers to concrete, the tensile force (relative movement of the particles) moves towards fibers and, ultimately, the strength increases. Irrespective of curing time and condition, specimens with 1% fibers achieved the highest UCS. The UCS for specimens F1-3-O, F1-7-O, F1-7-L, and F1-28-L increased by 25.4%, 15.9%, 19.2%, and 26.0% compared to the reference specimens without fibers.

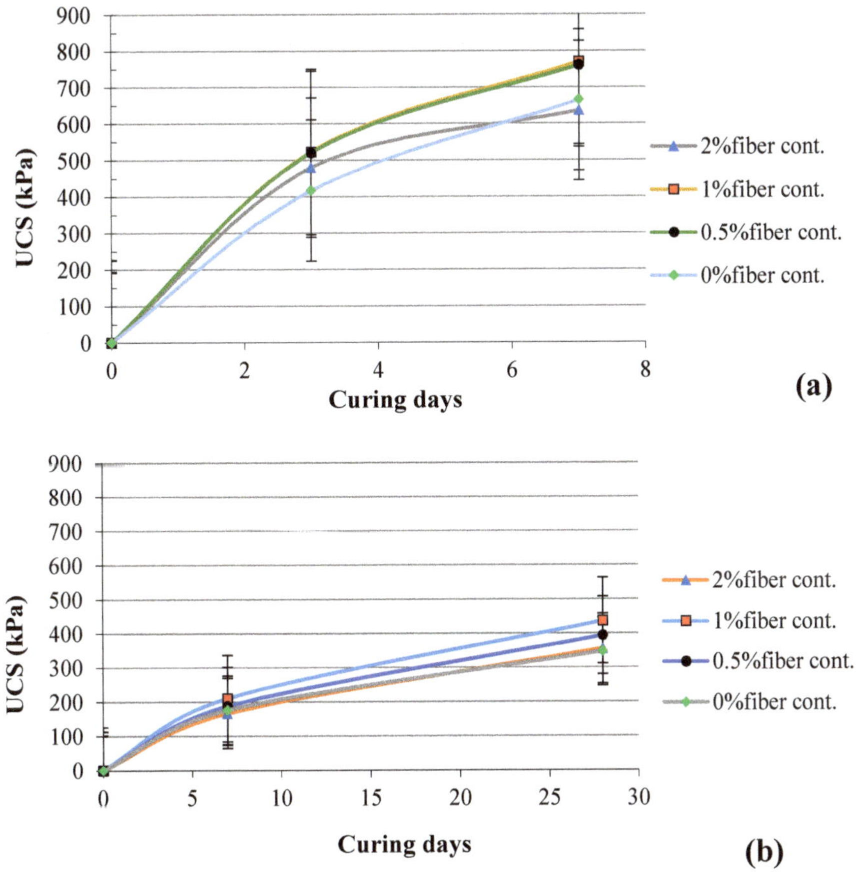

Figure 4. Peak UCS vs. curing time: (**a**) oven-cured samples, (**b**) ambient-cured samples.

The samples cured in an oven had higher UCS values compared with samples cured in an ambient condition. This is because hydration is enhanced in cementitious materials cured at high temperatures compared to ambient-cured materials [73–77]. The UCS of F1-7-O was more than 3.5 times that of F1-7-L. Therefore, by comparing the effects of all variables, it can be concluded that the curing condition has the highest impact on the UCS value. The significant impact of curing periods and saturation degrees on the compressive strength of LC has also been previously reported [13,15,78].

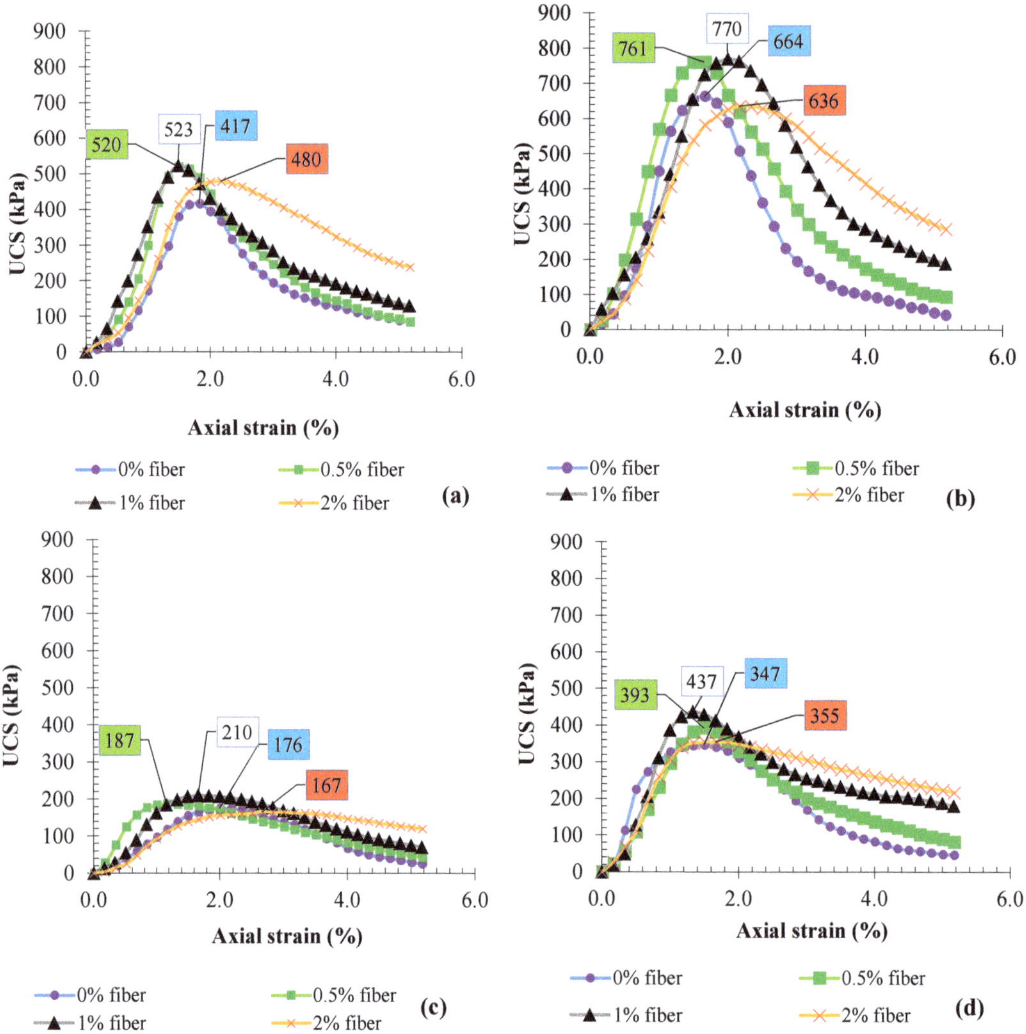

Figure 5. UCS vs. axial strain: (**a**) 3-day, oven-cured (3-O), (**b**) 7-day, oven-cured (7-O), (**c**) 7-day, lab-cured (7-L), (**d**) 28-day, lab-cured (28-L).

Saberian et al. [15] reported that the rate of increase in the UCS was negligible for specimens after a 28-day curing period. For instance, the UCS of LC at a 20% degree of saturation (S_r) in the first 28 days increased from 0 kPa to 259 kPa; however, in the next 32 days, it showed a further increase to 290 kPa. It was found that 89% of the maximum strength was achieved in the first 28 days. In the current study, all specimens achieved 68% of their 7-day strength in the first 3 days of their curing time in oven conditions; however, at ambient conditions, they only achieved 48% of their 28-day strength in the first 7 days of their curing time (see Figure 4).

Based on the experimental results shown in Figures 4 and 5, the relationships between UCS and the fiber content of the LCC specimens at different curing conditions and curing days were developed and summarized in Figure 6.

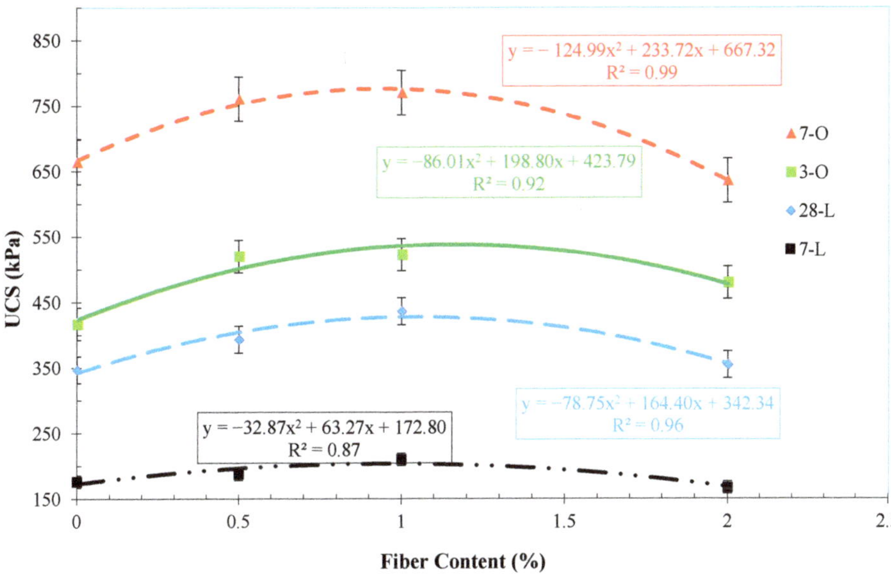

Figure 6. Relationship between UCS and fiber content at different curing conditions and curing days.

4.3. Mechanical Properties of Lime–Cement Concrete

A number of the mechanical properties of the LCC samples, including the failure strain, secant modulus (E_s), deformability index (I_d), resilient modulus (M_r), strain energy (SE), brittleness index (I_b), bulk modulus (K), shear modulus (G_s), strength ratio (R_{qu}), and the stiffness ratio (R_{Eu}) of specimens at different curing times, were determined by using the following equations and are summarized in Tables 4 and 5.

Table 4. Mechanical characteristics of oven-cured lime–cement concrete.

FC (%)	Curing Days	UCS (kPa)	ε_f (%)	E_s (MPa)	I_d	I_b	SE (kPa)	K (MPa)	G_s (MPa)	M_r (MPa)	R_{qu}	R_{Eu}
0	3	417	0.87	42	1.00	1.00	11	35	16	120	1.00	1.00
0.5	3	520	1.30	49	1.50	0.75	11	41	19	133	1.25	1.17
1	3	523	1.50	42	1.73	0.73	14	35	16	134	1.25	1.00
2	3	480	1.70	36	1.96	0.66	16	30	14	128	1.15	0.86
0	7	664	1.12	95	1.00	1.00	15	79	37	151	1.00	1.00
0.5	7	761	1.30	74	1.16	0.94	18	62	26	163	1.14	1.16
1	7	770	1.55	43	1.38	0.80	20	36	17	164	1.16	1.38
2	7	636	2.00	48	1.79	0.73	21	40	18	148	0.96	1.79

Table 5. Mechanical characteristics of ambient-cured lime–cement concrete.

FC (%)	Curing Days	UCS (kPa)	ε_f (%)	E_s (MPa)	I_d	I_b	SE (kPa)	K (MPa)	G_s (MPa)	M_r (MPa)	R_{qu}	R_{Eu}
0	7	176	1.83	12.14	1.00	1.00	4.87	10.12	5	91	1.00	1.00
0.5	7	187	1.06	29.26	0.58	0.78	6.15	24.38	11	92	1.06	2.41
1	7	210	1.29	21.73	0.70	0.67	7.00	18.11	8	95	1.19	1.79
2	7	167	1.44	12.18	0.79	0.61	7.16	10.15	5	89	0.95	1.00
0	28	347	1.92	66.13	1.00	1.00	9.51	55.11	25	112	1.00	1.00
0.5	28	393	1.00	35.89	0.52	0.83	10.35	29.91	14	118	1.14	0.54
1	28	437	1.03	52.46	0.54	0.83	11.57	43.72	20	123	1.26	0.79
2	28	355	1.17	45.04	0.61	0.79	13.67	37.53	17	113	1.02	0.68

The secant modulus was used to evaluate LCC resistance to deformation, which was determined from the UCS test results using Equation (1) [79].

$$E_S = \frac{\Delta\sigma}{\Delta\varepsilon} = \frac{\sigma_2 - \sigma_1}{\varepsilon_2 - \varepsilon_1} \quad (1)$$

where σ_2 is maximum stress in the elastic stage and σ_1 is minimum stress in the elastic stage; ε_1 and ε_2 are corresponding strains of σ_1 and σ_2.

The deformability index (I_D) is obtained from Equation (2) for measuring the deformation, brittleness, and ductility of soils and rocks [1,2].

$$I_D = \frac{\varepsilon_{ct}}{\varepsilon_{cu}} \quad (2)$$

where ε_{ct} is the axial strain at the maximum UCS of unreinforced LCC specimens (with 0% fiber content) and ε_{cu} is the axial strain at the maximum UCS of the fiber-reinforced LCC specimens.

The resilient modulus (M_r) is a measure of the soil elastic response under stress (AASHTO Test Method T307 2005), which is expressed as [3]:

$$M_r(\text{MPa}) = 0.124 \times \text{UCS}(\text{kPa}) + 68.8 \quad (3)$$

Bulk modulus (K) is the ratio of change in overall stress to the change in volumetric strain [80,81], which is determined as:

$$K = \frac{\sigma}{\frac{\Delta V}{V}} = \frac{\sigma}{\varepsilon_{xx} + \varepsilon_{yy} + \varepsilon_{zz}} = \frac{E_s}{3(1 - 2\vartheta)} \quad (4)$$

where $\frac{\Delta V}{V}$ is the relative volume change, σ is hydrostatic pressure, ε_{xx}, ε_{yy}, and ε_{zz} are direct strains, and υ is Poisson's ratio [81].

Shear modulus was measured using the following equation [82]:

$$G(\text{MPa}) = \frac{\sigma_{xy}}{(\varepsilon_{xy} + \varepsilon_{yx})} = \frac{\sigma_{xy}}{2\varepsilon_{xy}} = \frac{\sigma_{xy}}{\gamma_{xy}} = \frac{E_s(\text{kPa})}{2(1 + \vartheta)} = \frac{E_s(\text{kPa})}{3} \quad (5)$$

where σ_{xy} is shear stress, ε is shear strain, and $\gamma_{xy} = \varepsilon_{xy} + \varepsilon_{yx} = 2\varepsilon_{xy}$ [82].

Strain energy was measured as the area under the stress–strain curve [83,84]. The stiffness ratio and strength ratio for measuring the effects of fiber content (FC) on the stress–strain behavior of the soil and the failure mechanisms were obtained using Equations (6) and (7):

$$R_{q_u} = \frac{\text{UCS}_{(FC \neq 0)}}{\text{UCS}_{(FC = 0)}} \quad (6)$$

$$R_{E_u} = \frac{E_{s(FC \neq 0)}}{E_{s(FC = 0)}} \quad (7)$$

where UCS is the maximum unconfined compressive strength and Es is secant Young's modulus, which is the slope of the linear part of the stress–strain curve.

The brittleness index (I_b) is an indicator for the classification of soil strain softening severity and contractiveness. The brittleness index was determined using Equation (8) by calculating the variance between the peak and critical undrained shear strengths that are normalized by the peak undrained strength.

$$I_b = \frac{q_p - q_r}{q_p} \quad (8)$$

where q_p is the peak stress and q_r is the critical undrained shear strength obtained from UCS curves.

The results provided in Tables 4 and 5 show the relationship between the strain energy (energy absorption of LCC samples) and fiber content of samples cured for different days at both curing conditions. The testing outcomes show that an increase in the fiber content increased the area under the stress–strain curve. This effect can be seen in all specimens with different curing times and curing conditions. The strain energies of LCC specimens containing 0.5%, 1%, and 2% fiber contents in 3-day, oven-curing conditions increased by 6.54%, 35.73%, and 52.70%, respectively. This is because, by adding fibers to LCC specimens, the flexibility increased and the samples could resist greater strains.

Figure 7 shows the secant modulus versus the fiber content at different curing periods for ambient-cured and oven-cured samples, respectively. The results show that, at any curing time and curing condition, by increasing the fiber content, the secant modulus decreased because the contact behaviors between the cement and soil particles were decreased by the addition of fiber. This finding agrees with the conclusions drawn in previous studies [72,79,85,86].

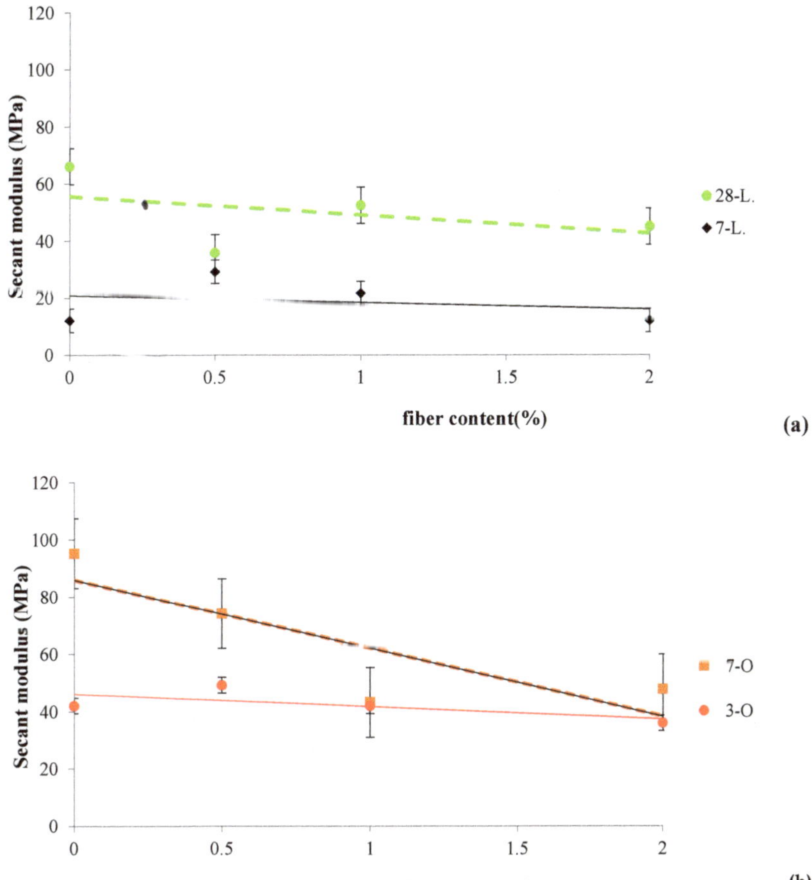

Figure 7. Relationship between secant modulus and fiber content of the LCC specimens in (**a**) ambient-curing and (**b**) oven-curing conditions.

Figure 8 shows the deformability index versus fiber content at different curing conditions. For samples under 3-day, oven-cured and 7-day, ambient-cured conditions, by increasing the fiber content, the deformability index increased. However, for specimens under the 7-day, oven-cured and 28-day, ambient-cured conditions, the deformability index increased by increasing the fiber content from 0% to 0.5%; moreover, it remained constant due to an increase in the fiber content from 0.5% to 2%. Therefore, using more than 0.5% fiber content did not affect the deformability index of the LCC specimens. The deformability of the fiber-reinforced LCC specimens increased regardless of the curing period and curing condition because the fiber increased the flexibility and deformability of the specimens.

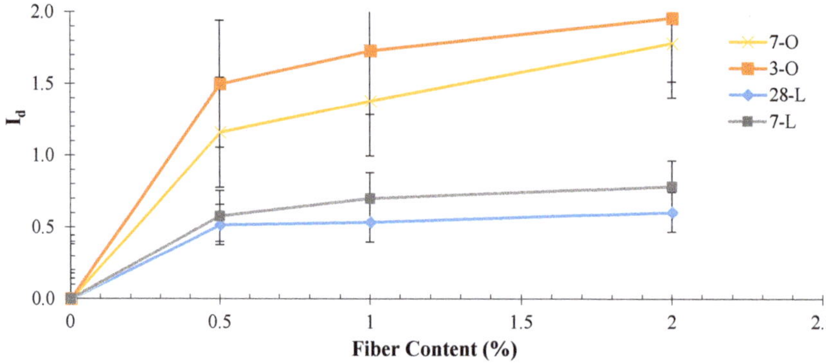

Figure 8. Relationship between deformability index and fiber content of specimens with different curing times and conditions.

According to Figure 9, it can be seen that with an increase in the amount of fibers and the age of the samples, the area under the stress-strain diagram, which is the energy absorbed by the material, is increased. Also, increasing the curing temperature is an important factor that increases the ability of materials to absorb energy. As can be seen in Figure 9, the samples that were cured at a higher temperature had a greater ability to absorb energy than the samples that were cured at the laboratory temperature. In other words, the ability of the samples to absorb energy was increased with increasing the curing temperature, age and fiber content.

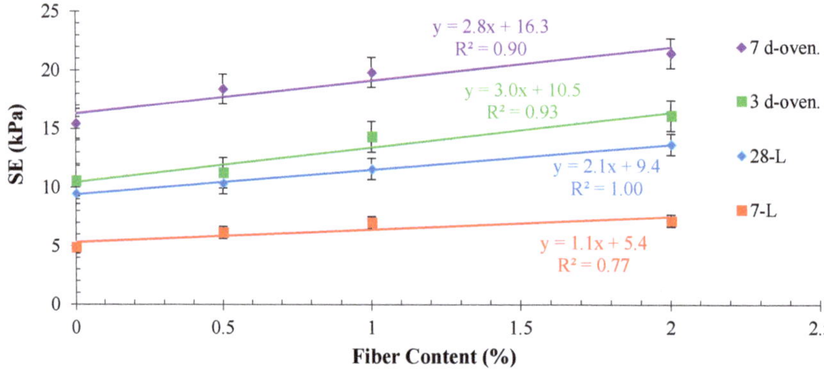

Figure 9. Relationship between strain energy and fiber content of specimens with different curing times and conditions.

5. Mathematical (Phenomenological) Model

According to the UCS test results, the mechanical properties of the LCC samples were affected significantly by the fiber content, curing condition, and curing time. Saberian et al. [4,15] and Cong et al. [87] used the power function expressed by Equation (9) to predict the results as this function provides excellent matches between the experimental and predicted results. Similar to Equation (9), a power function was introduced in the current study to predict the experimental results. In the proposed model, FC is the percentage of fiber content and CD is number of curing days. The brittleness index, deformability index, secant modulus, bulk modulus, resilient modulus, and shear modulus of fiber-reinforced LCC specimens, as well as the parameter of the FC/CD, were variables for predicting UCS.

$$UCS = a \times C^b \tag{9}$$

where C is equal to FC/CD; a and b (dimensionless parameter) are fitting parameters.

Based on the testing outcomes presented in Tables 4 and 5 and Equation (8), Figures 10–15 present the fitted curves for the relationships of compressive strength, E_s, K, G_s, M_r, and I_b versus the FC/CD of the LCC specimens. It can be seen that the fitted curves in Figures 10–15 have good accuracy.

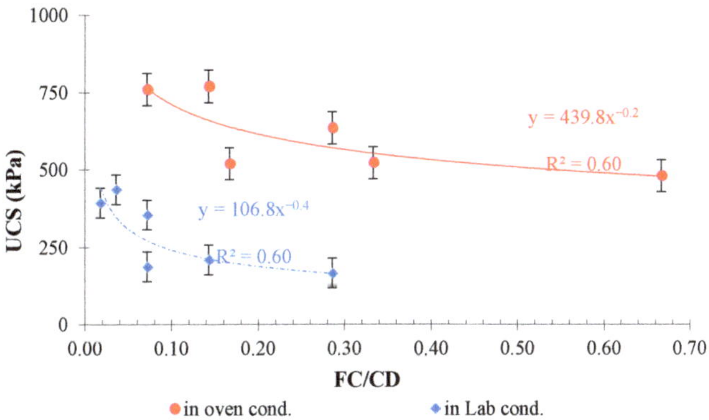

Figure 10. Fitted curves of UCS development as a function of FC/CD in oven- and ambient-cured conditions.

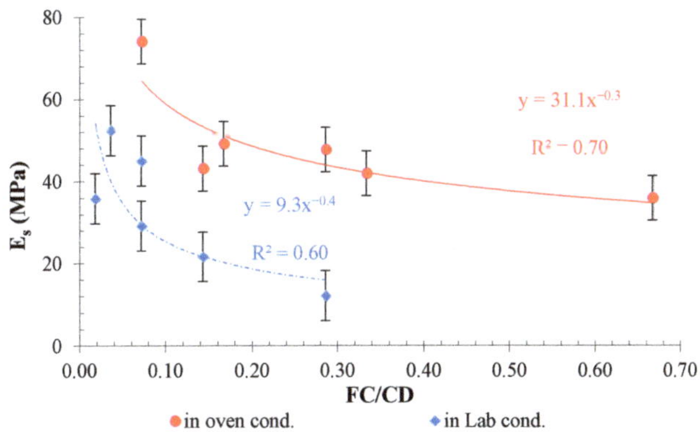

Figure 11. Fitted curves of Es development as a function of FC/CD in oven- and ambient-cured conditions.

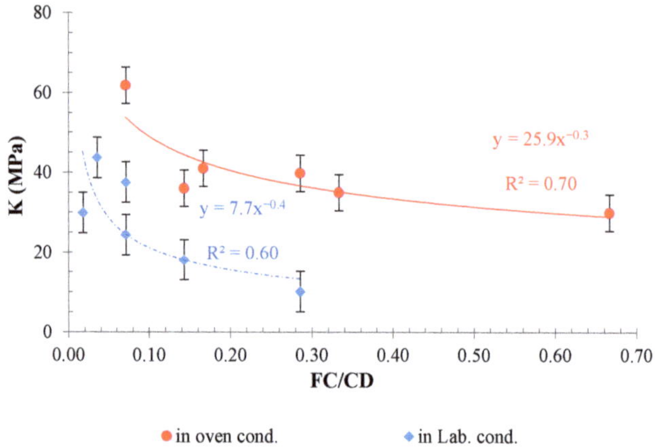

Figure 12. Fitted curves of bulk modulus development as a function of FC/CD in oven- and ambient-cured conditions.

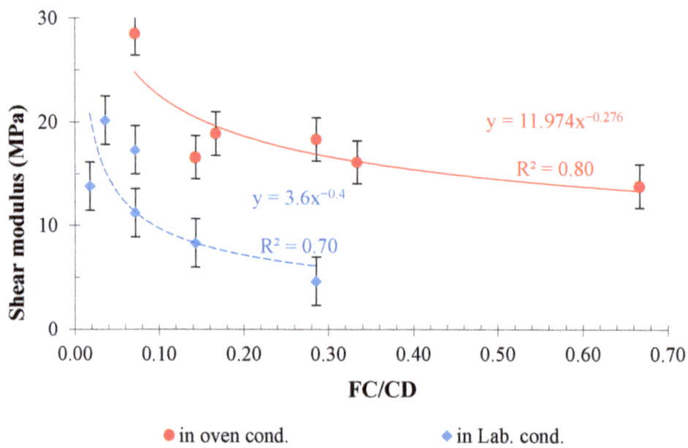

Figure 13. Fitted curves of G_s development as a function of FC/CD in oven- and ambient-cured conditions.

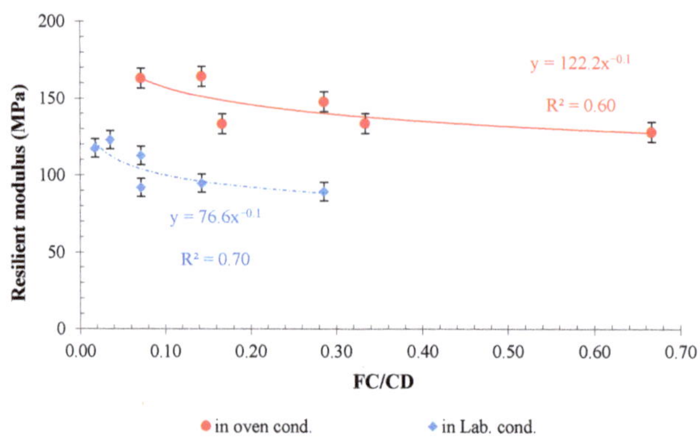

Figure 14. Fitted curves of resilient modulus development as a function of FC/CD in oven- and ambient-cured conditions.

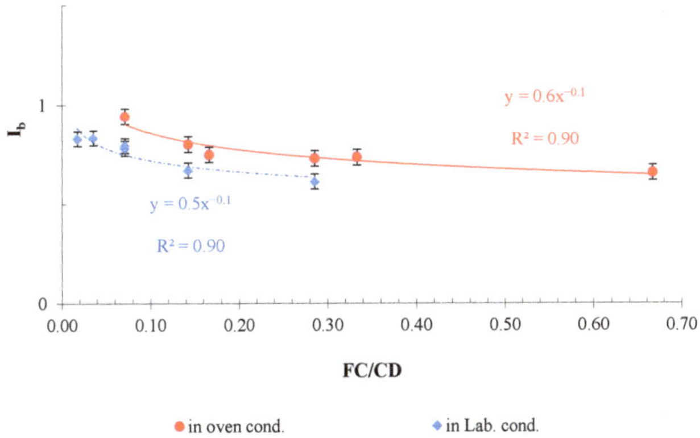

Figure 15. Fitted curves of brittleness index development as a function of FC/CD in oven- and ambient-cured conditions.

6. Conclusions

This research investigated the effects of curing times, curing conditions, and polyamide fiber contents on the mechanical characteristics of lime–cement concrete (LCC). According to the findings, the conclusions are summarized below:

- The optimum fiber content to increase the UCS of LCC is 1%. Increasing the fiber content from 1% to 2% led to a decrease in the UCS due to a reduction in cohesion;
- The energy absorption in LCC increased with increasing fiber content. In addition, LCC with a higher fiber content (i.e., over 1%) showed more ductile post-peak behavior compared to LCC with a lower fiber content;
- Curing times and conditions have significant effects on UCS values. Specimens cured in oven conditions showed higher UCS values compared to the ambient-cured specimens;
- The application of polyamide fibers, in general, showed a positive impact on improving the mechanical properties of LCC. However, the secant modulus of specimens for any curing condition and curing period decreased by increasing the fiber content;
- The deformability index of specimens for any curing condition and curing period increased by increasing the fiber content;
- Based on the laboratory test results, simple models were developed to predict the mechanical properties of LCC samples in relation to fiber content and curing days. Their prediction accuracy is reasonably good.

It is recommended that future studies be conducted to study the impact of fiber content on the flexural strength and workability of LCC specimens. Also, the durability of the fiber-reinforced LCC samples in exposure to wetting–drying and freezing–thawing conditions can be investigated.

Author Contributions: Conceptualization, M.M.J., S.J. and A.R.; methodology, M.M.J., S.J. and A.R.; validation, M.M.J., S.J. and A.R.; formal analysis, M.M.J., S.J. and A.R.; investigation, M.M.J., S.J. and A.R.; resources, H.R.; data curation, D.J.A.; writing—original draft preparation, M.M.J., S.J., T.O., H.R., D.J.A. and A.R.; writing—review and editing, T.O., S.J., H.R. and D.J.A., supervision, T.O., S.J., H.R. and D.J.A.; project administration, T.O., H.R. and D.J.A. All authors have read and agreed to the published version of the manuscript.

Funding: This research was funded by Technical and Vocational University (TVU) with Fund Number 25/400/96/161 and Letter Number 25/410/1616.

Institutional Review Board Statement: Not applicable.

Informed Consent Statement: Not applicable.

Data Availability Statement: The data presented in this study are available on request from the corresponding author.

Acknowledgments: The authors wish to acknowledge the technical support from Chem Concrete Pty Ltd. (www.chemconcrete.com.au).

Conflicts of Interest: The authors declare no conflict of interest.

References

1. Islam, M.S.; Hashim, R. Bearing Capacity of Stabilised Tropical Peat by Deep Mixing Method. *Aust. J. Basic Appl. Sci.* **2009**, *3*, 682–688.
2. Al-Swaidani, A.; Hammoud, I.; Meziab, A. Effect of Adding Natural Pozzolana on Geotechnical Properties of Lime-Stabilized Clayey Soil. *J. Rock Mech. Geotech. Eng.* **2016**, *8*, 714–725. [CrossRef]
3. Umar, M.; Kassim, K.A.; Chiet, K.T.P. Biological Process of Soil Improvement in Civil Engineering: A Review. *J. Rock Mech. Geotech. Eng.* **2016**, *8*, 767–774. [CrossRef]
4. Saberian, M.; Khabiri, M.M. Effect of Oil Pollution on Function of Sandy Soils in Protected Deserts and Investigation of Their Improvement Guidelines (case Study: Kalmand Area, Iran). *Environ. Geochem. Health* **2018**, *40*, 243–254. [CrossRef] [PubMed]
5. Zainorabidin, A.; Wijeyesekera, D.C. Geotechnical Challenges with Malaysian Peat. 2007. Available online: https://repository.uel.ac.uk/download/41d7bd3a3df6b74d6069261c0817e0b8f3d7dad24a5ff57f5dc3af0e34011ad3/363072/Zainorabidin%2C%20A%20%282007%29%20AC%26T%20252-61.pdf (accessed on 13 April 2023).
6. Puppala, A.J.; Musenda, C. Effects of Fiber Reinforcement on Strength and Volume Change in Expansive Soils. *Transp. Res. Rec.* **2000**, *1736*, 134–140. [CrossRef]
7. Saberian, M.; Rahgozar, M.A. Geotechnical Properties of Peat Soil Stabilised with Shredded Waste Tyre Chips in Combination with Gypsum, Lime or Cement. *Mires Peat* **2016**, *18*, 1–16. [CrossRef]
8. Rogers, C.D.; Boardman, D.I.; Papadimitriou, G. Stress Path Testing of Realistically Cured Lime and Lime/cement Stabilized Clay. *J. Mater. Civ. Eng.* **2006**, *18*, 259–266. [CrossRef]
9. Thyagaraj, T.; Zodinsanga, S. Swell–shrink Behaviour of Lime Precipitation Treated Soil. *Proc. Inst. Civ. Eng. Improv.* **2014**, *167*, 260–273. [CrossRef]
10. Jahandari, S. *Laboratory Study of Moisture and Capillarity Impact on Lime Concrete Resistance due to the Increase of Ground Water Level*; Faculty of Civil and Surveying Engineering, Department of Geotechnical Engineering, Graduate University of Advanced Technology: Kerman, Iran, 2015.
11. Ameri, M.; Kalantari, B.; Jahandari, S. Laboratory Study of Determination of Optimum Amount of Water and Clay in Mortar Made with Lime and Fly Ash. In Proceedings of the International Conference on Research in Science and Technology, Kualalumpur, Malaysia, 14 December 2015.
12. Jahandari, S.; Saberian, M.; Zivari, F.; Li, J.; Ghasemi, M.; Vali, R. Experimental Study of the Effects of Curing Time on Geotechnical Properties of Stabilized Clay with Lime and Geogrid. *Int. J. Geotech. Eng.* **2019**, *13*, 172–183. [CrossRef]
13. Jahandari, S.; Toufigh, M.M.; Li, J.; Saberian, M. Laboratory Study of the Effect of Degrees of Saturation on Lime Concrete Resistance due to the Groundwater Level Increment. *Geotech. Geol. Eng.* **2018**, *36*, 413–424. [CrossRef]
14. Jahandari, S.; Li, J.; Saberian, M.; Shahsavarigoughari, M. Experimental Study of the Effects of Geogrids on Elasticity Modulus, Brittleness, Strength, and Stress-Strain Behavior of Lime Stabilized Kaolinitic Clay. *GeoResJ* **2017**, *13*, 49–58. [CrossRef]
15. Saberian, M.; Jahandari, S.; Li, J.; Zivari, F. Effect of Curing, Capillary Action, and Groundwater Level Increment on Geotechnical Properties of Lime Concrete: Experimental and Prediction Studies. *J. Rock Mech. Geotech. Eng.* **2017**, *9*, 638–647. [CrossRef]
16. Firoozfar, A.; Khosroshiri, N. Kerman Clay Improvement by Lime and Bentonite to Be Used as Materials of Landfill Liner. *Geotech. Geol. Eng.* **2017**, *35*, 559–571. [CrossRef]
17. Costigan, A.; Pavía, S.; Kinnane, O. An Experimental Evaluation of Prediction Models for the Mechanical Behavior of Unreinforced, Lime-Mortar Masonry under Compression. *J. Build. Eng.* **2015**, *4*, 283–294. [CrossRef]
18. Pavlík, V.; Užáková, M. Effect of Curing Conditions on the Properties of Lime, Lime–metakaolin and Lime–zeolite Mortars. *Constr. Build. Mater.* **2016**, *102*, 14–25. [CrossRef]
19. de Bruijn, P.B.; Jeppsson, K.-H.; Sandin, K.; Nilsson, C. Mechanical Properties of Lime–hemp Concrete Containing Shives and Fibres. *Biosyst. Eng.* **2009**, *103*, 474–479. [CrossRef]
20. Walker, R.; Pavia, S.; Mitchell, R. Mechanical Properties and Durability of Hemp-Lime Concretes. *Constr. Build. Mater.* **2014**, *61*, 340–348. [CrossRef]
21. Thyagaraj, T.; Samuel, Z.; Kumar, K.S.R. Relative Efficiencies of Electrolytes in Stabilization of an Expansive Soil. *Int. J. Geotech. Eng.* **2016**, *10*, 107–113. [CrossRef]
22. Bartlett, S.; Farnsworth, C. Performance of Lime Cement-Stabilized Soils for the I-15 Reconstruction Project: Salt Lake City, Utah. *Transp. Res. Rec.* **2002**, *1808*, 58–66. [CrossRef]
23. Chand, S.K.; Subbarao, C. Strength and Slake Durability of Lime Stabilized Pond Ash. *J. Mater. Civ. Eng.* **2007**, *19*, 601–608. [CrossRef]
24. Malekpoor, M.; Toufigh, M. Laboratory Study of Soft Soil Improvement Using Lime Mortar-(well Graded) Soil Columns. *Geotech. Test. J.* **2010**, *33*, 225–235.

25. Toohey, N.M.; Mooney, M.A.; Bearce, R.G. Stress-Strain-Strength Behavior of Lime-Stabilized Soils during Accelerated Curing. *J. Mater. Civ. Eng.* **2013**, *25*, 1880–1886. [CrossRef]
26. Pakravan, H.R.; Ozbakkaloglu, T. Synthetic Fibers for Cementitious Composites: A Critical and in-Depth Review of Recent Advances. *Constr. Build. Mater.* **2019**, *207*, 491–518. [CrossRef]
27. Wang, L.; Tang, S. High-Performance Fiber-Reinforced Composites: Latest Advances and Prospects. *Buildings* **2023**, *13*, 1094. [CrossRef]
28. Wang, L.; Tang, S.; Chen, T.E.; Li, W.; Gunasekara, C. Sustainable High-Performance Hydraulic Concrete. *Sustainability* **2022**, *14*, 695. [CrossRef]
29. Wang, L.; Tang, S. High-Performance Construction Materials: Latest Advances and Prospects. *Buildings* **2022**, *12*, 928. [CrossRef]
30. Maher, M.H.; Ho, Y.C. Behavior of Fiber-Reinforced Cemented Sand under Static and Cyclic Loads. *Geotech. Test. J.* **1993**, *16*, 330–338.
31. Omine, K.; Ochiai, H.; Yasufuku, N.; Kato, T. Effect of Plastic Wastes in Improving Cement-Treated Soils. *Proc. 2nd Int. Congr. Environ. Geotech.* **1996**, 875–880.
32. Prabakar, J.; Sridhar, R.S. Effect of Random Inclusion of Sisal Fibre on Strength Behaviour of Soil. *Constr. Build. Mater.* **2002**, *16*, 123–131. [CrossRef]
33. Michalowski, R.L.; Čermák, J. Triaxial Compression of Sand Reinforced with Fibers. *J. Geotech. Geoenviron. Eng.* **2003**, *129*, 125–136. [CrossRef]
34. Cai, Y.; Shi, B.; Ng, C.W.W.; Tang, C.S. Effect of Polypropylene Fibre and Lime Admixture on Engineering Properties of Clayey Soil. *Eng. Geol.* **2006**, *87*, 230–240. [CrossRef]
35. Miraftab, M.; Lickfold, A. Utilization of Carpet Waste in Reinforcement of Substandard Soils. *J. Ind. Text.* **2008**, *38*, 167–174. [CrossRef]
36. Lovisa, J.; Shukla, S.K.; Sivakugan, N. Behaviour of Prestressed Geotextile-Reinforced Sand Bed Supporting a Loaded Circular Footing. *Geotext. Geomembr.* **2010**, *28*, 23–32. [CrossRef]
37. Mandal, J.N.; Murthi, M.V.R. Potential Use of Natural Fibres in Geotechnical Engineering. In Proceedings of the International Workshops on Geo-Textiles, Bangalore, India, 22 November 1989; pp. 22–29.
38. Sivakumar Babu, G.L.; Vasudevan, A.K. Strength and Stiffness Response of Coir Fiber-Reinforced Tropical Soil. *J. Mater. Civ. Eng.* **2008**, *20*, 571–577. [CrossRef]
39. Dang, L.C.; Fatahi, B.; Khabbaz, H. Behaviour of Expansive Soils Stabilized with Hydrated Lime and Bagasse Fibres. *Procedia Eng.* **2016**, *143*, 658–665. [CrossRef]
40. Kafodya, I.; Okonta, F. Effects of Natural Fiber Inclusions and Pre-Compression on the Strength Properties of Lime-Fly Ash Stabilised Soil. *Constr. Build. Mater.* **2018**, *170*, 737–746. [CrossRef]
41. Yixian, W.; Panpan, G.; Shengbiao, S.; Haiping, Y.; Binxiang, Y. Study on Strength Influence Mechanism of Fiber-Reinforced Expansive Soil Using Jute. *Geotech. Geol. Eng.* **2016**, *34*, 1079–1088. [CrossRef]
42. Azadegan, O.; Kaffash, E.A.; Yaghoubi, M.J.; Pourebrahim, G.R. Laboratory Study on the Swelling, Cracking and Mechanical Characteristics of the Palm Fiber Reinforced Clay. *Electron. J. Geotech. Eng.* **2012**, *17*, 47–54.
43. Akbulut, S.; Arasan, S.; Kalkan, E. Modification of Clayey Soils Using Scrap Tire Rubber and Synthetic Fibers. *Appl. Clay Sci.* **2007**, *38*, 23–32. [CrossRef]
44. Consoli, N.C.; Montardo, J.P.; Donato, M.; Prietto, P.D. Effect of Material Properties on the Behaviour of Sand—Cement—Fibre Composites. *Proc. Inst. Civ. Eng. Improv.* **2004**, *8*, 77–90. [CrossRef]
45. Chaduvula, U.; Viswanadham, B.V.S.; Kodikara, J. A Study on Desiccation Cracking Behavior of Polyester Fiber-Reinforced Expansive Clay. *Appl. Clay Sci.* **2017**, *142*, 163–172. [CrossRef]
46. Consoli, N.C.; Prietto, P.D.; Ulbrich, L.A. Influence of Fiber and Cement Addition on Behavior of Sandy Soil. *J. Geotech. Geoenviron. Eng.* **1998**, *124*, 1211–1214. [CrossRef]
47. Mukherjee, K.; Mishra, A.K. Hydro-Mechanical Properties of Sand-Bentonite-Glass Fiber Composite for Landfill Application. *KSCE J. Civ. Eng.* **2019**, *23*, 4631–4640. [CrossRef]
48. Patel, S.K.; Singh, B. Strength and Deformation Behavior of Fiber-Reinforced Cohesive Soil under Varying Moisture and Compaction States. *Geotech. Geol. Eng.* **2017**, *35*, 1767–1781. [CrossRef]
49. Gao, L.; Zhou, Q.; Yu, X.; Wu, K.; Mahfouz, A.H. Experimental Study on the Unconfined Compressive Strength of Carbon Fiber Reinforced Clay Soil. *Mar. Georesources Geotechnol.* **2017**, *35*, 143–148. [CrossRef]
50. Cui, H.; Jin, Z.; Bao, X.; Tang, W.; Dong, B. Effect of Carbon Fiber and Nanosilica on Shear Properties of Silty Soil and the Mechanisms. *Constr. Build. Mater.* **2018**, *189*, 286–295. [CrossRef]
51. Fatahi, B.; Khabbaz, H.; Fatahi, B. Mechanical Characteristics of Soft Clay Treated with Fibre and Cement. *Geosynth. Int.* **2012**, *19*, 252–262. [CrossRef]
52. Hejazi, S.M.; Sheikhzadeh, M.; Abtahi, S.M.; Zadhoush, A. A Simple Review of Soil Reinforcement by Using Natural and Synthetic Fibers. *Constr. Build. Mater.* **2012**, *30*, 100–116. [CrossRef]
53. Valipour, M.; Shourijeh, P.T.; Mohammadinia, A. Application of Recycled Tire Polymer Fibers and Glass Fibers for Clay Reinforcement. *Transp. Geotech.* **2021**, *27*, 100474. [CrossRef]
54. Jiang, H.; Cai, Y.; Liu, J. Engineering Properties of Soils Reinforced by Short Discrete Polypropylene Fiber. *J. Mater. Civ. Eng.* **2010**, *22*, 1315–1322. [CrossRef]

55. Consoli, N.C.; Vendruscolo, M.A.; Prietto, P.D.M. Behavior of Plate Load Tests on Soil Layers Improved with Cement and Fiber. *J. Geotech. Geoenviron. Eng.* **2003**, *129*, 96–101. [CrossRef]
56. Zaimoglu, A.S. Freezing–thawing Behavior of Fine-Grained Soils Reinforced with Polypropylene Fibers. *Cold Reg. Sci. Technol.* **2010**, *60*, 63–65. [CrossRef]
57. Miller, C.J.; Rifai, S. Fiber Reinforcement for Waste Containment Soil Liners. *J. Env. Eng.* **2004**, *130*, 891–895. [CrossRef]
58. Papayianni, I.; Stefanidou, M.; Pachta, V. Plastering the Prehistory: Marl as a unique material to cover, maintain and decorate the Neolithic walls of Catalhöyük. In Proceedings of the 4th Historic Mortars Conference HMC2016, Santorini, Greece, 10–12 October 2016.
59. Almerich-Chulia, A.; Fenollosa, E.; Martin, P. Reinforced Lime Concrete with FRP: An Alternative in the Restoration of Architectural Heritage. *Appl. Mech. Mater.* **2016**, *851*, 751–756. [CrossRef]
60. *ASTM C 618-19*; Standard Specification for Mortar for Coal Fly Ash and Raw or Calcined Natural Pozzolan for Use in Concrete. ASTM International: West Conshohocken, PA, USA, 2015.
61. Sirivitmaitrie, C.; Puppala, A.J.; Saride, S.; Hoyos, L. Combined Lime–cement Stabilization for Longer Life of Low-Volume Roads. *Transp. Res. Rec.* **2011**, *2204*, 140–147. [CrossRef]
62. *ASTM D 2487-06*; Standard Practice for Classification of Soils for Engineering Purposes (Unified Soil). ASTM International: West Conshohocken, PA, USA, 2010.
63. *ASTM D422-63*; Standard Method for Particle-Analysis of Soils. ASTM International: West Conshohocken, PA, USA, 2002.
64. *ASTM-D424-54*; Standard Method of Test for Plastic Limit. ASTM International: West Conshohocken, PA, USA, 1982.
65. *ASTM D423-66*; Standard Test Method for Liquid Limit of Soils. ASTM International: West Conshohocken, PA, USA, 1982.
66. Das, B.M. *Advanced Soil Mechanics*; CRC press: Boca Raton, FL, USA, 2019; ISBN 1351215167.
67. *ASTM D2487-17*; Standard Practice for Classification of Soils for Engineering Purposes (Unified Soil Classification System) 1. ASTM International: West Conshohocken, PA, USA, 2017.
68. *AASHTO T 180-10*; Standard Method of Test for Moisture-Density Relations of Soils Using a 4.54-Kg (10-Lb) Rammer and a 457-Mm (18-In.) Drop. American Association of State Highway and Transportation Officials: Washington, DC, USA, 2010.
69. *ASTM D 854-1*; Standard Test Methods for Specific Gravity of Soil Solids by Water Pycnometer. ASTM International: West Conshohocken, PA, USA, 2010.
70. *ASTM D 2166*; Standard Test Method for Unconfined Compressive Strength of Cohesive Soil. ASTM International: West Conshohocken, PA, USA, 2016.
71. Piro, N.S.; Mohammed, A.; Hamad, S.M.; Kurda, R. Artificial Neural Networks (ANN), MARS, and Adaptive Network-Based Fuzzy Inference System (ANFIS) to Predict the Stress at the Failure of Concrete with Waste Steel Slag Coarse Aggregate Replacement. *Neural Comput. Appl.* **2023**, *35*, 13293–13319. [CrossRef]
72. Duan, X.; Zhang, J. Mechanical Properties, Failure Mode, and Microstructure of Soil-Cement Modified with Fly Ash and Polypropylene Fiber. *Adv. Mater. Sci. Eng.* **2019**, *2019*, 9561794. [CrossRef]
73. Elkhadiri, I.; Palacios, M.; Puertas, F. *Effect of Curing Temperatura on Hydration Process of Different Cement*; Czech Academy of Sciences: Staré Město, Czech Republic, 2009.
74. Price, W.H. Factors Influencing Concrete Strength. *J. Proc.* **1951**, *47*, 417–432. [CrossRef]
75. Escalante-Garcıa, J.I.; Sharp, J.H. The Microstructure and Mechanical Properties of Blended Cements Hydrated at Various Temperatures. *Cem. Concr. Res.* **2001**, *31*, 695–702. [CrossRef]
76. Lothenbach, B.; Winnefeld, F.; Alder, C.; Wieland, E.; Lunk, P. Effect of Temperature on the Pore Solution, Microstructure and Hydration Products of Portland Cement Pastes. *Cem. Concr. Res.* **2007**, *37*, 483–491. [CrossRef]
77. Ballester, P.; Hidalgo, A.; Mármol, I.; Morales, J.; Sánchez, L. Effect of Brief Heat-Curing on Microstructure and Mechanical Properties in Fresh Cement Based Mortars. *Cem. Concr. Res.* **2009**, *39*, 573–579. [CrossRef]
78. Jahandari, S.; Saberian, M.; Tao, Z.; Mojtahedi, S.F.; Li, J.; Ghasemi, M.; Rezvani, S.S.; Li, W. Effects of Saturation Degrees, Freezing-Thawing, and Curing on Geotechnical Properties of Lime and Lime-Cement Concretes. *Cold Reg. Sci. Technol.* **2019**, *160*, 242–251. [CrossRef]
79. Tang, L.; Cong, S.; Geng, L.; Ling, X.; Gan, F. The Effect of Freeze-Thaw Cycling on the Mechanical Properties of Expansive Soils. *Cold Reg. Sci. Technol.* **2018**, *145*, 197–207. [CrossRef]
80. Duncan, J.M.; Bursey, A. Soil Modulus Correlations. In *Foundation Engineering in the Face of Uncertainty: Honoring Fred H. Kulhawy*; American Society of Civil Engineers: Reston, VA, USA, 2013; pp. 321–336.
81. Hobbs, D.W. The Dependence of the Bulk Modulus, Young's Modulus, Creep, Shrinkage and Thermal Expansion of Concrete upon Aggregate Volume Concentration. *Matériaux Constr.* **1971**, *4*, 107–114. [CrossRef]
82. Selvadurai, A.P.S.; Katebi, A. Mindlin's Problem for an Incompressible Elastic Half-Space with an Exponential Variation in the Linear Elastic Shear Modulus. *Int. J. Eng. Sci.* **2013**, *65*, 9–21. [CrossRef]
83. Ostadan, F.; Ostadan, F.; Arango, I. *Energy-Based Method for Liquefaction Potential Evaluation, Phase I-Feasibility Study*; US Department of Commerce, National Institute of Standards and Technology: Gaithersburg, MD, USA, 1996.
84. Green, R.A.; Mitchell, J.K.; Polito, C.P. An Energy-Based Excess Pore Pressure Generation Model for Cohesionless Soils. In *Proceedings of the John Booker Memorial Symposium, Sydney, Australia*; AA Balkema Publishers: Rotterdam, The Netherlands, 2000; Volume 3.

85. Ding, M.; Zhang, F.; Ling, X.; Lin, B. Effects of Freeze-Thaw Cycles on Mechanical Properties of Polypropylene Fiber and Cement Stabilized Clay. *Cold Reg. Sci. Technol.* **2018**, *154*, 155–165. [CrossRef]
86. Li, L.; Shao, W.; Li, Y.; Cetin, B. Effects of Climatic Factors on Mechanical Properties of Cement and Fiber Reinforced Clays. *Geotech. Geol. Eng.* **2015**, *33*, 537–548. [CrossRef]
87. Cong, M.; Longzhu, C.; Bing, C. Analysis of Strength Development in Soft Clay Stabilized with Cement-Based Stabilizer. *Constr. Build. Mater.* **2014**, *71*, 354–362. [CrossRef]

sustainability

Article

Factorial Mixture Design for Properties Optimization and Modeling of Concrete Composites Incorporated with Acetates as Admixtures

Ammar Ali Abed [1,2,*], Alireza Mojtahedi [2] and Mohammad Ali Lotfollahi Yaghin [2]

1 Department of Engineering Affairs, General Company for Ports of Iraq, Basrah 61004, Iraq
2 Department of Water Resources Engineering, Faculty of Civil Engineering, University of Tabriz, Tabriz 51368, Iran; a.mojtahed@tabrizu.ac.ir (A.M.); lotfollahi@tabrizu.ac.ir (M.A.L.Y.)
* Correspondence: eng.ammar@tabrizu.ac.ir

Abstract: Nowadays, admixtures are used with the aim to provide strength and durability to concrete with less water use. New and low-cost admixtures gained a large amount of consideration to mitigate the problems associated with concrete's durability and service life without upsetting its strength properties. The current work investigates the effect of three types of acetates on the workability, density, and compressive strength of concrete, which is used in structures of the Iraqi ports that suffer from corrosion damages and deterioration owing to the aggressive marine environments. Potassium acetate (KA), calcium acetate (CaA), and ethyl acetate (EA) are incorporated with different doses (1.38–5.6 wt.% of cement) in concrete mixtures using different water/cement ratios (0.48–0.54) based on an espoused central composite experimental design. The experimental results confirmed that the average workability increased with increasing the acetate dose, particularly with CaA. The density and compressive strength of 28 days of water-cured mixtures increased with increasing acetate dose following the order: Ca > K > Ethyl acetate and decreased with increasing w/c ratio. The high rise in compressive strength and workability linked to control mixtures was 30.8% and 77.3% as well as 15.7% and 64.3% for the mixtures incorporated with 5.6 wt.% CaA and KA, respectively. While it was 14.2% and 58.3% for the mixtures incorporated with 3.5 wt.% EA. RSM was employed to optimize and model the design and hardened properties of concrete mixtures. ANOVA results predicted the same trend, which was obtained from the experimental results. The mathematical models were valued with high-regression coefficients. The highest compressive strength of 42.68 MPa has been achieved for a concrete mixture of 0.48 w/c ratio by the incorporation of 5.1 wt.% CaA through a model with R^2 96.97%. The relatively low-cost acetate admixtures, particularly CaA, seemed promising for the fabrication of concrete with outstanding properties.

Keywords: acetates; compressive strength; concrete; optimization; response surface methodology; workability

Citation: Abed, A.A.; Mojtahedi, A.; Lotfollahi Yaghin, M.A. Factorial Mixture Design for Properties Optimization and Modeling of Concrete Composites Incorporated with Acetates as Admixtures. *Sustainability* **2023**, *15*, 10608. https://doi.org/10.3390/su151310608

Academic Editor: Ramadhansyah Putra Jaya

Received: 19 May 2023
Revised: 1 July 2023
Accepted: 3 July 2023
Published: 5 July 2023

1. Introduction

Iraq owns many commercial ports with distinguished rural locations. They occupy an effective position in the national economy. The ports are designated for commercial purposes (import and export). They are located in the Basra governorate, on the right bank of the Shatt al-Arab, such as the port of Al-Maqil, Abu Flus, Umm Qasr, and Khor al-Zubayr. Although the Iraqi ports have an important geographical location in world trade, as they are considered the ideal way for transit traffic trade between East and Southeast Asia and Europe, the ports have suffered many challenges including environmental stresses, resulting in corrosion damages and deterioration of the structures of the ports exposed to the aggressive marine environments, in particular, the high salinity of seawater with an average of about 40.5 g/kg [1]. Therefore, it became essential and economical to plan port activities in order to reduce port activity disruption by preserving port infrastructure, rather than demolishing and reconstructing the damaged facilities. This can be practiced by applying some concrete corrosion mitigation procedures using anti-corrosion materials as concrete admixtures.

It is well-confirmed that materials are one of the most important considerations for solving the shortcoming in concrete. In severe exposure conditions, the specification and/or use of unsuitable materials can lead to poor durability; therefore, using efficient materials in construction has become the most innovative method used for construction projects. In addition, using low water-to-cementitious materials ratio and supplementary materials will provide denser and more durable concrete equated with representative Portland cement concrete [2].

Several additives have been applied to obtain low permeability, high density, and load-carrying capacity, as well as consent for the building of durable and high-strength concrete structures. Among those additives are nanoparticles including nano-silica (nano-SiO_2), nano-alumina (nano-Al_2O_3), nano-ferric oxide (nano-Fe_2O_3), nano-titanium oxide (nano-TiO_2), carbon nanotubes (CNTs), graphene and graphene oxide [2–4]. Moreover, nanopolymers were found to participate in self-healing mechanisms [5,6], as well as rewarding an improved product layer on the steel surface and calcium content in the bulk matrix [7].

Among other additives are corrosion inhibitors which are added in slight amounts to extend the corrosion start time and avoid the onset of corrosion in reinforced concrete structures [8]. These include inorganic inhibitors, such as nitrates, chromates, molybdate, phosphate, carbonates, polyphosphates, and silicates [9–13], as well as organic inhibitors including amines and alkanol amines and their salts. The organic inhibitors can inhibit steel corrosion through an adsorption mechanism by forming a thin layer of shielding barrier film on the surface of the rebar [11,14]. Moreover, corrosion inhibitors may be embedded in microcapsules that could be added into fractured concrete and successfully passivated or maintained the inactiveness of the rebar when the concrete is damaged [15]. However, controlling the concrete, such as using low w/c ratios, increasing cement content in the concrete composition, and optimizing the inhibitor doze leads to better anticorrosion resistance [16].

In addition, the growth and operative application of chemical admixtures in concrete technology resulted in new advancements in the formulation and application of cement and concrete composites. The incorporation of a small amount of admixture could effectively adjust the performance of cement-based materials. The improvement includes the physical, mechanical, durability, rheology, and hydration properties [17]. The functional groups of admixtures are generally categorized into hydroxyl, carboxyl, amino, and sulfonic acid [18]. However, it was pointed out that modifying and improving the performance of hydroxyl- or carboxyl-containing admixtures could be carried out by esterification [19].

Superplasticizers are one of the admixtures that act to modify the concrete workability, and at the same time, increase the strength of concrete. They neutralize the surface attractions between cement particles resulting in a less open structure that requires less water/cement ratio; therefore, increasing the strength and durability of concrete. However, most of the superplasticizers used are synthetic and more expensive than other types of concrete admixtures [20].

On the other hand, new green and low-cost admixtures that can work efficiently in the presence of water are developed by some researchers. Those additives are added to mitigate the problems associated with concrete's durability and service life without upsetting its strength properties [21,22].

From a different perspective, RSM is an efficient statistical tool that has been widely used and successfully employed to set the experimental design and models that mathematically describe the effects of the operating parameters on the process's responses. Literature information on using RSM in the optimization and modeling of concrete mixture design and properties is highlighted in the methodology and results in parts of the current work.

There are limited studies regarding increasing the durability of concrete by incorporating acetates as an admixture. The literature studies are concentrated on using sodium acetate (CH_3COONa) [23–27]. However, the studies lack investigation and optimization of the effect of other types of metal acetates and organic acetates on the strength and physical characteristics of cementitious materials.

The objectives of the current work are to pave the way for the application of acetates as admixtures in the construction of Basrah port infrastructure by providing scientific knowledge about the standardized procedures used to fabricate and test the properties of reinforced concrete specimens incorporated with diverse kinds of acetates, as well as to determine the optimum features of acetates to achieve high strength and durability properties. The current part of the work focuses on using potassium acetate, calcium acetate as inorganic acetates, and ethyl acetate as organic acetate in unreinforced concrete. The research studies the effect of acetate type and amount, and the water-to-cement ratio on workability, density, and compressive strength. The effect of the variables is optimized and modeled using a two-factorial central composite experimental design.

2. The Experimental Part

2.1. Materials

Ordinary Portland cement (OPC) confirming Iraqi specifications (Iraqi standards 2019) [28], natural sand passing through a 2.36 mm sieve, gravel passing through a 20 mm sieve, and tap water were used as concrete main ingredients. Potassium acetate (KA), calcium acetate (CaA), and ethyl acetate (EA) with an assay of ≥99% are purchased from Fluka™ Chemicals and Reagents for Analytical Chemistry and used as admixtures. The mixture proportions (1:2:4) for concrete were used. Fresh concrete mixtures were prepared using 90% of the total amount of water needed for the concrete mix. The remaining amount of water was used to dissolve the predetermined amount of the acetate. The acetate solution was added to the fresh concrete mixture. All samples were mixed in a laboratory drum mixer. Concrete mixtures incorporated with different types (Ka, CaA, and EA) and doses (1.38–5.6 wt.%) in concrete mixtures using different water/cement ratios (0.48–0.54) based on an espoused central composite experimental design were formulated, molded, and cured in water at room temperature for 28 days and then tested. Three samples were formulated for each test taking into account the average value. Control samples without acetates were formulated, molded, cured, and tested similarly for comparison purposes. Figure 1 shows images of some aspects of the experimental work.

Figure 1. (**a**) Slump test, (**b**) mold preparation, (**c**) a typical casted specimen, (**d**) cured specimens.

2.2. Methodology

Response surface methodology (RSM) is a powerful methodology that uses statistical techniques based on the partial factorial design that is employed to optimize and model specific responses influenced by the process operating variables. The attractive feature of

RSM made the tool popular and has significant potential [14,29,30]. The importance of RSM could be summarized by its applications in various processes for designing simultaneous runs, which could be performed quickly ensuring savings in the process time, expense, and effort. In the current work, mathematical and statistical software was used as a tool to identify the relationships between the type and content of the acetate additives and the water/cement ratio of the concrete mixtures as independent operating parameters on workability, density, and compressive strength as dependent variables (responses), and to optimize and establish the mathematical models that describe the mentioned relations. The methodology used is carried out by using appropriate design and analysis of experiments in an empirical way based on a two-factorial central composite design followed by the response surface methodology.

The individual and interactive effects of the content of the acetates and water/cement ratio on the responses were studied through 10 experimental runs. The mathematical empirical model is fitted to evaluate the impact of each independent operating parameter on the responses. The empirical model is demarcated as:

$$Y = \beta0 + \beta1X1 + \beta2X2 + \beta11X1^2 + \beta22X2^2 + \beta12X1X2 \quad (1)$$

The response (Y) is correlated to the set of regression coefficients (β): The model intercept is (β0); β1 and β2 are linear terms; β12 is the interaction term; β11 and β22 are quadratic coefficient terms. The software portable statgraphics centurion 15.2.11.0.exe was used for the regression and graphical analysis of data.

The three types of acetates were incorporated with different doses (1.38–5.6 wt.%) in concrete mixtures using different water/cement ratios (0.48–0.54) based on the adopted central composite experimental design. Table 1 shows the experimental design with the actual operating parameter levels and the response values.

Table 1. The adopted experimental design includes the actual levels of the operating parameters as well as the average values of the responses.

The Operating variables		Exp. No.								
		1	2 *	3	4	5	6	7	8	9
(Acetate/cement) % wt.		3.50	3.50	2.00	5.00	2.00	5.60	5.00	1.38	3.50
W/C		0.48	0.51	0.49	0.49	0.53	0.51	0.53	0.51	0.54
Type of acetate	Concrete Property	Exp. No.								
		1	2 *	3	4	5	6	7	8	9
K acetate	Compressive strength (MPa)	38.90	32.45	35.80	37.00	25.29	33.61	29.12	29.80	25.02
	Density (g/cm^3)	2.39	2.37	2.37	2.41	2.38	2.408	2.376	2.393	2.365
	Slump (cm)	2.00	4.00	3.50	4.50	2.00	3.00	4.50	2.00	5.00
Ca acetate	Compressive strength (MPa)	42.88	40.50	39.98	41.97	35.82	40.89	37.22	38.31	33.60
	Density (g/cm^3)	2.39	2.37	2.37	2.40	2.37	2.385	2.38	2.36	2.37
	Slump (cm)	3.0	3.5	7.0	10.0	15.0	18.0	8.0	4.0	6.0
Ethyl acetate	Compressive strength (MPa)	35.20	33.9	32.10	29.70	27.29	26.10	32.90	28.99	27.30
	Density (g/cm^3)	2.43	2.37	2.42	2.43	2.39	2.397	2.405	2.393	2.399
	Slump (cm)	2.50	3.00	5.00	3.00	5.00	3.50	5.00	4.00	4.50
(Without acetate)	Compressive strength (MPa)	29.36			29.36		28.31	25.96		24.88

* Repetitions.

2.3. The Properties Investigated

The workability of fresh concrete mixtures is the measure of the ease with which the concrete composites flow and consolidate. Workability is a critical factor that affects the long-term performance of hardened concrete, which may complicate the compaction of cement composites due to the trapping of air spaces that highly affects the mechanical performance [31].

Standard flow tests for measuring the workability were conducted immediately after mixing freshly mixed concrete composites using a slump cone. The cone was filled in three layers and manually tamped 25 times. After filling all the layers and cleaning around the slump cone, the slump cone was removed within 3 to 5 s in a straight ascending direction. Slump reading is the difference in height of the slump cone and average of the height of the concrete specimen. The average of four measurements represented the test findings.

Dried concrete cubes of 150 × 150 × 150 mm that are cured purely in water for 28 days were tested for density. The volume of the cubes was accurately measured after weighing the cubes. The density of the concrete specimens was obtained as the average of the three test results.

A compressive strength machine control model 50-C23C02 with a 2000 KN load capacity was used to test the compressive strength of 150 × 150 × 150 mm concrete cubes. The load was applied at a constant rate until the specimen failed. The compressive strength of concrete was obtained as the record of the failure load. The average of the three test results represented the average compressive strength.

3. Results and Discussion

3.1. The Experimental Results

Compressive strength is the main property that affects the fundamental behavior of construction elements. The experimental findings concerning the average values of compressive strength of concrete mixtures cured for 28 days are shown in Figure 2. The mixtures were formulated with different w/c ratios and incorporated with different types and doses of acetates. The results confirmed that the compressive strength of concrete could be increased by the incorporation of the three acetates of admixtures compared to the control samples; however, the mixtures incorporated with calcium acetate seemed to exhibit the highest compressive strength followed by the mixtures with potassium acetate than ethyl acetate. In general, the compressive strength increases with increasing the dose of the three types of acetate; however, concrete mixtures incorporated with ethyl acetate exhibited maximum compressive strength at a content of 3.5% (Figure 2).

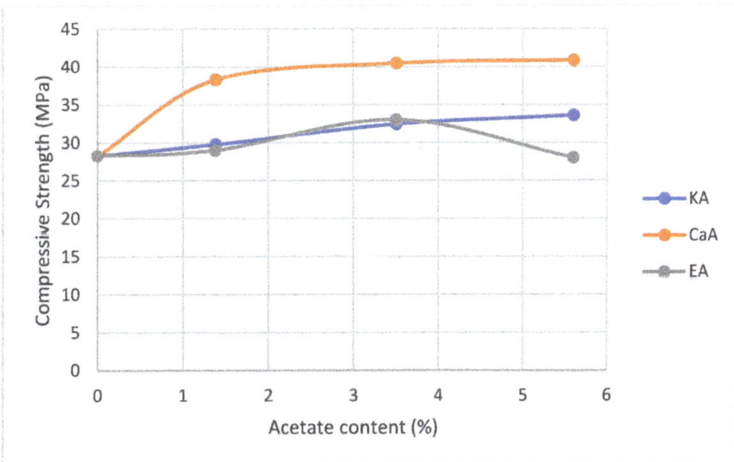

Figure 2. Compared to control specimens, the variation of compressive strength with acetates content for mixtures of (w/c = 0.51).

In general, the interaction between cement and admixtures is a surface phenomenon. Some evidence of chemical interactions between the cementitious phase and acetates was acknowledged and documented [18,32,33]. Acetates, such as KA, CaA, and EA that are used in the current work undergo hydrolysis when dispersed in an alkaline medium. The alkaline medium in the cement pastes is the $Ca(OH)_2$-saturated pore solution. The dissolution of cement grains leads to the release of Ca^{+2}. The alkaline hydrolysis of the acetate admixture leads to the release of acetate anion CH_3COO^-. The combination of the calcium and acetate ions resulted in the formation of the organic salt (calcium acetate). Figure 3 shows a scheme of the interactions described above.

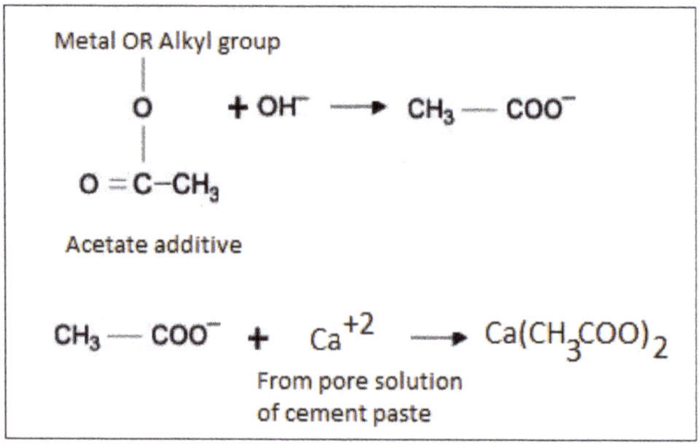

Figure 3. Scheme of the interactions of acetate admixtures with cement.

The chemical interaction could result in changes in the morphology and composition of the hydrated cement phases and increase the amounts of the hydrates which enhance the strength of the dried mixtures.

It indicated that the order of increasing the compressive strength of concrete mixtures was incorporated with Ca > K > Ethyl acetate. The situation may be attributed to the size of the acetate partner (Ca, K metal atoms, and the organic ethyl group) and its chemical nature, which are the factors that may play an important role in boosting the compressive strength. Regarding the size of the acetate partner, the smaller the size of the acetate partner, the more the enhancement in the tendency of linkages that lead to a more compacted concrete morphology. The small size of the calcium atom (Ca^{+2} radius = 114 pm) was compared to the K atom (K^{+1} radius = 152 pm) [34], and the larger size of ethyl group may be behind the aforementioned order of compressive strength of the mixtures which are formulated in the current study. Concerning the nature of the acetate partner, the stronger chemical bonding of the inorganic Ca and K metal ions with cement ingredients compared to the weaker physical linkages of the organic ethyl group is also one of the causes behind the higher compressive strength values of the mixtures incorporated with Ca and K acetates. The supreme increase in compressive strength linked to conventional mixtures was 30.8% and 15.7% for the mixtures incorporated with 5.6% CaA and KA, respectively. While it was 14.2% for the mixtures incorporated with 3.5% ethyl acetate.

The findings of the experimental work for the variation of compressive strength of concrete with w/c ratio revealed that strength rises with the declining w/c ratio using the three types of acetates. A typical illustration showing the tendency of decreasing the compressive strength of concrete mixtures using similar acetate content (3.5%) with the increasing w/c ratio is clarified in Figure 4.

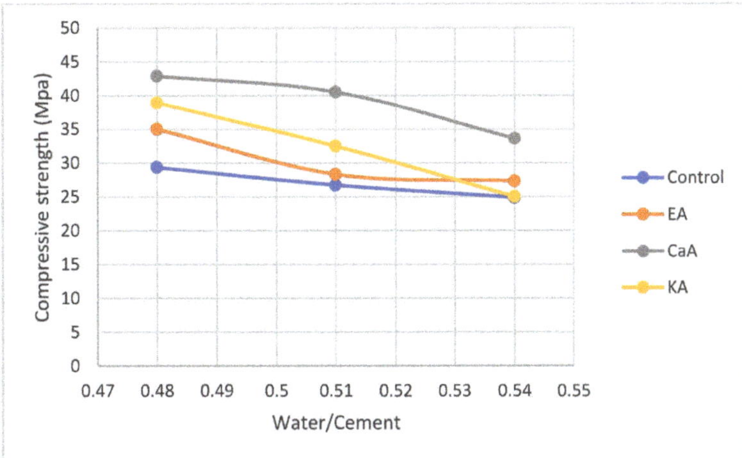

Figure 4. Compressive strength set against w/c ratio for concrete mixtures incorporated with 3.5% cement weight of the acetates.

A high w/c ratio causes aggregate segregation from the cement paste. Moreover, not consuming water by the hydration reaction may lead to microscopic pores (bleeding) and shrinkage resulting in declining the concrete strength.

Figure 5 shows the average workability results for the fresh concrete mixtures of w/c ratio of 0.51 containing different types and contents of acetates. It is noticed that increasing the content of acetates in the concrete mixtures augments its consistency. Notable compacted composites with no observable cracks were obtained.

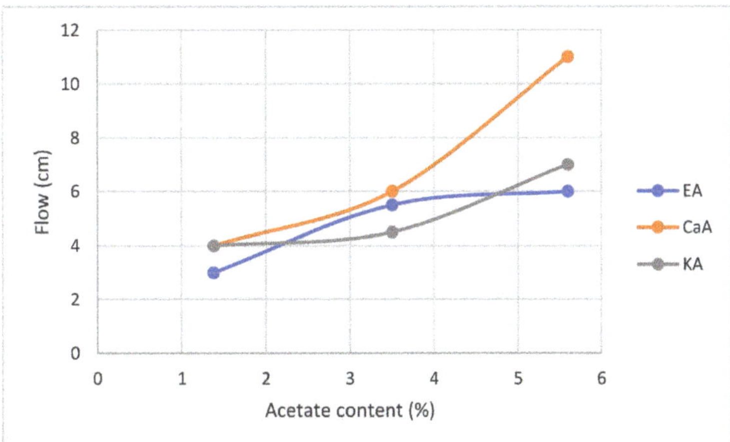

Figure 5. Average workability versus acetate content in concrete mixtures (w/c = 0.51) incorporated with different acetates.

The increase in workability is more announced for the mixtures incorporated with the metallic calcium and potassium acetates. The reason may be attributed to the fact that when the acetates dissociate in water, CH_3COO^- ions form. The metal ions already present in the cement, such as Na^+ and Ca^{+2} will react with the acetate ions. The product of this reaction comprises crystals of metal acetate. The dissociated K^+ and Ca^{+2} ions will react with free water and produce KOH and $Ca(OH)_2$. Moreover, acetic acid (CH_3COOH) is produced by the reaction of free water with CH_3COO^- ions. Suppose a large quantity of acetate admixture is added to the mixture. In that case, the formation of acetic acid in the mixture may delay the hydration process and increase the workability of the concrete mixtures. Similar findings were reported in [25]. The result of the dissociation of ethyl acetate also produces acetic acid that contributes to delaying the hydration reaction as well as increasing the mixture's workability.

The experimental results for the slump shown in Table 1 revealed that at a w/c ratio of up to 0.51, the increase in the calcium acetate content resulted in an increase in the slump. Moreover, the situation may be attributed to the adsorption of the hydrolyzed Ca ions from CaA on the surface of the cement particles owing to the concentration incline of Ca ions between the cement particles and liquid, in order that the cement particles become well dispersed due to mutual repulsion, causing the increase in slump. However, at a high w/c ratio (0.53 for experiment Nos. 5 and 7) the results revealed that increasing calcium acetate content (from 2% to 5%) gave rise to the decrease in slump. The decrease in slump may be attributed to the interaction effect of the two variables. The dilution effect of the high-water content may cause the desorption of Ca ions from the surface of cement particles leading to less-dispersed cement particles. Lowering the dispersity of the cement particles leads to a decrease in the slump values (workability). Therefore, it can be deduced that both the acetate content and w/c ratio govern the hydration process.

3.2. ANOVA Results

Analysis of variance (ANOVA) results for compressive strength are illustrated in Figures 6–8. The results confirmed that the acetate type, acetate amount, and w/c ratio remarkably affect compressive strength. The findings are reflected in the plots resulting from the analysis of the experimental compressive strength results by the RSM software. Pareto charts which show the absolute values of the properties show that the impact of the two operating parameters on compressive strength is significant (the effects extend past the reference line). However, the effect of the acetate content is more significant for EA (Figure 8a), while for CaA and KA, the w/c ratio seemed more significant (Figures 6a and 7a). The situation may be related to the chemical nature of the acetates. Ka and CaA are mineral acetates that have a good affinity to water, while EA is a non-metallic acetate that has an organic part of poor affinity to water. The mean effect plots indicated that compressive strength is highly affected by the w/c ratio, which decreases with increasing w/c ratio, in particular at higher w/c ratios. A dramatic decrease was observed for the mixtures with KA and CaA (Figures 6b and 7b), while mixtures with EA showed a flat maximum (Figure 8b). The reason for growing the strength by increasing the acetate amount may be due to the reaction of calcium hydroxide present in concrete with K and Ca in water. The product of this reaction is silica gel that will grow into solid crystals that adhere to the walls of the pores after hydration giving the concrete rigidity and strength. However, the compressive strength rises with increasing the content of EA up to a maximum and then decreases (Figure 8b), while for CaA and KA, the mixture of compressive strength rises with increasing the acetate content in a closely linear mode (Figures 6b and 7b). Decreasing the compressive strength of the mixtures incorporated with EA after the maxima may be attributed to the fact that at higher concentrations of EA, the repulsion forces between the organic EA which has low solubility in water and the polar constituents of concrete mixture will increase, leading to more porous and less compacted concrete microstructure. In agreement with our findings, it was reported that adding acetate could increase the content of ettringite and portlandite in the concrete [35].

Furthermore, the interaction plots display the mean effect of acetate content versus the w/c ratio at each level. No interaction effects of the operating variables for compressive strength were observed for the mixtures incorporated with CaA and KA (Figures 6c and 7c). The situation is reflected by the parallel lines of the effects. While the interaction plots reflected remarkable interaction effects for the compressive strength of the mixtures incorporated with EA, in particular at a w/c ratio above 0.48 (Figure 8c). The situation is visualized by the cross lines that suggest the interaction. The behavior above is more apparent throughout the contours shown in the two-dimensional (2D) plot (Figure 8g). The elliptical shapes of the contours indicate the variable's substantial interactions.

The normal probability plot presented in (Figures 6d, 7d and 8d) revealed the model's effectiveness by displaying the adjacent positions of the points to the straight lines. The three-dimensional (3D) response surface plots (Figures 6f, 7f and 8f) illustrate the conception of the response surface and the style of the compressive strength performance in terms of the two operating factors.

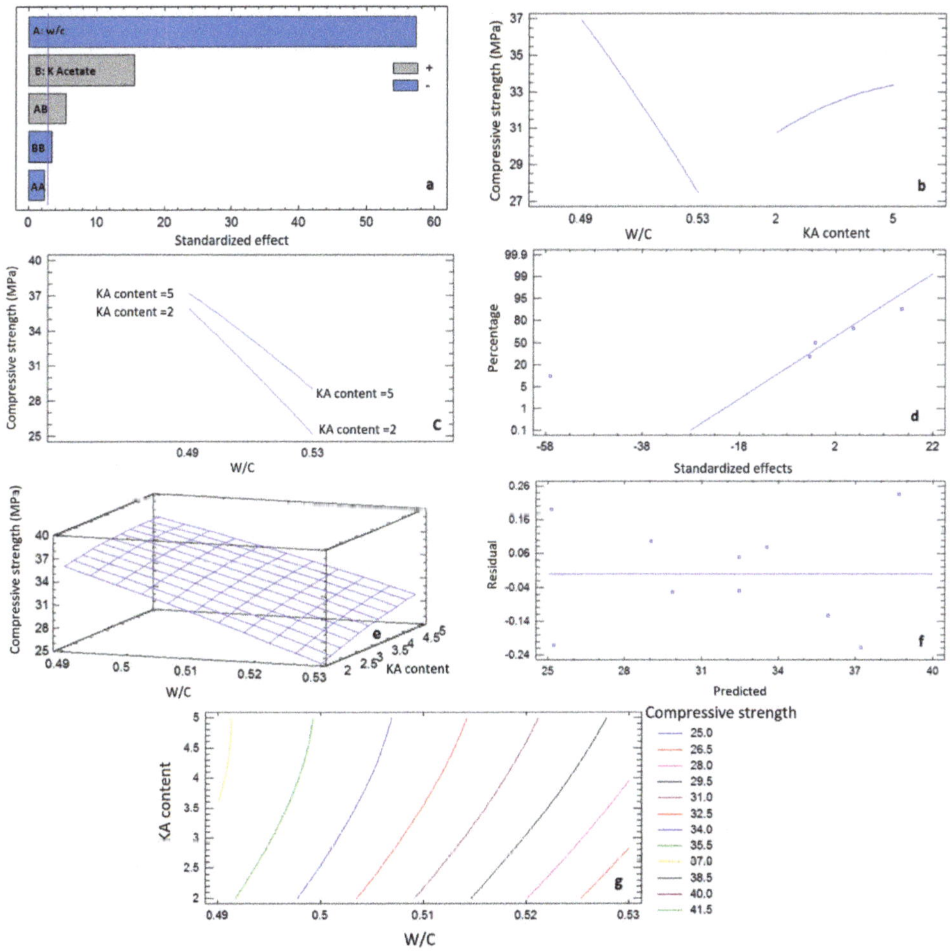

Figure 6. Plots of Pareto chart (**a**), main effects (**b**), interaction (**c**), normal probability (**d**), response surface (**e**), residual (**f**), and contour (**g**) for compressive strength of concrete mixtures incorporated with potassium acetate (KA).

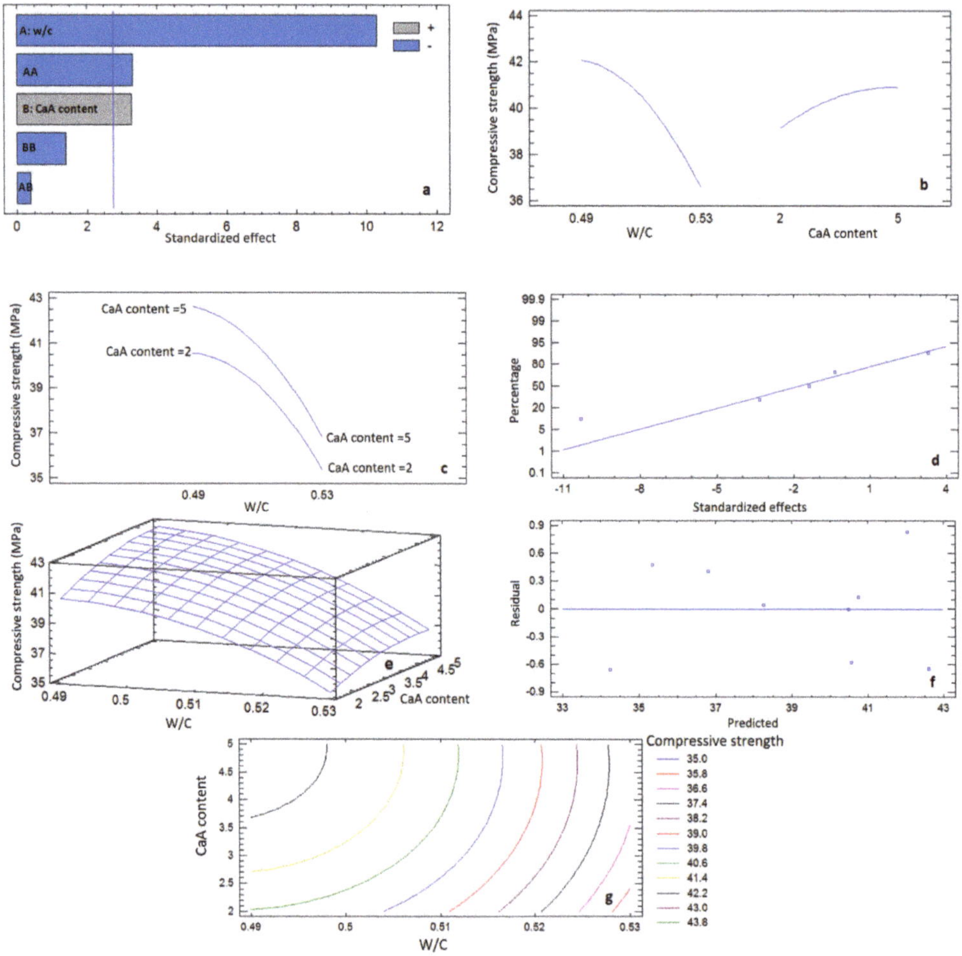

Figure 7. Plots of Pareto chart (**a**), main effects (**b**), interaction (**c**), normal probability (**d**), response surface (**e**), residual (**f**), and contour (**g**) for compressive strength of concrete mixtures incorporated with calcium acetate (CaA).

The contour plots show how varying the two operating variables affects the predicted compressive strength values, which are represented as colors. Figures 6g and 7g show the two-dimensional (2D) contour plots for KA and CaA mixture. The figures reflect less effective interactions between the variables. The non-elliptical shapes of the contours specify the situation.

The plots of the residuals (Figures 6e, 7e and 8e) verify the validity of the regression. The random scattering of residuals versus predicted reflects that the errors are independent. The fall of the points randomly on both sides of (0) demonstrates the normal distribution of the points. The results obtained affirm that the espoused model can be utilized to detect the optimum compressive strength and that it is appropriate for usage. The methodology was applied successfully to optimize and model the properties of concrete composites including compressive strength [36–39].

The empirical regression model equations, regression coefficients, and the optimum compressive strength values and operating variables for the mixtures formulated with different acetates are listed in Table 2. The R^2 is reasonable for fitting uniformity. The high-regression coefficient proves that the estimated models are accurate and can explain the experimental results successfully.

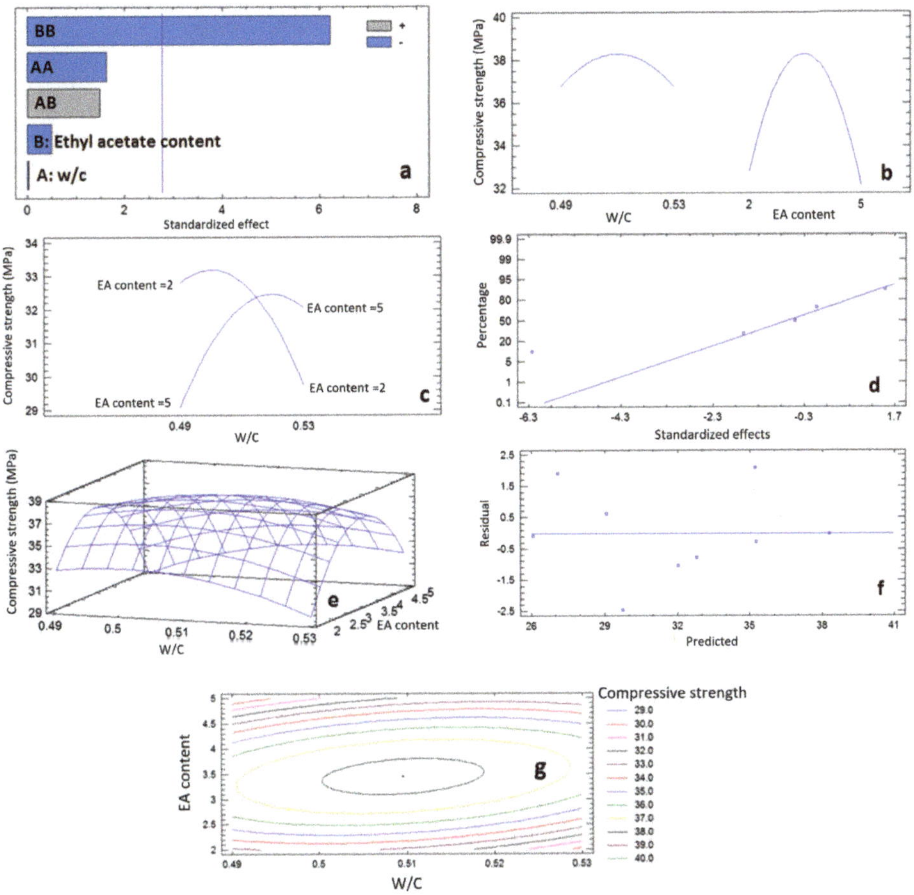

Figure 8. Plots of Pareto chart (**a**), main effects (**b**), interaction (**c**), normal probability (**d**), response surface (**e**), residual (**f**), and contour (**g**) for compressive strength of concrete mixtures incorporated with ethyl acetate (EA).

Table 2. The empirical regression model equations, regression coefficients, and the optimum compressive strength values and operating variables for the mixtures formulated with different acetates.

Acetate Type	The Generated Empirical Regression Model Equation	Regression Coefficient R^2 (%)	Optimized Compressive Strength Value (MPa)	Optimized Operating Variables
Ca	Compressive strength = $-665.62 + 2871.01\ W + 4.63\ A - 2932.86\ W^2 - 4.92\ W\ A - 0.22A^2$	96.97	42.68	w/c = 0.48 and CaA content = 5.1%
K	Compressive strength = $20.70 + 339.75\ W - 8.99\ A - 640.64\ W^2 + 21.67\ W\ K - 0.17\ A^2$	99.89	38.76	w/c = 0.48 and KA content = 4.24%
Ethyl	Compressive strength = $-898.76 + 3728.86\ W - 7.52\ A - 3828.22\ W^2 + 50.0\ W\ A - 2.62\ A^2$	91.41	38.31	w/c = 0.51 and EA content = 3.45%
Where W = W/C, A = acetate content (wt.%)				

ANOVA results for bulk density confirmed that the acetate type, acetate content, and w/c ratio remarkably influence density. However, the effect of acetate content is more significant (Figures 9a, 10a and 11a). The mean effect plots indicated that density declines with the rising w/c ratio and increases with increasing the acetate amount, particularly at high-acetate contents (Figures 9b, 10b and 11b). The rise in density may be ascribed to the decrease in air in the cement matrix as acetates fill the empty spaces within the pore structure. The higher density values may be due to the improvement in the hydration process and subsequently the increase in the hydration products was compared with those of control samples.

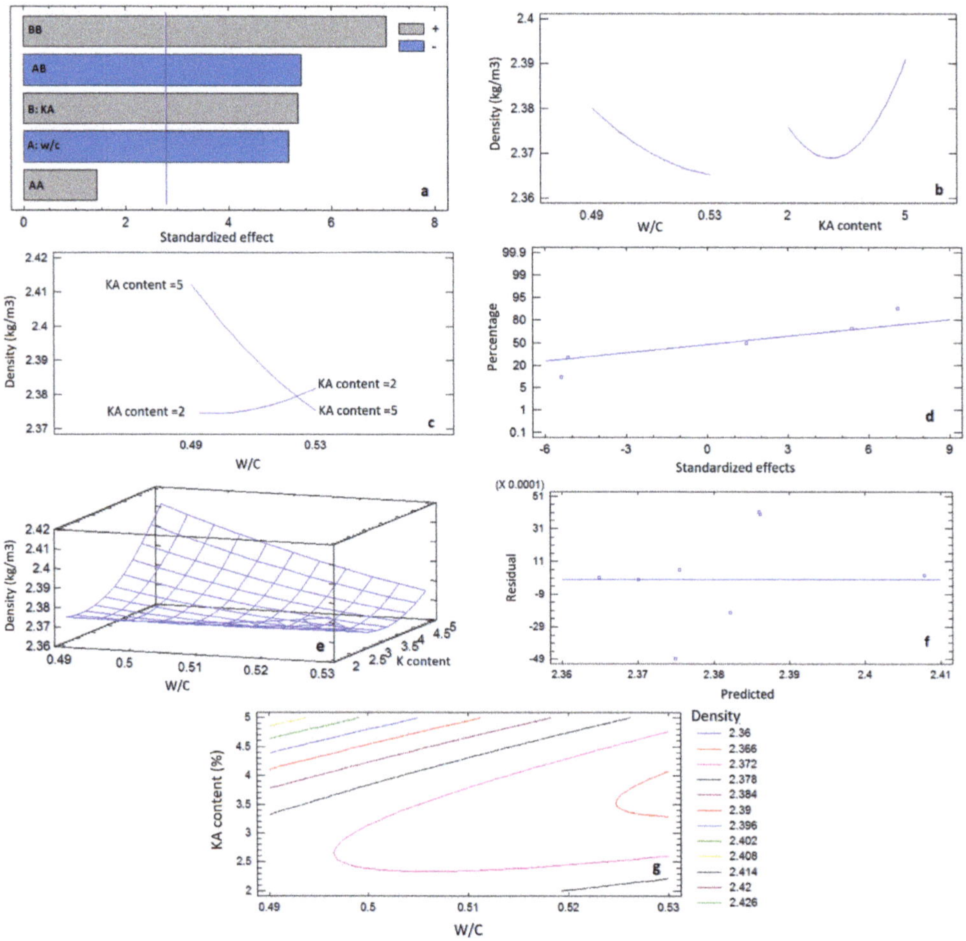

Figure 9. Plots of Pareto chart (**a**), main effects (**b**), interaction (**c**), normal probability (**d**), response surface (**e**), residual (**f**), and contour (**g**) for bulk density of concrete mixtures incorporated with potassium acetate (KA).

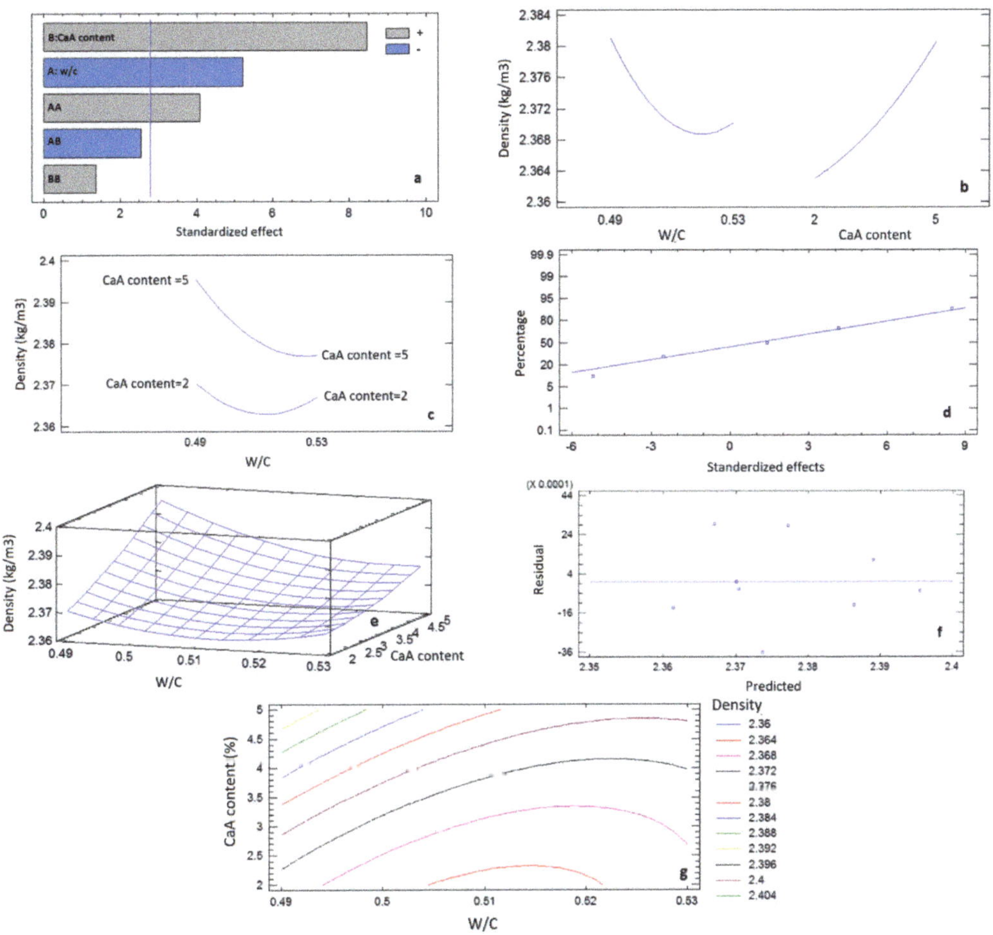

Figure 10. Plots of Pareto chart (**a**), main effects (**b**), interaction (**c**), normal probability (**d**), response surface (**e**), residual (**f**), and contour (**g**) for bulk density of concrete mixtures incorporated with calcium acetate (CaA).

The interaction plots reflected no remarkable interaction effects of the operating variables for density (Figures 9c, 10c and 11c). However, an interaction effect of the acetate content and w/c was observed for mixtures incorporated with KA at a higher w/c ratio (Figure 9c). The significant interactions are highlighted throughout the change in the contours to shape approximately the half ellipse and total eclipse as illustrated in (Figures 9g and 11g) representing the (2D) contour plots, respectively.

The normal probability plot presented in Figures 9d, 10d and 11d, and the plots of the residuals (Figures 9f, 10f and 11f) show the same trend established for compressive strength, which specifies the model's validity and precision describing the experimental results for bulk density.

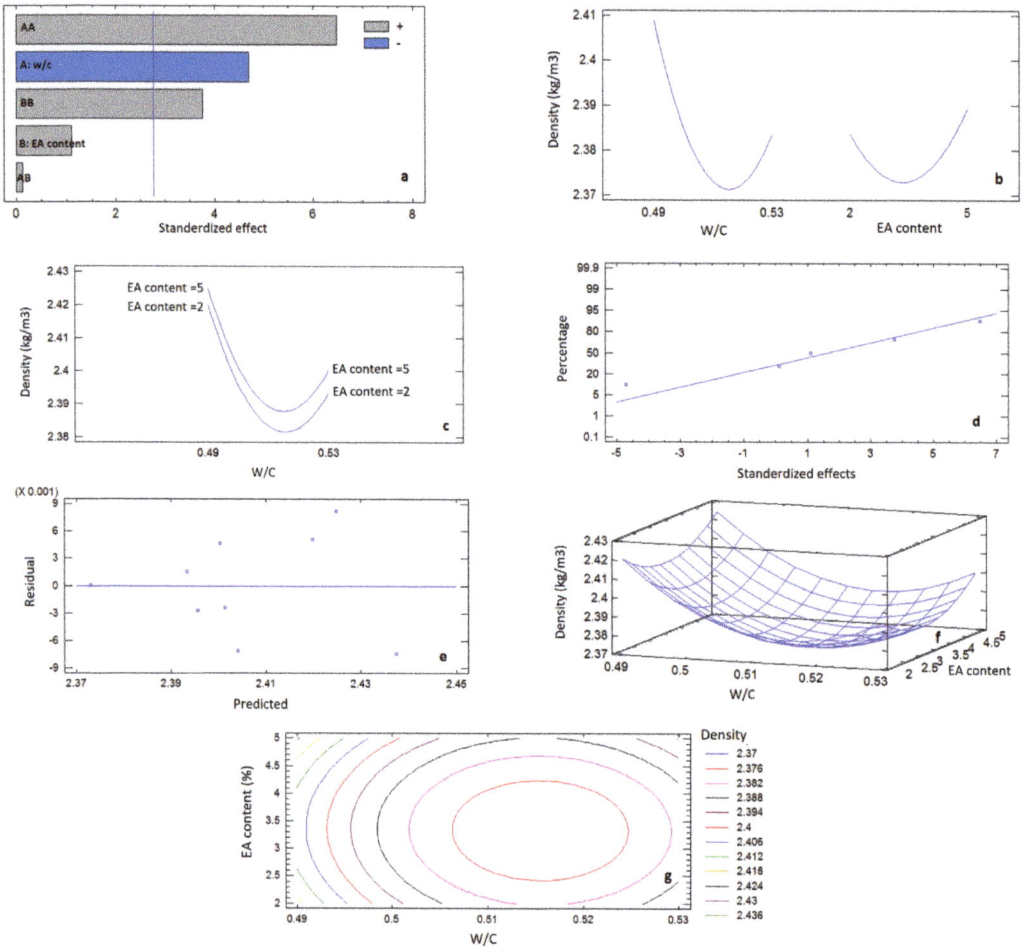

Figure 11. Plots of Pareto chart (**a**), main effects (**b**), interaction (**c**), normal probability (**d**), response surface (**e**), residual (**f**), and contour (**g**) for bulk density of concrete mixtures incorporated with ethyl acetate (EA).

The empirical regression model equations, regression coefficients, and the optimum bulk density values and operating variables for the mixtures formulated with different acetates are listed in Table 3. The high-regression coefficients of the estimated models confirm the model's precision. However, less significant differences in the optimum values were estimated for bulk density compared to those estimated for compressive strength.

Table 3. The empirical regression model equations, regression coefficients, and the optimum bulk density values and operating variables for the mixtures formulated with different acetates.

Acetate Type	The Generated Empirical Regression Model Equation	Regression Coefficient R^2 (%)	Optimized Density g/cm^3	Optimized Operating Variables
Ca	Density = 5.93225 − 14.18 W + 0.064 CaA + 14.06 W^2 − 0.13 W CaA + 0.0008 CaA^2	96.83	2.42	w/c = 0.48 and CaA content = 5.6 %
K	Density = 3.70748 − 5.94 W + 0.15 KA + 6.72 W^2 − 0.37 W KA + 0.006 KA^2	97.17	2.45	w/c = 0.48 and KA content = 5.6%
Ethyl	Density = 17.8716 − 59.82 W − 0.048 EA + 57.97 W^2 + 0.017 W EA + 0.006 EA^2	94.33	2.47	w/c = 0.48 and EA content = 5.6%

Where W = W/C, CaA = Calcium acetate content (wt.%), KA = Potassium acetate content (wt.%), EA = Ethyl acetate content (wt.%)

3.3. Microstructure Analysis

The interpretation of microstructure imaging has become well-recognized as a technique for studying the microstructure of cement and concrete composites. It is soundly established that the hydration of cement paste develops its microstructure. The hydrous cement grains form by the reaction of anhydrous cement grains with water. The products of the hydration reactions are the alite or tricalcium silicate, belite or dicalcium silicate, and calcium hydroxide. The alite and belite are accountable for the formation of the C-S-H gel. The calcium silicate hydrate is produced principally around the cement grains, while calcium hydroxide is deposited in the water-filled pores [40]. In this study, SEM images of different magnifications (the two tops are in lower magnification) were obtained for the hardened cementitious part of the control samples and acetate-modified samples. Typical SEM images for the control sample are shown in (Figure 12, top and bottom left). The figure indicates the formation of cement hydration products. Calcium hydroxide crystals are represented by areas of light gray with irregular shape, calcium silicate hydrate is signified by areas of dark gray, while the pores are displayed by black areas [41].

The top and bottom right of Figure 12 shows the typical SEM images for the hardened cementitious part of a sample incorporated with CaA. The figure illustrates less porous and denser microstructure resulting from the addition of the acetate. The denser microstructure may be owing to the high adhesion between CaA and the cement hydration products. The packed microstructure gives rise to a decrease in the volume of capillary pores in the sample incorporated with CaA. The modified microstructure enhanced by acetates enables the attainment of building materials with higher density, strength, and durability characteristics. Incorporating the acetates in the mixtures leads to reducing voids in concrete, thereby decreasing the porosity [42] and avoiding the diffusion of any fluids.

Figure 12. *Cont.*

Figure 12. Hardened cementitious part of control sample (**top** and **bottom left**), sample incorporated with CaA (**top** and **bottom right**).

4. Conclusions

The effective application of chemical admixtures in concrete technology resulted in a new development in the formulation and usage of cement concrete materials. Small amounts of admixture can efficiently adjust the cement-based materials' performance. This study is a part of research aimed at examining the effect of the incorporation of potassium acetate (KA), calcium acetate (CaA), and ethyl acetate (EA) in concrete as anti-corrosion admixtures, and its impact on other concrete properties, mainly workability and compressive strength. The current work focuses on studying the impact of the merging of admixtures on the workability, density, and compressive strength of concrete based on the acetate dose and water/cement ratio. The experimental and the RSM analysis results confirmed that the relatively low-cost acetate admixtures, in particular CaA seemed promising for the manufacturing of concrete with notable workability and mechanical strength. The high rise in compressive strength linked to control mixtures was 30.8% for the mixtures incorporated with 5.6 wt.% CaA. Optimum compressive strength of 42.68, 38.76, and 38.31 MPa, and density of 2.42, 2.45, and 2.47 g/cm^3 were estimated for concrete mixtures incorporated with CaA, KA, and EA, respectively using 0.48 w/c ratio. The promising results paved the way to continue the study in order to explore the impact of the admixtures on the water permeability and corrosion resistance of reinforced concrete incorporated with the optimum content of the acetate admixture, which could be used in Basrah marine construction applications.

Author Contributions: Conceptualization, A.M. and M.A.L.Y.; Methodology, A.A.A.; Software, A.A.A.; Formal analysis, A.A.A.; Investigation, A.A.A.; Resources, A.A.A.; Writing—original draft, A.A.A.; Writing—review & editing, A.M. and M.A.L.Y.; Visualization, A.A.A. All authors have read and agreed to the published version of the manuscript.

Funding: This research received no external funding.

Institutional Review Board Statement: Not applicable.

Informed Consent Statement: Not applicable.

Data Availability Statement: Not applicable.

Conflicts of Interest: The authors declare no conflict of interest.

References

1. Ibrahim, H.D.; Xue, P.; Eltahir, E.A.B. Multiple Salinity Equilibria and Resilience of Persian/Arabian Gulf Basin Salinity to Brine Discharge. *Front. Mar. Sci.* **2020**, *7*, 573. [CrossRef]
2. Dodds, W.; Goodier, C.; Christodoulou, C.; Austin, S.; Dunne, D. Corrosion risk assessment of structural concrete with coarse crushed concrete aggregate. *Proc. Inst. Civ. Eng.—Constr. Mater.* **2020**, *173*, 57–69. [CrossRef]

3. del Campo, J.M.; Negro, V. Nanomaterials in Protection of Buildings and Infrastructure Elements in Highly Aggressive Marine Environments. *Energies* **2021**, *14*, 2588. [CrossRef]
4. Bautista-Gutierrez, K.P.; Herrera-May, A.L.; Santamaría-López, J.M.; Honorato-Moreno, A.; Zamora-Castro, S.A. Recent Progress in Nanomaterials for Modern Concrete Infrastructure: Advantages and Challenges. *Materials* **2019**, *12*, 3548. [CrossRef]
5. Khudhair, N.A.; Al-Sammarraie, A.M.A. Enhancing of Corrosion Protection of Steel Rebar in Concrete Using TiO_2 Nanoparticles as Additive. *Iraqi J. Sci.* **2019**, *60*, 1898–1903. [CrossRef]
6. Amran, A.; Onaiz, A.M.; Fediuk, R.; Vatin, N.I.; Rashid, R.S.M.; Abdelgader, H.; Ozbakkaloglu, T. Self-Healing Concrete as a Prospective Construction Material: A Review. *Materials* **2022**, *15*, 3214. [CrossRef]
7. Koleva, D.A. An Innovative Approach to Control Steel Reinforcement Corrosion by Self-Healing. *Materials* **2018**, *11*, 309. [CrossRef]
8. Hossain, N.; Chowdhury, M.A.; Kchaou, M. An overview of green corrosion inhibitors for sustainable and environment friendly industrial development. *J. Adhes. Sci. Technol.* **2020**, *35*, 673–690. [CrossRef]
9. Topçu, I.B.; Eskişehir Osmangazi University; Uzunömeroğlu, A. Properties of corrosion inhibitors on reinforced concrete. *J. Struct. Eng. Appl. Mech.* **2020**, *3*, 93–109. [CrossRef]
10. Inamuddin, M.I.A.; Ahamed, M.I.; Luqman, M.; Altalhi, T. *Sustainable Corrosion Inhibitors, Ch.6*; Materials Research Forum LLC.: Millersville, PA, USA, 2019.
11. Lee, H.-S.; Saraswathy, V.; Kwon, S.-J.; Karthick, S. Corrosion Inhibitors for Reinforced Concrete: A Review. In *Corrosion Inhibitors, Principles and Recent Applications, Mahmood Aliofkhazraei*; IntechOpen: London, UK, 2017. [CrossRef]
12. Paulson, B.M.; Thomas, K.J.; Raphael, V.P.; Shaju, K.S.; Ragi, K. Mitigation of concrete reinforced steel corrosion by penta sodium triphosphate: Physicochemical and electrochemical investigations. *SN Appl. Sci.* **2020**, *2*, 1813. [CrossRef]
13. Reddy, V.S.; Prashanth, T.; Raju, S.P.; Prashanth, P. Effect of Organic and Inorganic Corrosion Inhibitors on Strength Properties of Concrete. *E3S Web Conf.* **2020**, *184*, 01112. [CrossRef]
14. Awolusi, T.; Oke, O.; Akinkurolere, O.; Sojobi, A. Application of response surface methodology: Predicting and optimizing the properties of concrete containing steel fibre extracted from waste tires with limestone powder as filler. *Case Stud. Constr. Mater.* **2018**, *10*, e00212. [CrossRef]
15. Castaneda, H.; Hassan, M.; Radovic, M.; Milla, J.; Karayan, A. *Self-Healing Microcapsules as Concrete Aggregates for Corrosion Inhibition in Reinforced Concrete*; Project No. 17CLSU08; Tran-SET Publication: Baton Rouge, LA, USA, 2018.
16. Karavokyros, L.; Katsiotis, N.; Tzanis, E.; Batis, G.; Sapalidis, A.; Chatzopoulos, A.; Sideris, K.; Beazi-Katsioti, M. The Effect of Mix-Design and Corrosion Inhibitors on the Durability of Concrete. *J. Mater. Sci. Chem. Eng.* **2020**, *8*, 64–77. [CrossRef]
17. Mohammed, I.U.; Ahmed, T.; Apurbo, S.M.; Mallick, T.A.; Shahriar, F.; Munim, A.; Awal, M.A. Influence of Chemical Admixtures on Fresh and Hardened Properties of Prolonged Mixed Concrete. *Adv. Mater. Sci. Eng.* **2017**, *2017*, 9187627. [CrossRef]
18. Li, D.-L.; Zheng, D.-P.; Wang, D.-M.; Zhao, J.-H.; Du, C.; Ren, C.-F. Influence of Organic Esters on Portland Cement Hydration and Hardening. *Adv. Mater. Sci. Eng.* **2018**, *2018*, 3203952. [CrossRef]
19. Mao, Q.; Zhang, J.; Ma, J.; Liu, H.; Wang, Z.; Cui, S. Improvement of Shrinkage Reduction and Superplasticity of Polycarboxylate Admixture by Ester and Silane Groups. *J. Mater. Civ. Eng.* **2022**, *34*, 10. [CrossRef]
20. Hassouna, F.M.A.; Abu-Zant, H. Effects of Superplasticizers on Fresh and Hardened Portland Cement Concrete Characteristics. *Int. J. Appl. Sci. Technol.* **2016**, *5*, 32–36. Available online: https://staff.najah.edu/media/published_research/2020/06/20/final.pdf (accessed on 9 February 2020).
21. Lai, G.; Liu, X.; Li, S.; Xu, Y.; Zheng, Y.; Guan, J.; Gao, R.; Wei, Z.; Wang, Z.; Cui, S. Development of chemical admixtures for green and environmentally friendly concrete: A review. *J. Clean. Prod.* **2023**, *389*, 136116. [CrossRef]
22. Fang, Y.; Wang, J.; Qian, X.; Wang, L.; Chen, P.; Qiao, P. A renewable admixture to enhance the performance of cement mortars through a pre-hydration method. *J. Clean. Prod.* **2022**, *332*, 130095. [CrossRef]
23. Kılınççeker, G.; Menekşe, C. The effect of acetate ions on the corrosion of reinforcing steel in chloride environments. *Prot. Met. Phys. Chem. Surfaces* **2015**, *51*, 659–666. Available online: https://link.springer.com/article/10.1134/S2070205115040176 (accessed on 9 February 2020). [CrossRef]
24. Ryu, H.-S.; Kim, D.-M.; Shin, S.-H.; Park, W.-J.; Kwon, S.-J. Steel-Corrosion Characteristics of an Environmental Inhibitor using Limestone Sludge and Acetic Acid. *Int. J. Concr. Struct. Mater.* **2018**, *12*, 13. Available online: https://ijcsm.springeropen.com/articles/10.1186/s40069-018-0243-x (accessed on 9 February 2020). [CrossRef]
25. Al-Kheetan, M.J.; Rahman, M.M. Integration of Anhydrous Sodium Acetate (ASAc) into Concrete Pavement for Protection against Harmful Impact of Deicing Salt. *Prop. Interfaced Mater. Film. JOM* **2019**, *71*, 4899–4909. Available online: https://link.springer.com/article/10.1007/s11837-019-03624-3 (accessed on 9 February 2020). [CrossRef]
26. Al-Kheetan, M.J.; Ghaffar, S.H.; Awad, S.; Chougan, M.; Byzyka, J.; Rahman, M.M. Microstructural, Mechanical and Physical Assessment of Portland Cement Concrete Pavement Modified by Sodium Acetate under Various Curing Conditions. *Infrastructures* **2021**, *6*, 113. [CrossRef]
27. Al-Kheetan, M.J.; Ghaffar, S.H.; Madyan, O.A.; Rahman, M.M. Development of low absorption and high-resistant sodium acetate concrete for severe environmental conditions. *Constr. Build. Mater.* **2020**, *230*, 117057. [CrossRef]
28. Iraqi Standard (IQ. S 5:2019) for CEM I (42.5 R). Available online: https://uowa.edu.iq/filestorage/file_1542638623.pdf (accessed on 9 February 2020).

29. Ahmed, S.M.; Kamal, I. Electrical resistivity and compressive strength of cement mortar based on green magnetite nanoparticles and wastes from steel industry. *Case Stud. Constr. Mater.* **2022**, *17*, e01712. [CrossRef]
30. Hameed, M.M.; AlOmar, M.K.; Baniya, W.J.; AlSaadi, M.A. Prediction of high-strength concrete: High-order response surface methodology modeling approach. *Eng. Comput.* **2021**, *38*, 1655–1668. [CrossRef]
31. Öztaş, A.; Pala, M.; Özbay, E.; Kanca, E.; Çağlar, N.; Bhatti, M.A. Predicting the compressive strength and slump of high strength concrete using neural network. *Constr. Build. Mater.* **2006**, *20*, 769–775. [CrossRef]
32. Silva, D.; Monteiro, P. Analysis of C3A hydration using soft X-rays transmission microscopy: Effect of EVA copolymer. *Cem. Concr. Res.* **2005**, *35*, 2026–2032. [CrossRef]
33. Silva, D.; Roman, H.; Gleize, P. Evidences of chemical interaction between EVA and hydrating Portland cement. *Cem. Concr. Res.* **2002**, *32*, 1383–1390. [CrossRef]
34. Peter, L.F.; Smith, B.C. Ionic radii for Group 1 and Group 2 halide, hydride, fluoride, oxide, sulfide, selenide and telluride crystals. *Dalton Trans.* **2010**, *39*, 7786–7791. Available online: https://pubs.rsc.org/en/content/articlelanding/2010/dt/c0dt00401d (accessed on 9 February 2020).
35. Cao, K.; Wang, L.; Xu, Y.; Shen, W.; Wang, H. The Hydration and Compressive Strength of Cement Mortar Prepared by Calcium Acetate Solution. *J. Adv. Civ. Eng.* **2021**, *2021*, 8817725. [CrossRef]
36. Abed, A.A.; Kamal, I.M. Factorial Design for Studying the Properties of Recycled Aggregate Concrete Exposed to Aggressive Media. *J. Eng. Res.* **2021**, *9*, 1–24. [CrossRef]
37. Abed, A.A.; Bas, Y.J.; Al-Hasani, A.; Kamal, I. Investigation on some properties of hardened cement-Biogenic ash composites. *AIP Conf. Proc.* **2022**, *2660*, 020054. [CrossRef]
38. Ma, H.; Sun, Z.; Ma, G. Research on Compressive Strength of Manufactured Sand Concrete Based on Response Surface Methodology (RSM). *Appl. Sci.* **2022**, *12*, 3506. [CrossRef]
39. Vishnupriyan, M.; Annadurai, R. A study on the macro-properties of PCB fiber-reinforced concrete from recycled electronic waste and validation of results using RSM and ANN. *Asian J. Civ. Eng.* **2023**, *24*, 1667–1680. [CrossRef]
40. Scrivener, K.L. Backscattered electron imaging of cementitious microstructures: Understanding and quantification. *Cement Concrete Compos.* **2004**, *26*, 935–945. [CrossRef]
41. Hassan, M.S. SEM-Backscattered Imaging Analysis of Cementitious Composite Matrix Incorporating Mineral Admixture. *Eng. Technol. J.* **2014**, *32*, 696–703.
42. Ozturk, A.U.; Baradan, B. Effects of admixture type and dosage on microstructural and mechanical properties of cement mortars. *KSCE J. Civ. Eng.* **2011**, *15*, 1237–1243. [CrossRef]

sustainability MDPI

Article

Experimental Study and Modeling of the Fracture Behavior, Mechanical Properties, and Bonding Strength of Oil Well Cement

Cumaraswamy Vipulanandan [1,*], Ahmed Salih Mohammed [2,3,*] and Praveen Ramanathan [4]

1 Center for Innovative Grouting Materials and Technology (CIGMAT), Department of Civil and Environmental Engineering, University of Houston, Houston, TX 77204, USA
2 Department of Civil Engineering, College of Engineering, University of Sulaimani, Sulaimaniyah 46001, Kurdistan Region, Iraq
3 Engineering Department, Civil Engineering, American University of Iraq, Sulaimani (AUIS), Sulaimaniyah 46001, Kurdistan Region, Iraq
4 Department of Civil and Environmental Engineering, University of Houston, Houston, TX 77204, USA
* Correspondence: cvipulan@central.uh.edu or cvipulanandan@uh.edu (C.V.); ahmed.mohammed@univsul.edu.iq (A.S.M.); Tel.: +964-770158865 (A.S.M.)

Abstract: This study aimed to analyze the outcomes of stress intensity factor (K_I) and new bond strength tests of oil well cement (class H) with a water-to-cement ratio (w/c) of 0.38. Mechanical properties of the cement paste, such as the compressive and flexural strengths, were tested and qualified at 1, 7, and 28 days of curing. The relationship between the elastic modulus and axial strain using the differential of the Vipulanandan p-q model for the cement paste was obtained. The stress intensity factor of the cement paste was between 0.3 and 0.6 MPa.m, and the crack tip opening displacement (CTOD) was between 2.798 and 6.254 µm at three different ratios between the initial notch height (a) and the thickness of the beam (d) (a/d = 0.3, 0.4, and 0.5). The nonlinear Vipulanandan p-q model was used to model the compressive and flexural stress–strain behavior of the cement at three curing times. The bonding strength between the cement and steel tube representing the casing in the borehole was 0.75, 1.89, and 2.59 MPa at 1, 7, and 28 days respectively.

Keywords: mechanical properties; fracture mechanics; curing age; bonding strength; models

Citation: Vipulanandan, C.; Mohammed, A.S.; Ramanathan, P. Experimental Study and Modeling of the Fracture Behavior, Mechanical Properties, and Bonding Strength of Oil Well Cement. *Sustainability* **2023**, *15*, 9566. https://doi.org/10.3390/su15129566

Academic Editors: José Ignacio Alvarez and Hosam Saleh

Received: 12 May 2023
Revised: 5 June 2023
Accepted: 6 June 2023
Published: 14 June 2023

1. Introduction

Each cement particle forms a type of growth on its surface during hydration. It gradually spreads until it adheres to the growth of other adjacent cement particles, resulting in progressive stiffening, hardening, and strength development [1]. Even if cement is a well-consolidated material, the chemistry of cement (and the chemistry inside cement) remains very complex and nonobvious. What is certain is that the hydration mechanism plays a pivotal role in the development of cement with specific final chemical compositions, mechanical properties, and porosities. This document provides a survey of the chemistry behind such inorganic materials. The text has been organized into five parts describing: (i) the manufacturing process of Portland cement, (ii) the chemical composition and hydration reactions involving Portland cement, (iii) the mechanisms of setting, (iv) the classification of the different types of porosities available in cement (with particular attention given to the role of water in driving the formation of pores), and (v) the recent findings on the use of recycled waste materials in cementitious matrices (with a particular focus on the sustainable development of cementitious formulations). In this study, the influence of water on the main relevant chemical transformations occurring in cement emerged with the formation of specific intermediates/products that might affect the final chemical composition of types of cement. Within the text, a clear distinction between setting and hardening is provided. Water's physical/structural role in influencing the porosities in cement is analyzed, providing a correlation between types of bound water and porosities [2,3]. In the petroleum industry, cement has been used in oil well operations. Cement is typically utilized to fill the annular space between the casing and rock formation by displacing the drilling fluid. In addition, cement will support the casing and protect it against corrosion and impact loading, restrict the movement of fluids between formations, and isolate productive and nonproductive zones. Oil well cement is used under different exposure conditions than the cement used in the conventional construction industry. The strength of oil well cement usually depends on factors such as time and conditions of curing, environmental conditions, slurry design and use of additives, and any additional treatments to the cement [4].

Cementing primarily aims to completely and irreversibly isolate the formation behind the casing. The success of a production operation of a well is almost entirely predicated on the quality of the primary cementing work. A full hydraulic seal is present between the formation and the casing across all the zones of interest [3]. Cementing also creates a hydraulic seal, which stops fluids from moving between producing zones in the borehole and escaping to the surface. In addition, it prevents the steel casing from rusting when exposed to formation fluids [5]. Over time, the cement is subjected to stresses due to pressure integrity tests, an increase in mud weight, casing perforation, well stimulation, the production of oil or gas, and an increase or fluctuation in the wellbore temperature [3,6]. Recent case studies on cementing damages found several problems that have caused varied cementing process delays [2,3]. These problems caused the cementing procedures to run longer than expected. Controlling fluid loss to rock and soil formation and effective well cementing have become essential challenges in oil well construction to maintain wellbore integrity [6]. This is because changing downhole environments threatens the integrity of the wellbore. Consequently, comprehensive monitoring and control of the entire oil-well-cementing process are necessary to ensure the cement maintains its integrity throughout its useful life [7,8].

The safety and effectiveness of the CO_2 injection procedure for geologic carbon storage depend on the integrity of the cement, which offers zonal isolation and mechanical support. This research focused on radial cracking in cement after CO_2 injection and interfacial debonding at wellbore contacts. It applied the energy release rate (ERR) definition to describe how fractures spread. The suggested model used the finite element approach and calculated the ERRs of both fractures using realistic wellbore layouts and injection settings. Additional parametric research revealed how the fracture geometry, the cement's mechanical and thermal characteristics, and the crack size affected crack propagation. According to simulation findings, with normal cement characteristics, interfacial and radial fracture ERRs would be more than 100 J/m^2. The Young's modulus, Poisson's ratio, and thermal conductivity of the cement were the next most significant influences on the ERR. Other crucial factors in regulating fracture propagation are the cracks' starting sizes and locations.

Furthermore, nonuniform in situ loads would accelerate fracture propagation at the interfaces. These important results might contribute to the improvement of cement sheath design and guarantee the long-term integrity of wells used for geological carbon storage. [1,9,10]: (i) cracking the cement sheath could allow fluids to migrate radially and vertically; and (ii) plastic deformation in the cement sheath could also allow fluids to migrate radially and vertically (Figures 1 and 2). A cement or concrete fracture manifests as cracking as an indication of internal stress or deterioration. Cement and concrete cracking is common, but they should never occur and usually mean more significant problems. Cement and concrete crack and break down into their respective stages, and cracking or fracturing in cement and concrete occur in three distinct phases. It is essential to distinguish between the mode of crack initiation and how this occurs at the microscopic level, the subsequent paths of propagation, and the ultimate macroscopic crack pattern. Strain and stress concentrations resulting from the incompatibility of the aggregate and paste components' elastic moduli are primarily responsible for forming small fissures or microcracks in fresh, hardened cement and concrete. The following are the different phases of concrete and cement cracking. In the first step, strain concentrations at the aggregate–

paste interface can occur before loading due to shrinkage or thermal movements in the cement or concrete [3,11,12].

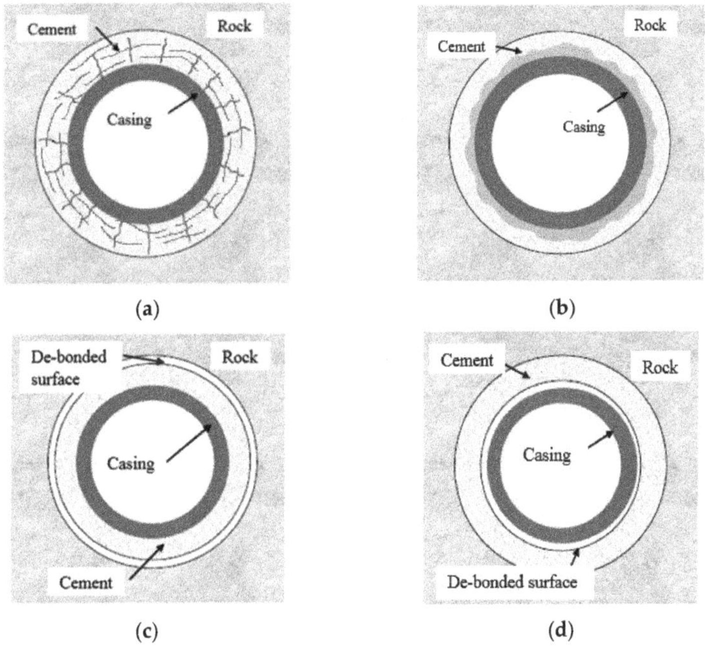

Figure 1. Debonding at Steel–Cement Interface. (**a**) Cracks in Cement Sheath, (**b**) Plastic Deformation in Cement Sheath, (**c**) De-bonding at Rock–Cement Interface, and (**d**) Debonding at Casing–Cement Interface [13].

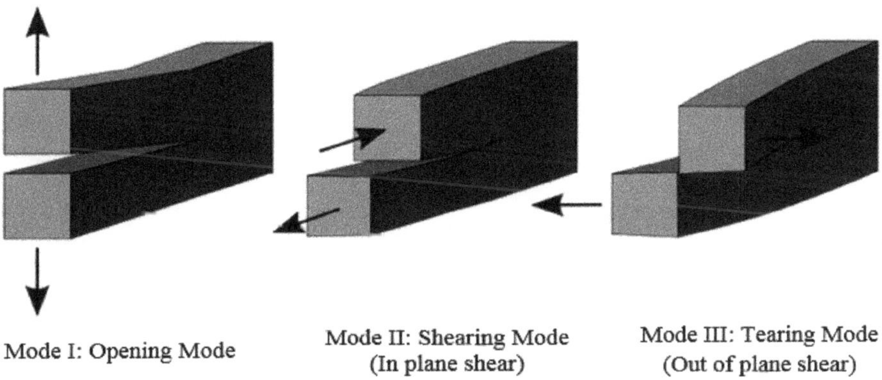

Figure 2. Fracture Modes of Materials. Tearing Mode [13].

In Phase I, cracks start microscopically at specific locations all over the specimen where the tensile strain is the highest. The lack of further crack growth at this load level demonstrates the stability of the cracks. Once the load is increased, initially stable cracks begin propagating in Phase II. Due to the likelihood of overlap between stable crack initiation and crack propagation, the transition between Stages I and II will be gradual (see Figure 2 for a visual demonstration). In Stage II, the crack system multiplies and spreads slowly and steadily, so propagation stops if loading stops and the stress level is constant [13]. The pressures acting from both sides of the sheath, the outside formation pore pressure, and the inside pressure transmitted from the casing and other axial loadings are the causes of the stresses induced in wellbore cement. Axial loadings can cause these stresses [14].

In most cases, it causes the material to transform into a different shape. The cement sheath can debond at the following interfaces [11]: cracking of the cement sheath, which could allow radial and vertical migration of fluids (Figure 1a); plastic deformation in the cement sheath, which could allow radial and vertical migration of fluids (Figure 1b); the rock–cement interface (Figure 1c); and the cement–casing interface (Figure 1d). In addition, the pressure and thermal cycle damage the cement sheath's integrity (sealing integrity and mechanical integrity), leading to gas leakage. The failure mode of a downhole cement sheath can be divided into four categories: (1) debonding/micro annulus at the interface of the cement sheath, (2) shear/tensile failure of the cement sheath, (3) radial cracks of cement sheath, and (4) plastic deformation of the cement sheath [10,13]. Researchers have used Linear Elastic Fracture Mechanics (LEFM) parameters and Elastic-Plastic Fracture Mechanics (EPFM) parameters in the literature to characterize the fracture resistance of cement paste, mortar, cement concrete, polymer concrete, and rocks. Figure 2 shows that the fracture mechanics field considers three distinct crack propagation modes. Mode I, the opening crack propagation model, is caused by a tensile stress field. The shearing or sliding mode that results from in-plane shear is Mode II. Mode III, also known as the tearing mode, is caused by shear outside the material's plane [15]. In engineering practice, the stress state near the tip of a sharp crack is more useful. In a linear elastic material, the stress intensity factor (K_I) characterizes the crack tip conditions [13–15]. If K_I is known, simple equations can be used to determine how the stress is distributed at the crack tip. No study was conducted on evaluating the fracture behavior of oil well cement at different heights of the initial crack.

Vipulanandan, C., and Dharmarajan [15,16] studied the Mode I fracture regarding the applicability of the critical stress intensity factor (K_{IC}) and/or critical crack-up opening displacement ($CTOD_c$) for epoxy and polyester polymer composites (PCs). Single-edge notched beams were used to study unreinforced and glass-fiber-reinforced polymer composite systems at room temperature in three-point and four-point bending. The crack extension during the pre-peak stress was calculated using the crack mouth opening displacement (CMOD) technique. The effective crack's tip was determined by the critical stress intensity factor, and the elastic crack tip opening displacement served as a proxy for the critical crack extension. The performance of this model was compared to other models provided for metals and cement concrete to calculate the elastic CTOD from the observed elastic CMOD. The test findings showed that the initial notch depth did not affect the two fracture characteristics. The two fracture characteristics predicted the notch sensitivity of polyester and epoxy polymer concrete. Resistance curves were created for a 4% glass-fiber-reinforced polyester PC based on the stress intensity factor and crack tip opening displacement [16]. Feng et al. [17] studied and thoroughly characterized the fracture mechanics of Portland cement mortars reinforced with multiwall carbon nanotubes and carbon nanofibers. To strengthen cement mortars with well-dispersed carbon nanotubes and carbon nanofibers, the critical values of the stress intensity factor, strain energy release rate, crack tip opening displacement, and critical crack length were experimentally found. A three-point closed-loop bending test was performed on prismatic notched specimens of neat mortars, mortars reinforced with 0.1 weight percent carbon nanofibers, and mortars reinforced with 0.1 and 0.2 weight percent multiwall carbon nanotubes. The crack mouth opening displacement was used as the feedback signal. The two-parameter fracture model was then used to calculate the fracture parameters of the nano-reinforced mortars [18]. The results of assessing parameters like compressive strength, modulus of elasticity, compressive toughness, flexural strength, flexural toughness, flexural residual strength, and fracture energy on steel and polypropylene fiber reinforced concrete that will be used in industrial ground floor slabs. The characterization approach included creating

cylindrical and prismatic specimens of fiber-reinforced concrete with nine different kinds of fibers, including four synthetic or polypropylene and five steel fibers. On the basis of a design for concrete with a 40 MPa flexural strength used for industrial ground floor slabs, non-reinforced concrete samples were also mixed as a reference. Based on the findings, it was concluded that the length and shape of fibers have an impact on the development of flexural strength as well as some other mechanical properties of concrete, such as ductility, since concrete samples made with steel fibers with hooks at their extremes exhibited better adhesion to concrete, while straight copolymer fiber and wavy polymer fiber displayed better overall performance. Finally, it was shown that the values of compression toughness and flexural toughness exhibited a strong association [19].

Cracks are commonly found in cement and concrete structures. They are undesirable features that may be brought about by environmental causes, workmanship, natural causes, as well as the age of the concrete element. It is, therefore, essential that the causes and consequences of cracking are well understood so that suitable remedial measures can be adopted. Once the cement sheath seal fails, it will lead to issues such as fluid leakage and sustained casing pressure. Cyclic loading will cause cumulative plastic strain and strength degradation of cement stone, a relatively dangerous working condition for the downhole cement sheath.

This study evaluated and modeled the mechanical behaviors of oil well cement, such as the stress intensity factor, compressing, and flexural stress–strain under cycling load.

Research Significance

This study aimed to assess the mechanical performance of an oil well cement, including its bonding, compressive, and flexural strengths. Among the specific objectives were the following:

i. Identifying the mechanical properties of oil well cement at various curing times.
ii. Researching Mode I and evaluating the fracture properties of oil well cement using three different a/d ratios.
iii. Using the nonlinear Vipulanandan p-q to test and model the stress–strain behavior of oil well cement under compression and flexural stresses.
iv. Constructing a new model to predict the modulus of elasticity of cement under varying strains.
v. Conducting a new protocol test to determine the bonding strength between the cement and steel casing in the oil well.

2. Methodology

2.1. Cement Characterization

The cement composition was analyzed at 25 °C using XRD (Figure 3). The sample holder was packed to within 3 mm of its capacity with 2 g of cement powder. The XRD was evaluated at intervals of 0.02 from 10 to 90°. Scanning electron microscopy (SEM) was used to obtain information about the cement's morphology (Figure 4).

2.2. Sample Mixture

The samples were prepared following API standards with a w/c of 0.38. The results were verified by testing at least three samples for each condition.

2.3. Compressive Strength Test (ASTM C 39)

Compression tests were conducted on cement samples after 1, 7, and 28 days of curing using a hydraulic compression testing machine. The strain gauge used in this study had a 60 mm length and a resistance of 120 Ω. A cylindrical mold measuring 50 mm in diameter and 100 mm in height. A stress–strain test was carried out at three different curing ages using a hydraulic compression testing system with a loading speed of 0.15 MPa/s. Specimens were instrumented with a strain gauge for lateral strain and an extensometer to measure axial strain to determine Poisson's ratio. The accuracy of the extensometer was checked with an axial strain gauge. Specimens were cast into appropriate molds for the intended tests and then cured at room temperature in a relative humidity box with a humidity of at least 95%. The samples were demolded after 24 h and stored in the same conditions until testing.

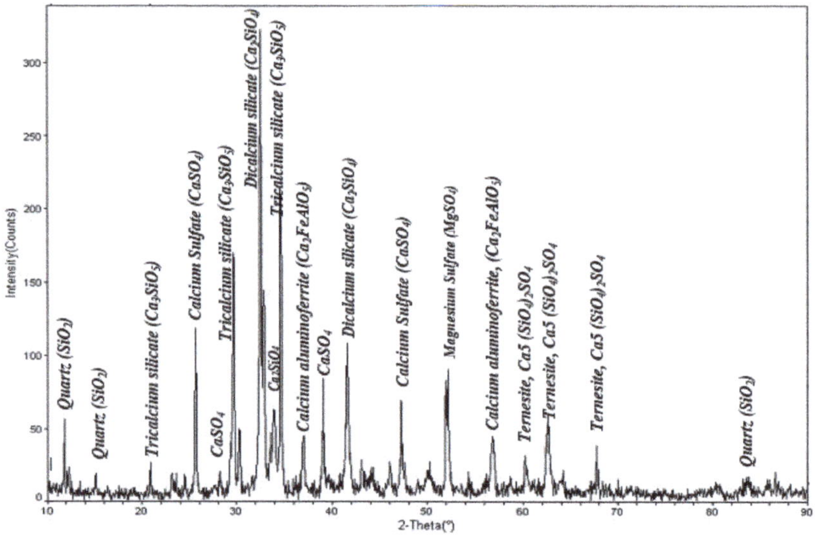

Figure 3. XRD for class H oil well cement.

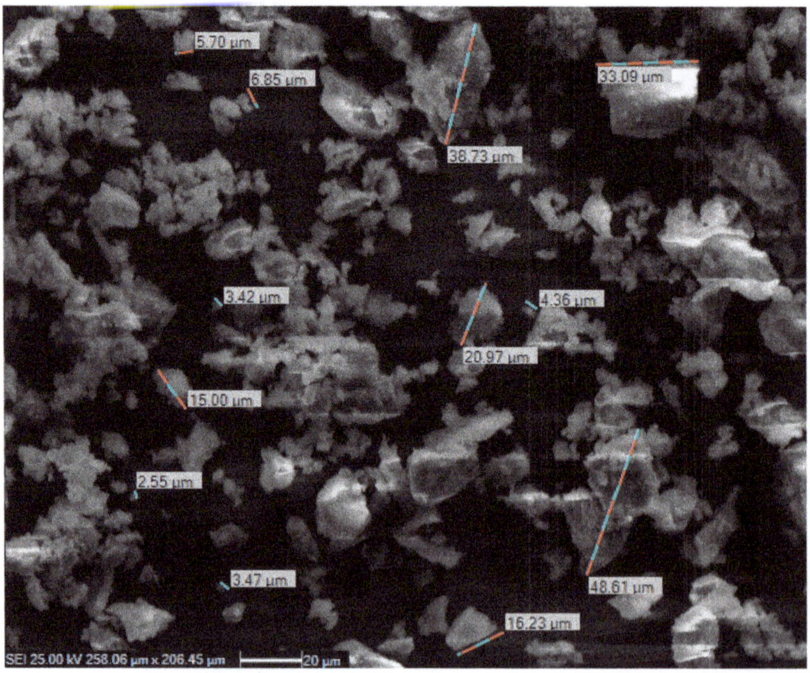

Figure 4. SEM for class H oil well cement.

2.4. Bonding between Steel Casing–Cement

Figure 5 depicts the experimental setup to evaluate the shear bonding between the steel casing and the cement. The steel casing was loaded up until it was flush with the cement. At a w/c of 0.38 and curing times of 1, 7, and 28 days, the applied load pushed the steel casing through the cement sheath. A portion of the steel enclosure was shifted aside (dia.: 45 mm; height: 74 mm). The load was applied to the steel casing until it reached the cement level (Figure 5b). The applied load pushed the casing through the cement sheath. Figure 5c shows the specimen at failure. The steel casing was moved to one side. After debonding from the cement sheath, the applied load acted against the friction between the steel–cement interface.

(**a**) (**b**) (**c**)

Figure 5. The specimen used to measure shear bonding between the steel casing and cement: (**a**) sample before the test, (**b**) specimen under the pull-out machine; (**c**) shear failure between casing and cement

2.5. Flexural Strength Test

The ASTM C293/C293M-10 standard [20] called for a three-point bending test. The dimensions of the specimen beam were 280 × 76 × 76 mm. A Vernier caliper was used to measure the specimen's dimensions for stress analysis. Strain gauges were installed on the tension and compression fibers near the midspan. Equation (1) was utilized in the strength calculation:

$$\sigma_f = \frac{3PL}{2bd^2} \quad (1)$$

where σ_f is the flexural stress, P is the applied load, L is the span length, b is the width of the specimen, and d is the depth of the sample. The strain was determined for each fiber by using strain gauges.

2.6. Fracture Toughness and CTOD

At least three samples were tested for each criterion, and the mean of those results is presented. A data-acquisition system was used to record the load (P), crack mouth opening displacement (CMOD), and load point deflection. The load cell's voltage output was used to determine the ideal value for the applied load. The strain was calculated from the two-wire DC resistance output of the strain gauge. Mode I fracture properties (Figure 2) were investigated using the beam specimen. As shown in Figure 6, a band saw was used to create a notch with a depth of 3 mm. A three-point loading setup was used to find the fracture toughness and crack tip opening displacement (CTOD) of different types of oil well cement. A clip on the CMOD gauge was used to measure the crack mouth opening displacement. Knife edges were used to clip the CMOD gauge to the pre-crack. At the same time, the resistance and pulse velocity was monitored to characterize the crack using a nondestructive method. Midspan deflection also was monitored using LVDT. Using a three-point loading setup, the oil well cement was tested for fracture toughness and crack tip opening displacement (CTOD). The crack mouth opening displacement was measured using a clip-on CMOD gauge. The CMOD gauge was clipped to the pre-crack with the blades of knives. Figure 7 shows the overall experimental setup, which utilized a clip-on CMOD gauge to derive K_I values. The schematic diagram of load vs. CMOD is presented in Figure 8.

Figure 6. Making a notch in the specimen using hand saw.

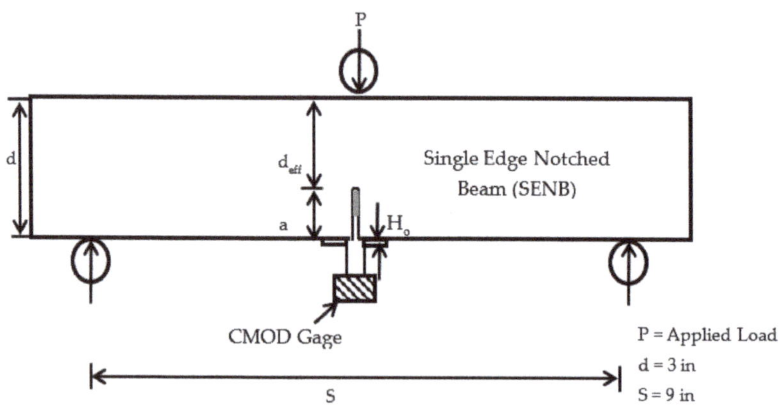

Figure 7. Schematic diagram of the experimental setup to find K_I.

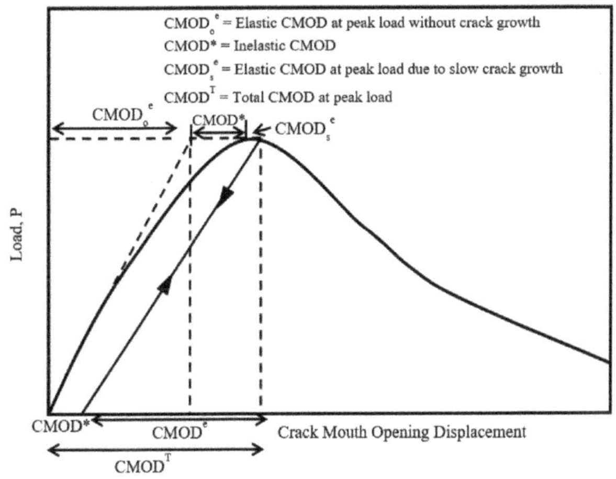

Figure 8. Schematic diagram of load versus CMOD with the components of CMOD^T.

For any geometries, Equation (2) provides the stress intensity factor K_I:

$$K_I = \sigma\sqrt{a}F(\alpha) \tag{2}$$

where σ is the bending stress calculated as shown in Equation (1), F(α) is a finite width and loading geometry correction factor, and "a" is the height of the crack. For the three-point loading bending test, F(α) is given in Equation (3):

$$F(\alpha) = \frac{\left[1.99 - \alpha(1-\alpha)\left(2.15 - 3.93\alpha + 2.7\alpha^2\right)\right]}{\left[(1+2\alpha)(1-\alpha)^{\frac{3}{2}}\right]} \tag{3}$$

where α is a factor that depends on the crack length. Equation (4) defines α as:

$$\alpha = \frac{(a + H_0)}{(d + H_0)} \tag{4}$$

where H_o is the clip gauge holder thickness, as shown in Figure 8. In Equation (4), 'a' must be calculated from a numerical iteration procedure using Equation (5):

$$a_e = a_0\left[\frac{C_U}{C_0}\right]\left[\frac{V(\alpha_0)}{V(\alpha_e)}\right] \tag{5}$$

where initial compliance Co is given by Equation (6):

$$Co = \frac{CMOD}{P} \tag{6}$$

Unloading compliance C_u was measured at about 95% of the peak load, and V(α) is given by Equation (7):

$$V(\alpha) = 0.76 - 2.28\alpha + 3.87\alpha^2 - 2.04\alpha^3 + \frac{0.66}{(1-\alpha)^2} \tag{7}$$

Variations in the load (P) versus the crack mouth opening displacement (CMOD) of the class H oil well cement in the experiment are shown in Figure 8 for the initial notch-to-depth ratios (a/d) of 0.3, 0.4, and 0.5 [18].

3. Results, Analysis, and Discussion

3.1. XRD and SEM

Major ingredients of the cement (Class H) included dicalcium silicate (Ca_2SiO_4), tricalcium silicate (Ca_3SiO_5), magnesium sulfate ($MgSO_4$), calcium aluminoferrite (Ca_2FeAlO_5), calcium sulfate ($CaSO_4$), and quartz (SiO_2) (Figure 3).

Figure 4 displays the results of the SEM analysis, which revealed that the diameter sizes of the cement particles ranged from 2.55 to 48.61 μm.

3.2. Vipulanandan p-q Model

The Vipulanandan p-q model was used to predict the stress–strain behavior of the soils, rocks, and piezoelectrical resistivity [21]. The ratio of the secant modulus to the initial tangential modulus is given by parameter q in Equation (8) (Table 1). The value of the model parameters was determined by minimizing the error in the stress–strain relationship. The model predicts a linear material up to peak stress when q equals 1. A smaller q value indicates a greater presence of nonlinear behavior. However, the post-peak behavior is controlled by the parameter p, while the pre-peak behavior is only moderately affected by it. Initially, the tangent and secant moduli were estimated to determine q. Then, the p values were adjusted to the lowest possible RMSE (Figure 9). Parameter q was defined as the ratio of the secant modulus at peak stress to the initial tangent modulus. Parameter p was obtained by minimizing the error in the predicted stress–strain relationship. The parameters p and q are influenced by the curing time of the cement. The shape of the stress–strain curve before and after the peak can be changed based on the p and q values, as shown in Figure 9.

$$\sigma = \frac{\frac{\varepsilon}{\varepsilon_f} * \sigma_f}{q + (1 - p - q)\frac{\varepsilon}{\varepsilon_f} + p\left(\frac{\varepsilon}{\varepsilon_f}\right)^{\frac{(p+q)}{p}}} \tag{8}$$

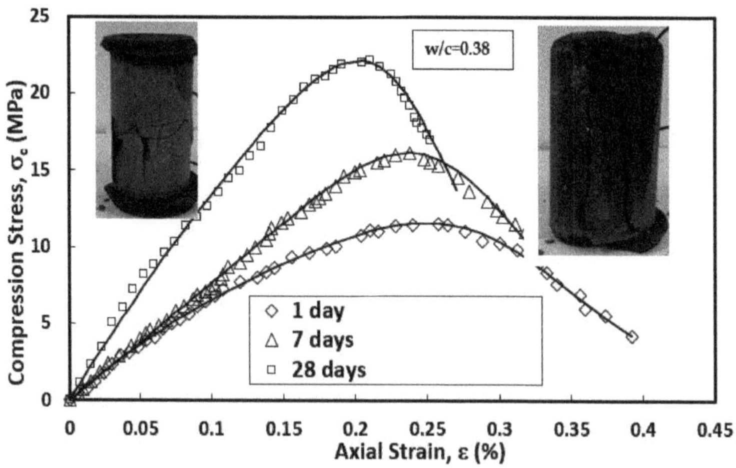

Figure 9. Compressive stress–strain behavior of the cement paste.

By differentiating Equation (8) and using the same p and q parameters for the stress-strain models in Equation (8), the modulus of elasticity of the cement paste versus strain can be predicted using Equation (9) (s is the axial stress, σ_f and ε_c are the stress at failure and the corresponding strain, and p and q are model variables (Table 1)):

$$\frac{d\sigma}{d\varepsilon} = \frac{q*\left[1-\left(\frac{\varepsilon}{\varepsilon_c}\right)^{\frac{(p+q)}{p}}\right]*\left(\frac{\sigma_c}{\varepsilon_c}\right)}{\left[\left[q*(1-p-q)*\frac{\varepsilon}{\varepsilon_c}\right]+p*\left(\frac{\varepsilon}{\varepsilon_c}\right)^{\frac{(p+q)}{p}}\right]^2} \quad (9)$$

The model evaluations in Equations (10) and (11) are the coefficient of determination and the root mean square error, respectively:

$$R^2 = 1 - \frac{\sum_{i=1}^{n}\left(y_{experimental} - y_{predicted}\right)^2}{\sum_{i=1}^{n}\left(y_{experimental} - mean\right)^2} \quad (10)$$

$$RMSE = \left(\frac{\sum_{i=1}^{n}\left(y_{experimental} - y_{predicted}\right)^2}{N}\right)^{0.5} \quad (11)$$

3.3. Compressive Strength

The compressive stress–strain relationship for cement was predicted using the Vipulanandan p-q stress-strain model. The compressive strength (σ_f) of the cement after 1, 7, and 28 days of curing was 10.6, 15.8, and 18.3 MPa, respectively (Figure 9). The axial strain of the samples at failure varied between 0.21 and 0.3% (Table 1). The coefficient of determination (R^2) was 0.99 (Table 1). Differentiating Equation (8), the modulus of elasticity of the cement versus axial strain can be calculated using Equation (9), as shown in Figure 10. The compression modulus of elasticity of the cement at 1 and 28 days of curing times is summarized in Table 1. The same p and q values were obtained to predict the stress–strain behavior of the cement shown in Figure 9 and were used to predict the modulus of elasticity of the cement shown in Figure 10.

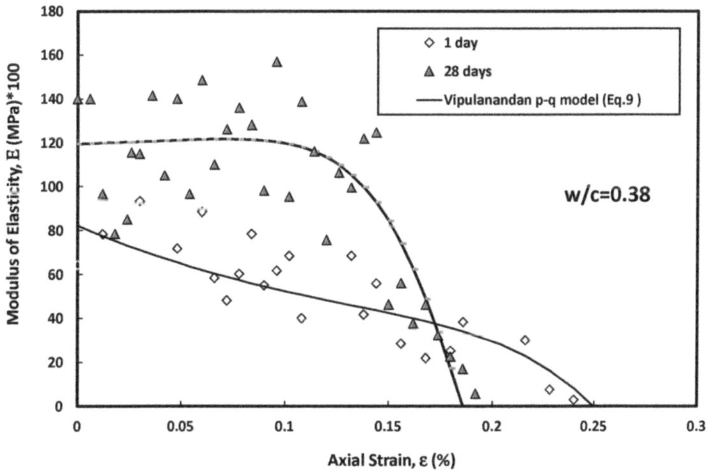

Figure 10. Relationship between axial strain and modulus of elasticity of the cement.

3.4. Fracture Properties of Cement

This study investigated the influence of utilizing a different a/d on the fracture behavior of cement at three different curing times and using the single-edge notch beam in three-point bending determined parameters such as the s K_I (mode 1), which has been frequently measured and widely reported for other mortars and concrete [13,22–24].

The experimental observations included the crack mouth opening displacement (CMOD), load versus load point displacement, and crack development due to applied loads.

3.4.1. Experimental Calculations

At different load levels (applied load), the following parameters could be calculated:

i. Crack length (a);
ii. Crack extension (Δa = crack length − initial crack length = $a - a_i$);
iii. Stress intensity factor (K_I);
iv. Crack tip opening displacement (CTOD).

It should be mentioned that at peak load, the stress intensity factor and crack length were called the critical stress intensity factor (K_I) and effective crack length, respectively.

3.4.2. Calculation Procedures

For a three-point bend test specimen, by using the concept of the linear elastic fracture mechanic (LEFM), the crack length can be obtained from the relationship between the elastic crack mouth opening displacement ($CMOD^e$) and the corresponding crack length (a), as shown in Equation (12).

$$CMOD^e = 4\sigma\, a \frac{V(\propto)}{\acute{E}} \quad (12)$$

where:

σ is the net stress, which is equal to $6\ M/bd^2$;
M is the applied pure bending moment;
$\propto$ is equal to $(a + H_\circ)/\ (d + H_\circ)$;
a is the crack length;
É for the plane, the stress is equal to E (Modulus);
É for the plane, the strain is equal to $E/\ (1 - v^2)$;
v is Poisson's ratio;
$V(\propto)$ can be calculated by using the following empirical formula:

$$V(\propto) = 0.8 - 1.7 \propto + 2.4 \propto^2 + 0.66\ /\ (1 - \propto)^2 \quad (13)$$

The elastic crack mouth opening displacement could be determined at various loading levels if the crack grew slowly. Accordingly, using Equation (15), the corresponding crack length (a) could be evaluated. For a given measured load (P), the crack length (a) could be evaluated by applying a numerical iterative procedure such that the crack length (a) could be assumed in Equation (15) to obtain the CMOD calculated and then be compared with the CMOD measured. The procedure had to be repeated until the measured and calculated values of CMOD were in agreement.

Since the effective crack length a_e was the sum of the initial notch plus an effective crack extension at the peak load, it can be mentioned that if the material behaved elastically up to the peak load without any crack extension, the relationship between the load and CMOD would be linear. However, suppose any displacement occurred due to unloading the specimen just before and immediately after the peak load. In that case, the displacement could be considered an inelastic displacement associated with the response of the notched beam. Studies have indicated that at peak load, the total CMOD ($CMOD^T$) is composed of the elastic displacement (no crack extension—$CMOD^e_\circ$), inelastic displacement ($CMOD^*$), and elastic displacement due to slow crack growth ($CMOD^e_s$).

The nonlinear displacement observed in the P-CMOD response could be attributed to creep and slow crack growth. To apply LEFM, the inelastic displacement ($CMOD^*$) had to be extracted from the total CMOD ($CMOD^T$) at peak load. The elastic ($CMOD^e$) at peak load was found by unloading the specimen immediately after peak load using the initial compliance (C_i) and the unloading compliance (C_u).

Where (C_i) is equal to (CMOD/P) and (C_u) is measured as about 95% of the peak load in Equation (15); by using the results, the effective crack length (a_e) could be determined using the following relationship:

$$a_e = a_i\left(\frac{C_u}{C_i}\right)\left[V_{(\propto_i)}\,V_{(\propto_e)}\right] \tag{14}$$

where a_i is the initial crack length.

A numerical iterative procedure had to be used to estimate the effective crack length using Equation (14); thus, for a beam of cross-section (b x d) with crack length (a), the stress intensity factor (K_I) for the four-point bending could be calculated by applying the equation as follows:

$$K_I = \sigma\sqrt{a}\,y\left(\frac{a}{d}\right) \tag{15}$$

where $y(a/d)$ is a correction factor for finite width and loading geometry. For three-point bend tests:

$$y(\propto) = \left[1.99 - \propto(1-\propto)\left(2.15 - 3.9\propto + 2.7\propto^2\right)\right]/\left[(1+2\propto)(1-\propto)^{\frac{3}{2}}\right] \tag{16}$$

Based on Equation (14) and by using the effective crack length (a_e), which was determined using Equation (15); instead of crack length (a) and peak load (P), the critical stress intensity factor (K_{Ic}) could be determined.

Figure 11 displays the load versus CMOD for each a/d at three curing times. The $CMOD^e$ was reduced by 39% when the a/d was increased from 0.3 to 0.5 (Figure 12a), and the value of the $CMOD^e$ increased by 11 to 36% when the curing time was increased from 1 to 28 days at the same a/d. Figure 11 depicts a linear relationship between the load and CMOD, assuming the material exhibited elastic behavior up to the peak load without any crack extension. This inelastic displacement (CMOD*) abruptly released the specimen after the peak load. The elastic displacement (no crack extension—$CMOD^e_o$), inelastic displacement (CMOD*), and elastic displacement due to slow crack growth at maximum load made up of the $CMOD^T$ ($CMOD^e_s$). The specimen was unloaded at 95% of the peak load to obtain the total elastic CMOD ($CMOD^e = CMOD^e_o + CMOD^e_s$). Figure 12a,b shows the relationships between curing time and $COMD^e$ and K_I of oil well cement at three different a/b, respectively.

3.5. Flexural Stress–Strain

Equation (1) was used to model the flexural stress–strain behavior of the cement with a w/c of 0.38 up to 28 days of curing. When the cement was left to cure for 7 days, there was a 12% increase in flexural strength compared with 1 day of curing. After 28 days, there was a 15% increase in flexural strength compared with 7 days of curing, as shown in Figure 13. The Vipulanandan p-q model performed well in predicting the experimental data, with R^2 and RMSE values between 0.99 and 0.98 and 0.03 and 0.03 MPa, respectively (Table 1).

Table 1. Model parameters for stress–strain for oil well cement (w/c = 0.38).

	Curing Age (Day)	σf (MPa)	εf (%)	Ei (MPa)	*p*	q	RMSE (MPa)	R^2	Figure No.
Compressive	1	10.6 ± 2	0.28 ± 0.01	9670 ± 21	0.034 ± 0.01	0.36 ± 0.02	0.128	0.99	Figure 9
	7	15.8 ± 1	0.3 ± 0.02	12,000 ± 25	0.03 ± 0.02	0.36 ± 0.02	0.02	0.99	
	28	18.3 ± 1.5	0.21 ± 0.02	19,360 ± 20	0.160 ± 0.03	0.87 ± 0.04	0.11	0.99	
Flexural	1	1.36 ± 0.02	0.041 ± 0.002	-	71,199.4 ± 100	0.316 ± 0.02	0.03	0.99	Figure 13
	7	2.32 ± 0.01	0.101 ± 0.001	-	71,214.6 ± 95	0.082 ± 0.01	0.05	0.99	
	28	3.0 ± 0.02	0.02 ± 0.002	-	71,138.8 ± 105	0.550 ± 0.03	0.10	0.99	

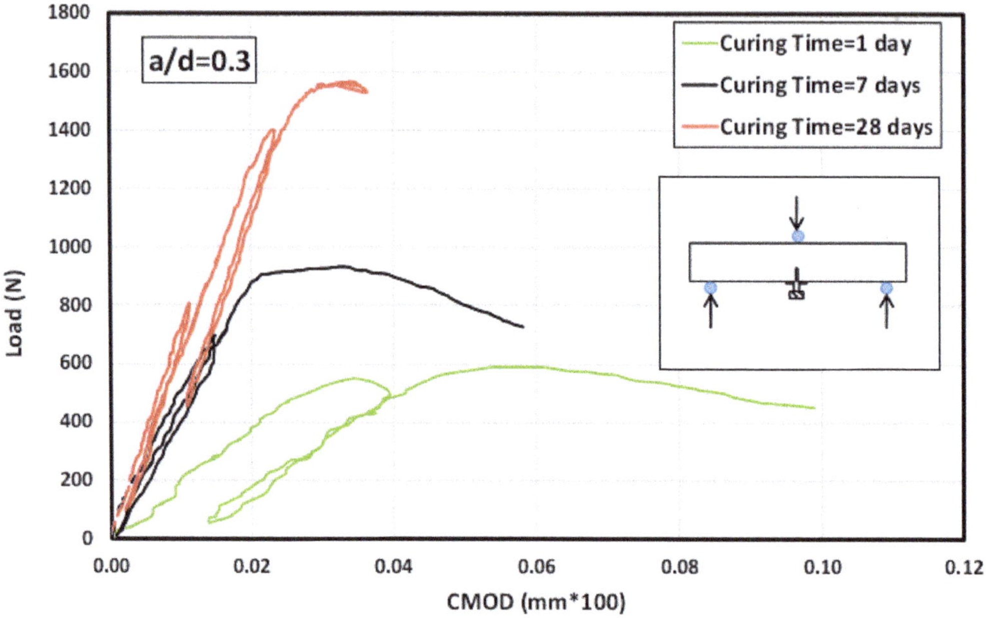

(a)

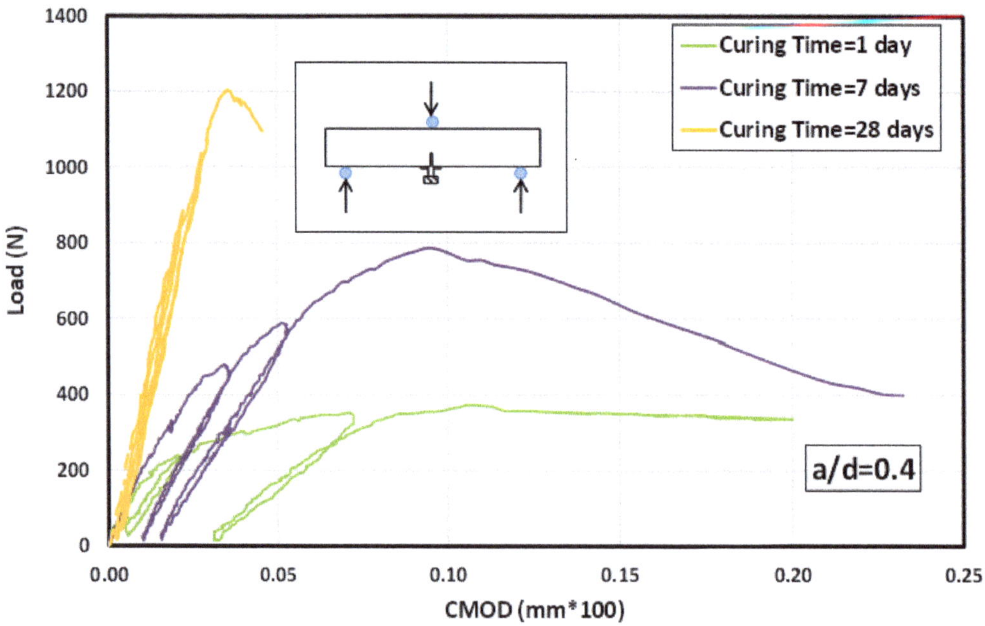

(b)

Figure 11. *Cont.*

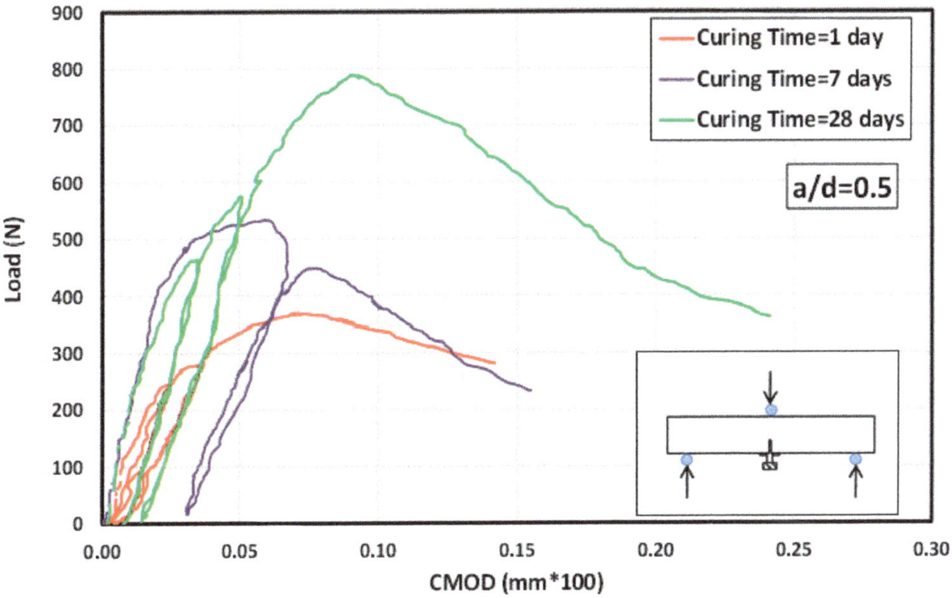

(c)

Figure 11. Relationship between axial strain and modulus of elasticity of cement: (**a**) a/b = 0.3 m; (**b**) a/b = 0.4; (**c**) a/b = 0.5.

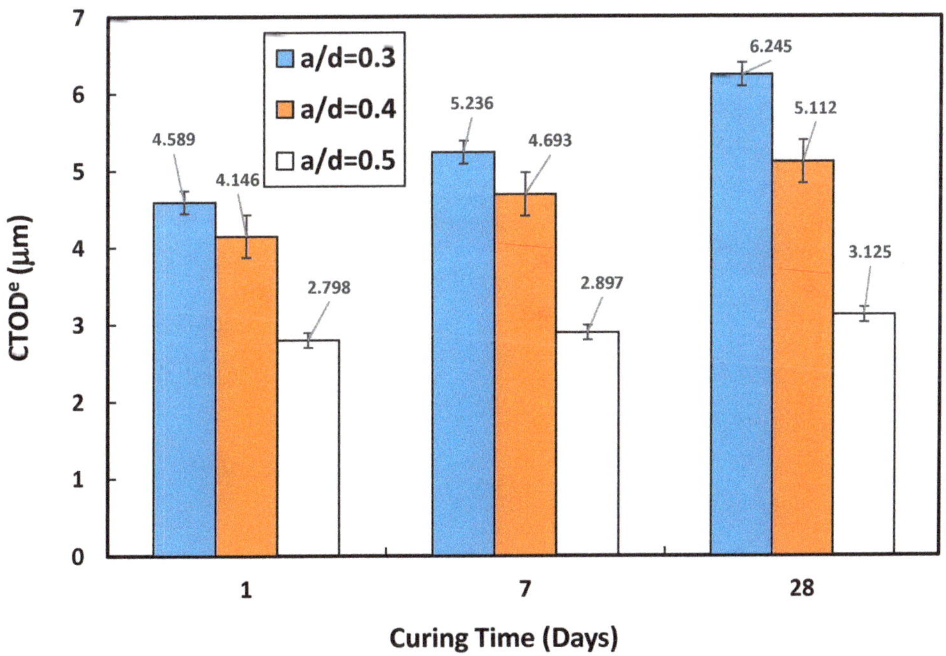

(a)

Figure 12. *Cont.*

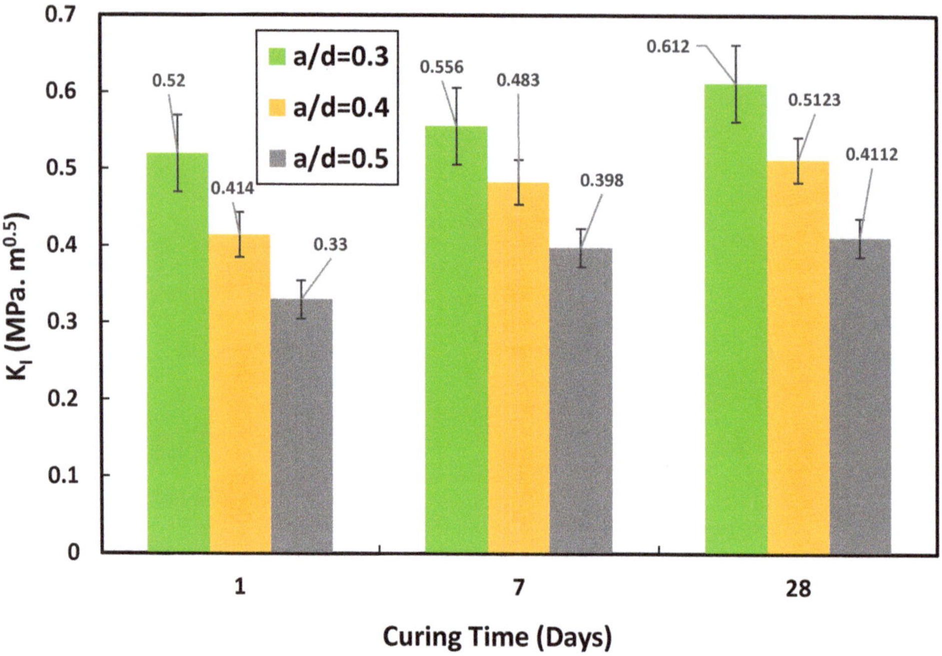

(**b**)

Figure 12. Variations in a/d ratio for oil well cement for (**a**) CTOD^e and (**b**) K_I.

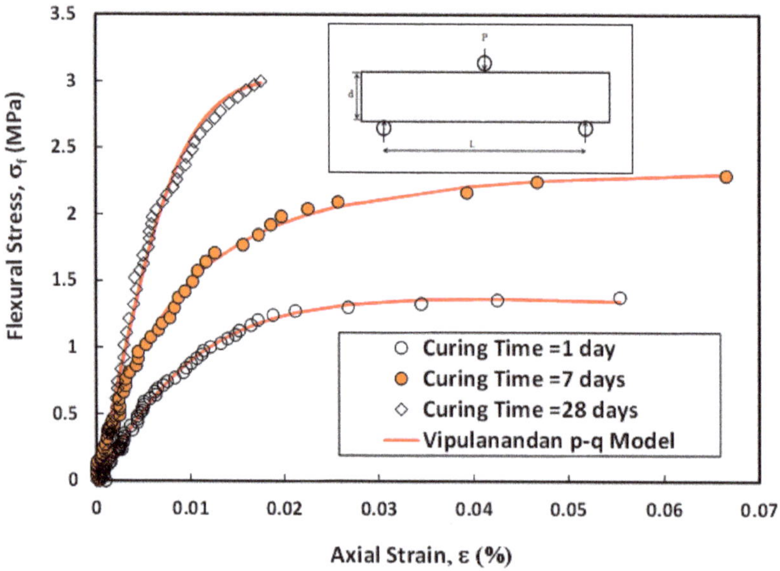

Figure 13. Flexural stress–strain for the oil well cement at three different curing times.

The K_I increased as the notch-to-depth ratio grew (Figure 14b). In this study, the K_I ranged from 0.33 to 0.612 MPa.m$^{0.5}$. The results of this study were consistent with previous reports that the K_{IC} of cement pastes, cement mortars, and cement concretes can range from 0.1 to 1.7 MPa.m$^{0.5}$. It was hypothesized that the wide variety of K_I values was due to the variations in the material that caused scatter in experimental results and the limitations in the applicability of fracture mechanics concepts. The K_I of the cement paste decreased with the ratio of a/d; also, the K_I increased from 0.33 at 1 day of curing to 0.398 and 0.4112 MPa.m$^{0.5}$ when the curing time was increased to 7 and 28 days, respectively.

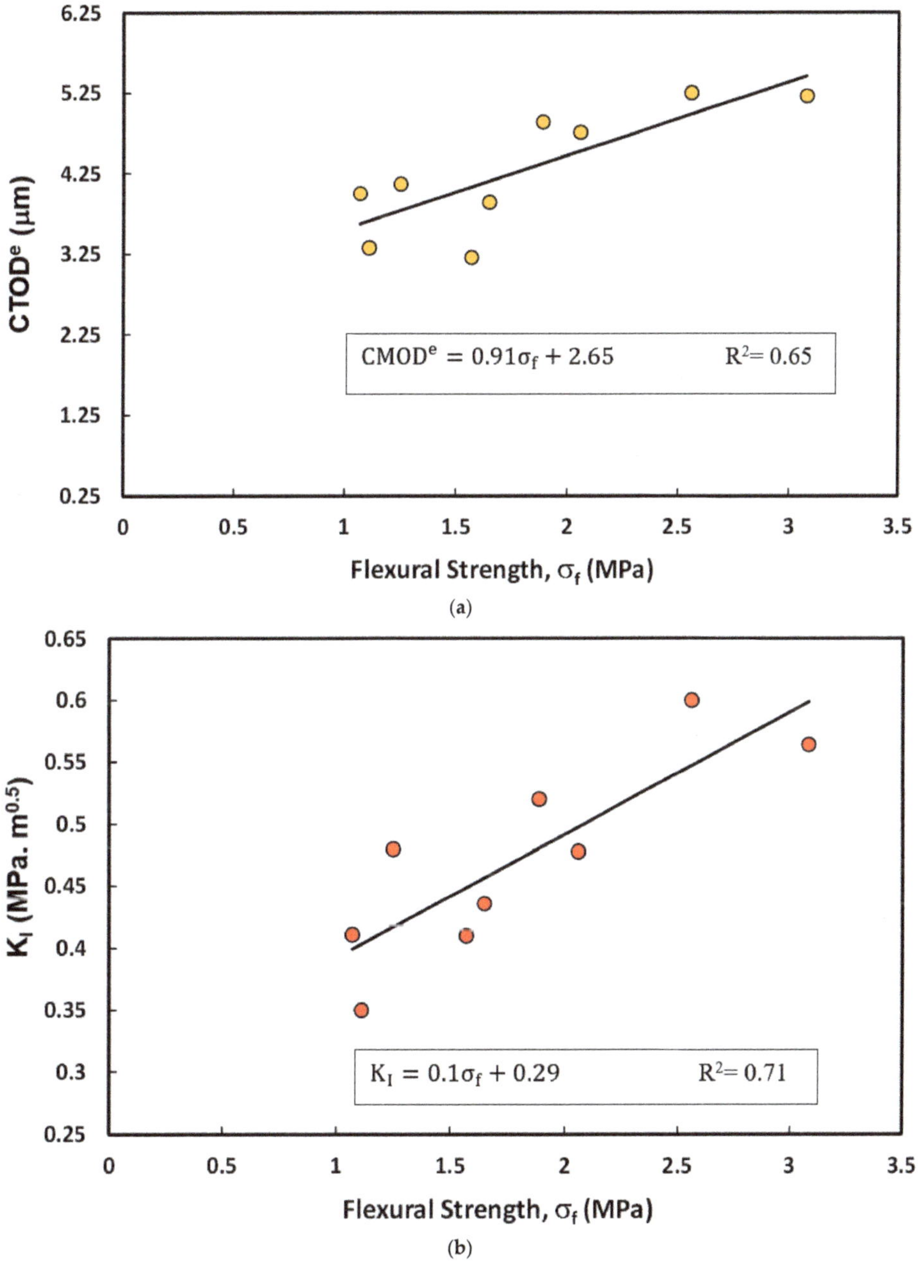

Figure 14. Correlations between flexural strength and (**a**) CTODe and (**b**) K_I.

The following correlations were developed between the CTOD^e and K_I with the flexural strength (σ_f) of the cement paste (Figure 14):

$$CMOD^e = 0.91\sigma_f + 2.65 \quad R^2 = 0.65 \tag{17}$$

$$K_I = 0.1\sigma_f + 0.29 \quad R^2 = 0.71 \tag{18}$$

3.6. Bonding Strength

It is essential to consider not only short-term well integrity factors such as cement quality and pumpability at the time of cementing operation but also the long-term integrity of the cement sheath, including the cement/casing bonding throughout the well's life and developing a high bonding strength between the cement and steel. The casing between the cement and the formation on the other side is an essential property determining the life of oil and gas wells. This new protocol test measured the bonding strength between the steel tube and the oil well cement at three curing times (1, 7, and 28 d). The respective bonding strengths were 0.35, 1.39, and 3.59 MPa for the three curing times (Figure 15). The steel casing was moved to one side. After debonding from the cement sheath, the applied load acted against the friction between the steel–cement interface. Mechanical properties were tested by conducting compression, indirect tension, flexural, fracture toughness, and steel casing–cement bonding tests.

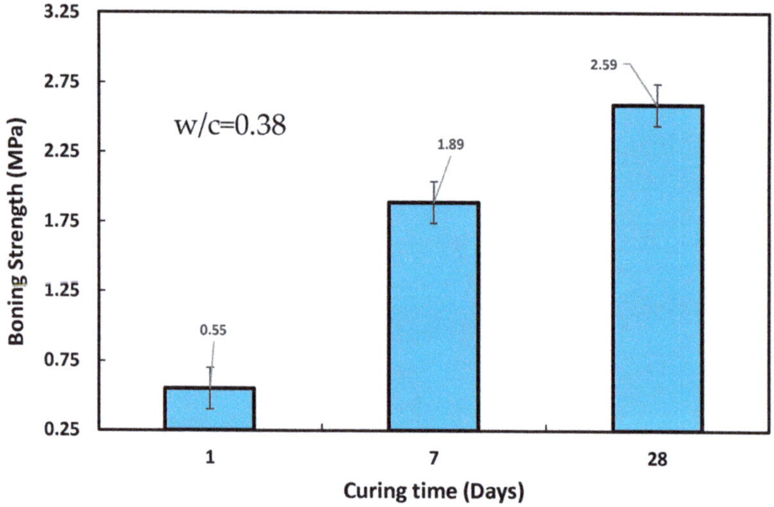

Figure 15. Relationship between bonding strength and curing time.

4. Conclusions

This study aimed to determine the mechanical properties of an oil well cement paste and its fracture behavior. The following conclusions were obtained from this study.

1. The compressive and flexural strengths for the oil well cement at a w/c of 0.38 increased from 10.6 to 18.3 MPa and from 1.36 to 3 MPa, respectively, when the curing time was increased from 1 to 28 days of curing.
2. Studies of fractures showed that the K_I values ranged from 0.3 to 0.6 MPa.m, and the CMOD values ranged from 2 to 6 m.
3. Increasing the ratio of the crack mouth depth and the thickness of the beam (a/d) from 0.3 to 0.5 decreased the CMOD^e and the K_I by 39–50% and 33–37% based on curing time, respectively. Based on the flexural strength of the oil well cement, the CMOD^e and K_I could be estimated from the developed linear relationship.

4. Based on the R^2 and RMSE, the Vipulanandan p-q model accurately predicted the tested oil well cement's compressive and flexural stress–strain behaviors.
5. More studies on the effect of the w/c on the bonding, compression, and flexural strengths of oil well cement should be conducted to justify the behavior of the oil well cement under different environmental conditions in an oil well.

Author Contributions: C.V. and A.S.M. collected the data, conducted planning, and wrote the manuscript; P.R. was responsible for the results and analysis; A.S.M. and C.V. were responsible for the conclusions and editing of the manuscript. All authors have read and agreed to the published version of the manuscript.

Funding: This research received no external funding.

Informed Consent Statement: Informed consent was obtained from all subjects involved in the study.

Data Availability Statement: The data supporting the conclusions of this article are included in the article.

Acknowledgments: Center for Innovative Grouting Materials and Technology (CIGMAT), and Texas Hurricane Center for Innovative Technology in the University of Houston, TX-USA, CIGMAT Laboratory supported this work.

Conflicts of Interest: The authors declare no conflict of interest.

References

1. Dou, H.; Dong, X.; Duan, Z.; Ma, Y.; Gao, D. Cement Integrity Loss due to Interfacial Debonding and Radial Cracking during CO_2 Injection. *Energies* **2020**, *13*, 4589. [CrossRef]
2. Lavagna, L.; Nisticò, R. An Insight into the Chemistry of Cement—A Review. *Appl. Sci.* **2023**, *13*, 203. [CrossRef]
3. Wu, H.; Hu, R.; Yang, D.; Ma, Z. Micro-macro characterizations of mortar containing construction waste fines as replacement of cement and sand: A comparative study. *Constr. Build. Mater.* **2023**, *383*, 131328. [CrossRef]
4. Labibzadeh, M.; Zahabizadeh, B.; Khajehdezfuly, A. Early-age compressive strength assessment of oil well class G cement due to borehole pressure and temperature changes. *J. Am. Sci.* **2010**, *6*, 1–7.
5. Vipulanandan, C.; Dharmarajan, N. Fracture properties of particle filled polymer composites. *J. Compos. Mater.* **1989**, *23*, 846–860. [CrossRef]
6. Vipulanandan, C.; Krishnamoorti, R.; Mohammed, A.; Boncan, V.; Narvaez, G.; Head, B.; Pappas, J.M. Iron Nanoparticle Modified Smart Cement for Real Time Monitoring of Ultra Deepwater Oil Well Cementing Applications. In Proceedings of the Annual Offshore Technology Conference 3, Houston, TX, USA, 4–7 May 2015; p. OTC-25842-MS.
7. McCarter, W.J.; Chrisp, T.M.; Starrs, G.; Blewett, J. Characterization and monitoring of cement-based systems using intrinsic electrical property measurements. *Cem. Concr. Res.* **2003**, *33*, 197–206. [CrossRef]
8. Han, B.; Zhang, L.; Zhang, C.; Wang, Y.; Yu, X.; Ou, J. Reinforcement Effect and mechanism of Carbon Fibers to Mechanical and Electrically Conductive Properties of Cement-Based Materials. *Constr. Build. Mater.* **2016**, *125*, 479–489. [CrossRef]
9. Choolaei, M.; Rashidi, A.M.; Ardjmand, M.; Yadegari, A.; Soltanian, H. The effect of nanosilica on the physical properties of oil well cement. *Mater. Sci. Eng. A* **2012**, *538*, 288–294. [CrossRef]
10. Li, H.; Xiao, H.G.; Ou, J.P. A study on mechanical and pressure-sensitive properties of cement mortar with nanophase materials. *Cem. Concr. Res.* **2004**, *34*, 435–438. [CrossRef]
11. Sorathiya, J.; Shah, S.; Kacha, M.S. Effect on Addition of Nano "Titanium Dioxide"(TiO_2) on Compressive Strength of Cementitious Concrete. *ICRISET2017. Sel. Pap. Civ. Eng.* **2017**, *1*, 219–225.
12. Compendex, S.; Elsevier, G. Influence of Zinc Oxide Nanoparticle on Strength and Durability of Cement Mortar. *Int. J. Earth Sci. Eng.* **2016**, *9*, 175–181.
13. Ramanathan, P. Mechanical, Piezoresistive and Fracture Behavior of Various Types of Smart Cements. Master's Thesis, University of Houston, Houston, TX, USA, 2014.
14. Teodoriu, C.; Amani, M.; Yuan, Z.; Schubert, J.; Kosinowski, C. Investigation of the mechanical properties of Class G cement and their effect on well integrity. *Int. J. Eng.* **2013**, *3*, 2305–8269.
15. Vipulanandan, C.; Dharmarajan, N. Critical crack tip opening displacement for polymer composites. *Eng. Fract. Mech.* **1989**, *33*, 409–419. [CrossRef]
16. Vipulanandan, C.; Dharmarajan, N. Effect of temperature on the fracture properties of epoxy polymer concrete. *Cem. Concr. Res.* **1988**, *18*, 265–276. [CrossRef]
17. Feng, G.; Liu, W.; Du, X.; Wang, J.; Li, X.; Zheng, Y. Crack evolution characteristics of cemented-gangue–fly-ash backfill with different proportions of fly ash and cement. *Constr. Build. Mater.* **2023**, *385*, 131498. [CrossRef]
18. Reis, J.M.L.; Ferreira, A.J.M. The influence of notch depth on the fracture mechanics properties of polymer concrete. *Int. J. Fract.* **2003**, *124*, 33–42. [CrossRef]

19. de Luna, A.M.; Lozano JÁ, O.; Jaime, L.P.; Martínez, J.P.; Bernal, J.J.S.; Martínez, S.H.R.; Flores, J.A. Experimental mechanical characterization of steel and polypropylene fiber reinforced concrete. *Rev. Tec. Fac. Ing. Univ. Zulia* **2014**, *37*, 106–111.
20. *ASTM C293/C293M–10*; Standard Test Method for Flexural Strength of Concrete. American Society for Testing and Materials: West Conshohocken, PA, USA, 2010.
21. Vipulanandan, C.; Mohammed, A.; Qu, Q. Characterizing the hydraulic fracturing fluid modified with nano silica proppant. In Proceedings of the AADE Fluids Technical Conference and Exhibition, Houston, TX, USA, 15–16 April 2014; pp. 15–16.
22. Wen, S.; Chung, D.D.L. Effect of moisture on piezoresistivity of carbon fiber-reinforced cement paste. *ACI Mater. J.* **2008**, *105*, 274.
23. Huang, Y.; Wang, W.; Sun, Z.; Wang, Y.; Liu, P.; Liu, C. A multilayered flexible piezoresistive sensor for wide-ranged pressure measurement based on CNTs/CB/SR composite. *J. Mater. Res.* **2015**, *30*, 1869–1875. [CrossRef]
24. Joel, O.F.; Ndubuisi, E.C.; Ikeh, L. Effect of cement contamination on some properties of drilling mud. In Proceedings of the Nigeria Annual International Conference and Exhibition, Lagos, Nigeria, 6–8 August 2012.

sustainability

MDPI

Article

Synergistic Effect of Micro-Silica and Recycled Tyre Steel Fiber on the Properties of High-Performance Recycled Aggregate Concrete

Muhammad Talha Amir [1,2], Sobia Riaz [3], Hawreen Ahmed [4,5,*], Syed Safdar Raza [3,*], Ahmed Ali A. Shohan [6] and Saleh Alsulamy [6]

1 Department of Civil Engineering, University of Engineering and Technology, Taxila 47050, Pakistan; ascepak@gmail.com
2 House of Property and Entrepreneurship (HOPE), Wallayat Complex, Plaza 74, Phase 7 Bahria Town, Islamabad 46000, Pakistan
3 Department of Civil Engineering, Faculty of Engineering, Bahauddin Zakariya University, Multan 66000, Pakistan; sobiariaz@bzu.edu.pk
4 Department of Highway and Bridge Engineering, Technical Engineering College, Erbil Polytechnic University, Erbil 44001, Iraq
5 Department of Civil Engineering, College of Engineering, Nawroz University, Duhok 42001, Iraq
6 Architecture & Planning Department, College of Engineering, King Khalid University (KKU), Abha 61411, Saudi Arabia; ashohan@kku.edu.sa (A.A.A.S.)
* Correspondence: hawreen.a@gmail.com (H.A.); safdarshah91@bzu.edu.pk (S.S.R.)

Abstract: The present research investigates the mechanical and physical properties of recycled aggregate concrete (RAC) modified with micro-silica (MS) and recycled tire steel fiber (RTSF). Natural coarse aggregates (NCA) were completely replaced by recycled coarse aggregates (RCA) to prepare RAC. High-strength RAC mixes were prepared by replacing 5% and 10% of Portland cement with MS. With each level of MS, RTSF was incorporated as 0%, 0.5%, 1 and 2% by volume fraction. In addition to mechanical properties, ultrasonic pulse velocity (UPV), electrical resistivity (ER), and water absorption (WA) of the mixes were also evaluated. The performance of modified RAC mixtures was also compared with plain natural aggregate concrete (PNAC). The experimental investigation revealed that RTSF substantially increased the tensile strength of RAC, whereas MS improved the durability of RTSF-reinforced RAC. RAC made with 1% RTSF and 10% MS showed 54% more splitting-tensile strength compared to the PNAC. The WA capacity of RAC incorporating 10% MS was 15–22% lower than that of the PNAC.

Keywords: waste tyres; construction waste; water absorption; supplementary binder; steel fiber; optimization; fibers and environment

Citation: Amir, M.T.; Riaz, S.; Ahmed, H.; Raza, S.S.; Shohan, A.A.A.; Alsulamy, S. Synergistic Effect of Micro-Silica and Recycled Tyre Steel Fiber on the Properties of High-Performance Recycled Aggregate Concrete. *Sustainability* **2023**, *15*, 8642. https://doi.org/10.3390/su15118642

Academic Editor: José Ignacio Alvarez

Received: 26 March 2023
Revised: 11 May 2023
Accepted: 22 May 2023
Published: 26 May 2023

1. Introduction

Many countries are suffering a scarcity of space to landfill construction and demolition (C&D) wastes. Uncontrolled urbanization has caused a massive boom in the construction industry. The state-of-the-art structures are replacing the insufficient and older ones and consequently, it led to the increased generation of C&D wastes. Due to the scarcity of landfilling space in urban regions, C&D waste is normally dumped along with domestic waste which causes severe environmental and social issues. Globally, around 40 major countries produce more than 3 billion metric tons of C&D waste per year [1]. The most effective method to deal with a massive quantity of C&D waste is to recycle it as construction aggregate and use them in the construction of buildings and roads [2]. Reusing C&D waste as construction aggregate is a win-win model to save both environment and humans from the adverse impacts of the construction industry.

In the past decade, researchers advanced their interest in appraising the performance of structural concrete produced with recycled coarse aggregates (RCA). Successful applications of RCA have been reported in rigid concrete pavements and building structures [3,4]. Hoffman et al. [5] (2012) assessed the performance of reinforced concrete (RC) elements with RCA. They reported that RC beams made with RCA show insignificant changes in shear strength compared to those made with natural coarse aggregates (NCA). The life cycle assessment studies have confirmed the environmental benefits of RCA application in structural concrete [2,6]. Not only recycled aggregate concrete (RAC) has a lower CO_2 footprint than conventional plain natural aggregate concrete (PNAC) [7], but it also helps in avoiding the potential increase in toxicity of soils due to uncontrolled disposal of C&D wastes [6,8,9].

Due to the existence of adhered mortar, RCA is weaker than NCA. Therefore, RAC has inferior properties compared to natural aggregate concrete (NAC) for the same volume of aggregate. To address the drawbacks of RAC, researchers have preferred using additional cementitious materials like silica fume/micro-silica, metakaolin, slag, fly ash, etc. [10–13] and fiber reinforcements (i.e., glass, steel, macro-synthetic fibers, etc.) [14–17]. SCMs are highly useful in enhancing the durability characteristics of RAC such as imperviousness against harmful fluids [14,18], acid attack resistance [14,19], chloride-ion permeability resistance [11,19], etc. High-performance admixtures like silica fume or micro-silica (MS) have shown positive effects on the strength enhancement of RAC, in addition to its phenomenal contribution to durability [11,20–22].

Fiber reinforcement overcomes the inherent brittleness of plain concrete. It is an excellent option to advance the tensile and flexural strength of RAC [23,24]. The effect of several types of fibers i.e., steel, glass, polypropylene, hybrid fibers, etc. have been studied on the performance of RAC [14,16,25–27]. Generally, fiber-reinforcement significantly enhanced the tensile ductility, shrinkage and abrasion resistance and impact toughness of concrete [28–30]. It has shown a positive role in some durability aspects, such as control over the degradation and abrasion of plain concrete surfaces [31,32]. The acid-attack durability of plain RAC has been reported to improve due to the addition of steel and glass fibers [14,33]. Therefore, the use of fibers can increase the life of a structure and minimize the maintenance cost.

Despite numerous benefits, especially in terms of improved ductility behavior and toughness, artificial fibers are expensive and possess a high CO_2 footprint compared to the SCMs and conventional constituents of concrete. For instance, artificial steel fiber (ASF) at a 1% volume fraction can increase the cost of concrete by twice, as shown by literature [34,35]. The carbon footprint of plain concrete increased by more than 50% at 1% volume of ASF [34,36]. Therefore, ASF, despite its phenomenal utilization ratio in tensile ductility [25], is an expensive option to upgrade the performance of RAC. Other than artificial fibers, cheap alternatives for fiber-reinforcements are recycled tyre steel fibers (RTSF) [37–39] and organic fibers [40,41] which are currently being examined for their effects on the mechanical and durability properties of concrete.

RTSF is recovered from the steel bead wires of scrap tyres, and it possesses high tensile strength and toughness comparable to ASF since new tyres are manufactured with high-grade tension steel wires [42]. The flexural behavior, residual strength and energy absorption capacity of RTSF-reinforced concrete and ASF-reinforced concrete is almost similar [39,43]. Hence, RTSF is a suitable, eco-friendly, and low-cost fiber reinforcement. Until now, very few studies [44,45] have appraised the performance of RAC incorporating RTSF. Existing studies showed [44,45] that using 0.5–1% volume RTSF notably increased the splitting-tensile strength ($f$ctm) and flexural strength of RAC, while it showed a marginal effect on the compressive strength ($f$cm).

The coupling effects of MS and RTSF on the performance of RAC have never been studied. Since modern structures require high strength, excellent durability, and ductility behavior, therefore, the performance of RAC must be studied with the combined incorporation of MS and RTSF. The durability and ductility benefits of SCMs and fiber, respectively can be combined by the simultaneous addition of MS and RTSF in RAC. Eventually, it would lead to the development of low-cost and sustainable high-performance RAC. Therefore, the present study is devoted to examining the mechanical and physical properties of RAC with different levels of MS (0, 5, and 10%) and RTSF (0, 0.5, 1, and 2%). Mechanical properties included *f*cm (7, 28, 90, and 180 days) and *f*ctm (28 and 90 days). Physical properties included ER, WA, and UPV were assessed at the age 28 and 90 days. Statistical correlations between mechanical and physical parameters were analysed.

2. Materials and Methods

2.1. Materials

2.1.1. Cement and Micro-Silica

Portland cement of '53 Grade' was utilized as the main binder to prepare all concrete mixes. The properties of Portland cement complied with ASTM C150 [46]. The properties of cement are given in Table 1. Commercially available high-performance MS was used as a partial replacement for cement. It has a silica oxide (SiO_2) content of approx. 98.5%. The specific gravity (Gs) and specific surface area of MS particles were 2.18 and 27,000 m^2/kg, respectively. These properties of MS came with technical datasheet. The particle size distribution of cement and MS samples is illustrated in Figure 1.

Table 1. Chemical, physical and mechanical properties of cement.

Chemical Properties	% Composition
Lime (CaO)	63.15
Alumina (Al_2O_3)	5.24
Silica (SiO_2)	19.6
Iron Oxide (Fe_2O_3)	4.36
Magnesia (MgO)	0.76
Loss in the ignition (LOI)	1.13
Physical Properties	**Result**
Specific gravity	3.11
Specific surface area (m^2/kg)	371
Initial setting time (h)	1.45
Final setting time (h)	2.93
7-days *f*cm	45.5
28-days *f*cm	52.4

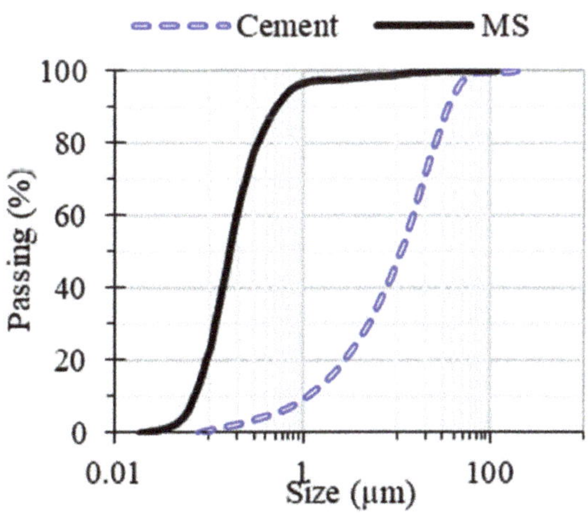

Figure 1. Gradation of binding materials.

2.1.2. Aggregates

For fine aggregates, siliceous sand from 'The Lawrancepur' quarry was used. This sand is used for the production of high-strength concrete in Pakistan. For NCA, crushed dolomitic sandstone of Kirana-Hills was used to prepare PNAC. The maximum aggregate size was 12.5 mm for NCA. The general properties of aggregates are given in Table 2. For the gradation of aggregates, ASTM C33 [47] was followed. For the determination of specific gravity (Gs) and WA, ASTM C127 [48] and ASTM C128 [49] were adopted for coarse aggregates and fine aggregates, respectively.

Table 2. Characteristics of aggregate samples.

Aggregate Type	Grain Size (mm)		24-h' WA (%)	Gs	FM
	Max.	**Min.**			
Sand	4.75	0.075	0.81	2.68	2.91
Crushed sandstone	12.5	2.36	0.93	2.73	-
RCA (Crushed-concrete waste)	12.5	2.36	3.56	2.57	-

RCA was derived from old high-strength concrete samples aged approx. 2 years. The aged samples were manually crushed to prepare. The samples were crushed and graded according to the size of NCA. The absorption capacity of RCA is almost four times higher than that of the NCA. Therefore, RCA was used in saturated surface dried (SSD) conditions to prepare concrete mixes. The properties of RCA are given in Table 1. Gradation charts of aggregates are shown in Figure 2.

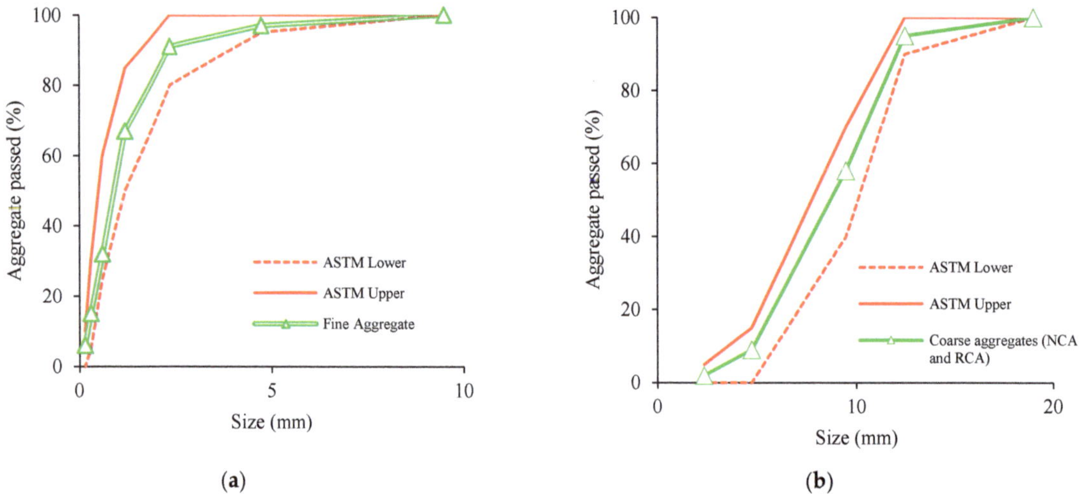

Figure 2. Aggregate gradation. (**a**) Fine aggregate. (**b**) Coarse aggregate.

2.1.3. Fiber Reinforcement

Steel chord wires of scrap tyres were manually shredded to prepare RTSF. The production of RTSF includes (1) manual extraction of chord wires from tyre waste, (2) burning of residual rubber particles to avoid the negative effect of weak rubber on the bond strength of RTSF, and (3) cleaned chord wires chopped into lengths varying between 25 mm and 37 mm. The density of RTSF is around 7700 kg/m^3. Due to the use of high-quality raw steel in the tyre, RTSF possesses good ductility and a high tensile strength comparable to an artificial steel fiber (ASF) [44]. The RTSF sample is shown in Figure 3.

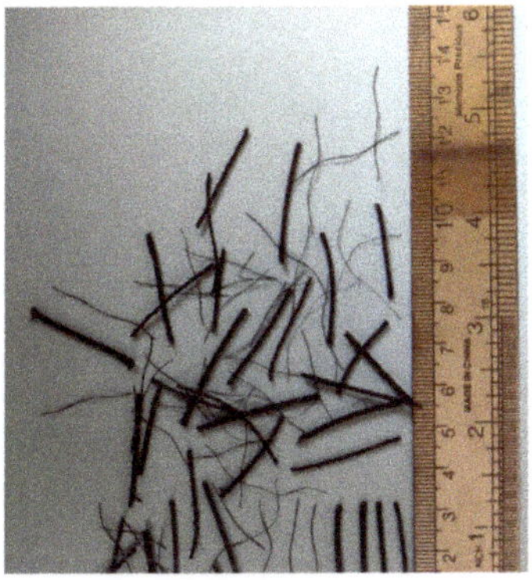

Figure 3. RTSF sample.

2.1.4. Plasticizer

For the mixing and curing of all mixtures, tap water was used. The effect of fibers on workability was minimized using a high-performance chemical admixture named Viscocrete 3110.

2.2. Details and Preparation of Concrete Mixtures

A total of thirteen mixes were investigated in this research, see Table 3. The first concrete mix "PNAC" was prepared with NCA to represent the control mix. After conducting trials, PNAC was designed as a high-performance concrete yielding slump of 200 ± 10 mm and *f*cm of 70 MPa at 28 days. RAC was produced by complete replacement of NCA with RCA, to maximize the recycling of C&D waste. The complete details about proportioning of concrete ingredients are given in Table 3. A total of twelve RAC mixes were designed with various contents of MS and RTSF. MS was used as 0, 5, and 10% by volume substitution of cement. The incorporation levels of MS were decided by simultaneously considering the performance of concrete in fresh and hardened states [11,50] and the economy. With each level of MS, RAC was reinforced with the four different volume fractions of RTSF i.e., 0, 0.5, 1, and 2%. These doses of RTSF were selected to evaluate the effect of a wide range of fiber content on the mechanical performance of RAC. Superplasticizer (SP) was employed to maintain the desired workability at a low water-binder ratio of 0.3.

Table 3. Design of concrete mixtures.

Mix ID	RTSF (%)	MS (%)	Cement (kg/m^3)	MS (kg/m^3)	RTSF (kg/m^3)	Fine Aggregate (kg/m^3)	Coarse Aggregate (kg/m^3)	Water (kg/m^3)	SP (kg/m^3)
PNAC	0	0	505	0	0	812	935	151.5	2
RAC/R0	0	0	505	0	0	812	860	151.5	2
RAC/R0.5	0.5	0	505	0	39	806	854	151.5	2
RAC/R1	1	0	505	0	78	799	847	151.5	2.5
RAC/R2	2	0	505	0	156	786	834	151.5	2.5
RAC/R0/M5	0	5	480	19	0	812	860	151.5	2.5
RAC/R0.5/M5	0.5	5	480	19	39	806	854	151.5	3
RAC/R1/M5	1	5	480	19	78	799	847	151.5	3
RAC/R2/M5	2	5	480	19	156	786	834	151.5	3.5
RAC/R0/M10	0	10	455	37	0	812	860	151.5	3
RAC/R0.5/M10	0.5	10	455	37	39	806	854	151.5	3
RAC/R1/M10	1	10	455	37	78	799	847	151.5	3.5
RAC/R2/M10	2	10	455	37	156	786	834	151.5	3.5

The mixing of fresh batches was completed in four stages (1) Firstly, all solid ingredients were dry blended at 40 rpm speed for 2 min; (2) In the second stage, half the amount of SP and water were added to the dry mix, mixing has proceeded at 40 rpm for 2 min. (3) In the third stage, the remaining halves of SP and water were added to the mixer, and the mixing speed was increased to 60 rpm, and it lasted for 2 min. Plain mixes were processed in the three stages; however, fiber-reinforced mixes have proceeded for the fourth stage. (4) In the final/fourth stage, required quantities of fibers were added gradually to the plain fresh concrete, while mixing continued at a speed of 80 rpm for 4 min. After the completion of mixing, Abram's cone slump test was performed on all mixes according to ASTM C143 [51]. RAC mixes incorporating 0% and 0.5% RTSF showed slump values between 200–220 mm. While highly reinforced RAC mixes showed slump values between 130–200 mm. Three replicate samples of all mixes were made for the determination of a property at a given age. Samples were cast in the steel molds and kept for 24 h to set and eventually immersed for curing in the tap water.

2.3. Testing Methods

Several performance indicators of concrete such as fcm, $fctm$, density, WA, ER, and UPV were evaluated to investigate the effects of MS and RTSF on RAC. For each type of parameter/property, three replicate samples of all mixes were tested at the specified ages, and their average result is reported in this research with standard deviation values. The fcm of each mix was evaluated at the age of 7, 28, 90, and 180 days. For this purpose, 100 mm cubical specimens of concrete were tested according to ASTM C39 [52]. The $fctm$ of all mixes was evaluated at the age of 28 and 90 days. For the splitting-tensile test, 100 × 200 mm cylindrical samples were prepared and tested according to ASTM C496 [53].

The density of each mix was measured to investigate the effect of RTSF and MS on the unit weight of RAC. For the evaluation of dry bulk density, 100 mm cubical samples were tested according to ASTM C948 [54]. The same samples were used to find out the WA capacity as per ASTM C948. The WA of each sample was measured at the age of 28 and 90-Days. To investigate the influence of RTSF and MS on the corrosion-risk potential of RAC, an ER test was performed on 100 mm cubical samples according to ASTM C1876 [55]. For the assessment of changes in the porosity of RAC due to the addition of RTSF and MS, an ultrasonic pulse velocity (UPV) test was conducted on 100 mm cubical samples according to ASTM C597 [56]. The ER and UPV of each mix were determined at the age of 28 and 90 days.

3. Results and Discussions

3.1. Density

The effect of MS and RTSF addition on the density of RAC is shown in Figure 4. As expected, the density decreased with the full replacement of NCA with RCA. This was because RCA had less dense or porous attached mortar which had a density lower than the natural aggregate. Therefore, RAC resulted in a 3.2% lower density than the PNAC. The addition of MS had shown a slight increase in the density of RAC. The particles of MS were smaller than that of the cement, therefore, MS can accommodate the spaces between cement particles and pores inside RCA. The filling effect of MS particles can improve the density of RAC [13]. Moreover, the addition of MS had the potential to reduce the free portlandite in the binder matrix and convert it into useful and dense calcium silicate hydrate (C-S-H) gel.

The addition of RTSF could cause a noticeable increment in the density of RAC for a given percentage of MS. At the addition of 2% RTSF, the density of RAC almost became equal to that of the PNAC. This was because the density of steel chords was about 3.5 times higher than that of plain concrete, therefore, a rising volume fraction of RTSF caused noticeable increments in the density. Therefore, RAC incorporating 10% MS and 2% RTSF exceeds in density compared to the PNAC mix.

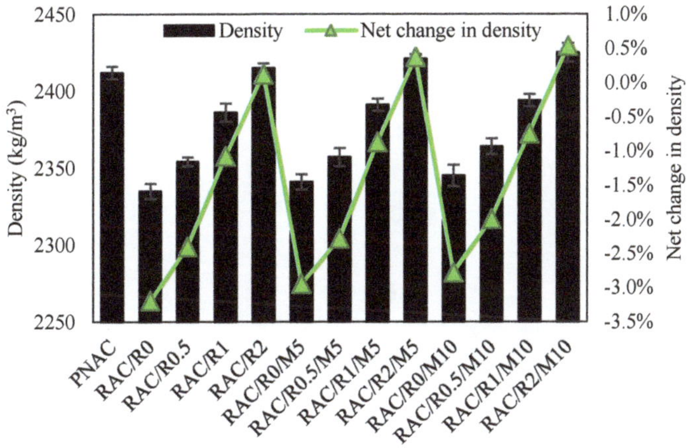

Figure 4. Effect of varying contents of MS and RTSF on the density of RAC.

3.2. Compressive Strength

The *f*cm results of all mixes at 7, 28, 90, and 180 days are shown in Figure 5. The *f*cm of RAC was around 15% lower than the PNAC. This is because plain RAC had a lower density than NAC. The voids present in the adhered mortar of RCA had a high amount of absorbed water in the saturated surface dry state. Therefore, the overall void content of RAC was high and could cause a reduction in the density and strength of concrete. Other than that, interfacial transition zones (ITZs) within RCA could played a negative role in reducing the strength of concrete [11], as ITZs acted as weak links in any concrete matrix under compressive loads. To elaborate, ITZs were regions where the properties of the RCA and the surrounding old cement paste were different. The ITZs were weaker than the rest of the concrete matrix, making them more prone to cracking or failure under stress.

MS addition caused notable improvements in the *f*cm of RAC at the age of 90 and 180 days. The *f*cm of RAC at 5% and 10% MS incorporation experienced an improvement of about 8 and 16%, respectively. At 10% MS incorporation, RAC showed *f*cm similar to the PNAC. MS addition promoted the growth of CSH-gel, which could cause an increase in the strength of RAC, especially at later ages. Past literature [13,50] had reported that MS overcame the strength deficit of plain RAC compared to the PNAC. The ITZs between RCA and the binder matrix could also strengthen due to cross-reactions at ITZs between portlandite present in RCA and micro-silica particles in the binder matrix. Moreover, pores present inside RCA can also be penetrated by fine silica particles, which could also lead to the strengthening of ITZs between RCA and the binder matrix. Hence, MS addition could be beneficial to the strength of RAC.

The increase in RTSF content from 0 to 2% showed a mixed effect on the *f*cm of RAC. The addition of 0.5 and 1% RTSF showed 4–6% increments in the *f*cm, while the 2% RTSF addition showed no notable change in the *f*cm. The changing behavior of *f*cm with the rising fiber content could be explained as fiber-reinforcement affected the *f*cm in two opposite ways (1) confinement effect and control over the premature cracking and brittle failure may contribute positively to the strength [23,26] (2) while poor dispersion may lead to the increase in air voids of concrete detrimental to the *f*cm [57]. It was hypothesized that at a high fiber volume, the accumulation of RTSF filaments could cause a reduction in the utilization of fibers. Previous studies [44,45] confirmed that RTSF incorporation of up to 1% volume fraction can increase the *f*cm by up to 5–8%.

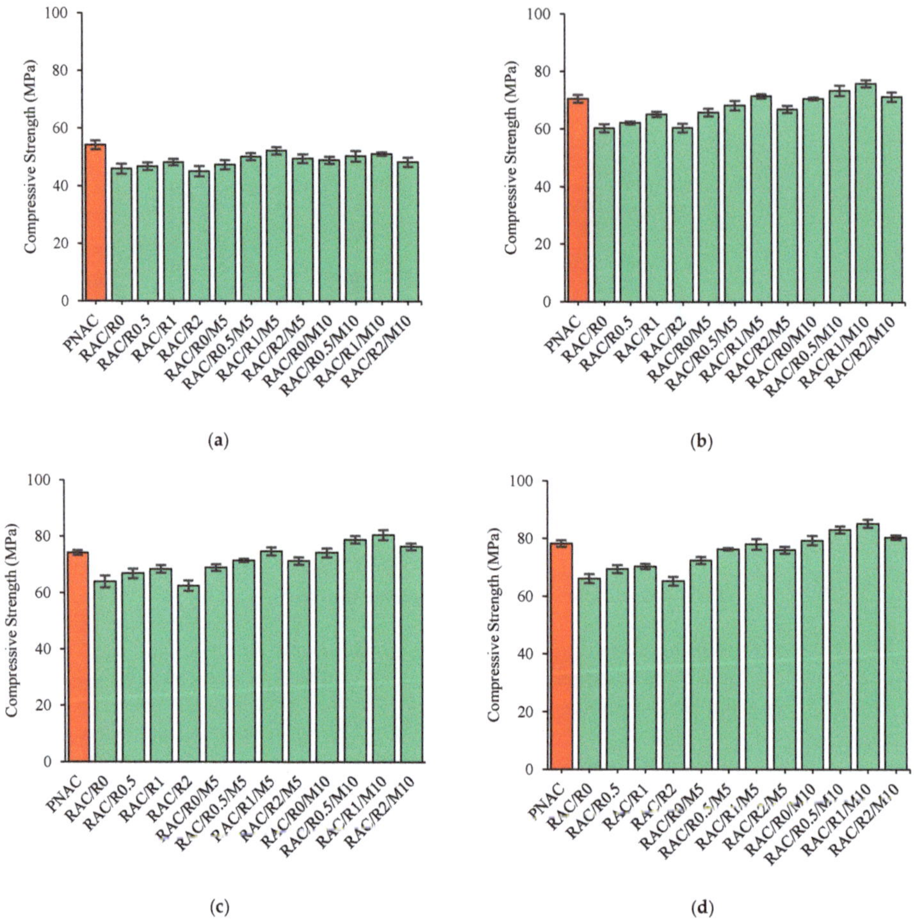

Figure 5. Compressive strength (*f*cm) of RAC with the varying content of MS and RTSF at the age of (**a**) 7-Days (**b**) 28-Days (**c**) 90-Days and (**d**) 180-Days.

Simultaneous incorporation of MS and RTSF led to noticeable improvements of 18–29% in the *f*cm of RAC compared to plain RAC (i.e., RAC/R0). Mixes containing 5% MS and 1% RTSF showed *f*cm similar to the PNAC. While mixes made with 10% MS and all RTSF (0.5, 1, and 1.5%) contents showed higher strength compared to PNAC. Among all mixtures, the maximum *f*cm, 7.6–8.5% higher than PNAC, was shown by RAC made with 10% MS and 1% RTSF. The MS incorporation seems to improve the strength of the matrix through the filler effect and pozzolanic reaction. while RTSF could improve the strength by offering crack resistance. The maximum contribution of MS in RAC was noticed at the age of 180 days. This could be because the pozzolanic reaction between silica particles and portlandite was slow and took a long duration. At the age of 180 days, the *f*cm of RAC increased by about 20% at the addition of 10%. Moreover, MS addition also seemed to increase the efficiency of RTSF. For example, mixes incorporating MS showed more net achievement in the *f*cm due to 1% RTSF than the mixes without MS. This can be related to the improvement in dispersion and bond of RTSF with plain matrix owing to improved strength and dense packing of binder particles with MS addition. The improvement in the efficiency of artificial steel fibers with MS addition was also noted in previous study [50]. The strengthening of the binder improved the bond performance of fiber filaments and plain matrix [58].

3.3. Splitting Tensile Strength

The *f*ctm of the PNAC mix and RAC with different contents of MS and RTSF is shown in Figure 6. Complete substitution of NCA with RCA showed a reduction of 13% in *f*ctm. This seemed to be the result of the low-density mortar present in RCA. Previous studies [26,44,45] reported a decline of around 15–20% in *f*ctm when RCA was used as a full replacement for NCA. MS led to notable increments in the *f*ctm of RAC. At 28 days, the tensile strength of RAC was increased by 9% and 8.3% at 5% MS and 10% MS incorporation, respectively. While, at 90 days, RAC experienced increments of 13% and 14.7% respectively due to 5% and 10% MS addition. The high percentage of MS (i.e., 10%) showed a major contribution to the tensile strength at 90 days due to the slow development of C-S-H in pozzolanic reactions. The plain mix of RAC containing 10% MS showed tensile strength comparable to that of the PNAC mix. MS addition could result in the strengthening of the plain matrix and improve the bond between the RCA and binder matrix of RAC [11]. This may have resulted in the improvement of the tensile strength of plain RAC.

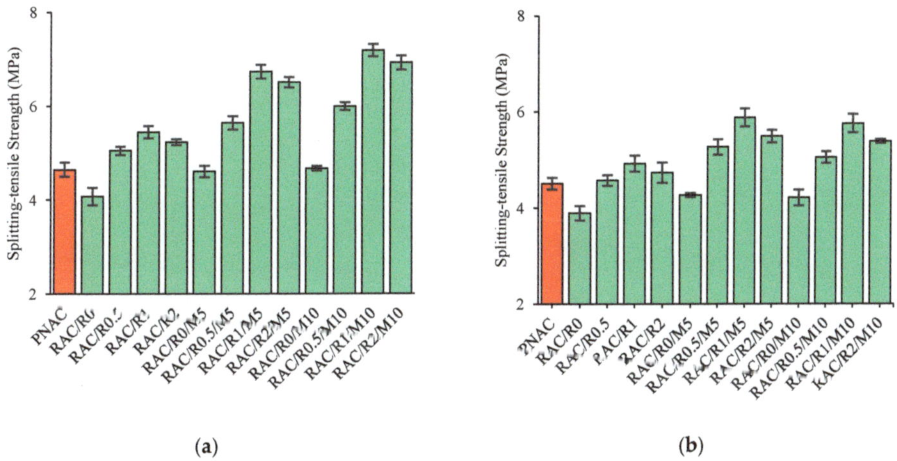

Figure 6. Effect of varying contents of MS and RTSF on the *f*ctm of RAC at the age of (**a**) 28 and (**b**) 90-days.

Compared to MS, RTSF showed a substantial increase in the *f*ctm of plain RAC. At 28 days, RTSF incorporation at 0.5, 1, and 1.5% volume fractions correspondingly caused increments of 17, 26, and 23% in the *f*ctm of plain RAC. While at 90 days, the efficiency of RTSF was further improved notably, for example, 0.5, 1, and 2% RTSF contents caused *f*ctm of RAC to increase by 23%, 33%, and 28%, respectively. As the concrete aged, it seemed to strengthen the binder matrix hence the bond between fibers and concrete could also strengthen, which may have resulted in the increased efficiency of RTSF. Available studies have [59,60] reported the tensile strength increase of about 20–30% at 1.5% incorporation of shredded RTSF. The efficiency of RTSF declined at a 2% volume fraction and it seemed to be related to the increase in porosity or air voids due to lack of proper dispersion of fiber filaments at a high fiber volume [50].

Figure 7 shows the *f*ctm of RAC with various levels of RTSF and MS relative to the PNAC. High-level net gains in the *f*ctm of RAC were noticed when RTSF was incorporated along with MS. For example, 1% RTSF addition showed about a 40% increase in the *f*ctm of RAC when used in conjunction with 5% MS, while it showed about 26% increase in the *f*ctm when used without MS. Similar improvements were noticed with other combinations of MS and RTSF as well. Thus, MS could increase the utilization of RTSF towards the ductility of RAC. The improvement in fiber efficiency with MS addition could be credited to the increased bond strength at the ITZs between fiber filaments and plain matrix. The strengthening of the bond seemed to be effective and ensured the high pull-out strength of fibers in binders containing MS [58]. The results of *f*ctm also highlighted the importance of RTSF alone, which can notably change the tensile strength even at 0.5% volume fraction. All RAC mixes incorporating RTSF showed higher tensile strength than PNAC at both ages of testing. Furthermore, conjunctive addition of MS with RTSF could lead to RAC with possibly high tensile strength than PNAC. As can be noted in Figure 8, RAC containing 1% RTSF and 5–10% MS outperformed PNAC by exhibiting 38–41% more *f*ctm.

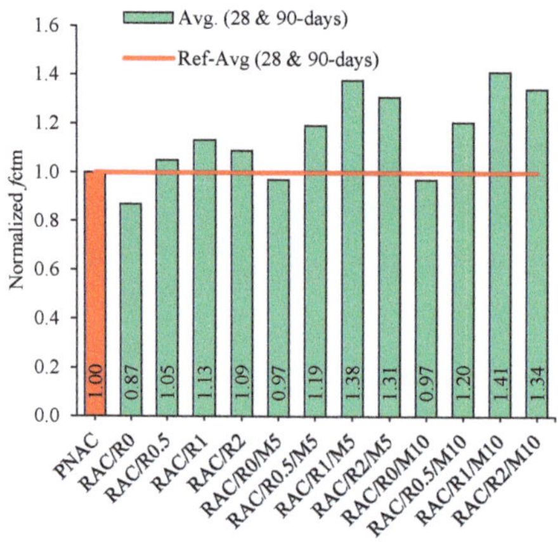

Figure 7. Variation of normalized *f*ctm (*f*ctm_mix/*f*ctm_PNAC) with changing RTSF and MS contents.

3.4. Water Absorption

The effect of MS and RTSF contents on the WA capacity of RAC is shown in Figure 8. WA measures the permeable porosity of concrete which is an indirect assessment of durability against the ingress of harmful substances into concrete. The results showed RAC had an 18–23% higher WA than that of the PNAC mix. This notable increase in the WA capacity could be attributed to the presence of adhered mortar on the surface of the RCA particles. The inclusion of RCA as a replacement for natural coarse aggregate could increase both pore connectivity and pore volume.

The inclusion of MS can notably reduce the WA capacity of RAC. For example, 5% and 10% MS addition caused 33% and 45% reductions in the WA capacity of RAC, respectively. As MS has extremely fine particles compared to cement, therefore, it could have efficiently increased the imperviousness of the plain matrix. Besides that, MS particles can also reduce the permeation of water along the weak ITZs between RCA and the binder matrix. The presence of fine particles of MS seems to be the tortuosity of permeable paths within a material, which ultimately may lead to the creation of an impermeable microstructure [18]. This is worth mentioning here, that RAC with 5% MS showed a 10% lower WA capacity than that of the PNAC mix at 90 days. Whereas RAC with 10% MS showed a 15% and 22% reduction in WA w.r.t PNAC mix at the age of 28 and 90 days, respectively. Thus, MS could notably control the negative effect of RCA on imperviousness.

The addition of RTSF without MS showed mixed effects on the WA capacity of RAC. Initially, WA was reduced at 0.5% RTSF, then absorption capacity started increasing with a further rise in RTSF compared to plain RAC and the maximum WA was observed at 2% RTSF. The positive effect of a small volume of fibers could be credited to the increased control of fibers over the aggregate sinking, and slurry flow which may eventually reduce cracks due to the drying shrinkage [61]. Alsaif et al. [62] reported that using artificial steel fibers could result in a 13% decline in the WA capacity of concrete. While at a high volume of fibers, the difficulty in compaction and improper dispersion issue could increase the void pockets in the concrete and it eventually caused the WA increase [19]. Mixes of RAC with combined incorporation of MS and RTSF showed smaller WA capacity as compared to PNAC mix except for the RAC with 5% MS and 2% RTSF (RAC/R2/M5). The influence of MS was found to be dominant to reduce the WA in the mixes with both MS and RTSF. RAC with 10% MS and all contents of RTSF can yield notably lower WA than that of the PNAC mix.

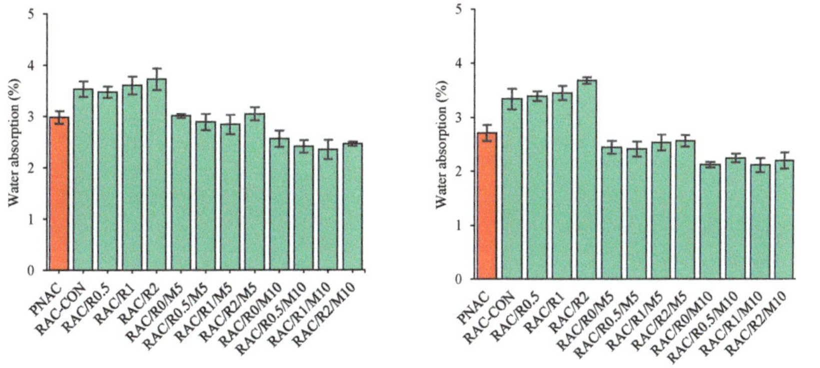

(**a**) (**b**)

Figure 8. Effect of RTSF and MS on WA capacity of RAC measured at (**a**) 28 days, and (**b**) 90-days.

3.5. *Ultrasonic Pulse Velocity (km/s)*

Changes in homogeneity and porosity of concrete can be assessed by the UPV test. The effect of RTSF and MS contents on the UPV of RAC was shown in Figure 9. All UPV values were pertaining to concretes of high strength and good quality as no experimental UPV value was below 3500 m/s. Since RCA had high porosity than NCA, RAC showed a lower UPV value than the PNAC mix. The possible porosity increase delayed the time of travel of pulse waves from a transmitter to the receiver poles of the UPV test apparatus. However, MS addition could cause an increase in the UPV of concrete. This indicated that the homogeneity and imperviousness of RAC can be increased with MS incorporation. Similar to the present study, Kou et al. [11] reported that UPV gain of RAC was increased with MS addition. Fine MS particles may consume free portlandite, strengthened the binder matrix, and bond between RCA and the binder matrix. The reduction in the distance between particles and improved packing of constituent particles could also cause gains of UPV in RAC mixes.

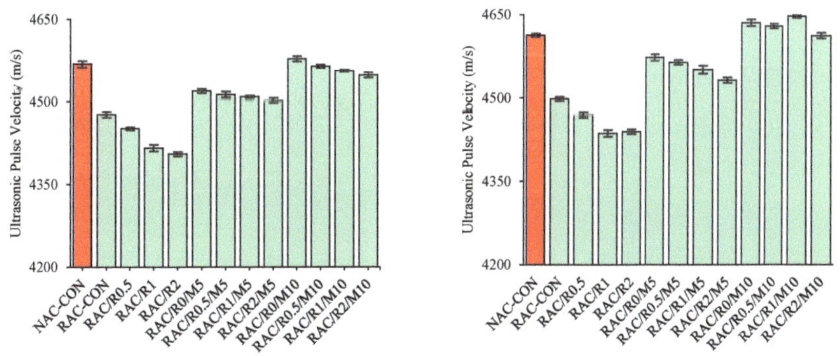

(**a**) (**b**)

Figure 9. Effect of varying contents of MS and RTSF on UPV of RAC at the age of (**a**) 28 and (**b**) 90-days.

Unlike MS, RTSF incorporation showed a decline in UPV. Although RTSF had a high density, it cannot notably change the porosity of the plain matrix. Besides that, fiber filaments seemed to introduce heterogeneity in the RAC matrix and could possibly cause UPV to decline. Fibers oriented in a different direction may also deflect the pulse waves slowing the propagation time of pulse waves. Yazici et al. [63] noted that the addition of steel fiber can cause a 1–9% decline in the UPV of concrete when fiber content was increased from 0.5 to 1.5%. This could have been caused by the increase in the porosity due to the increasing difficulty in compaction with the rise in fiber content. A major decline in UPV due to RTSF was noticed in RAC without MS. While minor UPV decline was observed in RAC with MS. The interaction of the binder matrix with RTSF may be poor when MS was not used. This may create more voids due to RTSF incorporation in mixes without MS. On the other hand, fine MS particles can ensure a strong interaction between the binder phase and fiber filaments. Thus, the intensity of the negative effect of RTSF on the void ratio could be minimized with the application of MS. A little UPV change due to RTSF variation in MS-containing mixes also indicated that the mineral admixture incorporation improved the dispersion of fiber filaments.

3.6. Electrical Resistivity

The durability of steel-reinforced concrete against corrosion can be estimated via different techniques. ER is a non-destructive evaluation that allows a simple assessment of the corrosion risk potential of concrete according to existing classifications [64]. The variation of ER of RAC against different contents of RTSF and MS is shown in Figure 10.

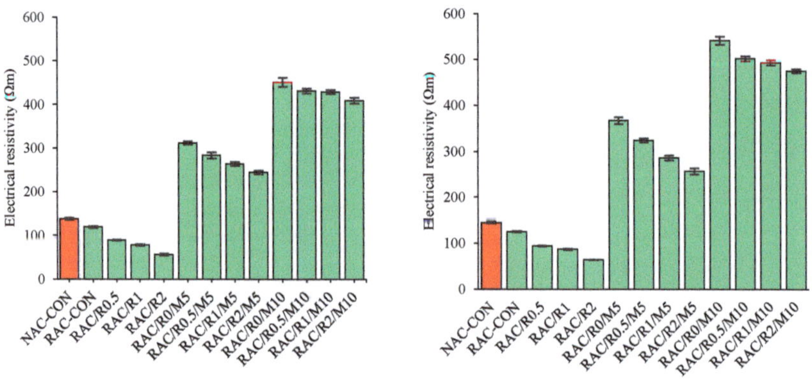

Figure 10. Effect of RTSF and MS contents on ER of RAC measured at (**a**) 28-days and (**b**) 90-days.

It was observed that RAC had a lower ER than that of the PNAC mix. This can be caused by the presence of high porosity in RCA which facilitates penetration of ions. Also, a high amount of moisture in RAC allowed faster penetration of electricity [65]. However, RAC showed ER values around 120 Ωm that pertains to concrete with 'no probability' of corrosion [64]. This may be credited to the low-water binder ratio used to develop the high-strength concrete in this study. While mixes with ER values of 120 to 50 Ωm and below 50 Ωm had a 'probable risk to rebar corrosion and 'vulnerable to rebar corrosion', respectively [64]. Thus, RAC containing RTSF without MS had a 'probable risk' of corrosion. The ER was reduced further, with the rise in RTSF content. This could be blamed on the high electrical conductivity of steel fibers. While a mild increase in the porosity of concrete due to high fiber volumes can also lead to an increase in the corrosion risk potential of concrete [66]. Afroughsabet et al. [25] reported that the ER value of concrete was reduced from 75 to 20 Ωm upon the addition of 1% volume of steel fiber.

The addition of 5% and 10% MS had proven extremely useful in increasing the ER of RAC. In other words, the corrosion-risk potential of RAC can be minimized considerably by incorporating MS in the binder. The pozzolanic influence of MS may reduce the free portlandite and produce secondary C-S-H gel. MS seemed to help in producing a dense microstructure that disrupted the free movement of ions present in pore solution triggered by electrical currents. Kou et al. [67] reported that the ingress of chloride ions was interrupted due to the incorporation of pozzolanic admixtures (i.e., MS, fly ash, and slag). The ER of RAC/R0/M5 and RAC/R0/M10 was about 2.2 and 3.3 times higher than that of the PNAC mix. Thus, the studied RAC made with 5% or 10% MS addition had no probable risk of corrosion.

The manufacture of steel fiber-reinforced RAC made with RTSF having a 'low' or 'no probable' risk of corrosion was achieved with the application of MS. The noticeable reduction in the porosity, as indicated by the results of WA, reduced the penetration of electrical current and compensates for the high electrical conductivity of steel fibers. It is also worth mentioning here that, although ER was a quick and convenient measure of the corrosion-risk potential of concrete, it can overestimate the influence of steel fibers on the corrosion risk of concrete [19]. Therefore, a reliable assessment of the corrosion risk potential of steel fiber-reinforced concrete can only be made by conducting a chloride ion diffusion test or immersion method [68]. As steel fibers caused little change in porosity [14], therefore, they are also anticipated to show little influence on the chloride-ion penetration resistance of concrete.

4. Conclusions

This paper studied the effect of different RTSF and MS contents on the mechanical and physical parameters of RAC at different ages. The following key findings can be drawn from experimental results:

- The overall density of RAC may be increased by the incorporation of both RTSF and MS. However, the increase in fiber content could increase the voids/pore connectivity in the plain concrete matrix. Seemingly, owing to the filling action and pozzolanicity, MS addition showed minor improvements in the density of concrete.
- The optimum dosage of RTSF can be taken as '1%' considering the maximum improvement in fcm and $fctm$. RAC made with 10% MS and 1% RTSF could show 7.6–8.5% higher fcm compared to PNAC.
- The $fctm$ of RAC with all contents of RTSF was found to be higher compared to the PNAC. At the age of 90 days, RAC made with 10% MS and 1% RTSF could result in 54% more $fctm$ than PNAC. MS inclusion could result in improvement in the utilization of RTSF towards the ductility of RAC. The increased fiber efficiency with MS addition seemed to be linked with the increased bond strength at the ITZs between fiber filaments and plain matrix.
- Even though 2% RTSF did not cause a noticeable improvement in mechanical performance but samples with 2% RTSF can yield showed smaller crack widths after peak load than samples with low levels of RTSF.
- The efficiency of RTSF can be increased by improving the mechanical properties with MS incorporation and increase in age.
- The UPV of RAC was slightly reduced due to the RTSF addition. Therefore, UPV could not be taken as a good measure to estimate the mechanical properties of steel fiber-reinforced concretes.
- Fine MS particles can ensure a strong interaction between the binder phase and fiber filaments. Thus, the intensity of the negative effect of RTSF on the void ratio can be minimized with the application of MS.
- The absorption capacity of RAC notably can be decreased due to MS addition. At 10% MS addition, RAC exhibited 15–23% lower WA than PNAC.
- The corrosion-risk potential of RAC could increase with RTSF incorporation. While MS could notably enhance the ER of RAC. The corrosion-risk potential of RAC made with 5 and 10% MS was 'not probable'. MS seemed to be effective in controlling the corrosion resistance of RAC made with RTSF.

Author Contributions: Conceptualization, M.T.A., H.A. and S.S.R.; methodology, H.A.; software, S.R.; validation, S.R., A.A.A.S. and S.A.; formal analysis, M.T.A.; investigation, S.R.; resources, S.R.; data curation, M.T.A.; writing—original draft preparation, M.T.A.; writing—review and editing, A.A.A.S.; visualization, S.A.; supervision, H.A., S.S.R. and S.A.; project administration, S.R.; funding acquisition, A.A.A.S. All authors have read and agreed to the published version of the manuscript.

Funding: The authors extend their appreciation to the Deanship of Scientific Research at King Khalid University for funding this work through large group Research Project under grant number R.G.P. 2/351/44.

Data Availability Statement: All data used is available in this research paper.

Acknowledgments: The authors extend their appreciation to the Deanship of Scientific Research at King Khalid University for funding this work through large group Research Project under grant number R.G.P. 2/351/44.

Conflicts of Interest: The authors declare no conflict of interest.

References

1. Akhtar, A.; Sarmah, A.K. Construction and Demolition Waste Generation and Properties of Recycled Aggregate Concrete: A Global Perspective. *J. Clean. Prod.* **2018**, *186*, 262–281. [CrossRef]
2. Kurda, R.; Silvestre, J.D.; de Brito, J. Life Cycle Assessment of Concrete Made with High Volume of Recycled Concrete Aggregates and Fly Ash. *Resour. Conserv. Recycl.* **2018**, *139*, 407–417. [CrossRef]
3. Şahmaran, M.; Li, V.C. Durability Properties of Micro-Cracked ECC Containing High Volumes Fly Ash. *Cem. Concr. Res.* **2009**,*39*, 1033–1043. [CrossRef]
4. Strieder, H.L.; Dutra, V.F.P.; Graeff, Â.G.; Núñez, W.P.; Merten, F.R.M. Performance Evaluation of Pervious Concrete Pavements with Recycled Concrete Aggregate. *Constr. Build. Mater.* **2022**, *315*, 125384. [CrossRef]
5. Hoffmann, C.; Schubert, S.; Leemann, A.; Motavalli, M. Recycled Concrete and Mixed Rubble as Aggregates: Influence of Variations in Composition on the Concrete Properties and Their Use as Structural Material. *Constr. Build. Mater.* **2012**, *35*, 701–709. [CrossRef]
6. Shi, X.; Mukhopadhyay, A.; Zollinger, D.; Grasley, Z. Economic Input-Output Life Cycle Assessment of Concrete Pavement Containing Recycled Concrete Aggregate. *J. Clean. Prod.* **2019**, *225*, 414–425. [CrossRef]
7. Kurad, R.; Silvestre, J.D.; de Brito, J.; Ahmed, H. Effect of Incorporation of High Volume of Recycled Concrete Aggregates and Fly Ash on the Strength and Global Warming Potential of Concrete. *J. Clean. Prod.* **2017**, *166*, 485–502. [CrossRef]
8. Kurda, R.; Silvestre, J.D.; de Brito, J. Toxicity and Environmental and Economic Performance of Fly Ash and Recycled Concrete Aggregates Use in Concrete: A Review. *Heliyon* **2018**, *4*, e00611. [CrossRef]
9. Maia, M.B.; De Brito, J.; Martins, I.M.; Silvestre, J.D. Toxicity of Recycled Concrete Aggregates: Review on Leaching Tests. *Open Constr. Build. Technol. J.* **2018**, *12*, 187–196. [CrossRef]
10. Kurda, R.; de Brito, J.; Silvestre, J.D. Influence of Recycled Aggregates and High Contents of Fly Ash on Concrete Fresh Properties. *Cem. Concr. Compos.* **2017**, *84*, 198–213. [CrossRef]
11. Kou, S.; Poon, C.; Agrela, F. Comparisons of Natural and Recycled Aggregate Concretes Prepared with the Addition of Different Mineral Admixtures. *Cem. Concr. Compos.* **2011**, *33*, 788–795. [CrossRef]
12. Lam, C.S.; Poon, C.S.; Chan, D. Enhancing the Performance of Pre-Cast Concrete Blocks by Incorporating Waste Glass—ASR Consideration. *Cem. Concr. Compos.* **2007**, *29*, 616–625. [CrossRef]
13. Dilbas, H.; Şimşek, M.; Çakır, Ö. An Investigation on Mechanical and Physical Properties of Recycled Aggregate Concrete (RAC) with and without Silica Fume. *Constr. Build. Mater.* **2014**, *61*, 50–59. [CrossRef]
14. Koushkbaghi, M.; Kazemi, M.J.; Mosavi, H.; Mohseni, E. Acid Resistance and Durability Properties of Steel Fiber-Reinforced Concrete Incorporating Rice Husk Ash and Recycled Aggregate. *Constr. Build. Mater.* **2019**, *202*, 266–275. [CrossRef]
15. Xie, J.; Fang, C.; Lu, Z.; Li, Z.; Li, L. Effects of the Addition of Silica Fume and Rubber Particles on the Compressive Behaviour of Recycled Aggregate Concrete with Steel Fibres. *J. Clean. Prod.* **2018**, *197*, 656–667. [CrossRef]
16. He, W.; Kong, X.; Fu, Y.; Zhou, C.; Zheng, Z. Experimental Investigation on the Mechanical Properties and Microstructure of Hybrid Fiber Reinforced Recycled Aggregate Concrete. *Constr. Build. Mater.* **2020**, *261*, 120488. [CrossRef]
17. Kazmi, S.M.S.; Munir, M.J.; Wu, Y.-F.; Patnaikuni, I.; Zhou, Y.; Xing, F. Axial Stress-Strain Behavior of Macro-Synthetic Fiber Reinforced Recycled Aggregate Concrete. *Cem. Concr. Compos.* **2019**, *97*, 341–356. [CrossRef]
18. Kurda, R.; de Brito, J.; Silvestre, J.D. Water Absorption and Electrical Resistivity of Concrete with Recycled Concrete Aggregates and Fly Ash. *Cem. Concr. Compos.* **2019**, *95*, 169–182. [CrossRef]
19. Ali, B.; Raza, S.S.; Kurda, R.; Alyousef, R. Synergistic Effects of Fly Ash and Hooked Steel Fibers on Strength and Durability Properties of High Strength Recycled Aggregate Concrete. *Resour. Conserv. Recycl.* **2021**, *168*, 105444. [CrossRef]
20. Çakır, Ö.; Sofyanlı, Ö.Ö. Influence of Silica Fume on Mechanical and Physical Properties of Recycled Aggregate Concrete. *HBRC J.* **2015**, *11*, 157–166. [CrossRef]
21. Sasanipour, H.; Aslani, F.; Taherinezhad, J. Chloride Ion Permeability Improvement of Recycled Aggregate Concrete Using Pretreated Recycled Aggregates by Silica Fume Slurry. *Constr. Build. Mater.* **2021**, *270*, 121498. [CrossRef]

22. Dimitriou, G.; Savva, P.; Petrou, M.F. Enhancing Mechanical and Durability Properties of Recycled Aggregate Concrete. *Constr. Build. Mater.* **2018**, *158*, 228–235. [CrossRef]
23. Ali, B.; Qureshi, L.A.; Khan, S.U. Flexural Behavior of Glass Fiber-Reinforced Recycled Aggregate Concrete and Its Impact on the Cost and Carbon Footprint of Concrete Pavement. *Constr. Build. Mater.* **2020**, *262*, 120820. [CrossRef]
24. Kazmi, S.M.S.; Munir, M.J.; Wu, Y.-F.; Patnaikuni, I. Effect of Macro-Synthetic Fibers on the Fracture Energy and Mechanical Behavior of Recycled Aggregate Concrete. *Constr. Build. Mater.* **2018**, *189*, 857–868. [CrossRef]
25. Afroughsabet, V.; Biolzi, L.; Ozbakkaloglu, T. Influence of Double Hooked-End Steel Fibers and Slag on Mechanical and Durability Properties of High Performance Recycled Aggregate Concrete. *Compos. Struct.* **2017**, *181*, 273–284. [CrossRef]
26. Das, C.S.; Dey, T.; Dandapat, R.; Mukharjee, B.B.; Kumar, J. Performance Evaluation of Polypropylene Fibre Reinforced Recycled Aggregate Concrete. *Constr. Build. Mater.* **2018**, *189*, 649–659. [CrossRef]
27. Ali, B.; Qureshi, L.A.; Shah, S.H.A.; Rehman, S.U.; Hussain, I.; Iqbal, M. A Step towards Durable, Ductile and Sustainable Concrete: Simultaneous Incorporation of Recycled Aggregates, Glass Fiber and Fly Ash. *Constr. Build. Mater.* **2020**, *251*, 118980. [CrossRef]
28. Mastali, M.; Dalvand, A. Use of Silica Fume and Recycled Steel Fibers in Self-Compacting Concrete (SCC). *Constr. Build. Mater.* **2016**, *125*, 196–209. [CrossRef]
29. Wang, L.E.I.; Guo, F.; Yang, H.; Wang, Y.A.N.; Tang, S. Comparison of Fly Ash, PVA Fiber, MgO and Shrinkage-Reducing Admixture on the Frost Resistance of Face Slab Concrete via Pore Structural and Fractal Analysis. *Fractals* **2021**, *29*, 2140002. [CrossRef]
30. Wang, L.; Zeng, X.; Li, Y.; Yang, H.; Tang, S. Influences of MgO and PVA Fiber on the Abrasion and Cracking Resistance, Pore Structure and Fractal Features of Hydraulic Concrete. *Fractal Fract.* **2022**, *6*, 674. [CrossRef]
31. Zarei, A.; Rooholamini, H.; Ozbakkaloglu, T. Evaluating the Properties of Concrete Pavements Containing Crumb Rubber and Recycled Steel Fibers Using Response Surface Methodology. *Int. J. Pavement Res. Technol.* **2022**, *15*, 470–484. [CrossRef]
32. Papachristoforou, M.; Anastasiou, E.K.; Papayianni, I. Durability of Steel Fiber Reinforced Concrete with Coarse Steel Slag Aggregates Including Performance at Elevated Temperatures. *Constr. Build. Mater.* **2020**, *262*, 120569. [CrossRef]
33. Ali, B.; Raza, S.S.; Hussain, I.; Iqbal, M. Influence of Different Fibers on Mechanical and Durability Performance of Concrete with Silica Fume. *Struct. Concr.* 2020; *in press*. [CrossRef]
34. Ali, B.; Qureshi, L.A.; Kurda, R. Environmental and Economic Benefits of Steel, Glass, and Polypropylene Fiber Reinforced Cement Composite Application in Jointed Plain Concrete Pavement. *Compos. Commun.* **2020**, *22*, 100437. [CrossRef]
35. Xie, J.; Zhang, Z.; Lu, Z.; Sun, M. Coupling Effects of Silica Fume and Steel-Fiber on the Compressive Behaviour of Recycled Aggregate Concrete after Exposure to Elevated Temperature. *Constr. Build. Mater.* **2018**, *184*, 752–764. [CrossRef]
36. Chan, R.; Santana, M.A.; Oda, A.M.; Paniguel, R.C.; Vieira, L.B.; Figueiredo, A.D.; Galobardes, I. Analysis of Potential Use of Fibre Reinforced Recycled Aggregate Concrete for Sustainable Pavements. *J. Clean. Prod.* **2019**, *218*, 183–191. [CrossRef]
37. Hu, H.; Papastergiou, P.; Angelakopoulos, H.; Guadagnini, M.; Pilakoutas, K. Mechanical Properties of SFRC Using Blended Recycled Tyre Steel Cords (RTSC) and Recycled Tyre Steel Fibres (RTSF). *Constr. Build. Mater.* **2018**, *187*, 553–564. [CrossRef]
38. Zhong, H.; Zhang, M. Experimental Study on Engineering Properties of Concrete Reinforced with Hybrid Recycled Tyre Steel and Polypropylene Fibres. *J. Clean. Prod.* **2020**, *259*, 120914. [CrossRef]
39. Revuelta, D.; Carballosa, P.; García Calvo, J.L.; Pedrosa, F. Residual Strength and Drying Behavior of Concrete Reinforced with Recycled Steel Fiber from Tires. *Materials* **2021**, *14*, 6111. [CrossRef]
40. Khan, M.; Rehman, A.; Ali, M. Efficiency of Silica-Fume Content in Plain and Natural Fiber Reinforced Concrete for Concrete Road. *Constr. Build. Mater.* **2020**, *244*, 118382. [CrossRef]
41. Islam, M.S.; Ahmed, S.J.U. Influence of Jute Fiber on Concrete Properties. *Constr. Build. Mater.* **2018**, *189*, 768–776. [CrossRef]
42. Liew, K.M.; Akbar, A. The Recent Progress of Recycled Steel Fiber Reinforced Concrete. *Constr. Build. Mater.* **2020**, *232*, 117232. [CrossRef]
43. Aiello, M.A.; Leuzzi, F.; Centonze, G.; Maffezzoli, A. Use of Steel Fibres Recovered from Waste Tyres as Reinforcement in Concrete: Pull-out Behaviour, Compressive and Flexural Strength. *Waste Manag.* **2009**, *29*, 1960–1970. [CrossRef] [PubMed]
44. Ahmadi, M.; Farzin, S.; Hassani, A.; Motamedi, M. Mechanical Properties of the Concrete Containing Recycled Fibers and Aggregates. *Constr. Build. Mater.* **2017**, *144*, 392–398. [CrossRef]
45. Moghadam, A.S.; Omidinasab, F.; Abdalikia, M. The Effect of Initial Strength of Concrete Wastes on the Fresh and Hardened Properties of Recycled Concrete Reinforced with Recycled Steel Fibers. *Constr. Build. Mater.* **2021**, *300*, 124284. [CrossRef]
46. *ASTM-C150*; Standard Specification for Portland Cement. ASTM International: West Conshohocken, PA, USA, 2018.
47. *ASTM C33/C33M-18*; Standard Specification for Concrete Aggregates. ASTM International: West Conshohocken, PA, USA, 2018.
48. *ASTM-C127*; Standard Test Method for Relative Density (Specific Gravity) and Absorption of Coarse Aggregate. ASTM International: West Conshohocken, PA, USA, 2015.
49. *ASTM-C128*; Standard Test Method for Density, Relative Density (Specific Gravity), and Absorption of Fine Aggregate. ASTM International: West Conshohocken, PA, USA, 2001.
50. Raza, S.S.; Ali, B.; Noman, M.; Fahad, M.; Elhadi, K.M. Mechanical Properties, Flexural Behavior, and Chloride Permeability of High-Performance Steel Fiber-Reinforced Concrete (SFRC) Modified with Rice Husk Ash and Micro-Silica. *Constr. Build. Mater.* **2022**, *359*, 129520. [CrossRef]

51. *ASTM-C143*; Standard Test Method for Slump of Hydraulic-Cement Concrete. ASTM International: West Conshohocken, PA, USA, 2015.
52. *ASTM C39/C39M-12*; Standard Test Method for Compressive Strength of Cylindrical Concrete Specimens. ASTM International: West Conshohocken, PA, USA, 2012.
53. *ASTM-C496*; Standard Test Method for Splitting Tensile Strength of Cylindrical Concrete Specimens. ASTM International: West Conshohocken, PA, USA, 2017.
54. *ASTM-C948*; Standard Test Method for Dry and Wet Bulk Density, Water Absorption, and Apparent Porosity of Thin Sections of Glass-Fiber Reinforced Concrete. ASTM International: West Conshohocken, PA, USA, 2016.
55. *ASTM C1876-19*; Standard Test Method for Bulk Electrical Resistivity or Bulk Conductivity of Concrete. ASTM International: West Conshohocken, PA, USA, 2019. Available online: https://www.astm.org/c1876-19.html (accessed on 5 September 2020).
56. *ASTM C597-16*; Standard Test Method for Pulse Velocity through Concrete. ASTM International: West Conshohocken, PA, USA, 2016. Available online: https://www.astm.org/c0597-16.html (accessed on 5 September 2020).
57. Xie, J.; Guo, Y.; Liu, L.; Xie, Z. Compressive and Flexural Behaviours of a New Steel-Fibre-Reinforced Recycled Aggregate Concrete with Crumb Rubber. *Constr. Build. Mater.* **2015**, *79*, 263–272. [CrossRef]
58. Wu, Z.; Shi, C.; Khayat, K.H. Influence of Silica Fume Content on Microstructure Development and Bond to Steel Fiber in Ultra-High Strength Cement-Based Materials (UHSC). *Cem. Concr. Compos.* **2016**, *71*, 97–109. [CrossRef]
59. Simalti, A.; Singh, A.P. Comparative Study on Performance of Manufactured Steel Fiber and Shredded Tire Recycled Steel Fiber Reinforced Self-Consolidating Concrete. *Constr. Build. Mater.* **2021**, *266*, 121102. [CrossRef]
60. Mastali, M.; Dalvand, A.; Sattarifard, A.R.; Abdollahnejad, Z.; Nematollahi, B.; Sanjayan, J.G.; Illikainen, M. A Comparison of the Effects of Pozzolanic Binders on the Hardened-State Properties of High-Strength Cementitious Composites Reinforced with Waste Tire Fibers. *Compos. Part B Eng.* **2019**, *162*, 134–153. [CrossRef]
61. Zhang, S.; He, P.; Niu, L. Mechanical Properties and Permeability of Fiber-Reinforced Concrete with Recycled Aggregate Made from Waste Clay Brick. *J. Clean. Prod.* **2020**, *268*, 121690. [CrossRef]
62. Alsaif, A.; Bernal, S.A.; Guadagnini, M.; Pilakoutas, K. Durability of Steel Fibre Reinforced Rubberised Concrete Exposed to Chlorides. *Constr. Build. Mater.* **2018**, *188*, 130–142. [CrossRef]
63. Yazıcı, Ş.; İnan, G.; Tabak, V. Effect of Aspect Ratio and Volume Fraction of Steel Fiber on the Mechanical Properties of SFRC. *Constr. Build. Mater.* **2007**, *21*, 1250–1253. [CrossRef]
64. Afroughsabet, V.; Biolzi, L.; Ozbakkaloglu, T. High-Performance Fiber-Reinforced Concrete: A Review. *J. Mater. Sci.* **2016**, *51*, 6517–6551. [CrossRef]
65. Singh, N.; Singh, S.P. Carbonation and Electrical Resistance of Self Compacting Concrete Made with Recycled Concrete Aggregates and Metakaolin. *Constr. Build. Mater.* **2016**, *121*, 400–409. [CrossRef]
66. Nili, M.; Afroughsabet, V. The Long-Term Compressive Strength and Durability Properties of Silica Fume Fiber-Reinforced Concrete. *Mater. Sci. Eng. A* **2012**, *531*, 107–111. [CrossRef]
67. Kou, S.-C.; Poon, C.-S.; Etxeberria, M. Influence of Recycled Aggregates on Long Term Mechanical Properties and Pore Size Distribution of Concrete. *Cem. Concr. Compos.* **2011**, *33*, 286–291. [CrossRef]
68. Qureshi, L.A.; Ali, B.; Ali, A. Combined Effects of Supplementary Cementitious Materials (Silica Fume, GGBS, Fly Ash and Rice Husk Ash) and Steel Fiber on the Hardened Properties of Recycled Aggregate Concrete. *Constr. Build. Mater.* **2020**, *263*, 120636. [CrossRef]

sustainability

MDPI

Article

Effect of Silica Fume on Engineering Performance and Life Cycle Impact of Jute-Fibre-Reinforced Concrete

Rawaz Kurda [1,2,3]

[1] Department of Highway and Bridge Engineering, Technical Engineering College, Erbil Polytechnic University, Erbil 44001, Iraq; rawaz.kurda@tecnico.ulisboa.pt
[2] Department of Civil Engineering, College of Engineering, Nawroz University, Duhok 42001, Iraq
[3] CERIS, Civil Engineering, Architecture and Georresources Department, Instituto Superior Técnico, Universidade de Lisboa, Av. Rovisco Pais, 1049-001 Lisbon, Portugal

Abstract: The brittleness of plain concrete (PC) is a result of its lack of tensile strength and poor resistance to cracking, which in turn limits its potential uses. The addition of dispersed fibres into the binding material has been demonstrated to have a positive impact on the tensile properties of PC. Nevertheless, using new or engineered fibres in concrete significantly increases the overall cost and carbon footprint of concrete. Consequently, the main obstacle in creating environmentally friendly fibre-reinforced concrete is the traditional design process with energy-intensive materials. This study investigated how the engineering properties and life cycle impact of concrete were influenced by varying the volume fractions of jute fibre (JF). The impact of incorporating silica fume (SF) as a partial replacement of Portland cement was also studied. The studied parameters included mechanical behaviour, non-destructive durability indicators, and the life cycle impact of concrete using JF and SF. The efficiency of JF in mechanical performance improved with the increase in age and with the addition of SF. When using both SF and 0.3% JF, there was an improvement of around 28% in the compressive strength (CS). When 0.3% JF was added, in the presence and absence of SF, the splitting tensile strength (STS) improvement was around 20% and 40%, respectively. The addition of JF improved the residual flexural strength (FS) and flexural ductility of PC. The SF addition overcame the drawbacks of the poor resistance of JF-reinforced concrete (JFRC) against water absorption (WA) and rapid chloride ion penetration (RCIP).

Keywords: cellulose fibre; cement replacement; fibre reinforcement; flexural ductility; life cycle; engineering properties

Citation: Kurda, R. Effect of Silica Fume on Engineering Performance and Life Cycle Impact of Jute-Fibre-Reinforced Concrete. *Sustainability* **2023**, *15*, 8465. https://doi.org/10.3390/su15118465

Academic Editor: Elena Lucchi

Received: 8 April 2023
Revised: 6 May 2023
Accepted: 15 May 2023
Published: 23 May 2023

1. Introduction

High-performance concrete (HPC) is created with the specific purpose of being more resilient and having greater strength compared to conventional PC systems. HPC achieves its high CS by utilizing a low water-cement (w/c) ratio, high cementitious material content, low aggregate volume, and "supplementary cementitious materials" (SCMs), resulting in a denser microstructure [1]. It is commonly understood that PC has a quasi-brittle nature, meaning that its brittleness increases as its strength increases [2]. Consequently, the primary concerns when using high-strength grades of PC are the increased brittleness and reduced ability to withstand cracking and spreading [3]. Cracking in PC can occur from various causes, including tensile loads and micro-cracks resulting from plastic and/or autogenous shrinkage during pre-hardening or drying shrinkage in the hardened state [4,5]. Inherent cracks allow harmful fluids to penetrate, compromising the durability of the concrete structure by subjecting it to detrimental chemicals [6,7]. Improving the tensile performance, crack resistance, and strain-hardening response of PC is a significant area of focus in the field of concrete technology [8].

Incorporating dispersed reinforcement/fibres into the PC matrix can address the problem of brittleness, resulting in a material with enhanced tensile strength, impact resistance, and energy/shock absorption capacity [9–12]. The impact of fibres on concrete's properties depends on multiple factors, including the material properties of the fibres, the size and shape of the fibres, the concentration and orientation of the dispersed fibres within the PC matrix, and the interfacial properties of the fibre–matrix interaction [13–16]. The impact of fibres on the characteristics of concrete is influenced by various factors, such as the fibre's material properties, as well as its size and shape [17,18]. Typically, engineered fibres, which are new or unused, are traditionally produced from materials such as glass, steel, polymers, and carbon fibres [19,20]. However, the manufacture and processing of these fibres require significant amounts of energy, resulting in substantial greenhouse gas emissions. There is a growing interest in evaluating natural plant-based fibres as an alternative to engineered fibres in concrete, given the environmental concerns associated with the manufacturing and processing of the latter [12,21,22]. Plant-based fibres are obtained from renewable, inexpensive, and recyclable sources, i.e., plants. Various kinds of plant fibres, such as sisal, coir, jute, wheat straws, hemp, etc., have been investigated as potential fibre reinforcements in concrete [12,23–25].

Jute fibre (JF) is regarded as an affordable and extremely durable natural fibre, among other plant-based fibres. Approximately 81% of the world's overall jute production is attributed to India and Bangladesh [26]. JF typically comprises plant components such as "cellulose" and "lignin". It possesses various beneficial properties, such as axial tensile strength, modulus of resilience, re-useability, a long life, and bio-degradation, among others. These properties make it a more advantageous option than other plant-based fibres [22,27].

Numerous studies have been conducted thus far to explore the characteristics of concrete reinforced with JF. Zakariya et al. [28,29] conducted a study on how different percentages and lengths of JF affect the properties of normal concrete. According to the research, the inclusion of JF at a smaller volume percentage led to an improvement in the strength parameters of the concrete, including CS, FS, and STS. The study also found that the optimal efficiency of JF occurred at fibre lengths of 10–15 mm. Ozawa and colleagues [30] discovered that the addition of 0.075% JF to HPC can rectify its brittle behaviour and serve as a protective measure against fire spalling. Sridhar et al. [31] reported that incorporating JF at a rate of 1.5% by weight of cement in normal-strength concrete (NSC) resulted in a 18% net increase in CS and a 25% increase in FS. Another study [22] reported that the highest enhancements in CS were observed at a volume percentage of 0.25% of JF, whereas the maximum improvements in flexural–tensile strength were observed at a volume percentage of 0.5% of JF. There have been numerous studies that have confirmed the favourable impact of JF inclusion on the mechanical characteristics of concrete. However, the effects of JF on parameters related to durability have not yet been established. According to Gulzar et al. [6], the inclusion of JF had unfavourable effects on the properties of ordinary concrete related to permeability. As the volume of JF increases, the porous nature of plant fibres leads to a decline in the concrete's ability to resist water absorption and chloride ion penetration. The inclusion of SCMs as a part binder in concrete, such as fly ash or SF, can help in controlling the worse impacts of plant fibres on the durability of material [6,7].

Previous research showed that the absorption-related characteristics of JF-reinforced concrete have not been adequately addressed. JF is a low-cost and greener alternative to synthetic fibres, and it offers high tensile strength. Moreover, no reliable information was found on the flexural response and life cycle impact of JFRC. It is widely recognized that the alkaline environment of cement can cause corrosion of the components of plant fibres [32]. The addition of a secondary binding material, such as SF or rice husk ash, to cement has been shown to mitigate the corrosive impact of an 'alkaline' environment on the plant fibres. The "pozzolanic reactions" convert free calcium-hydroxide (CH) into calcium-silicate-hydrate (C-S-H) [21,33]. There is limited research available on the modification of JF-reinforced concrete using SF.

This study aimed to examine the impact of various JF volume percentages on the mechanical, durability, and life cycle aspects of concrete, with and without SF as an SCM. The study examined the mechanical properties of the mixtures, including CS, flexural (load–deflection characteristics, ductility, residual strength), and tensile strength. This study conducted various tests to assess the durability behaviour of JFRCs, such as WA by immersion (%), RCIP/electric flux (Coulombs), ER, and UPV, to evaluate the durability of the modified JFRC mixtures. The findings of this research could provide useful knowledge on how to efficiently incorporate plant fibres in the design of concrete with improved flexural–tensile performance and resistance to weathering agents. JFRC possesses both economic and environmental significance. The use of JF as reinforcement reduces the cost of concrete production while potentially providing improved strength and durability. Additionally, JF is a renewable/reproduceable resource (unlike manufactured fibres), and their use in concrete helps to reduce carbon emissions, making it an eco-friendly alternative to traditional reinforcement materials. Besides, the natural synthesis of JF allows the absorption of CO_2 from the atmosphere. Therefore, plant-based/agro-fibres are suitable for a sustainable built environment.

2. Materials and Methods

2.1. Constituent Materials

As per ASTM C150 [34], Portland cement with a strength grade of 53 was utilized as the primary binder. The cement used in the study had a specific surface area (fineness) of 356 m^2/kg and an apparent density of 3.13 g/cm^3. Table 1 presents the oxide percentages of the cement used in the study. SF was also utilized as an SCM. Figure 1 displays an SEM image of the small particles of SF. The "apparent-density" and "surface-fineness" of SF was determined to be 2.33 g/cm^3 and 27,000 m^2/kg, respectively. Figure 2 shows the particle-size distribution of the binders used in the study.

Table 1. Oxide composition of OPC.

Oxide	(%)
SiO_2	17.44
CaO	65.17
Al_2O_3	4.61
MgO	2.16

Figure 1. SEM results of SF sample.

This research used fine and coarse, for the mixtures. Siliceous sand with a "bulk-density" of 1.63 g/cm^3 was used as the "fine" aggregate, while crushed dolomite aggregate with a "particle-density" of '2.68 g/cm^3' was used as the "coarse" aggregate. Aggregates that were "fine" and "coarse" had a 24 h WA capacity of "0.98%" and "1.12%", respectively. The gradation test results are depicted in Figure 3.

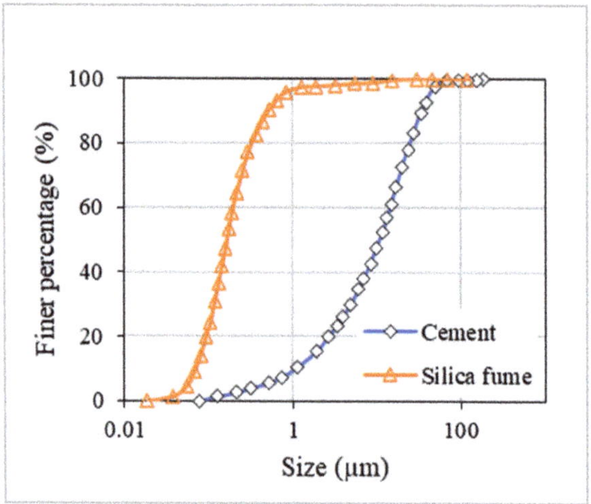

Figure 2. Particle size distribution of cement and SF.

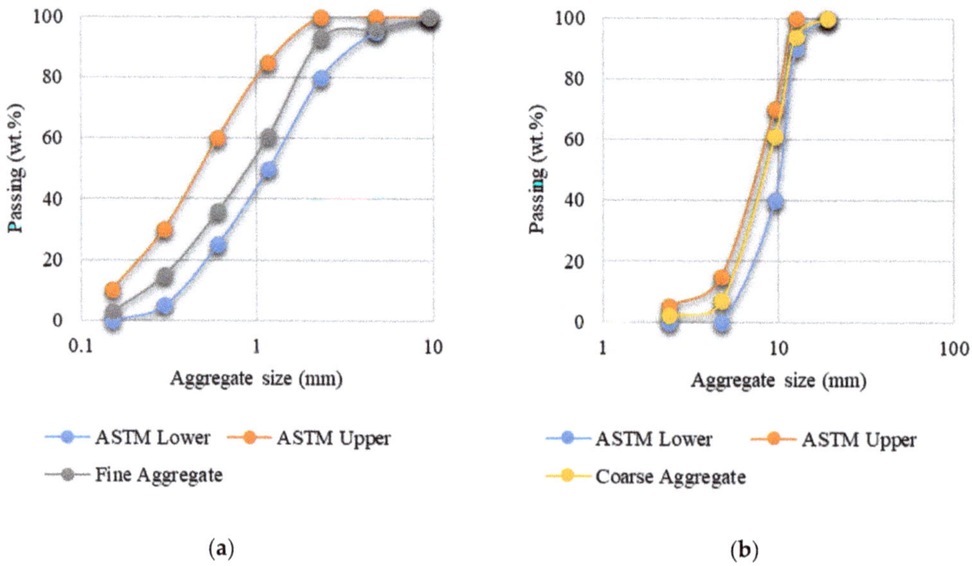

Figure 3. The particle size distribution of aggregates. (**a**) Fine; (**b**) coarse.

The research involved the production of JF using jute sacks that were no longer usable. The sacks were converted into fibres with varying lengths between 12 and 18 mm. As a result, the fibres utilized in the study were classified as recycled fibres. The average length of 15 mm was chosen as the optimal length for JF based on previous research, which demonstrated similar lengths of JF resulted in improved mechanical performance [28]. It was discovered that the average diameter of JF was approximately 0.1 mm. JF had an apparent density of 1.45 g/cm^3, and its tensile strength was 410 MPa. The general characteristics of JF are depicted in Figure 4. To reduce the amount of water used and produce concrete with a low water-cement ratio (w/c), a polycarboxylate-based plasticizer was utilized.

Figure 4. Jute fibre (JF).

2.2. Design and Preparation of Concrete Mixtures

Table 2 presents the names and proportions of the various components used in the mixture. Various trials were conducted to establish the optimal mixture ratio for the control concrete (JF 0) with the goal of attaining a CS of "60 MPa" and a slump range of 180 ± 30 mm. The mixtures had JF at three different volume percentages: 0.15%, 0.3%, and 0.5%. However, any volumes of JF higher than 0.5% were not considered as they have been reported to have negative effects on mechanical strength [22,28].

Table 2. Design of mixtures.

Mix Names	JF (%)	Type I Cement (kg/m^3)	SF (kg/m^3)	Fine Aggregate (kg/m^3)	Coarse Aggregate (kg/m^3)	Water (kg/m^3)	SP (kg/m^3)	JF (kg/m^3)
JF0	0	550	0	640	1073	182	3.03	0.00
JF0.15	0.15	-	0	-	-	-	-	2.16
JF0.3	0.3	-	0	-	-	-	-	4.32
JF0.5	0.5	-	0	-	-	-	-	7.20
JF0/SF	0	495	-	-	-	-	-	0.00
JF0.15/SF	0.15	-	-	-	-	-	-	2.16
JF0.3/SF	0.3	-	-	-	-	-	-	4.32
JF0.5/SF	0.5	-	-	-	-	-	-	7.20

Two sets of JFRCs were manufactured, with one using OPC only and the other replacing 10% of OPC with SF. The addition of SF to the mixture would serve two purposes: a "pozzolanic" effect and "mineral-filler" effect, as well as a reduction in the damaging impact of the highly alkaline cement on the plant fibre [35].

Figure 5 illustrates the process of mix preparation. The mechanical drum mixer was used to dry-mix the binders and aggregates for 4 min. During the next 4 min duration, water was added gradually to the dry mix along with the required amount of SP (at a rate of 3 litres per cubic meter). During the last stage of the mixing process, the fibre was slowly added to the wet mix over the course of 4 min. During all stages, the speed of mixing was around 20 rpm. After measuring the slump value of the fresh concrete, the mixture was then poured into moulds of different standard shapes. To ensure consistency across all mixtures, a vibrating table was used to compact the samples with a fixed vibration duration of 30 s. The compacted specimens were carefully shielded with a highly durable and waterproof membrane and then stored in a temperature-controlled environment for an exact duration of 24 h to ensure optimal setting conditions. After carefully removing the specimens from the moulds, they were placed in a water tank at a precisely controlled temperature of 25 ± 3 degrees Celsius to guarantee optimal curing conditions for the specified duration of testing.

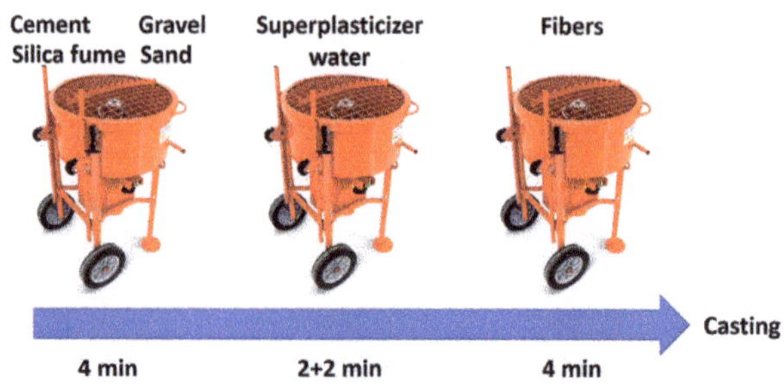

Figure 5. Mixing procedure.

2.3. Determination of Concrete Properties or Engineering Performance

Fresh concrete properties, namely fresh density, air content, and slump, were assessed. Air content and Abram's cone slump tests were conducted by following ASTM C231 [36] and ASTM C143 [37].

To determine the CS of the concrete specimens, "cubic" samples measuring 100 mm were tested according to the guidelines specified in the ASTM C39 standard [38]. The testing process involved applying a compressive force to the samples until they failed, and the maximum load that the specimens could bear before collapsing was recorded as the CS.

The STS of the concrete specimens was determined by testing "cylindrical" samples measuring "100 mm in diameter" and "200 mm in height" in accordance with the ASTM C496 standard [39]. The testing process involved applying a tensile force to the samples along the vertical axis until they failed. The maximum tensile load that the samples could withstand before failure was recorded as the STS or indirect tensile strength.

The FS of the concrete specimens was determined by testing "beam" samples measuring "100 mm in width", "100 mm in depth", and "350 mm in length", in accordance with the ASTM C1609 [40]. The testing process involved applying a load to the centre of the beam until it failed, and the maximum load that the beam could bear before collapsing was recorded as the FS. The load vs. deflection behaviour was also noted for specific mixtures (JF0.3, JF0.5, JF0.3/SF, and JF0.5/SF).

After 28 days of curing, the WA capacity of the concrete specimens was determined by immersing "disc-shaped" samples measuring "100 mm in diameter" and "50 mm in height" in water for a 24 h period in accordance with ASTM C948 [41]. The immersion method involved weighing the dry samples before and after immersion in water to determine the amount of water absorbed by the samples.

The 28-day RCIP values of the concrete specimens were determined by testing disc-shaped samples measuring "100 mm in diameter" and "50 mm in height", in accordance with ASTM C1202 [42]. This test is used to assess the ability of the concrete to resist the penetration of chloride ions, which can cause corrosion of reinforcing steel and affect the durability of the concrete structure. The RCIP test involves applying a voltage across the concrete sample and measuring the electrical charge/flux (Coulombs) passing through it.

The ER of the concrete mixtures was determined by testing cubic samples measuring "100 millimetres" on each side in accordance with ASTM C1876 [43]. The ER test is used to assess the ability of the concrete to resist the flow of electric current and is related to various properties such as permeability, durability, and corrosion resistance. In this test, a small electrical potential (60 V) is applied to the sample, and the resulting current is measured.

To evaluate the quality of the concrete, a UPV test was conducted. UPV values are known to be indicative of the concrete's strength, porosity, internal defects, and permeability. For this test, "100 mm cubic" samples were used, and the procedure followed ASTM C597 [44].

2.4. Life Cycle Assessment or Environmental Performance

The study evaluated the environmental impact (EI) of various concrete mixtures by analysing their global warming potential (GWP) resulting from carbon dioxide (CO_2) emissions. To assess the EI, the functional unit considered was one cubic meter (m^3) of concrete. Therefore, the analysis focused on determining the EI of producing one cubic meter of concrete.

The research investigated the EI of the materials used to make concrete, from when they are extracted from the Earth to when they are delivered to the concrete plant. Additionally, the study assessed the EI of the concrete production process, including the "extraction" of raw materials (B1), the "transportation" of processed materials (B2), and the actual "manufacturing" process (B3). In other words, the study aimed to analyse the environmental impact of the entire life cycle of concrete production, from start to finish. This kind of analysis is important for understanding the potential environmental impact of construction materials and processes and for identifying ways to reduce that impact.

To assess the EI resulting from the "processing" of "raw materials" (B1), the study utilized a database containing the GWP values (kg-CO_2/kg). These values were obtained from a previous study conducted by Braga et al. [45]. The EI of SF used in the concrete mixtures was obtained from a study by Hájek et al. [46]. The data for the EI of the raw materials used in the concrete mixtures are presented in Table 3. The collected data were adapted based on the cradle-to-gate scenario. The EI of JF was taken as similar to that of the recycled coconut fibre, since, in this research, JF was considered as a recycled material [47]. The study considered the environmental impact of transporting the concrete mixtures by lorry, which was calculated to be 6.57×10^{-5} kg-CO_2 per kilogram per kilometre. Table 4 provides information on the distances between the "concrete plant" and the "sources of raw materials", which were used to determine the EI of transporting the raw materials. The distances were utilized to evaluate the EI of the constituent materials' transportation.

Table 3. EI of raw materials (kg-CO_2/kg).

Material	GWP (kg-CO_2/kg)
OPC	0.898
Siliceous sand	0.002
Dolomite sandstone	0.053
SP	0.002
Water	0.000
SF	0.0011
JF	0.36
Concrete preparation	4.65
Transportation impact	0.0000657

Table 4. Transportation distances.

Material	Distance from Mixing Plant (km)
OPC	397
SF	178
Siliceous sand	456
Dolomitic sandstone	246
SP	177
JF	15
Water	0

3. Results and Discussion

3.1. Engineering Performance

3.1.1. Air Content

The effect of JF and SF incorporation on the air content of fresh concrete is illustrated in Figure 6. It can be noticed that, for both SF- and non-SF-containing mixtures, the "air content" increased with the increasing percentage of JF. The addition of plant-based fibres to concrete can increase its air content due to the interaction and cohesion between the fibres and the fresh concrete mixture. The fibres act as obstacles, hindering the movement or flow of the fresh concrete mixture, which in turn can create voids in the mix. These voids can trap air and increase the overall air content of the concrete. The lack of workability and balling effect caused by the presence of JF increases the air content in the compacted concrete. Similar findings have been reported by the use of coconut fibre in fresh concrete [48]. The inclusion of SF as the cement replacement material marginally decreases the air content of fresh concrete. This can be attributed to the filling effect of extremely fine SF particles that occupy the free spaces between cement particles, thus resulting in a more compacted concrete.

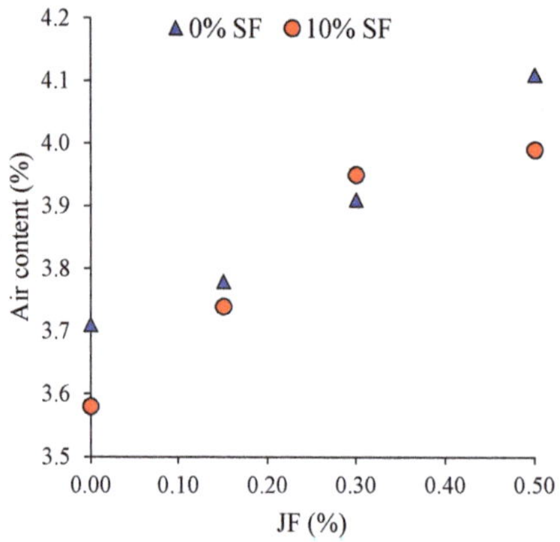

Figure 6. Effect of JF and SF incorporation on the air content of fresh concrete.

3.1.2. Fresh Density

The results of the fresh density tests are presented in Figure 7. The fresh density of concrete is of great practical significance since it is widely used to assess the quality and yield. It can be noted that plant-based JF reduces the fresh density of concrete with the increasing percentage. Up to a 1% decline in the fresh density of concrete was observed when the JF content was used up to a 0.5% concentration. This is primarily linked to the low density of JF filaments as compared to the binder matrix. Since low-density JF filaments replace the dense cementitious matrix in the concrete, it is not possible to control the loss in density due to the incorporation of the lightweight fibres.

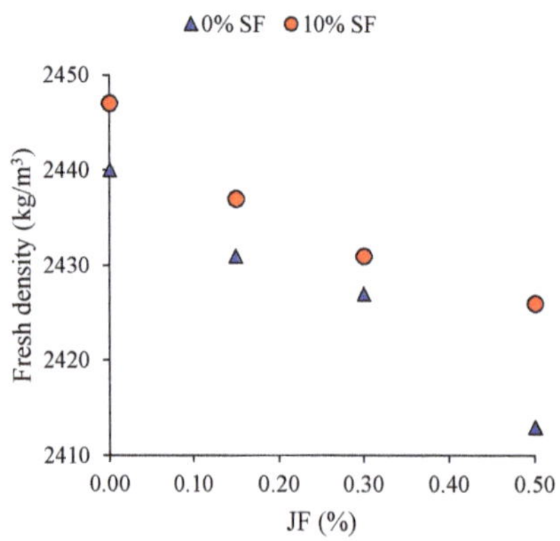

Figure 7. Effect of JF and SF incorporation on the density of fresh concrete.

Contrary to JF incorporation, SF addition increased the fresh density of concrete by 7 kg/m^3. As SF particles are extremely fine as compared to OPC particles, therefore, these can occupy the volume between OPC particles, resulting in increased compaction and densification. SF is a highly reactive pozzolan that can improve the packing density of the concrete particles, leading to a denser mixture [49]. The fine particles of silica fume can fill the voids between the larger particles of the other components, resulting in a more compact mixture.

3.1.3. Slump

The effect of JF and SF addition on the workability of fresh concrete is depicted in Figure 8. The addition of JF to a concrete mixture can generally make it more viscous and less fluid, reducing its workability due to possible clustering of thin JF filaments. JF incorporation can increase the friction between concrete particles, making it harder to mix and move the concrete. However, it can be noted that, despite the decline in the workability, the slump value of all fibre-reinforced mixes remained within the target slump range of 150–210 mm. This is because the control fresh concrete was produced with a carefully designed dosage of a high-range water reducer, which allowed for the possible workability loss due to the incorporation of JF. The concrete yielding a slump in the range of 150–230 mm can be used as a pumpable concrete.

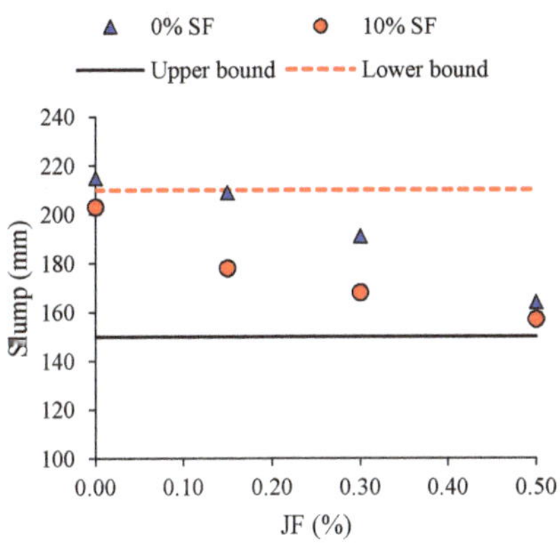

Figure 8. Effect of JF and SF incorporation on the slump of fresh concrete.

The incorporation of SF led to a small decline in the workability of the fresh concrete. The addition of SF to a concrete mixture can have a negative impact on workability due to its high surface area and fineness as compared to OPC particles, which can cause the concrete to become thicker, cohesive, and less fluid. SF particles can absorb water, resulting in a reduction in the water content available for cement hydration and increasing the demand for water. Therefore, the use of SP provides control over the significant loss of workability due to SF addition. SP does not only control the negative effects of SF and JF incorporation on the workability, it also helps in the dispersion of SF particles and JF filaments to ensure more homogenous and enhanced properties of the concrete [50].

3.1.4. Compressive Strength

Figure 9 depicts how the CS of concrete is influenced by inputs such as the presence of varying percentages of JF and SF. The impact of different volumes of JF on the CS of concrete was not consistent. At 28 days, there was a slight increase of "4.0%" and "4.8%" in CS when "0.15%" and "0.3%" volumes of JF were added, respectively. On the other hand, the CS of concrete at 91 days showed a higher increase. There was an increment of "5.5%" and "9.7%" when "0.15%" and "0.3%" of JF were added, respectively. The reason for the increase in CS when 0.15% to 0.3% of JF was added is due to the enhancement of the "transverse" distortion resistance of the plain concrete [22,51]. There is also a theory that the moist JFs can aid in the internal curing of the concrete's microstructure, which may lead to an increase in its strength [3]. Due to the lower density and higher surface area of JFs, their incorporation at higher rates may have a detrimental effect on the density of concrete. Therefore, when a 0.5% volume of JF was added, there was no significant change observed in the CS at both 28 and 91 days of testing.

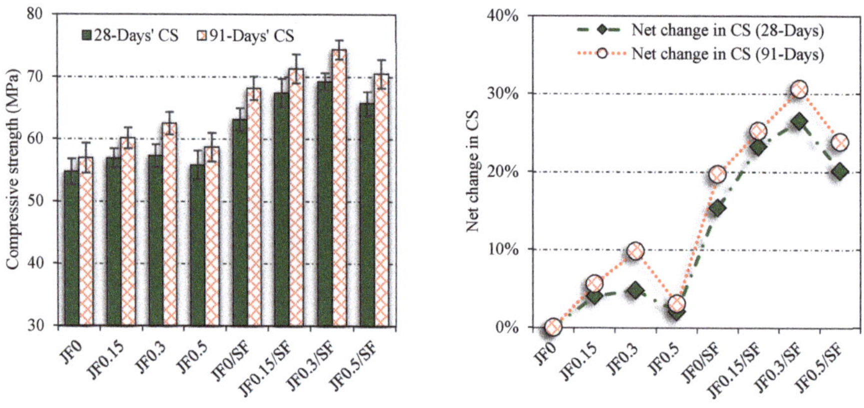

(a) (b)

Figure 9. Compressive strength results: (**a**) CS results with SF and varying JF content; (**b**) change in CS with respect to JF0.

When a 10% volume of SF was added to the concrete, there was a significant increase in the CS. At 28 and 91 days, there was an increase of 15.4% and 19.7% in CS, respectively. The reason behind this increase in strength was attributed to the pozzolanic reaction between the portlandite (CH) and SF particles, which led to the additional growth of C-S-H in the concrete. SF contributes to the increase in strength, not only through the pozzolanic reaction, but also by reducing the pores between cement particles, which leads to the solidification and strengthening of the microstructure.

When JF and SF were incorporated together, their combined effect showed a synergistic increase in the CS of the concrete. When a 0.3% volume of JF and a 10% volume of SF were added individually, there was a net increase of 5% and 15%, respectively, in CS at 28 days. However, when they were added together, there was a synergistic effect, which resulted in a much greater net increase of 26% in CS. When a 10% volume of SF and a 0.3% volume of JF were added together, there was a significant increase in CS compared to the control mixture. At 28 and 91 days, there was a net increase of 26% and 31%, respectively. The observation of similar synergistic effects between SF and other volumes of JF suggests that the efficiency of JF as a fibre reinforcement increases as the strength grade of PC improves. The enhancement in the fibre–matrix interfacial strength due to the incorporation of SF in the binder matrix may explain this phenomenon [52]. An important point to mention is that the increase in CS resulting from the addition of JF also increased as the concrete aged. For example, with the addition of 0.3% JF, the strength of the concrete increased by 4.8% and 9.8% (compared to the control mixture) at 28 and 91 days, respectively. This indicated that the strength improvement due to JF incorporation continues to increase with the aging of the concrete (not indefinitely).

3.1.5. Splitting Tensile Strength

The impact of adding JF and SF on the STS of concrete mixtures is presented in Figure 10. The figure shows that the inclusion of JF resulted in significant increases in the STS value of concrete. At different volumes of incorporation of JF, there were notable increases in the STS value of concrete, as seen in Figure 10. Specifically, there were net increases of "8.6%", "18.4%", and "19.8%" in the STS value of the concrete at "0.15%", "0.3%", and "0.5%" JF incorporation, respectively. The strength values of the concrete also showed notable improvements with an increase in JF content, particularly in the 91-day STS. Islam and Ahmed [22] reported that the addition of 0.5% of JF led to a net increase of approximately 20% in the splitting tensile strength of normal-strength concrete. Zakaria and colleagues [28] reported that the addition of 0.15–0.25% of JF resulted in a 16–20% improvement in the STS of PC. The incorporation of JF in concrete is believed to enhance the resistance of concrete against the onset and propagation of cracks. This can be explained by the fact that JFs can bridge the micro-cracks in the concrete matrix, which can lead to an increase in the crack-bridging capacity of the concrete. As a result, the stress concentration around the cracks is reduced, which can prevent the further growth and development of cracks. This mechanism ultimately leads to an improvement in the tensile strength of the concrete. Additionally, the improved interfacial bonding between the fibres and the cement matrix due to the pozzolanic reaction can also contribute to the increased tensile strength of the concrete [53]. The study found that the highest STS value was observed in the concrete containing a 0.5% volume of JF. However, it was noted that the percent difference in the STS value between the concrete mixtures containing a 0.3% and a 0.5% volume of JF was not significant. In other words, while increasing the volume percentage of JF resulted in higher STS values, the improvement in STS between 0.3% and 0.5% was not as significant as the improvement seen between lower percentages of JF. This suggests that the optimal volume percentage of JF may depend on various factors, such as the specific application and required strength of the concrete. Based on the results, it can be concluded that incorporating 0.3% vol. of JF in the mixture can provide optimum benefits in terms of the efficient utilization of fibre material and maintaining the workability of the fresh mixture. Therefore, this volume percentage can be considered as the ideal amount of JF to be added to the concrete mixture.

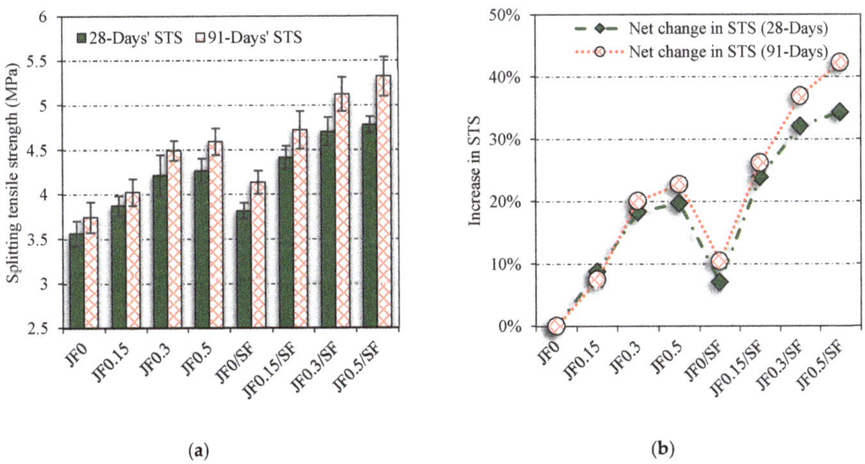

Figure 10. Splitting tensile test results: (**a**) STS results with SF and varying JF content; (**b**) change in STS with respect to JF0.

At 28 and 91 days, the strength of concrete was improved by 7% and 10.4%, respectively, compared to the control mixture, when SF was added to it. When both JF and SF were added to the concrete mixture, their combined effect showed a coupling and synergetic effect on the STS of the concrete. This means that the combined effect was greater than the sum of the individual effects of JF and SF. In other words, the use of JF and SF together resulted in a greater improvement in the STS of the concrete than using either one of them alone. This highlights the potential benefits of using multiple materials to enhance the properties of concrete. The combination of JF and SF in the concrete mixture resulted in a coupling and synergetic effect on the STS. Specifically, when 0.3% JF and 10% SF were added together, there was a net improvement of 32% in STS (which was 7% higher than the improvement achieved by using SF alone) at 28 days, and a net improvement of 42.2% in STS (which was 9% higher than the improvement achieved by using SF alone) at 91 days. This showed that the combined use of JF and SF can lead to significant improvements in the short-term and long-term strength of concrete. Adding SF to the concrete mixture enhances the pull-out strength of fibre filaments [52]. The higher STS of the concrete was a result of the formation of C-S-H and an improved JF–matrix interface, which led to an increase in the micro-hardness and a stronger bond between the fibres and the matrix.

3.1.6. Load versus Deflection

The effect of JF and SF incorporation on the load–deflection behaviour of the concrete is shown in Figure 11. It is distinct that the incorporation of SF had a slight influence on the peak load and no contribution towards the post-peak deflection behaviour of the concrete. Adding SF contributed to the strength improvement through pozzolanic reactions by creating more C-S-H gels. However, the incorporation of JF significantly changed the load–deflection behaviour of the concrete. JF0.3 and JF0.5 generally demonstrated greater ductility compared to traditional PC, allowing them to withstand more extensive deformations before collapsing. This is because the JF filaments distributed within the concrete help to distribute loads more evenly and bridge small cracks that could form under external loads [54].

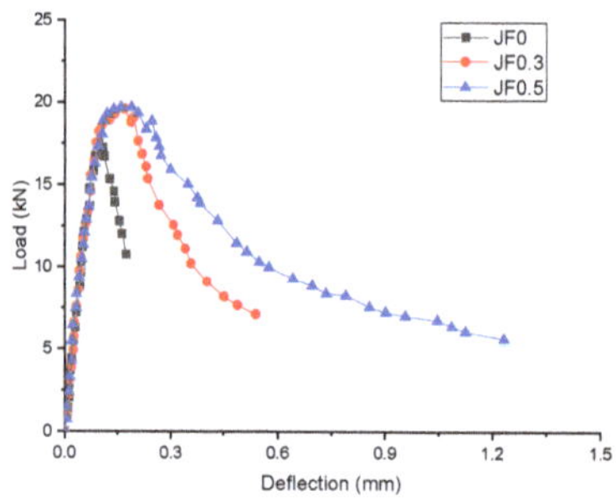

(**a**)

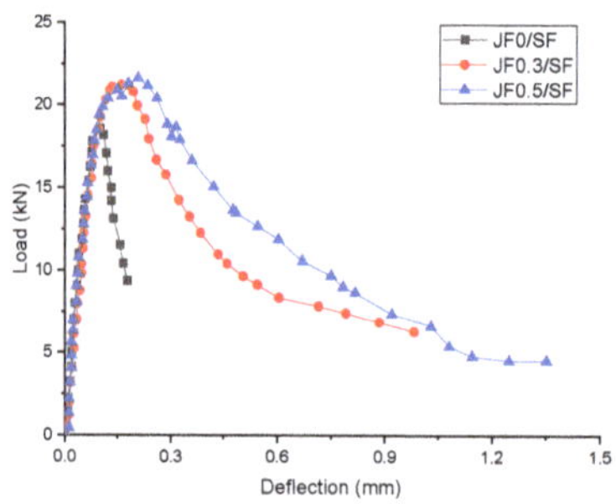

(**b**)

Figure 11. Load deflection behaviour of concrete with (**a**) different JF contents and 0% SF and (**b**) different JF contents and 10% SF.

The load–deflection curve of JFRC may exhibit plateaus or yielding, which correspond to the formation of new cracks and fibre pull-out events. The final failure of JFRC may occur due to the rupture of JF filaments or the complete separation of JF filaments from the matrix. The increase in JF content had a minor improving effect on the peak load sustained by JFRC. However, the increase in the fibre content ensured the delay in the failure of JFRC. This is because, at higher fibre contents, more fibres are present to restrict the crack movements and allow for a more ductile response than concretes with low fibre contents.

The incorporation of SF into JFRCs led to a further increase in the peak load and improved the post-peak deflection behaviour. The longer stems of the post-peak load–deflection were observed for JFRC mixtures incorporating SF. This can be attributed to the improvement in the bond performance of JF due to the solidification and densification of the concrete matrix with SF addition. Thus, SF can help improve the overall load–deflection behaviour of JFRCs.

3.1.7. Flexural Strength or Modulus of Rupture

Figure 12 presents a visual representation of how incorporating JF into concrete, with and without SF, affects the FS of the material. Adding a 0.3% and a 0.5% volume of JF to the concrete resulted in a net increase of approximately 16% in the FS of the material after 28 days. The addition of a 0.3–0.5% volume of JF to the concrete resulted in a net improvement of approximately 20% in the flexural strength (FS) of the material after 91 days. Concrete samples showed further improvements in strength at later ages. As the concrete aged, the chemical reactions within the mixture continued to occur, resulting in the formation of stronger bonds between the cement particles and other materials including the fibre reinforcement within the mixture. This finding corroborates the findings of a prior investigation [55], which revealed that the increase in FS resulting from the addition of JF can be attributed primarily to the reinforcement of the concrete against the initiation and propagation of cracks. The incorporation of JFs helps to distribute stress more uniformly throughout the concrete matrix, which reduces the likelihood of the formation of cracks and other forms of damage. By strengthening the concrete and reducing the risk of cracking, the JFs helped to increase the FS.

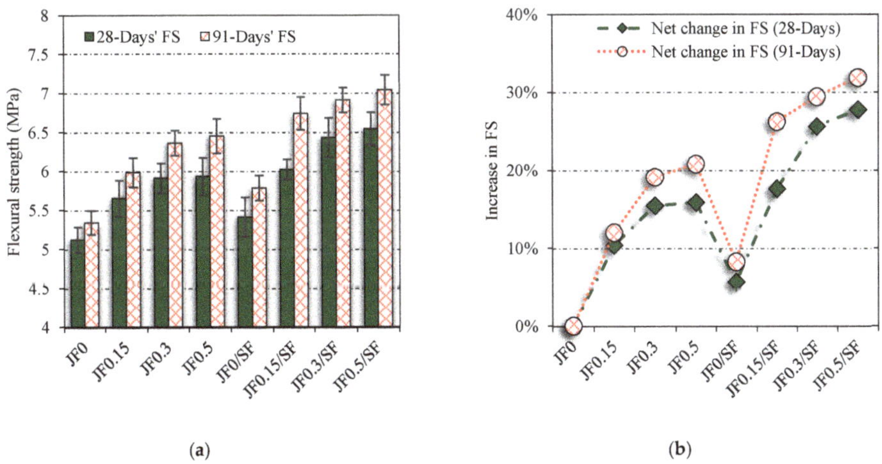

Figure 12. Flexural test results: (**a**) FS results with SF and varying JF content; (**b**) change in FS with respect to JF0.

It is worth noting that the difference in the percentage of improvement in FS resulting from the addition of a 0.3% and a 0.5% volume of JF was not significant. Hence, a 0.3% volume of JF may be regarded as the optimum amount to be added to the concrete mixture for maximum FS enhancement.

The FS of the concrete exhibited a net improvement of 5–10% as a result of incorporating SF as a partial substitute for OPC. In addition to enhancing the performance of OPC, the inclusion of SF also improved the impact of JF on the FS of the concrete. The joint addition of SF and a 0.3% volume of JF resulted in net improvements of 26% and 30% in FS at 28 days and 91 days, respectively, compared to the control mixture. The combined inclusion of JF and SF in the concrete mixture exhibited both a coupling effect and a synergetic effect on the FS of the material. The addition of SF had the effect of refining the bond between the fibre and binder matrix, which increased the pull-out strength of the fibre filaments. This, in turn, led to additional enhancements in the tensile and flexural strength of the concrete. Previous research has demonstrated the synergistic effects of incorporating SF into the concrete mixture with both micro and macro fibres. These findings suggest [56,57] that the inclusion of SF in combination with other reinforcing materials, such as JF, can lead to significant improvements in the strength of concrete, making it more resistant to cracking and other forms of damage.

3.1.8. Residual Strength and Flexural Toughness

The residual flexural strength (RS) and flexural toughness (FT) of concrete mixtures are presented in Figures 13 and 14, respectively. RS refers to the flexural strength of FRC after the initial cracking of the concrete. It is usually measured from the loads corresponding to 0.5 mm, 1 mm, and 2 mm deflections. However, in this study, JFRCs failed before 2 mm deflection; therefore, the RS measurement was limited to 0.5 mm and 1 mm only. The addition of JF filaments helps to bridge micro-cracks, preventing the cracks from propagating further and leading to a more gradual loss of strength after cracking. As a result, JFRC exhibited a higher RS compared to JF0. Due to the addition of 0.3% and 0.5% JF, the RS (0.5 mm) were 2.30 and 3.28 MPa as compared to 0 MPa RS of the plain JF0, respectively. Despite a nominal effect on the peak FS, the incorporation of 0.5% JF had a prominent effect on the RS at 1 mm deflection. JF0.3 and JF0.3/SF failed before 1 mm deflection, while JF0.5 and JF0.5/SF retained a notable RS at 1 mm deflection.

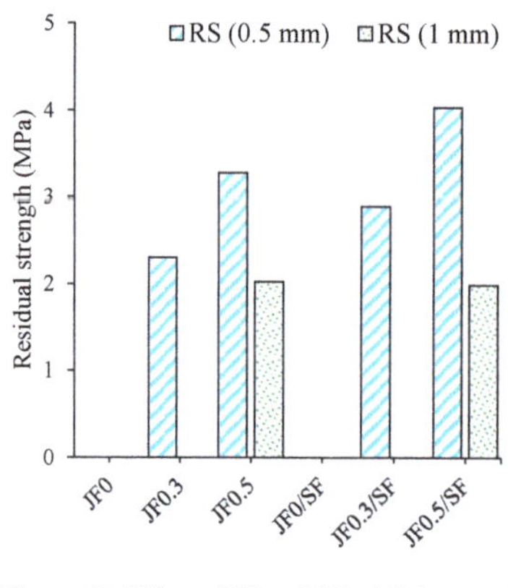

Figure 13. Effect of JF and SF addition on the residual strength.

FT refers to the ability of a concrete to absorb energy under "flexure" or "bending" and is an important property for many structural applications. JFRCs exhibit increased FT due to the bridging effect of the fibres, which redistributes stress around the crack and prevents further crack propagation. This results in a higher energy absorption capacity and increased ductility, allowing the JFRC to deform more before failing. For instance, JF0.3 and JF0.5 yielded 3.22- and 6.4-times higher FT as compared to JF0, respectively. The incorporation of SF also increased the FT of JFRCs. For instance, JF0.3/SF and JF0.5/SF showed FT values 5.35- and 7.31-times higher as compared to JF0, respectively.

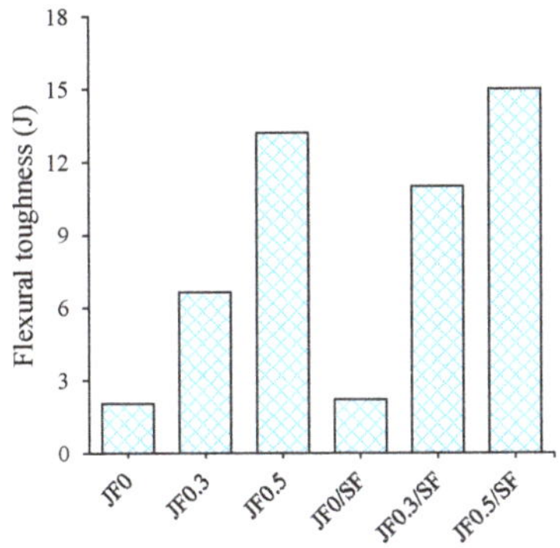

Figure 14. Effect of JF and SF addition on the flexural toughness.

3.1.9. Water Absorption Capacity

The durability of concrete is impacted by the movement of fluids through the pores that are connected to the concrete's surface. The volume of permeable voids, known as WA, can be used as an estimate of the durability of concrete. Essentially, the more permeable voids there are, the less durable the concrete is likely to be. Therefore, controlling the number of permeable voids in the concrete can help improve its durability. A test was carried out to measure WA, which represents the volume of permeable voids in concrete, at two different time intervals: 28 and 91 days. The results of the test are presented in a graph and figure labelled as Figure 15. This figure provides a visual representation of how WA changes over time due to the mixture design inputs such as JF volume and SF incorporation.

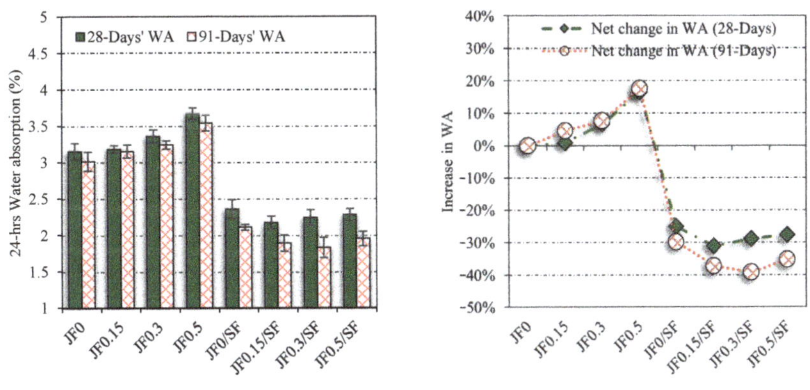

Figure 15. Water absorption test results: (**a**) WA results with SF and varying JF content; (**b**) change in WA with respect to JF0.

The observation was made that, when the volume of JF was increased from 0.15% to 0.5%, the volume of permeable voids in the concrete (represented by WA) increased by approximately 16%. This suggests that the addition of JF may have a negative impact on the durability of the concrete by increasing its permeability. The increase in WA observed when JF was added to concrete is believed to be caused by an increase in the connectivity of the microstructure of the concrete with its outer surface. In other words, the addition of JF may lead to a more porous concrete microstructure, allowing fluids to move more easily through the material and increasing its permeability [12]. As the number of fibre filaments in the concrete increased, more access points were created on the surface of the material, which can facilitate the movement of water into the concrete matrix.

The addition of SF resulted in a significant decrease in the volume of permeable voids, as represented by the WA capacity, in both plain and fibre-reinforced concrete mixtures. Specifically, the WA capacity decreased by 25.1% and 29.9% at the ages of 28 and 91 days, respectively. These results demonstrated that the incorporation of SF can be an effective way to improve the durability of JFRCs by increasing their resistance to the transport of fluids through their microstructure. The micro-particles of SF are believed to arrange themselves between the cement particles in the concrete mixture, which helps to densify the microstructure of the material. This can lead to a reduction in the volume of permeable voids and a corresponding decrease in the concrete's permeability. Additionally, the increased growth of calcium-silicate-hydrate (C-S-H) that occurred when SF was added to the mixture can also contribute to a reduction in the connectivity between capillary channels, further reducing the permeability of the concrete. Specifically, the WA values of the fibre-reinforced concrete containing JF (JFRC) mixtures were 30–40% lower than those of the control mixture.

3.1.10. Rapid Chloride Ion Permeability

Figure 16 displays the results of rapid chloride ion permeability (RCIP) testing conducted on various concrete mixtures. It can be observed that the RCIP values of the control concrete fall within the range of 1000 to 2000 Coulombs. RCIP testing is often used to assess the durability of concrete by measuring its resistance to the penetration of chloride ions, which can contribute to corrosion and other forms of damage in concrete structures. The moderate range of chloride permeability exhibited by the control concrete, as indicated by the RCIP values falling within the range of 1000 to 2000 Coulombs, is typically associated with concrete that has a low water-binder ratio, typically less than 0.4 [58]. These RCIP values are typically associated with high-strength and high-performance concretes [59]. The results indicated that there was a notable increase in RCIP with the increase in JF content. Specifically, when a 0.5% volume of JF was incorporated into the mixture, there was an observed increase in the RCIP value of approximately 43% compared to the control mixture. The observed increase in chloride permeability with the addition of JF can be attributed to an increase in the permeable porosity of the concrete, which allows for faster movement of chloride ions. As previously discussed, plant-based fibres such as JF are inherently porous in nature, and their incorporation into the concrete matrix can negatively impact its imperviousness [7,48].

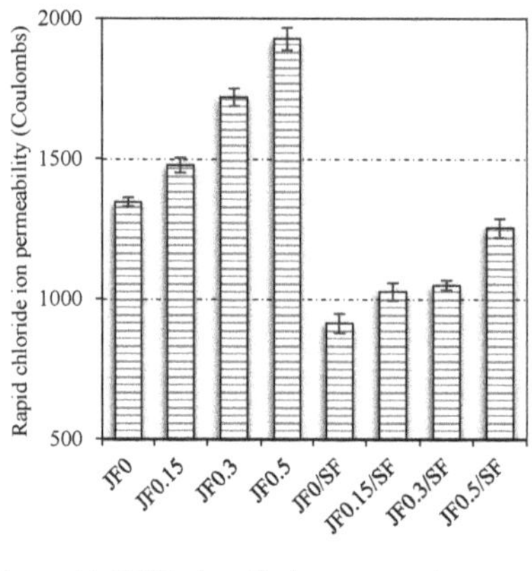

Figure 16. RCIP of studied concrete mixtures.

The addition of SF was found to be effective in reducing the RCIP value of the concrete. The observed reduction in the RCIP value with the incorporation of SF can be attributed to its superior filler effect, which helps to control the free movement of chloride ions in the concrete matrix. This is because SF has been shown to enhance the chloride binding capacity of the microstructure, effectively reducing the permeability of the concrete to chloride ions. The results indicated that the addition of SF led to a significant decline in the RCIP value of the concrete, with a reduction of approximately 32% compared to the control mixture without SF. Previous research [59,60] has shown that SCMs such as SF and rice husk ash have a significant effect on the permeability-related properties of concrete. The study conducted by Kou et al. [59] reported a reduction of approximately 25% in the RCIP value of concrete with the incorporation of 10% SF. The negative impact of JF on the chloride permeability resistance of concrete can be mitigated by incorporating SF into the mixture. By adding SF to the JFRC mixtures, lower RCIP values were obtained compared to the control mixture. This implies that the use of SF can be an effective strategy for improving the durability and resistance of fibre-reinforced concrete to chloride ion penetration, which can enhance its lifespan and performance in various applications.

3.1.11. Ultrasonic Pulse Velocity

UPV, or ultrasonic pulse velocity, is a non-destructive technique used to evaluate the quality and integrity of concrete. It is used to detect any cracks or defects present in the concrete and to determine the homogeneity of the hardened concrete. UPV is a valuable tool for assessing the overall condition of concrete structures without causing any damage. The range of UPV values between 3.5 km/s and 4.5 km/s is generally associated with normal-strength or medium-strength concretes of good quality [61]. These values indicated a relatively homogeneous and sound concrete structure without any significant cracks or defects. To clarify, UPV values above 4.5 km/s are typically associated with high-strength concrete and ultra-high-performance concretes (UHPCs) that have a low water-binder ratio, indicating excellent quality. The UPV test results are presented in Figure 17, and as expected, the incorporation of JF had a negative impact on the UPV value. For example, when the JF content was increased from 0.15 to 0.5%, the UPV value decreased by 2.31%. The decrease in UPV due to the addition of plant-based fibres is a common phenomenon as these fibres have a lower density and cellular microstructure, which lead to a reduction in the overall density of concrete [12,62]. The UPV test results confirmed that the high fibre contents had a negative impact on the WA, RCIP, and CS results of the concrete. This is because the presence of plant-based fibres in concrete reduces its density and homogeneity, leading to a decline in the UPV values.

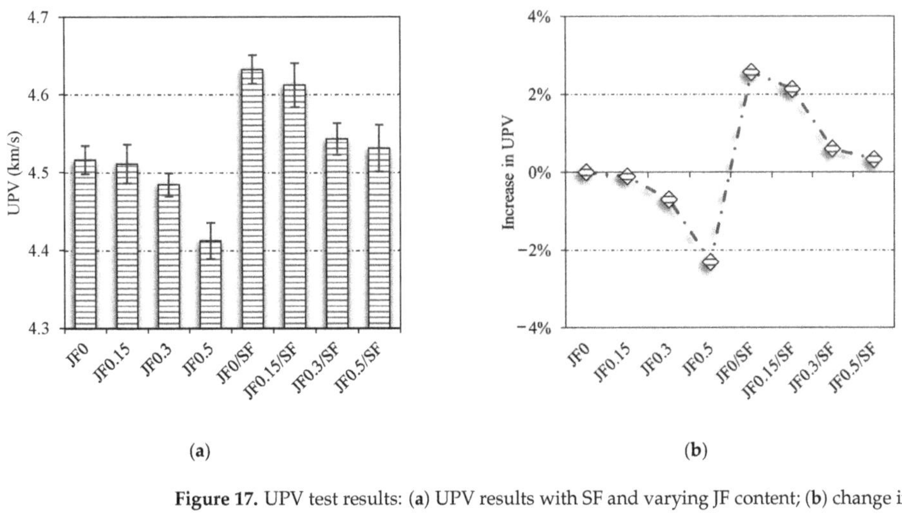

Figure 17. UPV test results: (**a**) UPV results with SF and varying JF content; (**b**) change in UPV with respect to JF0.

Adding SF increased the UPV value of the concrete by 2.6%. This increase can be attributed to the filler effect of SF, which helps improve the connectivity and homogeneity of the concrete. This, in turn, leads to a higher UPV value, which is above 4.5 km/s and is typically associated with excellent-grade concrete systems that have a low water-binder ratio, such as high-strength concrete. The compactness and reduction in pore size of the cementitious matrix around the fibres offsets the reduction in overall density caused by the presence of fibres in concrete [63].

3.1.12. Electrical Resistivity

ER is a crucial factor in determining the durability of concrete since the risk of corrosion in reinforced concrete is largely dependent on it. Figure 18 shows the ER measurements of all the concrete mixtures. Adenaert [64] suggested that, if the ER value of concrete is lower than 50 Ωm, then the steel reinforcement is at high risk of corrosion. However, if the ER value ranges between 50 and 120 Ωm, then the concrete is likely to be affected by corrosion in the long run. ER values above 120 Ωm are indicative of concrete with very low corrosion risk. The results indicate that all concrete mixtures had ER values greater than 50 Ωm. However, the PC mixtures exhibited higher ER values than their respective fibre-reinforced mixtures. The results showed that both plain concrete mixtures, with and without SF, are considered as durable and capable of protecting the steel reinforcement against corrosion. It should be noted that the incorporation of 10% SF into PC led to a 60% increase in ER. The increase in ER values of concrete containing SF can be explained by the filling of pores with micro-silica particles and hydration products. The C-S-H gel, which contributes significantly to the strength of concrete, reduces the formation of capillary pores, increases the volume of solid phases inside the microstructure, and ultimately, enhances the durability of concrete, including its resistance to corrosion [65]. The results indicated that the incorporation of JF has a negative impact on the corrosion-resistance potential of concrete, as evidenced by the results of WA and RCIP testing. Fibre-reinforced mixtures are more susceptible to corrosion due to the ability of the pore-solution to flow through the porous cellular internal structure of the fibres. Concrete mixtures with a 0.15% volume of JF, both with and without SF, have ER values above 120 Ωm, which are considered safe against corrosion. The results obtained from the durability tests, namely WA, RCIP, and ER, highlight the importance of SF in preserving the overall durability of JFRC.

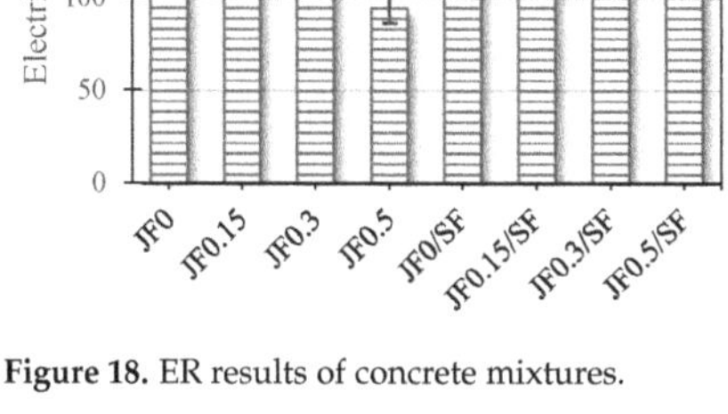

Figure 18. ER results of concrete mixtures.

3.2. Life Cycle Assessment Results

The global warming potential (GWP) of all studied concrete mixtures is illustrated in Figure 19. It can be observed that the incorporation of JF had a slight impact on the cradle-to-gate GWP of the concrete. Negligible reductions were noticed in GWP due to the increasing incorporation of JF. This is because JF possesses a very low GWP itself, and it replaces some volume of aggregates in concrete, which results in the marginal decline of

GWP. Contrary to JF inclusion, the incorporation of 10% SF as a cement substitution resulted in the 8% decline of the total GWP. Replacing OPC with SF in concrete can have a positive EI in several ways. SF is a by-product of the production of silicon and ferrosilicon alloys, and using it in concrete can help to reduce waste and limit the need for disposal. Additionally, the production of SF requires significantly less energy compared to cement production, resulting in reduced carbon emissions and energy consumption. By replacing a portion of the OPC with SF in concrete, the overall cement content of the mixture can be reduced. This can reduce the GWP of concrete as cement production is a significant contributor to greenhouse gas emissions (GHEs). Furthermore, SF can improve the durability and strength of concrete, allowing for thinner and lighter structures, which can further reduce the EI associated with construction.

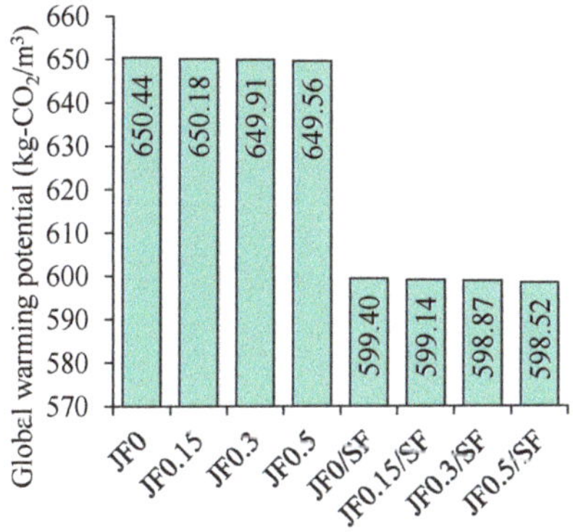

Figure 19. GWP of concrete mixtures with JF and SF incorporation.

The EI per unit CS, STS, and FS of each mixture is presented in Figure 20. These results show how much GWP is caused by the unit strength gained by each mixture studied. According to the results, it can be observed that the GWP per unit strength reduced with the rising JF content. GWP/STS and GWP/FS were highly sensitive to the JF addition, whereas GWP/CS was mildly influenced by the presence of JF. This is because JF yielded more benefit in the case of STS and FS, and it was mildly effective at enhancing the CS. On the other hand, SF addition was highly effective at reducing the GWP of concrete for unit CS, FS, and STS. The addition of 10% SF reduced the STS- and FS-related GWP by 16% and 14%, respectively, whereas JF0.5/SF yielded STS- and FS-related GWP corresponding to being 34% and 30% lower as compared to JF0.

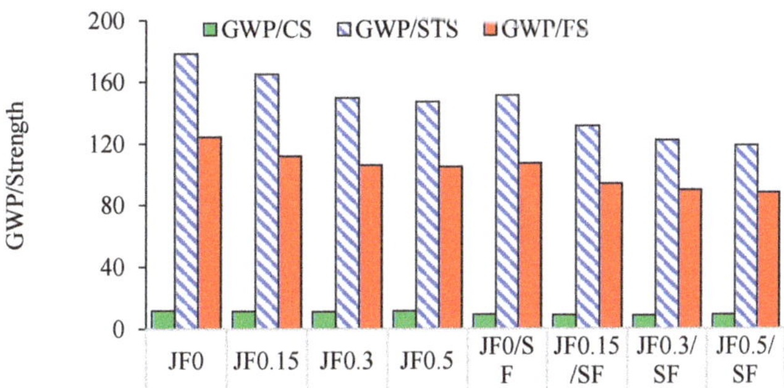

	JF0	JF0.15	JF0.3	JF0.5	JF0/SF	JF0.15/SF	JF0.3/SF	JF0.5/SF
GWP/CS	11.65	11.11	10.85	11.35	9.13	8.64	8.34	8.79
GWP/STS	178.20	164.81	149.40	146.79	150.98	131.25	121.97	118.52
GWP/FS	124.37	111.81	105.94	104.94	107.13	93.91	89.79	88.15

Figure 20. GWP per unit strength of mixtures.

4. Conclusions

Based on the results and discussion presented in this study, the following significant conclusions can be drawn regarding the impact of JF addition on various mechanical and durability properties of concrete, with and without SF:

- The inclusion of JF is harmful to the workability, and it increased the air content of the fresh concrete. The use of SF can be helpful in improving the fresh density and air content of fresh JFRC mixtures.
- The addition of 0.3% JF resulted in an overall improvement in compressive strength of 5–10%. The combination of 10% SF and 0.3% JF led to a remarkable improvement of 26.5% and 30.6%, respectively, compared to the control sample.
- The addition of SF in the binder further improved the STS gain from adding fibre. When a 0.3% JF volume and 10% SF were added, the STS increased by 32% and 37%, respectively, compared to the control sample.
- The increase in JF content improved the overall flexural response of the concrete. The flexural ductility and post-peak deflection behaviour were improved with the increase in the JF content from 0.3% to 0.5%. JFRCs with high fibre content also showed notable yielding before experiencing the strain-softening response.
- The best improvements in both STS and FS were achieved with a 0.5% JF dosage, but the difference in the net improvement between 0.3% JF and 0.5% JF was not significant. Therefore, it can be concluded that a 0.3% JF dosage is optimal in terms of both mechanical and economic performance.
- At both 28 and 91 days, the addition of 0.5% JF and 10% SF increased the FS of the concrete by 27% and 32%, respectively. The inclusion of 0.5% JF improved the FT of the concrete by 6.4-times and 7.3-times with 0% and 10% SF addition, respectively.
- Increasing the volume of JF from 0.15% to 0.5% resulted in a 16.5% increase in the water absorption (WA) capacity of the concrete. The addition of SF helped JFRCs achieve lower WA values compared to the control concrete.
- JFRCs showed a higher level of RCIP compared to the control concrete, regardless of the fibre content. Nonetheless, the incorporation of SF helped to decrease the chloride permeability in JFRCs. This is likely due to the additional growth of C-S-H and micro-filler action, which intercept the movement of chloride ions within the microstructure of JFRCs.
- The UPV findings indicated that the density of JFRCs is lower than that of traditional concrete, which could be attributed to the presence of light-density and porous fibres. Nonetheless, incorporating SF can enhance the quality of JFRCs and result in a superior final product.
- The study found that the risk of corrosion in JFRCs increased with higher volumes of JF, and incorporating 0.5% JF resulted in a 40% reduction in electrical resistivity (ER). However, adding SF significantly improved the ER of JFRCs.
- The use of JF is slightly useful in reducing the volume-related emissions of concrete. However, it can cause significant reductions in STS and FS-related emissions. Up to a 30% net reduction in the STS-related emissions was observed due to the addition of 0.5% JF and 10% SF.

Based on the extensive experimental campaign, the optimum dosage of JF can be 0.3%, which provides the maximum mechanical strength (peak CS, STS, and FS). However, more ductility, toughness and post-peak crack resistance in flexural loading can be attained by using a high JF volume. The durability was compromised with the increasing JF content; however, conjunctive use of JF with mineral admixtures such as SF is recommended to avoid probable durability losses. It is recommended to investigate the pull-out bond behaviour of JF in cement pastes with various water-cement ratios (w/c) and mineral admixtures. The elevated temperature performance and autogenous and drying shrinkage behaviour of JFRCs still need to be examined.

Furthermore, JF is susceptible to volumetric changes, due to the absorption of moisture; therefore, the drying-wetting behaviour of concrete with JF reinforcement must be understood and investigated in future research for applications where concrete is exposed to moisture or outdoor conditions.

Funding: This research received no external funding.

Institutional Review Board Statement: Not applicable.

Informed Consent Statement: Not applicable.

Data Availability Statement: All data shown in the manuscript. In that case, feel free to write whatever statement necessary.

Conflicts of Interest: The author declares no conflict of interest.

References

1. Afroughsabet, V.; Biolzi, L.; Ozbakkaloglu, T. High-performance fiber-reinforced concrete: A review. *J. Mater. Sci.* **2016**, *51*, 6517–6551. [CrossRef]
2. Mardalizad, A.; Caruso, M.; Manes, A.; Giglio, M. Investigation of mechanical behaviour of a quasi-brittle material using Karagozian and Case concrete (KCC) model. *J. Rock Mech. Geotech. Eng.* **2019**, *11*, 1119–1137. [CrossRef]
3. Lee, G.-W.; Choi, Y.-C. Effect of abaca natural fiber on the setting behavior and autogenous shrinkage of cement composite. *J. Build. Eng.* **2022**, *56*, 104719. [CrossRef]
4. Hashmi, A.F.; Shariq, M.; Baqi, A. An investigation into age-dependent strength, elastic modulus and deflection of low calcium fly ash concrete for sustainable construction. *Constr. Build. Mater.* **2021**, *283*, 122772. [CrossRef]
5. Afroughsabet, V.; Biolzi, L.; Monteiro, P.J.M. The effect of steel and polypropylene fibers on the chloride diffusivity and drying shrinkage of high-strength concrete. *Compos. Part B Eng.* **2018**, *139*, 84–96. [CrossRef]
6. Gulzar, M.A.; Ali, B.; Barakat, O.; Azab, M.; Najemalden, A.M.; Salih Mohammed, A.; Alashker, Y. Influence of Jute Fiber on Tensile, Electrical, and Permeability Characteristics of Slag Concrete: A Better, Cheaper, and Eco-Friendly Substitute for Conventional Concrete. *J. Nat. Fibers* **2023**, *20*, 2170947. [CrossRef]
7. Ali, B.; Fahad, M.; Ullah, S.; Ahmed, H.; Alyousef, R.; Deifalla, A. Development of Ductile and Durable High Strength Concrete (HSC) through Interactive Incorporation of Coir Waste and Silica Fume. *Materials* **2022**, *15*, 2616. [CrossRef]
8. Lao, J.-C.; Huang, B.-T.; Xu, L.-Y.; Khan, M.; Fang, Y.; Dai, J.-G. Seawater sea-sand Engineered Geopolymer Composites (EGC) with high strength and high ductility. *Cem. Concr. Compos.* **2023**, *138*, 104998. [CrossRef]
9. Khan, M.; Cao, M.; Ai, H.; Hussain, A. Basalt Fibers in Modified Whisker Reinforced Cementitious Composites. *Period. Polytech. Civ. Eng.* **2022**, *66*, 344–354. [CrossRef]
10. Khan, M.; Cao, M.; Xie, C.; Ali, M. Hybrid fiber concrete with different basalt fiber length and content. *Struct. Concr.* **2022**, *23*, 346–364. [CrossRef]
11. El Ouni, M.H.; Shah, S.H.A.; Ali, A.; Muhammad, S.; Mahmood, M.S.; Ali, B.; Marzouki, R.; Raza, A. Mechanical performance, water and chloride permeability of hybrid steel-polypropylene fiber-reinforced recycled aggregate concrete. *Case Stud. Constr. Mater.* **2022**, *16*, e00831. [CrossRef]
12. Ali, B.; Azab, M.; Ahmed, H.; Kurda, R.; El Ouni, M.H.; Elhag, A.B. Investigation of physical, strength, and ductility characteristics of concrete reinforced with banana (Musaceae) stem fiber. *J. Build. Eng.* **2022**, *61*, 105024. [CrossRef]
13. Ali, B.; Kurda, R.; Herki, B.; Alyousef, R.; Mustafa, R.; Mohammed, A.; Raza, A.; Ahmed, H.; Fayyaz Ul-Haq, M. Effect of Varying Steel Fiber Content on Strength and Permeability Characteristics of High Strength Concrete with Micro Silica. *Materials* **2020**, *13*, 5739. [CrossRef] [PubMed]
14. Ali, B.; Qureshi, L.A. Combined Effect of Fly Ash and Glass Fibres on Mechanical Performance of Concrete. *NED Univ. J. Res. Mech.* **2018**, *15*, 91–100.
15. Raza, S.S.; Qureshi, L.A. Effect of carbon fiber on mechanical properties of reactive powder concrete exposed to elevated temperatures. *J. Build. Eng.* **2021**, *42*, 102503. [CrossRef]
16. Raza, S.S.; Qureshi, L.A.; Ali, B.; Raza, A.; Khan, M.M. Effect of different fibers (steel fibers, glass fibers and carbon fibers) on mechanical properties of reactive powder concrete (RPC). *Struct. Concr.* **2021**, *22*, 334–346. [CrossRef]
17. Hussain, I.; Ali, B.; Akhtar, T.; Jameel, M.S.; Raza, S.S. Comparison of mechanical properties of concrete and design thickness of pavement with different types of fiber-reinforcements (steel, glass, and polypropylene). *Case Stud. Constr. Mater.* **2020**, *13*, e00429. [CrossRef]
18. Kizilkanat, A.B.; Kabay, N.; Akyüncü, V.; Chowdhury, S.; Akça, A.H. Mechanical properties and fracture behavior of basalt and glass fiber reinforced concrete: An experimental study. *Constr. Build. Mater.* **2015**, *100*, 218–224. [CrossRef]
19. Ali, B.; Qureshi, L.A.; Kurda, R. Environmental and economic benefits of steel, glass, and polypropylene fiber reinforced cement composite application in jointed plain concrete pavement. *Compos. Commun.* **2020**, *22*, 100437. [CrossRef]
20. Chan, R.; Santana, M.A.; Oda, A.M.; Paniguel, R.C.; Vieira, L.B.; Figueiredo, A.D.; Galobardes, I. Analysis of potential use of fibre reinforced recycled aggregate concrete for sustainable pavements. *J. Clean. Prod.* **2019**, *218*, 183–191. [CrossRef]
21. Khan, M.; Rehman, A.; Ali, M. Efficiency of silica-fume content in plain and natural fiber reinforced concrete for concrete road. *Constr. Build. Mater.* **2020**, *244*, 118382. [CrossRef]

22. Islam, M.S.; Ahmed, S.J.U. Influence of jute fiber on concrete properties. *Constr. Build. Mater.* **2018**, *189*, 768–776. [CrossRef]
23. Onuaguluchi, O.; Banthia, N. Plant-based natural fibre reinforced cement composites: A review. *Cem. Concr. Compos.* **2016**, *68*, 96–108. [CrossRef]
24. Assaedi, H.; Shaikh, F.U.A.; Low, I.M. Characterizations of flax fabric reinforced nanoclay-geopolymer composites. *Compos. Part B Eng.* **2016**, *95*, 412–422. [CrossRef]
25. Satyanarayana, K.G.; Sukumaran, K.; Mukherjee, P.S.; Pavithran, C.; Pillai, S.G.K. Natural fibre-polymer composites. *Cem. Concr. Compos.* **1990**, *12*, 117–136. [CrossRef]
26. Townsend, T. World natural fibre production and employment. In *Handbook of Natural Fibres*; Elsevier: Amsterdam, The Netherlands, 2020; pp. 15–36.
27. Affan, M.; Ali, M. Experimental investigation on mechanical properties of jute fiber reinforced concrete under freeze-thaw conditions for pavement applications. *Constr. Build. Mater.* **2022**, *323*, 126599. [CrossRef]
28. Zakaria, M.; Ahmed, M.; Hoque, M.M.; Islam, S. Scope of using jute fiber for the reinforcement of concrete material. *Text. Cloth. Sustain.* **2016**, *2*, 11. [CrossRef]
29. Zakaria, M.; Ahmed, M.; Hoque, M.; Shaid, A. A comparative study of the mechanical properties of jute fiber and yarn reinforced concrete composites. *J. Nat. Fibers* **2020**, *17*, 676–687. [CrossRef]
30. Ozawa, M.; Kim, G.-Y.; Choe, G.-C.; Yoon, M.-H.; Sato, R.; Rokugo, K. Thermal properties of jute fiber concrete at high temperature. *J. Struct. Fire Eng.* **2016**, *7*, 182–192. [CrossRef]
31. Sridhar, J.; Gobinath, R.; Kırgız, M.S. Comparative study for efficacy of chemically treated jute fiber and bamboo fiber on the properties of reinforced concrete beams. *J. Nat. Fibers* **2022**, *19*, 12224–12234. [CrossRef]
32. Teixeira, F.P.; de Andrade Silva, F. On the use of natural curauá reinforced cement based composites for structural applications. *Cem. Concr. Compos.* **2020**, *114*, 103775. [CrossRef]
33. Wei, J.; Meyer, C. Utilization of rice husk ash in green natural fiber-reinforced cement composites: Mitigating degradation of sisal fiber. *Cem. Concr. Res.* **2016**, *81*, 94–111. [CrossRef]
34. *ASTM-C150*; Standard Specification for Portland Cement. ASTM International: West Conshohocken, PA, USA, 2018.
35. Ren, G.; Yao, B.; Huang, H.; Gao, X. Influence of sisal fibers on the mechanical performance of ultra-high performance concretes. *Constr. Build. Mater.* **2021**, *286*, 122958. [CrossRef]
36. *ASTM-C231*; Testing Air Content of Concrete with a Type B Pressure Meter. ASTM International: West Conshohocken, PA, USA, 2003.
37. *ASTM-C143*; Standard Test Method for Slump of Hydraulic-Cement Concrete. ASTM International: West Conshohocken, PA, USA, 2015.
38. *ASTM-C39*; Standard Test Method for Compressive Strength of Cylindrical Concrete Specimens. ASTM International: West Conshohocken, PA, USA, 2015.
39. *ASTM-C496*; Standard Test Method for Splitting Tensile Strength of Cylindrical Concrete Specimens. ASTM International: West Conshohocken, PA, USA, 2017.
40. *ASTM-C1609*; Standard Test Method for Flexural Performance of Fiber-Reinforced Concrete (Using Beam with Third-Point Loading). ASTM International: West Conshohocken, PA, USA, 2019.
41. *ASTM-C948*; Standard Test Method for Dry and Wet Bulk Density, Water Absorption, and Apparent Porosity of Thin Sections of Glass-Fiber Reinforced Concrete. ASTM International: West Conshohocken, PA, USA, 2016.
42. *ASTM C1202-19*; Standard Test Method for Electrical Indication of Concrete's Ability to Resist Chloride Ion Penetration. ASTM International: West Conshohocken, PA, USA, 2012.
43. *ASTM C1876-19*; Standard Test Method for Bulk Electrical Resistivity or Bulk Conductivity of Concrete. ASTM International: West Conshohocken, PA, USA, 2019.
44. *ASTM C597-16*; Standard Test Method for Pulse Velocity Through Concrete. ASTM International: West Conshohocken, PA, USA, 2016.
45. Kurda, R.; Silvestre, J.D.; de Brito, J. Life cycle assessment of concrete made with high volume of recycled concrete aggregates and fly ash. *Resour. Conserv. Recycl.* **2018**, *139*, 407–417. [CrossRef]
46. Hájek, P.; Fiala, C.; Kynčlová, M. Life cycle assessments of concrete structures—A step towards environmental savings. *Struct. Concr.* **2011**, *12*, 13–22. [CrossRef]
47. Shah, S.H.A.; Amir, M.T.; Ali, B.; El Ouni, M.H. Mechanical performance and environmental impact of normal strength concrete incorporating various levels of coconut fiber and recycled aggregates. *Environ. Sci. Pollut. Res.* **2022**, *29*, 83636–83651. [CrossRef]
48. Ali, B.; Farooq, M.A.; El Ouni, M.H.; Azab, M.; Elhag, A.B. The combined effect of coir and superplasticizer on the fresh, mechanical, and long-term durability properties of recycled aggregate concrete. *J. Build. Eng.* **2022**, *59*, 105009. [CrossRef]
49. Kumar, B.G.; Muthu, M.; Chajec, A.; Sadowski, Ł.; Govindaraj, V. The effect of silica fume on the washout resistance of environmentally friendly underwater concrete with a high-volume of siliceous fly ash. *Constr. Build. Mater.* **2022**, *327*, 127058. [CrossRef]
50. Alyousef, R.; Ali, B.; Mohammed, A.; Kurda, R.; Alabduljabbar, H.; Riaz, S. Evaluation of Mechanical and Permeability Characteristics of Microfiber-Reinforced Recycled Aggregate Concrete with Different Potential Waste Mineral Admixtures. *Materials* **2021**, *14*, 5933. [CrossRef]

51. Das, C.S.; Dey, T.; Dandapat, R.; Mukharjee, B.B.; Kumar, J. Performance evaluation of polypropylene fibre reinforced recycled aggregate concrete. *Constr. Build. Mater.* **2018**, *189*, 649–659. [CrossRef]
52. Wu, Z.; Shi, C.; Khayat, K.H. Influence of silica fume content on microstructure development and bond to steel fiber in ultra-high strength cement-based materials (UHSC). *Cem. Concr. Compos.* **2016**, *71*, 97–109. [CrossRef]
53. Yang, Y.; Wu, C.; Liu, Z.; Wang, H.; Ren, Q. Mechanical anisotropy of ultra-high performance fibre-reinforced concrete for 3D printing. *Cem. Concr. Compos.* **2022**, *125*, 104310. [CrossRef]
54. Raza, S.S.; Amir, M.T.; Azab, M.; Ali, B.; Abdallah, M.; El Ouni, M.H.; Elhag, A.B. Effect of micro-silica on the physical, tensile, and load-deflection characteristics of micro fiber-reinforced high-performance concrete (HPC). *Case Stud. Constr. Mater.* **2022**, *17*, e01380. [CrossRef]
55. Razmi, A.; Mirsayar, M.M. On the mixed mode I/II fracture properties of jute fiber-reinforced concrete. *Constr. Build. Mater.* **2017**, *148*, 512–520. [CrossRef]
56. Xie, J.; Huang, L.; Guo, Y.; Li, Z.; Fang, C.; Li, L.; Wang, J. Experimental study on the compressive and flexural behaviour of recycled aggregate concrete modified with silica fume and fibres. *Constr. Build. Mater.* **2018**, *178*, 612–623. [CrossRef]
57. Ali, B.; Raza, S.S.; Kurda, R.; Alyousef, R. Synergistic effects of fly ash and hooked steel fibers on strength and durability properties of high strength recycled aggregate concrete. *Resour. Conserv. Recycl.* **2021**, *168*, 105444. [CrossRef]
58. McDonald, D.B. The Rapid Chloride Permeability Test and Its Correlation to the 90-Day Chloride Ponding Test. *PCI J.* **1994**, *39*, 38–47.
59. Kou, S.; Poon, C.; Agrela, F. Comparisons of natural and recycled aggregate concretes prepared with the addition of different mineral admixtures. *Cem. Concr. Compos.* **2011**, *33*, 788–795. [CrossRef]
60. Qureshi, L.A.; Ali, B.; Ali, A. Combined effects of supplementary cementitious materials (silica fume, GGBS, fly ash and rice husk ash) and steel fiber on the hardened properties of recycled aggregate concrete. *Constr. Build. Mater.* **2020**, *263*, 120636. [CrossRef]
61. *BIS 13311-1*; Method of Non-Destructive Testing of Concret, Part 1. BIS: Basel, Switzerland, 1992.
62. Jena, B.; Patra, R.K.; Mukharjee, B.B. Influence of incorporation of jute fibre and ferrochrome slag on properties of concrete. *Aust. J. Civ. Eng.* **2022**, *20*, 13–30. [CrossRef]
63. Hong, G.; Oh, S.; Choi, S.; Chin, W.-J.; Kim, Y.-J.; Song, C. Correlation between the compressive strength and ultrasonic pulse velocity of cement mortars blended with silica fume: An analysis of microstructure and hydration kinetics. *Materials* **2021**, *14*, f2476. [CrossRef] [PubMed]
64. Adenaert, K. Transport Mechanismen in Zelfverdichtend Beton in Relatie Met Carbonatatie en Chloride Penetratie. Ph.D. Thesis, Ghent University, Ghent, Belgium, 2006.
65. Afroughsabet, V.; Ozbakkaloglu, T. Mechanical and durability properties of high-strength concrete containing steel and polypropylene fibers. *Constr. Build. Mater.* **2015**, *94*, 73–82. [CrossRef]

sustainability

MDPI

Article

Rheological Characterization of Ground Tire Rubber Modified Asphalt Binders with Parallel Plate and Concentric Cylinder Geometries

Salih Kocak

Construction Management, University of West Florida, Pensacola, FL 32514, USA; skocak@uwf.edu

Abstract: Recently, scrap tire rubber-modified asphalt binders and pavements have been the preferred choice of state DOTs and parties involved due to the desirable engineering, as well as economic and environmental impacts. Rheological and mechanical properties of rubber modifications have been the main focus of researchers for the last couple of decades. This paper investigates the rutting potential, fatigue cracking resistance, and continuous performance grade (CPG) changes of waste tire rubber-modified, original, and aged asphalt binders. The CPG of asphalt binders is determined at high, intermediate, and low temperatures. A Delta T Critical comparison of the binder was carried out to establish a relationship between measured parameters. Linear amplitude sweep (LAS) tests at equi-stiffness temperatures were conducted to discover the fatigue life of all binders while the multiple stress creep recovery test is performed to assess the high-temperature rutting performance of asphalt binders as per the Superpave performance grading system at accepted regional (58 °C) as well as high PG temperatures. In addition, parallel-plate geometry and concentric cylinder geometry were used with the Multiple Stress Creep Recovery (MSCR) test to discover the impact of discrete particles available in crumb/ground tire rubber-modified asphalt binders as per standards. The results show that rubber modifications improved the base binder's rutting resistance and continuous PGs without adversely affecting the fatigue cracking resistance. Based on the mathematical expressions developed, 2.71%, 7.82%, 12.94%, and 18.05% (by weight of binder), GTR modifications improved the high PG of the modified binders one, two, three, and four grade bumps, respectively. Similar linear correlations with R^2 0.872 and 0.6 were established for continuous low and intermediate PGs, respectively. MSCR test results revealed that both 9% and 20% GTR modifications were achieved to enhance the H-grade traffic level of the original binder to E-grade.

Keywords: ground tire rubber; parallel plate geometry; concentric cylinder geometry; MSCR; LAS; continuous performance grade; Delta T Critical

Citation: Kocak, S. Rheological Characterization of Ground Tire Rubber Modified Asphalt Binders with Parallel Plate and Concentric Cylinder Geometries. *Sustainability* **2023**, *15*, 2880. https://doi.org/10.3390/su15042880

Academic Editor: Ahmed Salih Mohammed

Received: 12 January 2023
Revised: 2 February 2023
Accepted: 3 February 2023
Published: 5 February 2023

1. Introduction

Asphalt modifications to enhance the load and weather-related performance of pavements have been a common practice used by agencies nowadays. Ground tire rubber (GTR) has been one of the most applied asphalt modifiers even though there are many virgin and recycled asphalt modifiers in the market. The modified asphalt costs more than the original one, and the actual price changes depending on the amount as well as on the type of the modifier. Based on the literature, rubber-modified asphalts can cost 10% to 30% higher than conventional ones [1–4] while polymer-modified asphalts can cost twice as much [5]. However, the life cycle cost of modified asphalts, in general, is lower due to enhanced performance and fewer maintenance cycles required [6,7]. Rubberized asphalt can achieve as good as or even better performance than pricey polymer-modified binders, making GTR a cost-effective alternative to polymers [8–10].

Around one billion scrap tires are generated globally every year [11]. The state of California produces over 50 million waste tires by itself [12]. When they are not properly discarded in landfills, the scrap tires are stockpiled or illegally dumped on the land, creating fire hazards or housing for disease-carrying mosquitos, insects, and rodents [13]. To overcome such issues, state DOTs are searching for alternative ways to handle waste scrap tires. In one attempt, California state law obligates Caltrans to use GTR in 35% of all its asphalt paving projects. Moreover, it requires at least 20% rubber in the asphalt binder used in surface courses [12]. As a result of all these requirements, it is estimated by the California Department of Resources Recycling and Recovery (CalRecycle, Sacramento, CA, USA) that more than 35,000 tons of crumb rubber were integrated into paving activities in 2018. In addition, less than 25,000 waste scrap tires are anticipated in the stockpiles of California [14].

Other than the above-mentioned engineering, economic, and environmental benefits, some other added-value benefits of using GTR in asphalt pavements can be found in the literature. Some of them can be listed as reducing tire pavement interaction noise [15,16], decreasing tire wear [17], providing better ride quality [18], increasing longevity [19], softening stiff binders in high RAP pavements [20], improving aging properties [21], decreasing ice retention [22], and increasing resistance to reflective cracking [23]. This study aims to fill the gap in the literature by investigating the rheological and mechanical properties of wet-process GTR-modified asphalt binders at various percentages using the CPG and recently introduced binder performance tests such as Delta T Critical, LAS, and MSCR. As the state DOTs have been in the transition phase to a new binder grading system, this paper intends to establish an understanding between the old Superpave high-temperature specification AASHTO M320 and the new MSCR specification AASHTO M322 for GTR-modified asphalt binders. It also examined one of the main concerns regarding the use of parallel plate geometry for binders with discrete particles, such as rubber-modified binders, by comparing it with concentric cylinder geometry. The findings and correlations between various parameters studied can serve as the basis for future researchers and binder modifiers in this field.

2. Objectives and Scope

There were a few objectives of this study. The major objective was to evaluate the rutting potential and fatigue cracking resistance of GTR-modified asphalt binders along with the continuous low-, intermediate-, and high-performance grades. Another objective was to establish correlations between the percent GTR modifier and all three continuous grading temperatures. Yet, another objective was to investigate the impact of testing geometry for binders with discrete particles as per specifications. For this purpose, MSCR tests of modified and original binders were conducted by using both testing geometries, namely parallel-plate and concentric cylinder. The scope of the study covered 3%, 6%, and 9% GTR modification of the original PG58-28 binder along with the commonly used 20% GTR modification in rubberized asphalt pavements. The performance of the modified asphalt binders relative to the original binder was tested using Delta T Critical, LAS, and MSCR tests along with fundamental Superpave binder tests.

3. Materials and Methods

Asphalt binders with an original PG58-28 commonly used in the Midwest region of the United States was selected as the base binder for all modifications and testing. Basic information regarding the original asphalt binder is provided in Table 1.

GTR particles are produced at ambient temperature with the cracker mill process (CMP), which is the most common GTR production technique. CMP uses shredded tire pieces instead of full-size tires. Before shredding and further size reduction, fiber reinforcements and steel belting are removed from the tire bodies using a series of fiber and steel separators. In this process, scrap tire pieces pass between rotating corrugated steel drums for size reduction. The spacing and the differential speed of the drum pairs control the tearing of the scrap tires. GTR produced with this process has irregular shapes with larger surface areas. Particle sizes can be achieved over a range of 425 microns to 4.75 mm. Mesh size #20 acquired with the cracker mill process at ambient temperature is utilized in this study. According to the sieve analysis results, almost 100% of the GTR particles passed the #16 sieve, a little less than 50% were able to pass the #30 sieve, and only 2.3% of the rubber particles were retained in the #100 sieve.

Table 1. Basic properties of the original/neat/base binder.

Property	Value	Unit
Continuous High Grade	60.1	°C
Continuous Intermediate Grade	17	°C
Continuous Low Grade	−29.1	°C
Viscosity at 135 °C & 170 °C	285 & 68	cP
$\|G^*\|$ at 10 °C & 70 °C	3,470,167 & 238	Pa
Phase Angle at 10 °C & 70 °C	62.6 & 88.8	degrees
Flash Point	312	°C

Ground Tire Rubber Modification Process

There are various GTR modification methods in the literature. The modification method applied in this study is adapted from one of the GTR manufacturers in the industry, which is similar to the ones researchers have been using in the literature as well [24]. In this process, only low-shear mixing is used. It was conducted using a benchtop blender with a boat motor-type propeller as shown in Figure 1a–c, which illustrates the 20-mesh size ground tire rubber particles used in this research and a sample of a GTR-modified asphalt binder in a concentric cylinder geometry, respectively.

Figure 1. (**a**) Low-shear mixer with heating mantle. (**b**) GTR particles. (**c**) GTR-modified asphalt binder in a CC geometry.

Rubber particles are slowly added into the preheated asphalt binder at 180 °C while the mixer is set to rotate at 1000 rpm. The addition of the GTR is completed within 5 min. the binder–rubber blend is mixed for 60 min at the same rpm value unless a vortex is formed on the surface of the mixture. Otherwise, the speed is adjusted to ensure vortex-free surface mixing. The temperature of the mixing system is kept at around 180 °C by using an adjustable heating mantle for a gallon can. Circulatory heating oil baths and hot plates can be used as well depending on the size of the containers. It is important to leave some space in the containers to house the swollen rubber particles. Once the mixing process is completed, the GTR-modified asphalt binder is transferred to an oven at 163 °C for 16 h in an oxygen-free condition to complete the reaction between binder and rubber particles. This process is also known as static aging. At least two batches were prepared per modification for the repeatability of the modification process. The calculations showed the consistency of the modification process by providing minimal variations.

4. Tests, Results, and Discussion

PG, ΔTc, LAS, and MSCR tests and calculations were conducted to understand the relative performance of GTR-modified asphalt binders compared to each other and the original/neat binder. Statistical methods were applied to the test results to establish relationships between modifications and certain useful parameters such as continuous performance and traffic grades. All tests were carried out on a minimum of two replicates per binder batch. Hence, at least four samples per binder (either original or modified) were tested for each test type. The basic descriptive statistical values for the test results were computed and presented along with the data.

4.1. Continuous High-, Intermediate-, and Low-Performance Grade Tests and Results

The major focus of this research was to determine the change in all continuous grading temperatures of GTR-modified asphalt binders relative to the base binder as the PG is a basic step for binder classification based on AASHTO M320. The original/neat/base binder was denoted with "0%" in the test results since there were no modifiers incorporated. Similarly, the aged-base/original binder was denoted with "A-0%". The aged-base binder was prepared by only exposing the original binder to the modification process (i.e., heating and low-shear mixing) without adding any modifier to illustrate the aging effect during the binder modification process. Continuous PGs are selected in lieu of discrete PGs since they provide more information and precise data about the modifications. Additionally, they are a better fit for research that includes modifications. Asphalt binder producers, especially, use CPG values to achieve cost-effective binder modifications. In the literature, there are researchers who apply the CPGs to determine the required amount of asphalt modifiers, such as styrene-butadiene-styrene (SBS) polymers, devulcanized rubber, or their combinations to acquire the asphalt binders with desired PG values [25,26]. In addition, some other researchers benefitted the CPG concept to compare the accuracy of testing equipment [27]. The linear interpolation method stated at ASTM D7643 was implemented to determine the continuous grades even though there were other methods such as nonlinear, parabolic, and exponential curve fitting, used by researchers to calculate the continuous grades in the literature. Based on the information provided in ASTM D7643, a 3% GTR-modified binder has continuous grades of 66.2–31.32 (15.3) where 66.2 °C is the continuous high grade (CHPG), −31.32 is the continuous low grade (CLPG), and 15.3 is the continuous intermediate grade (CIPG).

The linear interpolation between the absolute highest passing and lowest failing temperatures as per ASTM D7643 was applied to the data to find the critical temperatures. There are two equations provided in the standards to determine the continuous grades. Both equations are developed based on the two-point linear relations between the test results and temperatures. Equation (1) uses the log_{10} scale for the test results and the arithmetic scale for test temperatures, and it is applicable for all test results other than the m-value, whereas Equation (2) uses the arithmetic scale for both parameters, and it is only applied to determine the m-value-based critical temperature for CLPG:

$$T_c = T_1 + \left(\frac{log_{10}(P_s) - log_{10}(P_1)}{log_{10}(P_2) - log_{10}(P_1)} \right) \times (T_2 - T_1) \quad (1)$$

$$T_c = T_1 + \left(\frac{P_s - P_1}{P_2 - P_1}\right) \times (T_2 - T_1) \quad (2)$$

where:

T_c = Continuous grading temperature, °C;
T_1 = Lower test temperature, °C;
T_2 = Higher test temperature, °C;
P_s = Required value for failing criteria based on the specifications;
P_1 = Test result at T_1;
P_2 = Test result at T_2.

Table 2 summarizes the process of determining the continuous high-grading temperature by linear interpolation method for 3% GTR-modified asphalt binder. The left part of the table gives information about the original binder measurements while the right part shows the test results for RTFO-aged binder along with some basic descriptive statistical values such as average (AVR), standard deviation (Stdev), and coefficient of variation (COV). Lastly, the bottom row provides the final continuous grading temperature (T_C-final) in degrees Celsius, which is the lower of T_C-neat and T_C-rtfo. To determine the CLPGs, RTFO-aged binders were further aged inside the pressurized aging vessel (PAV) to simulate the long-term/oxidative aging that occurs under field conditions. Before preparing the bending beam rheometer (BBR) samples, PAV-aged binders were degassed in a vacuum oven to minimize/eliminate the air bubbles introduced during the aging process. The BBR test provides the low-temperature relaxation and stiffness properties of asphalt binders. Those properties are used to examine the thermal cracking resistance of the asphalt binders. Based on the low PG of the unmodified binder, which was −28 °C, BBR tests were conducted at −12 °C, −18 °C, and/or −24 °C until the failure took place. As per the specifications, the failure occurred when either the stiffness was greater than 300 MPa or the m-value was smaller than 0.300 at the 60th second.

Table 2. Determining the continuous high grading temperature of 3% GTR-modified binder.

	GTR 3% Original			GTR 3% RTFO-Aged			
	Average of Run 1	**Average of Run 2**	**AVR [1]**		**Average of Run 1**	**Average of Run 2**	**AVR [1]**
T_C-neat (°C)	64.28	63.64	63.96	T_C-rtfo (°C)	64.75	65.01	64.87
T_1 (°C)	64.0	58	Stdev [2]	T_1 (°C)	64.0	64.0	Stdev [2]
T_2 (°C)	70.0	64.0	0.46	T_2 (°C)	70.0	70.0	0.17
P_1 (kPa)	1.03	1.97	COV [3]	P_1 (kPa)	2.4	2.47	COV [3]
P_2 (kPa)	0.548	0.956	0.71%	P_2 (kPa)	1.2	1.23	0.26%
P_s (kPa)	1.0	1.0		P_s (kPa)	2.2	2.2	
T_C-final (°C)				63.96			

AVR [1]: average, Stdev [2]: standard deviation, COV [3]: coefficient of variation.

Table 3, which illustrates the process of calculating the CLPG, presents the stiffness and m-value at the 60th second recorded during BBR testing along with the application of Equations (1) and (2). The last row in Table 3 shows the continuous low temperatures (T_C), which are the absolute smallest of T_C-s and T_C-m. It is important to note that continuous low temperatures (T_C, T_C-s, and T_C-m) are not the same as continuous low PG temperatures. The CLPGs can be acquired by adding −10 °C to the calculated continuous low temperatures as stated at AASHTO M320.

The third continuous performance grade calculation was performed to discover the CIPG of the binders. Even though the CIPG is not a part of PG designation, it is found to evaluate the fatigue cracking behavior of the binders. CIPG is known to be the weakest chain of the Superpave binder performance grading. It lacks either a theoretical or a practical basis. Intermediate PG was established as the arithmetic average of the high and low PG difference plus 4 °C according to AASHTO MPI-93. As an example, for the PG 58−28 grade binder, the intermediate PG can be calculated as (58−28) + 4 = 19 °C, which does not have any mechanistic foundations. Unlike high- and low-performance grades, the intermediate PG test temperatures change at 3 °C intervals. The lower the intermediate PG is, the more flexible the asphalt binder is, and it performs better in fatigue cracking resistance.

Table 3. Continuous low-temperature grade construction of 3% GTR-modified binder.

GTR 3% Stiffness (S-Value) Based				GTR 3% Slope (m-Value) Based			
	Average of Run 1	**Average of Run 2**	**AVR**		**Average of Run 1**	**Average of Run 2**	**AVR**
T_C-s (°C)	−20.03	−20.04	−20.04	T_C-m (°C)	−20.59	−20.36	−20.48
T_1 (°C)	−18	−18	Stdev	T_1 (°C)	−18	−18	Stdev
T_2 (°C)	−24	−24	0.46	T_2 (°C)	−24	−24	0.16
P_1 (MPa)	231	230	COV	P_1 (MPa)	0.319	0.313	COV
P_2 (MPa)	501	503	1.53%	P_2 (MPa)	0.275	0.280	0.53%
P_s (MPa)	300	300		P_s (MPa)	0.300	0.300	
T_C (°C)				−20.04			

Table 4 provides the results of continuous high-, intermediate-, and low-performance grades of neat and modified binders in a summary form. It is worth re-emphasizing that the A-0% binder underwent the heating and mixing process with no modifier to simulate the impact of aging that occurred during the modifications. While it improved the CHPG by 1.8 °C, it, as expected, worsened both CIPG and CLPG by 0.9 °C and 2.2 °C, respectively. Based on the measured data, the impact of the aging was strong enough to change the PG of the base binder from 58–28 to 58–22. Regardless of the modification type and amount, the original binder DSR testing protocol became the governing testing to determine the CHPG of the binders over the RTFO-aged DSR testing protocol. Similarly, m-value-based continuous low-grading temperatures were dominant for all asphalt binders tested other than the neat and 3% GTR-modified ones.

Table 4. Results of continuous high-, intermediate-, and low-performance grades.

Modification Type	**Continuous High PG (°C)**			**Continuous Inter. PG (°C)**	**Continuous Low PG (°C)**				**CHPG-CLPG**
	CH PG	**Original PG**	**RTFO Aged**	**CIPG PAV Aged**	**CLPG**	**S-Based**	**m-Value Based**	**Δ*Tc***	**(°C)**
0%	60.1	60.1	61.2	17.0	−29.1	−29.1	−30.6	1.5	89.2
A-−0%	61.9	61.9	62.5	17.9	−26.9	−28.4	−26.9	−1.5	88.8
3% GTR	64.0	64.0	64.9	15.3	−30.0	−30.0	−30.5	0.5	94.0
6% GTR	68.6	68.6	69.3	14.2	−30.7	−31.3	−30.7	−0.6	99.3
9% GTR	72.3	72.3	74.3	14.6	−30.2	−32.4	−30.2	−2.2	102.5
20% GTR	83.7	83.7	86.4	13.9	−31.9	−34.8	−31.9	−2.9	115.6

Figure 2a demonstrates the impact of the GTR modification on CHPG. As the amount of GTR modifier increased, the CHPG of the modified binders improved as well. The 3% and 6% GTR modifications achieved a one-grade bump while the 9% and 20% GTR modifications accomplished two-grade and four-grade bumps, respectively.

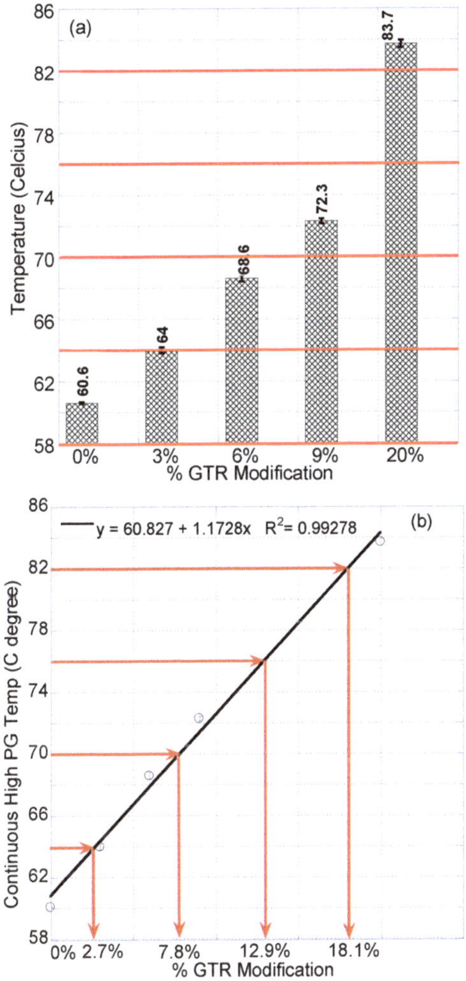

Figure 2. (**a**) Impact of GTR modification on CHPG. (**b**) Correlation between modifications and CHPG.

To determine the exact amount of GTR modifier to attain the desired grade bump, the correlation established in Figure 2b between the percent modifier and CHPG can be used. Linear correlation with R^2 of 0.993 implies a strong relationship between the parameters investigated. Based on the mathematical expression formulated during the linear curve fitting process, around 7.82% (by weight of binder) GTR is required to acquire two-grade bumps, and 12.94% rubber is needed to reach three-grade bumps.

Different modifiers are generally used to decrease the low PG or to increase the high PG of the asphalt binders. Multiple modifiers can be used simultaneously in the same asphalt binder to enhance both the high and low PGs. Even though GTR modification of the binder is generally conducted to increase the high PG of the binder to provide better resistance to rutting without adversely affecting the low and intermediate PGs, the results reveal that any GTR modification improved not only CHPGs but also CIPGs and CLPGs with a clear trend. The linear relationships between the percent modifiers and CIPG-CLPG were established to determine useful mathematical expressions.

Figure 3a,b illustrates the relationship between GTR modifications versus CLPG and CIPG, respectively. Linear correlation between percent GTR and CLPG was established with R^2 equal to 0.872. Similarly, a linear correlation between the percent GTR and CIPG was established with R^2 equal to 0.6. Despite being simple, yet reasonable, correlations for the asphalt binders, the relationship can be better represented by using different curve fitting options and by enhancing the testing matrix. Compared to the intermediate PG of the base binder, which was 19 °C, all modified binders resulted in one grade bump by lowering the intermediate PG to 16 °C. The highest improvement was attained at 20% GTR modification by lowering the CIPG 3.1 °C compared to the original binder.

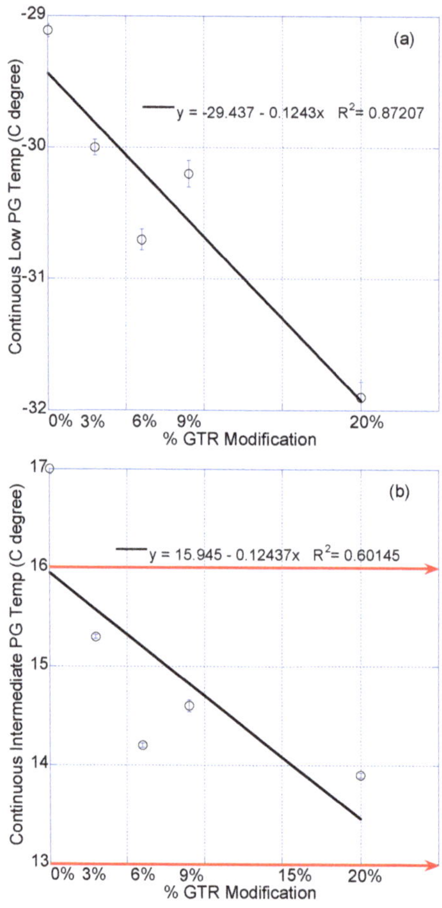

Figure 3. Linear correlations between (**a**) GTR modifications and CLPG; (**b**) GTR modifications and CIPG.

4.2. Delta T Critical (Delta Tc or ΔTc) Analysis

Delta T_C (ΔT_C-Delta T critical) was another parameter calculated during the CLPG determinations. It is used as an indicator of the impact of aging and modifiers on binder rheology [27]. It is a parameter state DOTs either have implemented or have been considering implementing into their specifications as another PG+ binder test. ΔT_C is used to understand the asphalt binders' relaxation properties, which are a sign of nonload-related cracking or aging-related embrittlement distress in the flexible pavements. The parameter is computed using the creep stiffness and rate from BBR testing for any asphalt binder including neat binders, binders with additives, modified binders, and recovered binders although it was primarily developed for neat asphalt binders [28]. However, researchers should be cautious against using ΔT_C for polymer-modified binders since it is generally accepted that polymer-modified asphalt binders show reduced laboratory aging (RTFO and PAV) due to their higher viscosities. The reduced aging can create a favorable condition for stiffness by causing lower T_C-s values [29]. There is ongoing research to investigate polymer-modified asphalt binders and their relationship with ΔT_C. Further research is needed to establish a better understanding between GTR-modified asphalt binders and ΔT_C as well since the aging characteristics, especially RTFO aging, of GTR-modified binders can show slight variations as described in the literature.

First conceptualized by Anderson et al. in 2011, ΔTc is a relatively new parameter that has been gaining popularity due to its ability to quantify the aging propensity of asphalt binders as well as to correlate it directly with certain types of pavement distress, such as block cracking [28–31]. The same procedure explained in the computation of the continuous low-grading temperature section, which is presented in Table 3, is followed to find the critical temperatures based on creep stiffness (*Tc-s*) and creep rate (*Tc-m*) at the end of the 60th second. Once these parameters are calculated, Equation (3) is used to calculate the ΔTc:

$$\Delta T_c = T_{C-S(60s)} - T_{C-m(60s)} \tag{3}$$

Depending on the *Tc-s* and *Tc-m* values, ΔTc can be positive or negative. $+\Delta Tc$ is an indicator of the fact that the binder's low PG is governed by the creep stiffness while $-\Delta Tc$ indicates that the low PG of the binder is controlled by the creep rate. In addition, the magnitude of the parameter provides information about the level of dominatΔTcion by stiffness or creep rate. In addition, a warning limit of −2.5 °C and a failure limit of −5.0 °C are suggested for 20-h PAV-aged asphalt binders. These limits are strongly related to the basic understanding of the parameter. As binders age, *Tc-m* increases quicker than *Tc-s* and makes the binders more m-controlled. The m-controlled binders are more brittle and less able to relax the applied stresses, hence more prone to cracking distress. Anderson et al. showed that there is a strong correlation between fatigue cracking (top-down cracking) and some other pavement distresses related to poor relaxation properties and higher negative values of ΔTc, although the ΔTc parameter was derived from the BBR testing, which was developed for thermal cracking [28,31].

ΔTc values for modified as well as the aged and original binders are provided in Table 4. As the amount of aging and GTR modification increased, the ΔTc parameter became more and more m-value controlled. While the ΔTc of the original binder 0% was 1.5 °C, it became −1.5 °C for the A-0% sample and 0.5 °C, −0.6 °C, −2.2 °C, and −2.9 °C for 3%, 6%, 9%, and 20% GTR modifications, respectively. Figure 4 illustrates two linear correlations between ΔTc and percent GTR modification. The only difference between these two correlations is the inclusion of 20% GTR modification results. The reason to construct relationships with and without the 20% GTR is the fact that not all binders have linear m-value or log-linear S-value with temperature. Moreover, as the amount of rubber modifier increases, the stiffness values become smaller. In the case of 20% GTR, the stiffness value was less than 300 MPa at any BBR temperature when Tc-m < 0.300 or Tc-m > 0.300, which was a different behavior than any other modified and original binders. Hence, it was decided to provide two linear correlations with and without a 20% GTR-modified binder. It is worth noting that the author established linear correlations for simplicity by assuming linear relationships between the m-value and temperature. Although both linear relations have strong correlations, a better curve fitting other than a linear one may be possible with the inclusion of 20% GTR modification. As can be seen from Figure 4 the first correlation with R^2 of 0.987 omitted the 20% GTR modification, and the second one with R^2 of 0.845 included it.

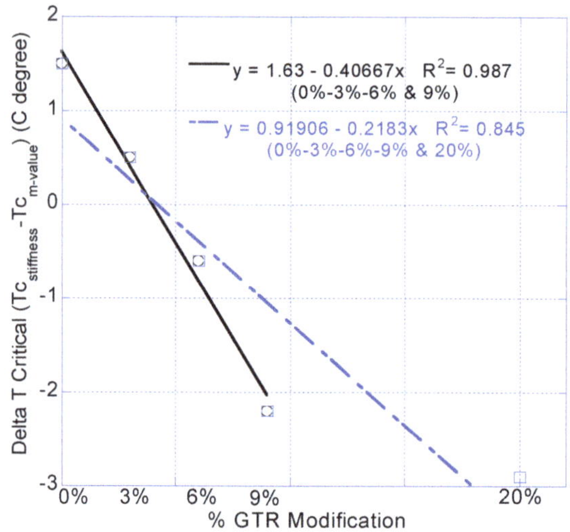

Figure 4. Linear correlations between ΔTc and percent GTR modifications.

4.3. Fatigue Cracking Resistance Test and Results

Superpave binder tests and the PG system were initially developed to find out the climatic and load-related performance of asphalt binders for certain pavement distresses. While high PG was developed for the rutting resistance, low PG was established for the thermal cracking of asphalt pavement. Similarly, the intermediate PG of the binders was created to provide information about the fatigue life of the asphalt binders. Although low and high PGs quite successfully estimated the relative performance of unmodified binders for resistance to low-temperature cracking and rutting, respectively, intermediate PG was not as successful to determine the fatigue cracking resistance of asphalt binders since it lacked any theoretical and practical fundamentals. Moreover, the recent changes in binder production practices as well as the need for binder modifications to prevent the early distress accumulation in the asphalt pavements as a result of heavier traffic volumes developed parallel to the exponentially growing world population in recent decades made Superpave binder tests obsolete and gave rise to the development of newer test methods to better understand their mechanistic behavior. In an effort, a viscoelastic continuum damaged (VECD) mechanics-based linear amplitude sweep (LAS) test was developed to determine asphalt binders' resistance to fatigue cracking. It is designed to evaluate the fatigue resistance ability of asphalt binders under cyclic loading with increasing amplitudes. LAS test can estimate the number of cycles to fatigue failure (N_f) as a function of strain developed in the asphalt pavement at a 35% reduction in initial modulus.

Theoretically, the LAS test has two parts. The first part, frequency sweep tests, is performed to acquire the undamaged rheological properties of the asphalt binders. In this stage, the sample was tested at a set of standard specified frequencies between 0.2 Hz and 30 Hz at 0.1% strain amplitude. This stage is important to derive the damage analysis parameter alpha (α). The second part of the tests is performed to determine the fatigue characteristics of asphalt binders under oscillatory shear strain-controlled mode with linearly increasing strain amplitude between 0.1% to 30% at a constant 10 Hz frequency for a total of 3100 loading cycles. The same binder specimen is used in both stages since the frequency sweep test does not create damage to the sample. LAS can be performed with 8-mm parallel-plate geometry and a 2-mm working gap using readily available DSR equipment on either RTFO- or PAV-aged asphalt binders as per AASHTO- T391 (previously TP101) specifications. In this study, PAV-aged, degassed asphalt binders were used. The same specification recommends testing temperature as the intermediate pavement temperature based on the PG grades of the binder. However, this suggestion lacks a mechanistic basis as stated in detail at CIPG calculations. According to this recommendation, base binder, 3% GTR-, 6% GTR-, 9% GTR-, and 20% GTR-modified binders would have been tested at 19 °C, 22 °C, 22 °C, 25 °C, and 28 °C, respectively. Many researchers experienced the same problem with the LAS testing temperature. They applied different methods and mechanistic approaches to establish a relation for testing temperature. Some researchers suggested using a certain temperature, such as 25 °C, as it corresponds to the local intermediate temperature of the region of interest [32], while some of them selected a typical representative intermediate temperature such as 20 °C [33]. Some others recommended the testing temperature as the average climatic PG minus 4 °C based on the linear viscoelastic range of the asphalt binders [34]. Further studies showed that the linear viscoelastic approach was suitable to select the LAS temperatures. It was suggested that LAS temperatures should be selected between the temperatures corresponding to $|G^*| < 60$ MPa at 10 Hz (approximately 2.5 MPa at 10 rad/s) to prevent excessive brittleness and adhesive failures between parallel-plate geometry of DSR and binder specimen and to $|G^*| > 60$ MPa at 10 Hz (about 25 MPa at 10 rad/s) to prevent the bulging and geometry changes [35]. Yet, some other researchers proposed using temperatures at which the iso-stiffness condition occurs, such as $|G^*| \times \sin(\delta) = 6.5$ MPa [36]. Typically, the LAS results at 2.5% and 5.0% strain levels are presented considering the pavement layer stiffness. The general approach is to use a 5.0% strain level for asphalt layers thinner than 4 inches and a 2.5% strain level for asphalt layers thicker than 4 inches [35].

In this study, the authors selected the LAS temperatures at which $|G^*| \times \sin(\delta) = 5000$ kPa. These temperatures are also known as the equi-stiffness temperatures or the temperatures equal to the CIPGs of the binders. LAS temperatures determined using this approach satisfied the linear viscoelastic range of asphalt binder tests. Visual inspections also confirmed that there was no bulging or excessive brittleness during the testing.

Figure 5 illustrates the findings of the LAS tests of GTR-modified asphalt binders as well as the original binder at different strain levels. Data analysis of the LAS test results was performed at 1%, 3%, 5%, and 8% applied strain levels. The aim of selecting numerous strain levels was to better understand the fatigue performance of thinner overlays, such as 2-inch, and thicker asphalt pavements, such as 18-inch-thick multi-lift pavements, other than only focusing on the 4-inch thickness. Analysis of LAS results revealed that 3% GTR modification shortened the fatigue life of the asphalt binder at all strain levels tested. However, as the strain level increased from 1% to 8%, the difference in fatigue life diminished. At 8% strain level, both original and 3% GTR-modified were statistically indifferent. The 6% GTR modification improved the fatigue life at smaller strain levels compared to the 3% GTR-modified binder; however, it did not enhance the fatigue life at higher strain levels. Similarly, 9% GTR modification improved fatigue life at 1% and 3% strain levels compared to 6% GTR modification. Ultimately, 20% GTR modification almost reached three times more fatigue life at a 1% strain level than the original binder. Both 20% GTR modified and original binders had almost the same fatigue life cycles at a 3% strain level. However, at higher strain levels, 20% GTR modification did not provide better fatigue life than the original binder. It can be summarized that as the GTR percentage increased, the fatigue life of the asphalt binders enhanced considerably at lower strain levels and worsened at higher strain levels.

Figure 6a–d illustrates the relationship between the number of fatigue cycles to failure and the percent GTR modification for 1%, 3%, 5%, and 8% strain levels, respectively. Linear correlations were established between the parameters investigated at four different strain levels. R^2 values of the linear relationships for all strain levels changed between 0.9 and 0.99 implying strong correlations. While there was a positive correlation between the fatigue live and GTR percentages at 1% and 3% strain levels, a negative correlation occurred between the same parameters at 5% and 8% strain levels.

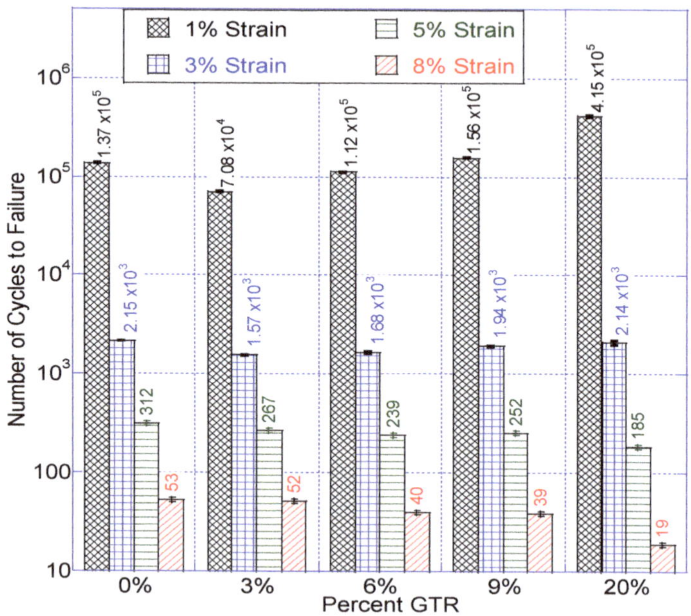

Figure 5. LAS test results for different strain levels at CIPG temperatures.

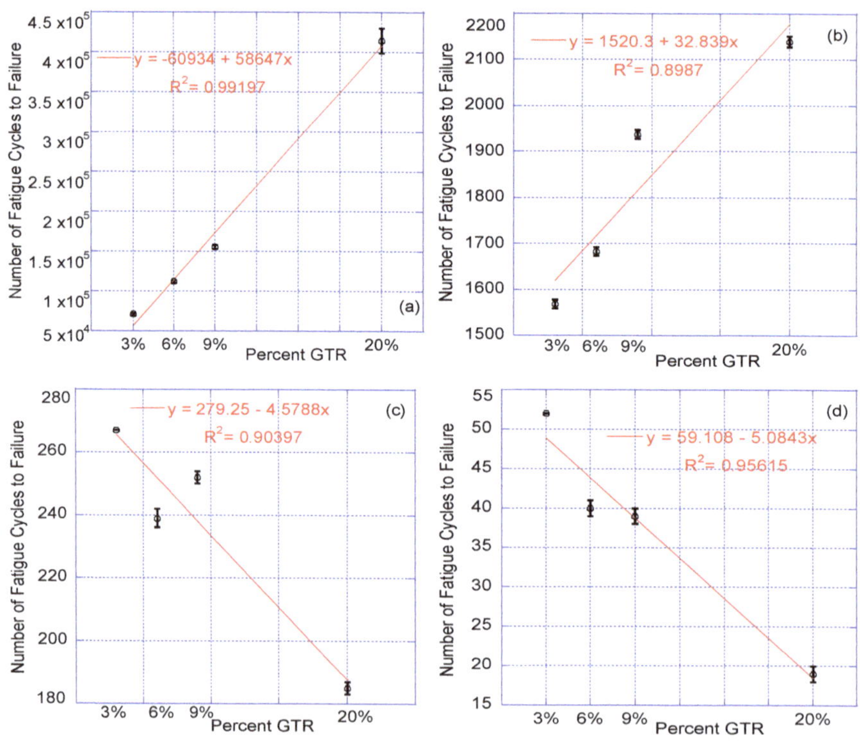

Figure 6. Fatigue life vs. GTR modification at (**a**) 1% strain, (**b**) 3% strain, (**c**) 5% strain, (**d**) 8% strain.

4.4. MSCR Rutting Resistance Test Results with Parallel Plate and Concentric Cylinder Geometries

The original Superpave performance grading system used a dissipated strain energy approach to determine the rutting performance of the asphalt binders. In this approach, the $|G^*|/\sin(\delta)$ parameter for original and RTFO-aged binders was limited to calculate the high PG temperature grade. The $|G^*|/\sin(\delta)$ parameter became inadequate to accurately assess the rutting potential of asphalt binders with the recent changes in the binder production practices and introduction of modified binders. Deficiencies related to the $|G^*|/\sin(\delta)$ parameter and corresponding test methods resulted in proposing novel test methods such as zero-shear viscosity, low-shear viscosity, and nonrecoverable creep compliance [37]. Further studies demonstrated that the nonrecoverable creep compliance is better correlated with the rutting performance of modified and unmodified binders [38–40]. As an alternative to PG plus tests, the MSCR test, which provides nonrecoverable creep compliance and percent recovery at multiple stress levels, was proposed to determine the high-temperature performance of the asphalt binders. Following a path with four phase evolutions between 2001 and 2014 (phase I-NCHRP Project 9-10, phase II-original MSCR test, phase III-standard MSCR test, and phase IV-modified MSCR test), the MSCR test protocol was documented at AASHTO T350 and AASHTO M332 standard specifications [37]. Similar to the LAS test to evaluate the fatigue cracking resistance of asphalt binders, the MSCR test is another mechanistic binder performance test developed to better understand the permanent deformation behavior of the original and modified asphalt binders. The test is conducted using DSR equipment with parallel plate geometry and a 1 mm gap setting at short-term aged asphalt binders. According to the MSCR testing protocol, the asphalt binder is loaded at 0.1 kPa and 3.2 kPa stress levels. The loading pattern follows a 9 s recovery after a 1 s shear creep in both stress levels. This loading pattern is repeated for 20 cycles for 0.1 kPa and 10 cycles for 3.2 kPa. The %R-percent recovery and Jnr-nonrecoverable creep compliance are the two main parameters recorded at the end of the test. MSCR grade bumping is carried out based on the traffic level, and traffic grade selection is performed using the MSCR nonrecoverable creep compliance value obtained at 3.2 kPa shear level, $J_{nr3.2kPa}$. Traffic grade is included in the binder PG by placing one of the four-grade letters next to the high PG temperature. The traffic letter grade can be S, H, V, or E according to the following criteria:

- Standard Traffic: S-grade, which occurs when $2.0\ kPa^{-1} < J_{nr@3.2kPa} < 4.5\ kPa^{-1}$;
- Heavy Traffic: H-grade, which occurs when $1.0\ kPa^{-1} < J_{nr@3.2kPa} < 2.0\ kPa^{-1}$;
- Very Heavy Traffic: V-grade, which occurs when $0.5\ kPa^{-1} < J_{nr@3.2kPa} < 1.0\ kPa^{-1}$;
- Extremely Heavy Traffic: E-grade, which occurs when $J_{nr@3.2kPa} < 0.5\ kPa^{-1}$.

The biggest changes in the last two phases of the MSCR tests were made to the number of creep and recovery cycles as well as the testing temperatures. While the latest version of the standards states that the testing should be performed at regional high temperatures, which is 58 °C for this study, the earlier version (AASHTO MP19 and AASHTO TP70) suggested the test be conducted at the high PG of the asphalt binders. This temperature change was necessary since the modified asphalt binders would never reach the high PG temperature in regions with lower high temperatures. In this study, for a comparison reason between previous and current MSCR protocols, the tests were performed at both the regional and high PG temperatures. Figure 7a demonstrates the change in $J_{nr@3.2kPa}$ with percent GTR for both high PG temperature and regional temperature. Even though there was not a clear trend for the results acquired at high PG temperatures, the nonrecoverable creep compliance values decreased as the percent GTR modifier increased at the regional high temperature of 58 °C. Thus, the comparison hereinafter focuses on the test results obtained at 58 °C. All modifications reduced the nonrecoverable creep compliance values compared to the neat binder. While the original binder with no modifications was at the H-traffic level, both 9% and 20% GTR modifications achieved two grade bumps and reached the E-traffic level. GTR 3% modification was not able to make a difference in the traffic level, and 6% GTR modification accomplished one grade bump by bringing the traffic grade to V-level. Figure 7b demonstrates the linear correlation developed between nonrecoverable creep compliance at 3.2 kPa and the percent GTR modifications.

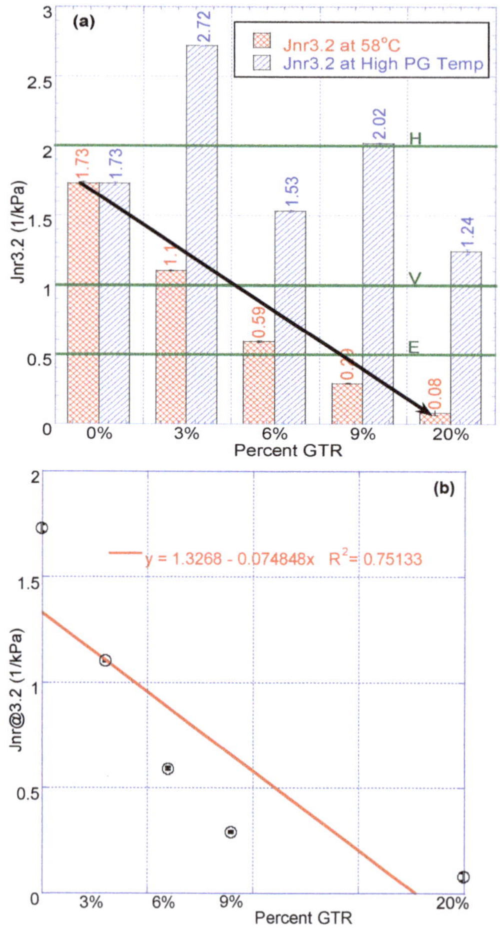

Figure 7. (**a**) Change in Jnr at 3.2 kPa as GTR modification increases at regional and high PG temperatures. (**b**) Correlation between Jnr at 3.2 kPa and GTR modifications at 58 °C.

One of the objectives of this research was to study the impact of testing geometry on the MSCR parameters. In the literature, there has been a concern regarding the testing of asphalt binders with discrete particles such as crumb rubber-modified asphalt binders, and the use of concentric cylinder geometry (a.k.a bob and cup) was promoted. In addition, AASHTO MP19 stated that the standards were not applicable for asphalt binders with undissolved distinct particles larger than 250 microns inside. To evaluate the impact of testing geometry on GTR-modified asphalt binders, MSCR tests were carried out with parallel plate (PP) and concentric cylinder (CC) setups. Figure 8a shows the correlation between nonrecoverable creep compliance values obtained at 3.2 kPa stress level for PP and CC geometries. There is almost a perfect linear fit between measured values with R^2 equal to 0.98 (0.993 with zero intercepts). This implies that either testing geometry can be used to determine the nonrecoverable creep compliance of mesh size 20 GTR-modified asphalt binder.

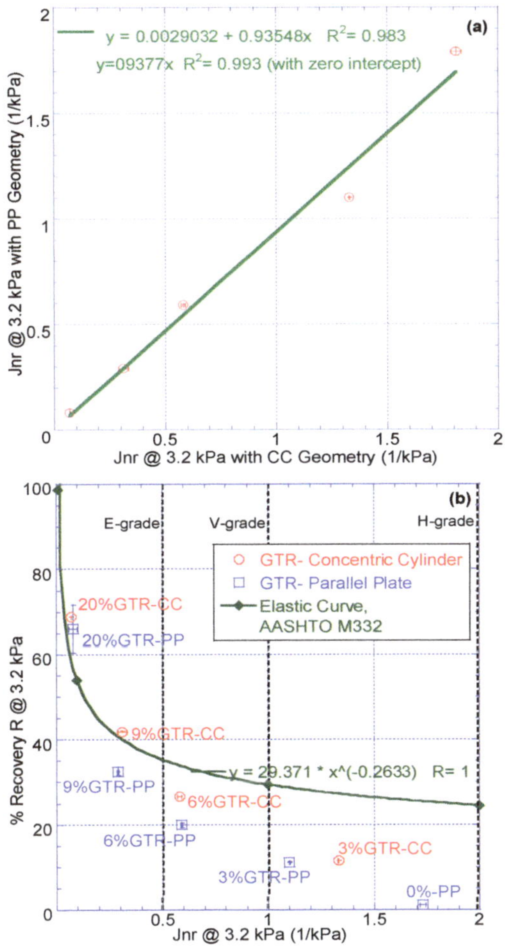

Figure 8. (**a**) Linear correlation between Jnr at 3.2 for PP and CC testing geometries. (**b**) Nonrecoverable creep compliance versus percent recovery graph for GTR modifications.

Figure 8b illustrates the nonrecoverable creep compliance and percent recovery values attained at 3.2 kPa stress level for PP and CC geometries along with the AASHTO M332 elastic curve. This graph is also known as MSCR elastic curve graph, which is mainly developed to ensure that there are enough elastic modifiers in the modified asphalt binders. According to the AASHTO T350, binders reaching above the curve have the desired quantity of elastic modifiers. $J_{nr3.2}$ values measured with PP and CC geometries were almost the same for all modified binders other than 3% GTR, and percent recovery results obtained from the CC setup were slightly higher than the PP geometry measurements for all tested binders. This resulted in 9% GTR CC measurement reaching above the elastic curve while 9% GTR PP measurement fell below the curve while it did not affect any other test outcomes. Regardless of the testing geometry, 20% of GTR modifications were able to pass the elastic curve, and none of the other modifications had enough elastic modifier to reach above the elastic curve. In addition to providing valuable information regarding the existence of elastic modifiers, the graph is an efficient way to show the traffic grade as well. Testing geometries had no impact on the traffic grade of the asphalt binders tested.

5. Discussion

PG of the asphalt binders has been major selling point for manufacturers for years based on the Superpave Binder Performance Grading system. Results of binder modification on binder performance grade studies mostly remained confidential. In general, each modifier is used to improve a certain property of the binders such as high PG, low PG, elasticity, ductility, viscosity, etc. The common practice is to use combined modifiers to enhance multiple properties, such as high and low PGs, of asphalt binders simultaneously. GTR modifications have been practiced improving the high PG of the asphalt binders mainly due to their preferable high-temperature/rutting performances [41]. On the other hand, some researchers showed that GTR modifications had an insignificant impact on the low-temperature properties [6–43]. Moreover, some other studies showed a significant decrease in the low-temperature performance of GTR-modified binders [42]. This study showed that GTR modification has a positive impact on both low and intermediate CPGs of the binders. The use of continuous PGs instead of discrete ones, which were assessed mainly in GTR modification studies in the literature, made it possible to reveal improvements in low and intermediate PGs even though the enhancements were smaller and easy to neglect when compared to high PGs. As the amount of GTR increased gradually from 3% to 20%, it was observed that all three continuous PGs improved proportionally. This study suggests further research with GTRs higher than 20%, different rubber gradations, and modification methods that have a great influence on the swelling or degradation behavior of the GTR [44,45].

ΔTc was another parameter evaluated in this study for GTR-modified asphalt binders. ΔTc is an indicator of the aging/nonload-related embrittlement in the asphalt pavements. ΔTc was initially developed for neat binders, and it has strong correlations with certain thermal crack types. However, the reduced laboratory aging of the modified binders due to higher viscosities prevents establishing strong correlations for modified binders. At this moment, the research to establish the relationship between ΔTc and polymer-modified asphalt binders has been ongoing [28,29,46]. This applies to GTR-modified asphalt binders as well. True aging of GTR-modified asphalt binders in laboratory conditions is challenging to determine, and there is further research needed to establish the guidelines. It should be noted that relations and data provided in this research for ΔTc are based on the so-called/artificial laboratory RTFO and PAV aging of GTR-modified asphalt binders. Further research may include increased aging time, pressure, temperature, or combinations thereof.

Fatigue cracking resistance of GTR-modified asphalt binders were analyzed using a VECD-based LAS test at different strain values. The fatigue life of modified binders showed an increasing trend at lower strain levels whereas they decreased at higher strain levels. This means that GTR-modified asphalt binders can perform better at low-volume or less traffic-loading roads better than high-volume roads such as interstates. The results do not necessarily follow the literatures' findings. There are many research studies showing that the addition of rubber particles improved the fatigue life of the asphalt binders [42,47–49]. However, it was also mentioned that the rubber type, size, production method, and mixing conditions have a considerable impact on the fatigue life [50,51]. It should be noted that most of the research in the literature was performed using a single LAS test temperature. In this study, LAS temperatures changed for each mixture type based on the equi-stiffness temperatures, which complies with early literature.

MSCR test results followed the results of high CPG closely as supported by earlier studies [52]. As the percent GTR increased, the nonrecoverable creep compliance values at both stress levels showed a decreasing trend, which conforms to the literature findings [52–54]. On the other hand, percent recovery values increased with growing GTR percentage. MSCR elastic curve, which is an indirect method of evaluating the elasticity of the binders, showed that only 20% GTR modification was able to provide enough elasticity to the binder. It also implies that as the rubber amount increases, binders become more elastic. One of the major concerns related to use of MSCR test with GTR modified asphalt binder was that the modified binder had larger than 250-micron discrete particles inside.

Since MSCR uses parallel-plate geometry, the gap between the plates may hinder the actual performance of rubber-modified binders under loading. In order to overcome this concern, tests were conducted using concentric cylinders in addition to parallel-plate geometry. Results showed a strong correlation between the two testing geometries by suggesting the use of either geometry for up to 20% GTR modifications. Additional research is recommended for higher than 20% GTR modifications. Furthermore, this study suggests improvements to MSCR specifications for a Jnr difference limit of 75%. Although the Jnr difference was lower than 75% for up to 9% GTR modifications, it was around 177% for 20% GTR modification. This was mainly due to the fact of insignificant nonrecoverable creep compliance at 0.1 kPa loading for heavily modified binders such as 20% GTR modification, which complies with literature findings as well [52].

6. Conclusions

Based on the data analysis and results obtained in this document, the following conclusions were compiled:

- The aging during the modification process (A-0% samples) resulted in a change in the PG of the original binder from 58–28 to 58–22. While it improved the CHPG by 1.8 °C, it worsened both CIPG and CLPG by 0.9 °C and 2.2 °C, respectively.
- A linear correlation with R^2 of 0.993 was established between CHPG and percent GTR modifications. The 3% GTR modifications achieved a one-grade bump while 9% and 20% GTR modifications accomplished two-grade and four-grade bumps. Similar to the linear correlation of CHPG, linear correlations between percent GTR and CIPG-CLPG were established as well. While the relationship for CLPG had an R^2 equal to 0.872, it was 0.6 for CIPG.
- The ΔTc parameter reduced as the artificial aging and percent GTR modification increased. Strong linear correlations between ΔTc and percent GTR modifications with R^2 value up to 0.987 were established.
- Based on the LAS test results, as the GTR percentage increased, the fatigue life of the asphalt binders enhanced considerably at lower strain levels and worsened at higher strain levels. The improvements compared to the original binder either were insignificantly small or did not exist.
- At regional high-temperature measurements, all modifications reduced the nonrecoverable creep compliance values compared to the neat binder. Compared to the original binder with no modifications, which was at the H-traffic level, both 9% and 20% GTR modifications achieved two grade bumps and reached the E-traffic level.
- Although the percent recovery values obtained with CC geometries were slightly higher than ones acquired with PP testing geometries, the difference between nonrecoverable creep compliance values was insignificant. A linear correlation between PP and CC geometries was established with R^2 equal to 0.98 (0.993 with zero intercepts), implying either testing geometry can be used to determine the nonrecoverable creep compliance.
- Only 20% of GTR modifications tested with PP and CC geometries and 9% of GTR modifications tested with CC geometry were able to pass the AASHTO M322-elastic curve. This indicates that low-percent GTR modifications do not provide enough elasticity to the asphalt binder. Moreover, the type of testing geometry may have an impact on the selection decision of the GTR-modified asphalt binders.

Funding: This research received no external funding.

Institutional Review Board Statement: Not applicable.

Informed Consent Statement: Not applicable.

Data Availability Statement: Not applicable.

Acknowledgments: The author is grateful for the resources and administrative support provided by M.E. Kutay at Michigan State University.

Conflicts of Interest: The author declares no conflict of interest.

References

1. Kaloush, K.E.; Biligiri, K.P. *Asphalt-Rubber Standard Practice Guide*; Links: West Devon, UK, 2011.
2. Kocak, S. Interaction between Recycled Tire Rubber, Polymers, and High Amount of Reclaimed Asphalt Pavements. Ph.D. Dissertation, Michigan State University, East Lansing, MI, USA, 2016.
3. Picado-Santos, L.G.; Capitao, S.D.; Neves, J.M.C. Crumb rubber asphalt mixtures: A literature review. *Constr. Build. Mater.* **2020**, *247*, 118577. [CrossRef]
4. Riekstins, A.; Haritonovs, V.; Straupe, V. Economic and environmental analysis of crumb rubber modified asphalt. *Constr. Build. Mater.* **2022**, *335*, 127468. [CrossRef]
5. Behnood, A.; Modiri, G.M. Morphology, rheology, and physical properties of polymer-modified asphalt binders. *Eur. Polym. J.* **2019**, *112*, 766–791. [CrossRef]
6. Bahia, H.U.; Davies, R. Effect of crumb rubber modifiers on performance related properties of asphalt binders. *J. Assoc. Asph. Paving Technol.* **1994**, *63*, 414–449.
7. Roque, R.; Birgisson, B.; Drakos, C.; Sholar, G. *Guidelines for Use of Modified Binders*; Project No: 4910-4504-964-12; Florida Department of Transportation: Tallahassee, FL, USA, 2005.
8. Willis, R.J. Use of ground tire rubber in a dense-graded asphalt mixture on US 231 in Alabama: A case study. In Proceedings of the 2013 Airfield and Highway Pavement Conference, Los Angeles, CA, USA, 9–12 June 2013.
9. Kocak, S.; Kutay, M.E. Fatigue performance assessment of recycled tire rubber modified asphalt mixtures using viscoelastic continuum damage analysis and AASHTOW are pavement ME design. *Constr. Build. Mater.* **2020**, *248*, 118658. [CrossRef]
10. Behnood, A.; Olek, J. Rheological properties of asphalt binders modified with styrene-butadiene-styrene (SBS), ground tire rubber (GTR), or polyphosphoric acid (PPA). *Constr. Build. Mater.* **2017**, *151*, 464–478. [CrossRef]
11. Global Tire Recycling Market Analysis 2025. Opportunity, Demand, Growth and Forecast 2017–2025 Edition 2020. Available online: https://www.goldsteinresearch.com/report/global-tire-recycling-industry-market-trends-analysis (accessed on 1 January 2023).
12. McElvery, R. Is the road to sustainable asphalt paved with tires. *Am. Chem. Soc. CS Cent. Sci.* **2020**, *6*, 2120–2122. [CrossRef]
13. Valentini, F.; Pegoretti, A. End-of-life options of tyres. A review. *Adv. Ind. Eng. Polym. Res.* **2022**, *5*, 203–213. [CrossRef]
14. CalRecycle. Tire Management. 2022. Available online: https://calrecycle.ca.gov/tires/ (accessed on 1 January 2023).
15. Kocak, S.; Kutay, M.E. Relationship between material characteristics of asphalt mixtures and highway noise. *Transp. Res. Rec.* **2012**, *2295*, 35–43. [CrossRef]
16. Xu, L.; Ni, H.; Zhang, Y.; Sun, D.; Zheng, Y.; Hu, M. Porous asphalt mixture use asphalt rubber binders: Preparation and noise reduction evaluation. *J. Clean. Prod.* **2022**, *376*, 134119. [CrossRef]
17. Noel, B. Five Major Benefits of Rubber Modified Asphalt. 2022. Available online: https://www.forconstructionpros.com/asphalt/article/22458257 (accessed on 1 January 2023).
18. Buttlar, W.G.; Rath, P. *State of Knowledge Report on Rubber Modified Asphalt*; US Tire Manufacturing Association: Washington, DC, USA, 2021.
19. Calabi-Floody, A.; Mignolet-Garrido, C.; Valdes-Vidal, G. Study of the effect of the use of asphalt binders modified with polymer fibres from end-of-life tyres (ELT) on the mechanical properties of hot mix asphalt at different operating temperatures. *Materials* **2022**, *15*, 7578. [CrossRef] [PubMed]
20. Kocak, S.; Kutay, M.E. Use of crumb rubber in lieu of binder grade bumping for mixture with high percentage of reclaimed asphalt pavement. *Road Mater. Pavement Des.* **2017**, *18*, 116–129. [CrossRef]
21. Papagiannakis, A.T. *A Review of Crumb-Rubber Modified Asphalt Concrete Technology*; Research Project T9902-09 Rubber-Asphalt Study; Washington State Transportation Center: Pullman, WA, USA, 1995.
22. Hassan, N.A.; Airey, G.D.; Jaya, R.P.; Mashros, N.; Aziz, M. A review of crumb rubber modification in dry mixed rubberised asphalt mixtures. *J. Teknol.* **2014**, *70*, 127–134.
23. Nuha, S.M.; Asim, H.A.; Mohamed, R.K.; Mahrez, A. An overview of crumb rubber modified asphalt. *Int. J. Phys. Sci.* **2012**, *7*, 166–170.
24. Xia, C.; Chen, M.; Geng, J.; Liao, X.; Chen, Z. Swelling and degradation characteristics of crumb rubber modified asphalt during processing. *Math. Probl. Eng.* **2021**, *2021*, 6682905. [CrossRef]
25. Kocak, S.; Kutay, M.E. Effect of devulcanized rubber modification on performance grade, fatigue cracking resistance and rutting resistance of asphalt binders. *J. Mater. Civ. Eng.* **2021**, *33*. [CrossRef]
26. Kocak, S.; Kutay, M.E. Combined effect of SBS and devulcanized rubber (DVR) modification on performance grade and fatigue cracking resistance of asphalt binders. In *8th RILEM International Conference on Mechanisms of Cracking and Debonding in Pavements*; Springer: Berlin/Heidelberg, Germany, 2016; Volume 13, pp. 269–274.

27. Zeng, Z.A.; Underwood, B.S.; Castorena, C. Low-temperature performance grade characterisation of asphalt binder using the dynamic shear rheometer. *Int. J. Pavement Eng.* **2022**, *23*, 811–823. [CrossRef]
28. Anderson, R.M.; King, G.N.; Hanson, D.I.; Blankenship, P.B. Evaluation of the relationship between asphalt binder properties and non-load related cracking. *J. Assoc. Asph. Paving Technol.* **2021**, *80*, 615–664.
29. Federal Highway Administration (FHWA). *Technical Brief on Delta Tc Binder Specification Parameter*; FHWA-HIF-21-042:29; Federal Highway Administration (FHWA): Washington, DC, USA, 2021.
30. Asphalt Institute (AI). *State -of-the-Knowledge: Use of the Delta Tc Parameter to Characterize Asphalt Binder Behavior*; AI-IS-240; Asphalt Institute (AI): Lexington, AL, USA, 2019.
31. Reinke, G. The relationship of binder delta Tc (ΔTc) and other binder properties to mixture fatigue and relaxation. In Proccedings of the Binder ETC Meeting, Fall River, MA, USA, 10 May 2018.
32. Jafari, M.; Babazadeh, A. Evaluation of polyphosphoric acid-modified binders using multiple stress creep and recovery and linear amplitude sweep tests. *Road Mater. Pavement Des.* **2016**, *17*, 859–876. [CrossRef]
33. Xie, W.; Castorena, C.; Wang, C.; Kim, Y. A framework to characterize the healing potential of asphalt binder using the linear amplitude sweep test. *Constr. Build. Mater.* **2017**, *154*, 771–779. [CrossRef]
34. Safaei, F.; Castorena, C. Temperature effects of linear amplitude sweep testing and analysis. *Transp. Res. Rec.* **2016**, *2574*, 92–100. [CrossRef]
35. Teymourpour, P.; Bahia, H. Linear amplitude sweep test: Binder grading specification and field validation. In Proccedings of the Binder Expert Task Group Meeting, Baton Rouge, LA, USA, 16 September 2014.
36. Hintz, C.; Bahia, H. Simplification of linear amplitude sweep test and specification parameter. *Transp. Res. Rec.* **2013**, *2370*, 10–16. [CrossRef]
37. Liu, H.; Zeiada, W.; Al-Khateeb, G.G.; Shanableh, A.; Samarai, M. Use of the multiple stress creep recovery (MSCR) test to characterize the rutting potential of asphalt binders: A literature review. *Constr. Build. Mater.* **2021**, *269*, 121320. [CrossRef]
38. D'Angelo, J.; Dongré, R. Practical use of multiple stress creep and recovery test: Characterization of styrene–butadiene–styrene dispersion and other additives in polymer-modified asphalt binders. *Transp. Res. Rec.* **2009**, *2126*, 73–82. [CrossRef]
39. D'Angelo, J. The relationship of the MSCR test to rutting. *Road Mater. Pavement Des.* **2009**, *10*, 61–80. [CrossRef]
40. Laukkanen, O.V.; Soenen, H.; Pellinen, T.; Heyrman, S.; Lemoine, G. Creep recovery behavior of bituminous binders and its relation to asphalt mixture rutting. *Mater. Struct.* **2015**, *48*, 4039–4053. [CrossRef]
41. Ahmed, T.A.; Hajj, E.Y.; Bahzad, D.S.; Piratheepan, M.; Warrag, A. Investigation of the rheological and bonding characteristics of crumb rubber-modified asphalt binders mixed with warm mix asphalt additive and antistrip agent. *Int. J. Pavement Res. Technol.* **2022**, *15*, 509–524. [CrossRef]
42. Zhu, H.; Zhang, M.; Li, Y.; Zou, Y.; Chen, A.; Wang, F.; Liu, L.; Gu, D.; Zhou, S. Swelled mechanism of crumb rubber and technical properties of crumb rubber modified bitumen. *Materials* **2022**, *15*, 7987. [CrossRef]
43. Haibin, L.; Cui, C.; Temitope, A.; Feng, Z.; Zhao, G.; Guo, P. Effect of SBS and crumb rubber on asphalt modification: A review of the properties and practical application. *J. Traffic Transp. Eng.* **2022**, *9*, 836–863.
44. Ren, S.; Liu, X.; Lin, P.; Wang, H.; Fan, W.; Erkens, S. The continuous swelling-degradation behaviors and chemo-rheological properties of waste crumb rubber modified bitumen considering the effect of rubber size. *Constr. Build. Mater.* **2021**, *307*, 124966. [CrossRef]
45. Ren, S.; Liu, X.; Lin, P.; Erkens, S. Influence of swelling-degradation degree on rheological properties, thermal pyrolysis kinetics, and emission components of waste crumb rubber modified bitumen. *Constr. Build. Mater.* **2022**, *337*, 127555. [CrossRef]
46. Baumgardner, G.L. *Delta Tc Asphalt Binder Specification Parameter, FHWA Development and Deployment of Innovative Asphalt Pavement Technologies Coop Agreement with University of Nevade, Reno*; Federal Highway Administration (FHWA): Washington, DC, USA, 2021.
47. Li, J.; Xiao, F.O.; Amirkhanian, S.N. Storage, fatigue and low temperature characteristics of plasma treated rubberized binders. *Constr. Build. Mater.* **2022**, *209*, 454–462. [CrossRef]
48. Chen, Z.; Wang, T.; Pei, J.; Amirkhanian, S.; Xiao, F.; Ye, Z. Fan Low temperature and fatigue characteristics of treated crumb rubber modified asphalt after a long term aging procedure. *J. Clean. Prod.* **2019**, *234*, 1262–1274. [CrossRef]
49. Shi, C.; Zhou, W.; Wang, T.; Cai, X.; Wang, H.; Yang, J.; Hong, J.; Gong, M. Fatigue performance characterization of styrene-butadiene-styrene and crumb rubber composite modified asphalt binders with high crumb rubber contents. *J. Clean. Prod.* **2022**, *331*, 129979. [CrossRef]
50. Niu, D.; Xie, X.; Zhang, Z.; Niu, Y.; Yang, Z. Influence of binary waste mixtures on road performance of asphalt and asphalt mixture. *J. Clean. Prod.* **2021**, *298*, 126842. [CrossRef]
51. Li, F.; Zhang, X.; Wang, L.; Zhai, R. The preparation process, service performances and interaction mechanisms of crumb rubber modified asphalt (CRMA) by wet process: A comprehensive review. *Constr. Build. Mater.* **2022**, *354*, 129168. [CrossRef]
52. Badri, M.R.; Alkaissi, Z.A.; Sutanto, M. Physical, rheological and morphological characterization of modified asphalt binder with differing crumb rubber contents. *Mater. Today Proc.* **2021**, *42*, 3028–3034. [CrossRef]

53. Chen, Y.; Ji, C.; Wang, H.; Su, Y. Evaluation of crumb rubber modification and short-term aging on the rutting performance of bioasphalt. *Constr. Build. Mater.* **2018**, *193*, 467–473. [CrossRef]
54. Zhang, L.; Xing, C.; Gao, F.; Li, T.; Tan, Y. Using DSR and MSCR tests to characterize high temperature performance of different rubber modified asphalt. *Constr. Build. Mater.* **2016**, *127*, 466–474. [CrossRef]

sustainability

Article

Influence of Heat–Cool Cyclic Exposure on the Performance of Fiber-Reinforced High-Strength Concrete

Ibrahim Hakeem [1,*], Md. Akter Hosen [2], Mana Alyami [1], Shaker Qaidi [3,4,*] and Yasin Özkılıc [5]

1 Department of Civil Engineering, College of Engineering, Najran University, Najran 55461, Saudi Arabia
2 Department of Civil and Environmental Engineering, College of Engineering, Dhofar University, Salalah 211, Oman
3 Department of Civil Engineering, College of Engineering, University of Duhok, Duhok 42001, Iraq
4 Department of Civil Engineering, College of Engineering, Nawroz University, Duhok 42001, Iraq
5 Department of Civil Engineering, Faculty of Engineering, Necmettin Erbakan University, Konya 42000, Turkey
* Correspondence: iyhakeem@nu.edu.sa (I.H.); shaker.abdal@uod.ac (S.Q.)

Abstract: Sometimes civil engineering infrastructures have been constructed in hot and cold weathering regions such as desert areas. In such situations, the concrete is not only smashed by hot and cold processes but also spoiled by shrinkage cracking. Therefore, this study intends to examine the influence of heat–cool cycles on high-strength concrete comprising various fibers, such as natural date palm, polypropylene, and steel fibers, and their different volume percentages. The most popular technique for improving the structural behavior of concrete is fiber insertion. Fibers decrease cracking occurrences, enhance early strength under impact loads, and increase a structure's ability to absorb additional energy. The main goal is to examine the effects of three different types of fibers on regular concrete exposed to heat–cool cycles. For each type of fiber, three dosages of 0.2%, 0.6%, and 1% were used to create high-strength concrete. After 28 days of regular water curing and six months of exposure to heat-and-cold cycles, all specimens were tested. The heat–cool cycles entailed heating for two days at 60 °C in the oven and cooling for another two days at room temperature. The results of the experiment showed that fiber reinforcement in concrete improves its strength and durability. The flexural strength was substantially improved by increasing the date palm, polypropylene, and steel fibers into the high-strength concrete with and without heat–cool cycles. Adding increments of date palm, polypropylene, and steel fibers into high-strength concrete revealed a significant improvement in energy absorption capacity in both cases, i.e., with or without the implementation of heat–cool cycles. Therefore, the natural date palm fibers might be utilized to produce sustainable fibrous high-strength concrete and be applicable in severe weathering conditions.

Keywords: fibrous concrete; date palm fiber; engineering characteristics; durability; thermal cycles; energy absorption capacity

Citation: Hakeem, I.; Hosen, M.A.; Alyami, M.; Qaidi, S.; Özkılıc, Y. Influence of Heat–Cool Cyclic Exposure on the Performance of Fiber-Reinforced High-Strength Concrete. *Sustainability* **2023**, *15*, 1433. https://doi.org/10.3390/su15021433

Academic Editor: Enzo Martinelli

Received: 8 December 2022
Revised: 27 December 2022
Accepted: 5 January 2023
Published: 12 January 2023

1. Introduction

Concrete structures are typically exposed to a variety of diverse environmental conditions throughout their lifespan [1–3]. Hence, a concrete structure's resilience is measured by how well it can resist certain exposure conditions without needing extensive maintenance or rehabilitation [4,5]. Concrete is said to be a composite material that can sustain itself for many years, or even centuries, with little to no maintenance [6]. Without reinforcing, plain cement concrete is made up of cement, fine, and coarse aggregate. To produce different types of concrete that are appropriate for diverse structural loading and environmental conditions, changes can be made to the ingredients used to generate plain cement concrete (PCC) and its mix design. Several performance-related difficulties are presented in order to emphasize the typical concrete's poor performance even further [7,8].

"The advent of ultra-high-performance concrete (UHPC) in the 1990s was a technological improvement in the creation of concrete [9,10]. High compressive strength, high tensile strength, and a good degree of fracture toughness and ductility were the distinguishing characteristics of this innovative concrete [11–13]. Additionally, fibers are being employed as a discrete three-dimensional reinforcement to overcome PCC's shortcomings and as a substitute for UHPC [14–16]. In order to increase its resistance to loads, fiber-reinforced concrete (FRC) inserts fiber into its composition. Different kinds of FRC have been developed, each with unique and significant benefits. FRCs have a wide range of applications because of their many benefits—good tensile strength, ductility, and fatigue resistance—which include building pavements, industrial floors, tunnel linings, slope stabilization, and impact-resistant constructions, among others" [17,18].

The initiation and spread of cracks in concrete under tensile and compressive loads can be prevented or delayed by using the right fiber type. Commercially accessible reinforcements come in a variety of categories and have features that make them suitable for particular applications. Examples include "carbon fiber [19,20], steel fiber (SF) [21,22], glass fiber [23], polypropylene fiber [24,25], organic fibers [26,27], carbon nanotubes [28], basalt fiber [29] and more. In comparison to other industrial fibers, SF is by far the better fiber when it comes to the mechanical performance of concrete. SF has a high elastic modulus of roughly 200 GPa, and a high tensile strength of over 1200 MPa. The literature has established a paradigm that supports the viability of SF as an excellent reinforcement material that ensures satisfactory tensile, compressive, flexural, and shear strength qualities [30,31]. Azad, A. K. et al. [32] expressed the experimental outcomes of the flexural test, the specimens can support more loads once they reach the cracking load; however, once they reach the peak load, a softening mode of collapse is seen, exhibiting great ductility. Additionally, it has been demonstrated that the use of steel fibers increases the reinforced concrete beams' resistance to shear failure, negating the need for stirrups" [33–35].

Mishra, S. et al. [36] have utilized numerous chemically altered sisal fibers as reinforcement, in addition to glass fibers, in the polyester matrix to improve the mechanical properties of the hybrid composites. According to the experimental findings, hybrid composites absorb less water than unhybridized composites. Mohanty, A. K et al. [37] demonstrated that these concrete specimens' water absorption rates were low when compared to unpalmed and categorized composites. When compared to well-known composites, such as glass and palm, bamboo and palm, and glass manufactured using the same techniques, an analysis of the tensile, flexural, and dielectric properties of composites revealed comparable results for characteristics, such as tensile strength. Priya, S. P. et al. [38] examined the tensile strength of these palm and glass composites and determined that adding more fabric to these composites improved their mechanical qualities. It was discovered that the matrix and the reinforcement had strong interfacial bonding and chemical resistance. Althoey, F. et al. [39] investigated the engineering characteristics of date palm fiber-infused high-strength concrete as well as the performance of conventional steel and polypropylene fibers. The concrete samples were fabricated using 0.0%, 0.20%, 0.60%, and 1.0% volumes of date palm, steel, and polypropylene fibers. The results revealed that 1% of date palm, steel, and polypropylene fibers boosted the splitting tensile strength by 17%, 43%, and 16%, respectively. For date palm, steel, and polypropylene fibers, the flexural strength was enhanced from 60% to 85%, 67% to 165%, and 61% to 79% respectively, by adding 1% fiber in comparison to the reference sample.

The main aim of this paper is to investigate the influence of three different types of fibers (date palm, polypropylene, and steel) in creating high-strength concrete under the heat–cool exposure cycles.

2. Research Significance

One of the most promising and cost-effective solutions is to replace traditional steel-reinforced concrete with fiber-reinforced concrete for structural applications. This study focuses on high-strength concrete comprising date palm fibers, which are agro-waste in the Arabian gulf region, and its properties compared to conventional polypropylene and steel fibers. Durability is the key concern for new date palm fiber-reinforced high-strength concrete. Concrete's resilience and service life are largely influenced by environmental variables including heating–cooling and freezing–thawing cycles. One of the most prevalent and harmful variables for concrete in service is the heat–cool cycle, which is found in many parts of the world, including the Arabian Gulf, Northwest China, and southern California in the United States. Therefore, this study focuses on the influence of heat–cool cyclic exposure on the performance of high-strength concrete comprising date palm fibers and conventional polypropylene and steel fibers.

3. Materials and Methods

3.1. Materials

3.1.1. Cement

In this investigation, the high-strength date palm, polypropylene, and steel fiber concrete specimens were made using regular Portland cement Type-I. The cement had a fineness of 4100 cm^2/g and a specific gravity of 3.15. According to the manufacturer [40] and verified by ASTM C 150 [41], the cement included 59% C3S, 12.10% C2S, 10.60% C3A, and 10.4% C4AF. Table 1 includes information about the cement's chemical arrangement.

Table 1. Chemical arrangement of cement.

Chemical Composite	Cao	Al_2O_3	Fe_2O_3	MgO	SiO_2	SO_3	LOI	K_2O	Insoluble
Mass (%)	63.83	6.25	3.45	0.97	19.70	2.25	1.52	1.08	0.95

3.1.2. Aggregates

Natural dune sand was utilized as a fine aggregate, with the majority of its particles passing through a 4.75 mm sieve [42]. The high-strength date palm, polypropylene, and steel fibrous concrete was made utilizing crushed stone with a maximum size of 20 mm as the coarse aggregate. Table 2 demonstrates the physical characteristics of the fine and coarse aggregates.

Table 2. Physical characteristics of the aggregates.

Characteristics	Type of Aggregate	
	Fine	Coarse
Bulk Density (kg/m^3)	1535.74	1630.00
Specific Gravity	2.67	2.77
Fineness Modulus	2.23	7.34
Water Absorption (%)	1.31	0.69

3.1.3. Superplasticizer and Water

For the production of high-strength date palm, polypropylene, and steel fibrous concrete, super plasticizers (SP) are renowned as excellent water reducers. In this study, Glenium® 110M, which is based on polycarboxylate ether, was applied as an SP when fabricating the fibrous concrete.

In both cases, filtered tap water was a key component in the production and curing of high-strength fibrous concrete. The requirements for producing high-strength fibrous concrete with ASTM C1602/C1602M [43] have been related to the qualities of water.

3.1.4. Date Palm Fibers

The date palm fibers were sourced from date palm trees in and around Najran, Saudi Arabia, that were 15 to 25 years old. These trees represent one of the best accessible diversities and are responsible for a sizable amount of agricultural waste products. Using a manual process, the date palm fibers were manually removed and collected from the palm trees. The bidirectional date palm fibers are positioned around the trunk of the tree and consist of two or three layers that are packed and overlaid. The different lengths and diameters of the raw date palm fibers were collected from the agro-farm, as shown in Figure 1.

Figure 1. Collecting procedure of date palm fibers as a waste material.

Chemical Treatment of Date Palm Fibers

"To remove any potential contaminants from the surface of the fibers and to improve the compatibility of the fibers with other concrete ingredients, the date palm fibers were chemically cured using varying concentrations of Sodium Hydroxide (NaOH). The major modification caused by this treatment is the disruption of hydrogen bonding within the network structure, here with increasing surface roughness. Here, aqueous NaOH is used to remove the lignin, wax, and oils from the cell walls. Thus, alkaline treatment often affects the cellulosic fibril, the degree of polymerization, and therefore the extraction of lignin and other non-cellulosic compounds. The treatment of date palm fibers was performed by NaOH solution immersion. The fibers were treated by immersing them individually in 1.5%, 3.0%, and 6.0% of NaOH solution. The fibers were immersed in the solution for 24 h at room temperature. Based on the effect on the fibers, a treatment with 3% NaOH was chosen due to the highest tensile strength of the fibers". Table 3 catalogs the physical characteristics of date palm fibers.

Table 3. Date palm fiber's physical characteristics.

Type of Date Palm Fibers	Diameter, (mm)	Length (mm)	Elongation (%)	Strain	Tensile Strength (MPa)
Raw Fibers	0.90	92	4	0.044	100
1.5% NaOH Treated	0.65	79	6	0.058	174
3.0% NaOH Treated	0.61	79	6	0.062	234
6.0% NaOH Treated	0.69	80	5	0.055	181

3.1.5. Polypropylene Fibers

Polypropylene fiber was applied to manufacture the high-strength fibrous concrete compared with date palm and steel fibrous concrete. The manufacturer provided the polypropylene fiber's physical characteristics, which are presented in Table 4. Figure 2 shows the polypropylene fibers that were utilized in this investigation.

Table 4. Polypropylene fiber's physical characteristics.

Length (cm)	Diameter (cm)	Density (g/cm^3)	Young Modulus (GPa)	Elongation at Breaking (%)	Tensile Strength (MPa)
1.20	0.0025	0.91	5.4	30	550

Figure 2. Polypropylene fibers.

3.1.6. Steel Fibers

The steel fibers were bundled with adhesive and hooked at both ends, as shown in Figure 3. The high-strength fibrous concrete was produced using these bundled steel fibers. Table 5 lists the physical characteristics of the steel fibers. Table 6 reveals the range of fibers used to manufacture the fibrous concrete, which ranges from 0% to 1.0% of the volume of the concrete.

Figure 3. Bundled steel fibers.

Table 5. Steel fiber's physical characteristics.

Length (cm)	Diameter (cm)	Aspect Ratio	Density (g/cm^3)	Tensile Strength (MPa)
6.0	0.075	80	7.85	625

Table 6. Fibrous concrete mix proportion.

Mix ID	Fibers				Aggregates		Cement	Water	SP
	(%)	Date Palm	PP	Steel	Fine	Coarse			
		Kilogram per Cubic Metre							
REF	0	-	-	-	736.93	1105.40	400.00	176.40	2.0
DF0.2	0.2	8	-	-					
DF0.6	0.6	24	-	-					
DF1.0	1.0	40	-	-					
PF0.2	0.2	-	8	-					
PF0.6	0.6	-	24	-					
PF1.0	1.0	-	40	-					
SF0.2	0.2	-	-	8					
SF0.6	0.6	-	-	24					
SF1.0	1.0	-	-	40					

3.2. Methodology

3.2.1. Mix Design and Specimens Preparation

Various tests that focus on specific aspects of the date palm, polypropylene, and steel fibrous high-strength concrete were used to evaluate the performance of the fibers. Different fibers (date palm, polypropylene, and steel) are blended with concrete in varying amounts. In this study, 0.2%, 0.6%, and 1.0% of fibers by volume of concrete have been applied for manufacturing the high-strength fibrous concrete. A total of ten mixtures were developed using date palm fiber (DF), polypropylene fiber (PF), and steel fiber (SF) as shown in Table 6.

The experimental program was conducted to calculate the various hardened features of the high-strength fibrous concrete using 100 mm cubes, 150 mm diameter × 300 mm height cylinders, and 100 mm × 100 mm × 500 mm prisms. Each mixture of high-strength fibrous concrete contained three specimens, and average values are presented in this study.

3.2.2. Thermal Cycles Procedure

The specimens were exposed to heat–cool cycles after 28 days of water curing by placing them into an oven. A single heating and cooling process cycle consisted of two days of heating at 60 °C and then two days of cooling specimens at room temperature 25 ± 5 °C, repeated for 180 days. This heat–cool cycle was chosen to simulate several areas of the Kingdom of Saudi Arabia's daily variation of ambient temperature, which is prevalent in the summer. Sufficient space between specimens was maintained to allow a uniform flow of hot air during heating and easy dissipation of heat during cooling. The specimens were placed at an adequate distance from the source of hot air so that the heat would not concentrate only on the surface of the specimens, as shown in Figure 4. After 180 days of exposure, the hardened density, compressive strength, flexural strength, ultrasonic pulse velocity (UPV), and water absorption of the specimens were determined.

Figure 4. Fiber reinforced concrete specimens under heat–cool cycles.

3.2.3. Investigation of Structural Properties

Compressive Strength Test

The capacity of concrete to carry compressive loads till failure is known as compressive strength. Based on ASTM C109 [44], the compressive strength test for high-strength date palm, polypropylene, and steel fiber reinforced concrete was carried out. The 100 × 100 × 100 mm cube specimens were evaluated after being hydrated for 28 days with ordinary drinkable water. The following formula was used to determine the compressive strength of the high-strength fibrous concrete specimens:

$$f_c' = \frac{F}{A_c} \tag{1}$$

where f_c' is compressive strength in MPa; F is the maximum compressive load at failure in N; A_c is the cross-sectional area of the specimens in mm^2.

Flexural Strength Test

According to ASTM C293 [45], a center-point loading arrangement was employed to test the flexural strength of high-strength date palm, polypropylene, and steel fiber concrete prism specimens. This test was performed using a Universal Instron machine with a 400 kN loading capacity and a constant loading rate of 0.0167 mm/s. This experiment evaluated the high-strength date palm, polypropylene, and steel fibrous concrete's modulus of rupture (MOR). This method is frequently used to succeed in high-strength fibrous concrete's flexural strength. A linear variable displacement transducer (LVDT) installed at the center of 100 mm × 100 mm × 500 mm high-strength fiber reinforced concrete prism specimens were used to assess the displacement. While conducting the experiment on the prism specimens, the applied load and displacement were automatically recorded in the data logger. To study the load-displacement curves, the data logger's recorded results were transferred to a computer.

Density Test

After a 28-day curing period, the density of the high-strength date palm, polypropylene, and steel fiber concrete specimens was measured with reference to ASTM C138 [46]. Before performing the mechanical compression test, this experiment was performed on the specimens. The weight and volume of the fiber concrete specimens were assessed throughout these studies.

Water Absorption Capacity Test

The presence of enclosing little holes that are changed by excessive water is a sign of concrete's greater quality. As a result, "concrete quality measurements like density, stiffness, and durability are typically computed using the experiment to determine water absorption capacity. After meeting the requirement of the curing period of 28 days, the water absorption test for high-strength date palm, polypropylene, and steel fiber reinforced concrete was carried out in accordance with BS 2011 Part 122 [47] using cylindrical specimens with sizes of 75 mm in diameter and 150 mm in height. The high-strength fiber reinforced concrete specimens were initially dried for the first 72 h in an electric power oven at a constant temperature of 105 °C. The specimens were then removed from the oven, allowed to cool for 24 h in a dry environment, and weighed. The specimens were placed right away in a water tank at a temperature of 20 °C. The specimens were submerged in water for 30 min with the specimen's longitudinal axis kept horizontal. Following the collection of the specimens from the water, the clothing was dried to obtain a saturated surface state before being reweighed". The increase in weight caused by immersion in water, which was shown as a percentage of the dry weight of the specimen, was used to determine the water absorption capacity of fiber reinforced concrete specimens.

Ultrasonic Pulse Velocity (UPV) Test

The ultrasonic pulse velocity (UPV) test was used to validate the integrity and homogeneity of the high-strength date palm, polypropylene, and steel fiber reinforced concrete specimens [48]. The test was performed using the high-strength fiber reinforced concrete specimens in accordance with ASTM C597 [49].

Energy Absorption Capacity

The energy retained by the unit cross-sectional area at any displacement terminal point is used to represent the high-strength date palm, polypropylene, and steel fiber-reinforced concrete specimens' ability to absorb energy [50]. By using the area under the load vs. deflection graphs up to the specimens' rupture, the energy absorption capacity of the specimens was calculated.

4. Results and Discussions

4.1. Compressive Strength

The experimental outcomes of the compressive strength for date palm, polypropylene, and steel fiber-reinforced concrete specimens with and without implementing heat–cool cycles are shown in Figure 5. The compressive strength gradually enhanced with increasing the fiber content for date palm, polypropylene, and steel fiber-reinforced concrete specimens without applying the heat–cool cycles compared with the reference specimen, as shown in Figure 5a. At the same time, the compressive strength for specimens that underwent the heat–cool cycles was slightly reduced as the volume of fibers increased for date palm and polypropylene fiber-reinforced concrete specimens, but not steel fiber specimens, compared with reference specimens. Therefore, the compressive strength might predominantly rely on the strength of the aggregates [51], whereas the impact of heat–cool cycles on the strength may be trivial.

Figure 5b–d show "the correlation between the compressive strength and date palm, polypropylene, and steel fibers contents, respectively for the application of heat-cool cycles and without heat-cool cycles. This correlation demonstrated strong R^2 values for date palm, polypropylene, and steel fiber-reinforced high-strength concrete specimens". The predicted equations for high-strength date palm, polypropylene, and steel fiber-reinforced concrete specimens for the implementation of heat–cool cycles and without heat–cool cycles are specified by:

$$f'_{cDF} = 0.72v_f + 69.84 \; for \; date \; palm \; fibers \; with \; heat - cool \; cycles \tag{2}$$

$$f'_{cDF} = 5.99v_f + 63.01\ for\ date\ palm\ fibers\ without\ heat - cool\ cycles \quad (3)$$

$$f'_{cPF} = 5.13v_f + 67.13\ for\ polypropylene\ fibers\ with\ heat - cool\ cycles \quad (4)$$

$$f'_{cPF} = 3.32v_f + 65.54\ for\ polypropylene\ fibers\ without\ heat - cool\ cycles \quad (5)$$

$$f'_{cSF} = 0.44v_f + 73.69\ for\ steel\ fibers\ with\ heat - cool\ cycles \quad (6)$$

$$f'_{cSF} = 7.57v_f + 62.51\ for\ steel\ fibers\ without\ heat - cool\ cycles \quad (7)$$

where f'_{cDF}, f'_{cPF}, and f'_{cSF} are the compressive strength (MPa) of date palm, polypropylene, and steel fiber-reinforced concrete specimens, respectively and v_f is the fibers content (%) in the concrete specimens.

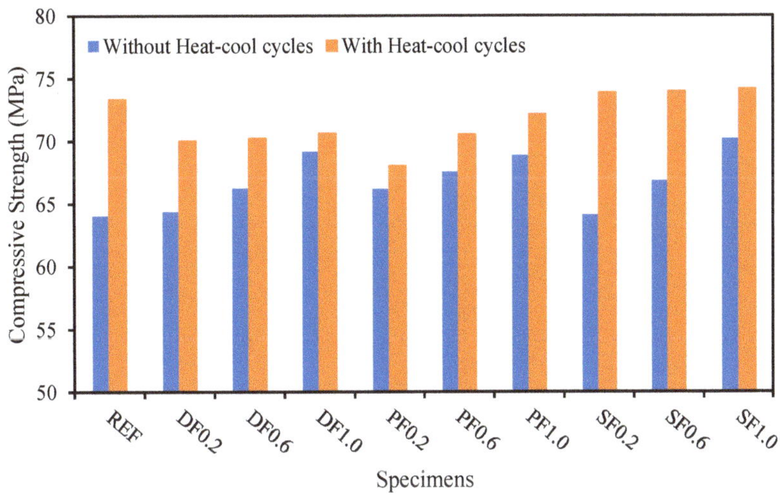

(**a**)

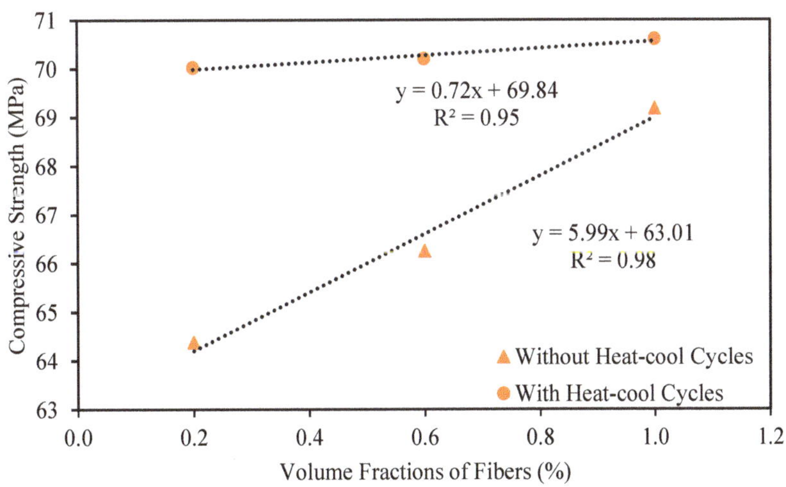

(**b**)

Figure 5. *Cont.*

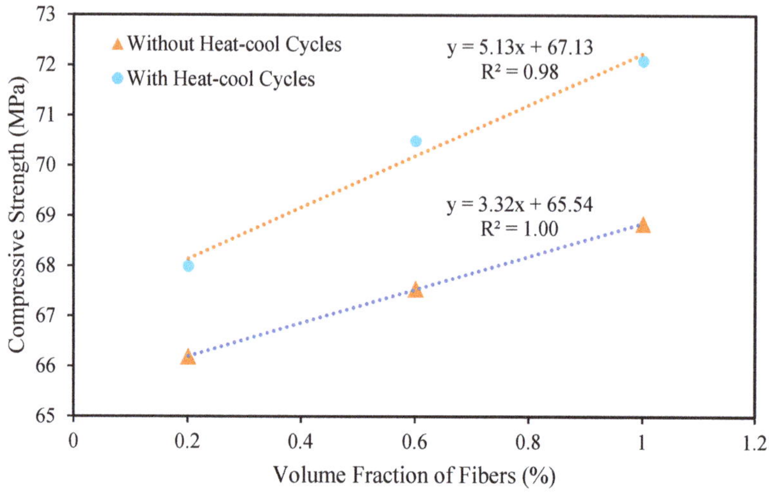

(c)

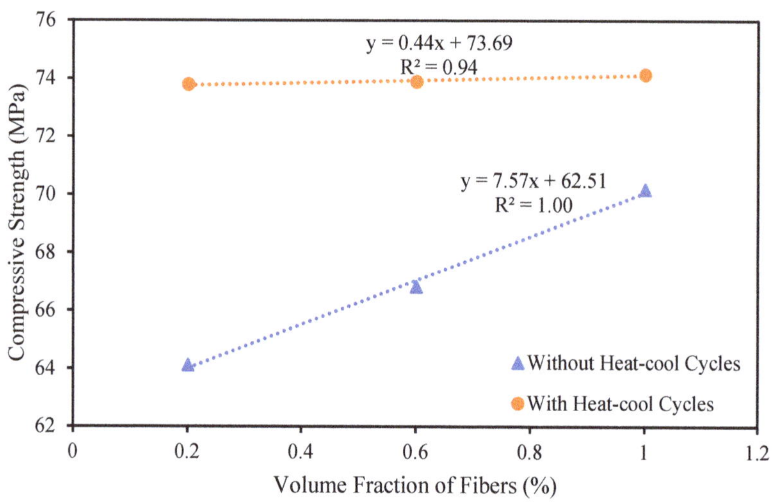

(d)

Figure 5. Influence of heat–cooling cycles on the compressive strength of fibrous concrete. (**a**) Assessment of compressive strength for heat–cool cycles; (**b**) Correlation between the compressive strength and amount of date palm fibers; (**c**) Correlation between the compressive strength and amount of polypropylene fibers; (**d**) Correlation between the compressive strength and amount of steel fibers.

4.2. Flexural Strength

The flexural performance of fiber-reinforced concrete is crucial for safeguarding infrastructures against severe weathering actions, such as freeze–thaws and extreme temperature [52–54]. The flexural strength of high-strength concrete comprising date palm, polypropylene, and steel fibers and its improvement are presented in Figure 6a. The addition of date palm, polypropylene, and steel fibers from 0% to 1% into the high-strength concrete significantly enhanced the flexural strength up to 85%, 79%, and 165%, respectively, compared with the reference specimen, without the implementation of heat–cool cycles. This was increased to 4%, 2%, and 34% for date palm, polypropylene, and steel fibers containing specimens, respectively, compared with the reference specimen when heat–cool cycles were applied on the specimens. In both cases, the steel fibers significantly enhanced the flexural strength compared with the date palm and polypropylene. The steel fibers might withstand or postpone the arising initial cracks in the cross-section of the specimens because of their higher flexural rigidity and higher capability to resist the severe weathering action. On the other hand, date palm fibers exhibited better flexural performance over polypropylene fibers because of the larger length of fibers.

Whereas, Kriker, A. et al. [55] investigated the mechanical and physical characteristics of four different varieties of date palm fibers. The attributes of date palm fiber-reinforced concrete are also provided as a function of curing in water and in a hot, dry climate, including strength, continuity index, toughness, and microstructure. When hot, in both dry and water curing, it was discovered that increasing the length and percentage of fiber reinforcement improved the post-crack flexural strength and toughness coefficients but decreased the first crack and compressive strengths.

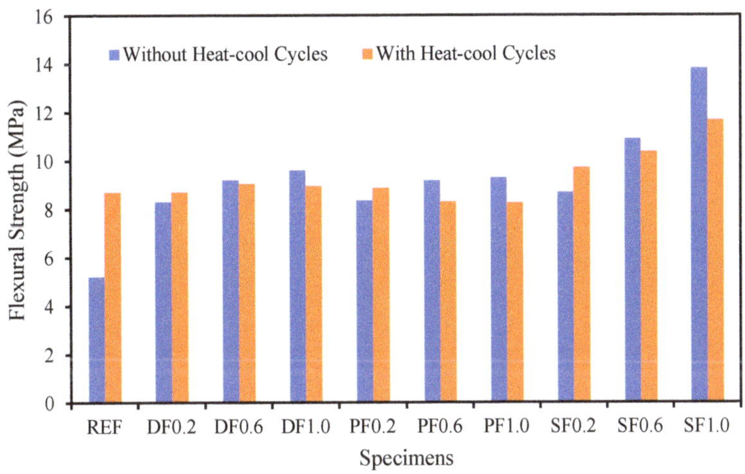

(a)

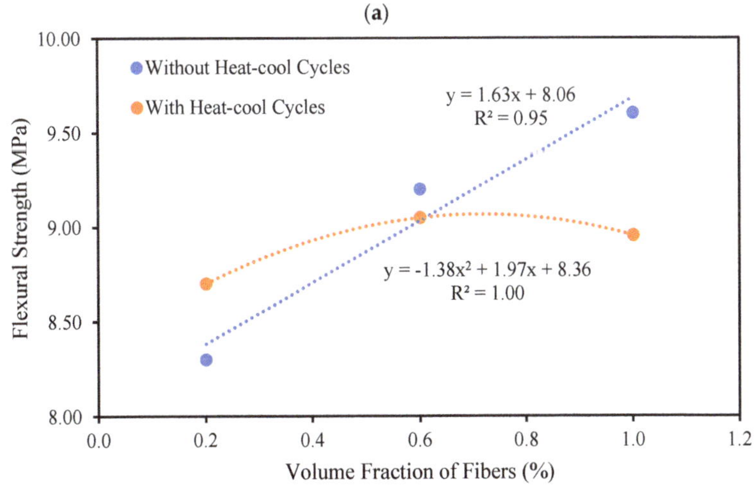

(b)

Figure 6. *Cont.*

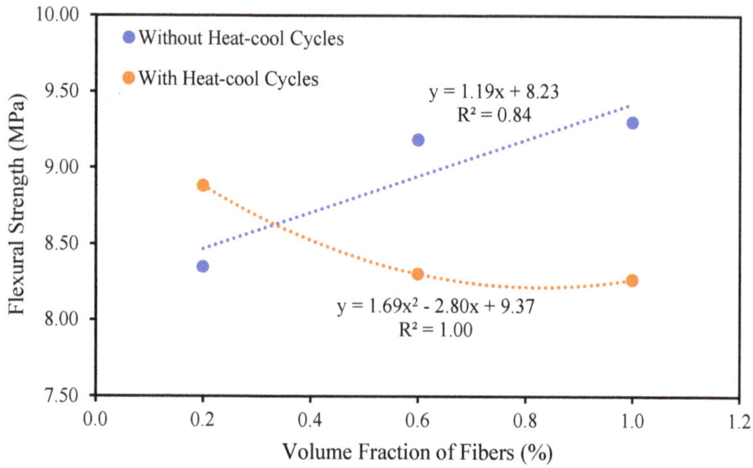

(c)

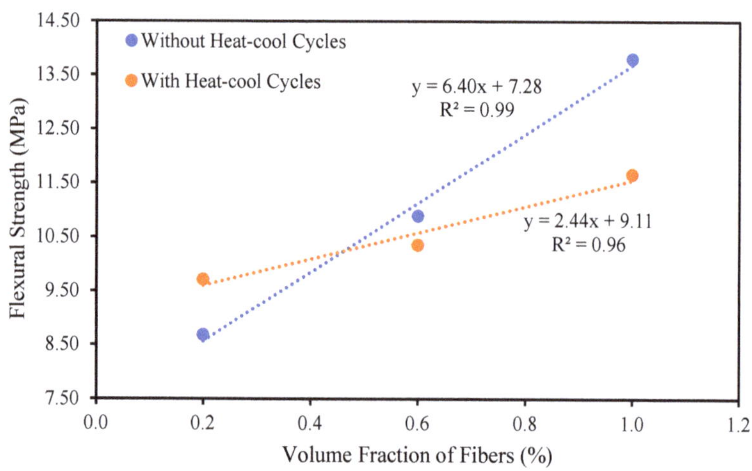

(d)

Figure 6. Impact of heat–cool cycles on the flexural strength of fibrous concrete. (**a**) Flexural performance of fibrous concrete changed due to heat–cool cycles; (**b**) Correlation between the flexural strength and amount of date palm fibers; (**c**) Correlation between the flexural strength and amount of polypropylene fibers; (**d**) Correlation between the flexural strength and amount of steel fibers.

"The relationship between the flexural strength and volume fraction of date palm, polypropylene, and steel fibers with applying the heat-cool cycles and without heat-cool cycles are displayed in Figure 6b–d, respectively. This correlation demonstrated linearly enhancing flexural strength for date palm, polypropylene, and steel fiber-reinforced high-strength concrete specimens without applying heat-cool cycles. By contrast, only steel fibers comprising specimens exhibited predominantly improving flexural strength under the heat-cool cycles". The flexural strength predicted equations for high-strength date palm, polypropylene, and steel-fiber-reinforced concrete specimens for the implementation of heat–cool cycles and without heat-cool cycles are specified by:

$$f_{rDF} = -0.32v_f^2 + 1.97v_f + 8.36\ for\ date\ palm\ fibers\ with\ heat - cool\ cycles \quad (8)$$

$$f_{rDF} = 1.63v_f + 8.06\ for\ date\ palm\ fibers\ without\ heat - cool\ cycles \quad (9)$$

$$f_{rPF} = 1.69v_f^2 + 2.8v_f + 9.37\ for\ polypropylene\ fibers\ with\ heat - cool\ cycles \quad (10)$$

$$f_{rPF} = 1.19v_f + 8.23\ for\ polypropylene\ fibers\ without\ heat - cool\ cycles \quad (11)$$

$$f_{rSF} = 2.44v_f + 9.11\ for\ steel\ fibers\ with\ heat - cool\ cycles \quad (12)$$

$$f_{rSF} = 6.40v_f + 7.28\ for\ steel\ fibers\ without\ heat - cool\ cycles \quad (13)$$

where f_{rDF}, f_{rPF}, and f_{rSF} are the flexural strength (MPa) of date palm, polypropylene, and steel fiber-reinforced concrete specimens, respectively.

4.3. Density

Density is an essential characteristic for high-strength fiber-reinforced concrete. "The density of fiber-reinforced concrete relying on the ingredients used to manufacture it [56]. The density of high-strength concrete containing of date palm, polypropylene, and steel fibers under application of the heat-cool cycles (with and without) as presented in Figure 7a. The density of fibrous high-strength concrete progressively reduces with the increasing the date palm and polypropylene fibers without applying the heat-cool cycles, whereas the density predominantly improves for steel fibers due to the heavier unit of steel fibers over the date palm and polypropylene fibers. In contrast, the densities were gradually reduced by adding different amounts of fiber for date palm, polypropylene, and steel fibers except for the SF1.0 specimen in the application of the heat-cool cycles on the specimens". The density increased up to 3% and 4% by adding steel fibers in the high-strength concrete incorporating heat–cool cycles and without heat–cool cycles, respectively.

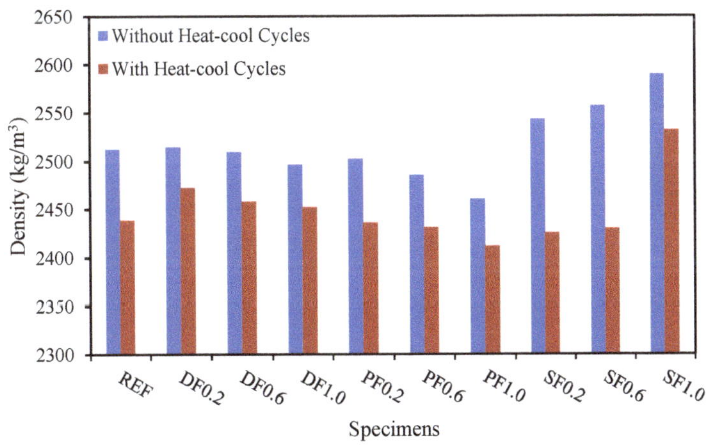

(a)

Figure 7. *Cont.*

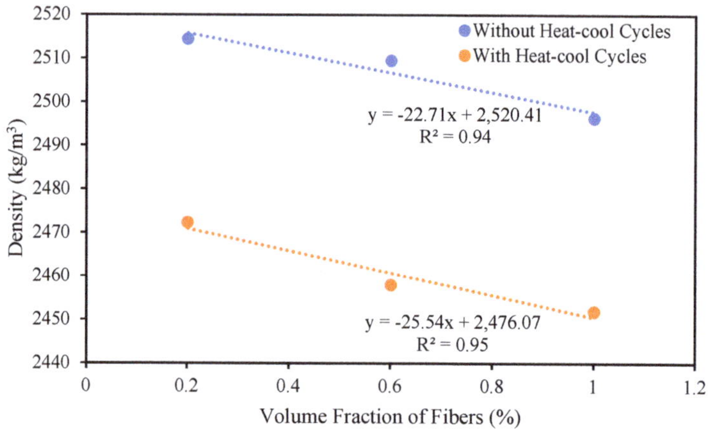

(b)

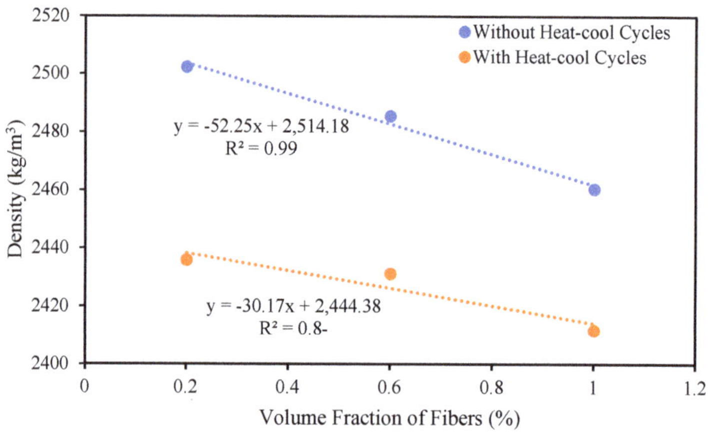

(c)

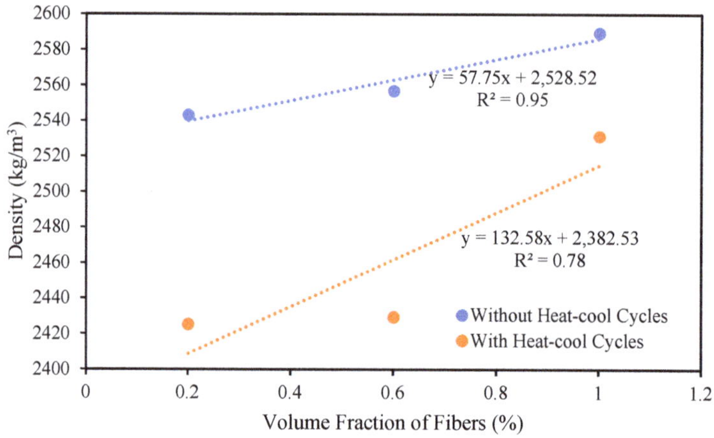

(d)

Figure 7. Effect of heat–cooling cycles on the density of high-strength fiber-reinforced concrete.

(**a**) Density of fiber reinforced concrete reflecting due to heat–cool cycles; (**b**) Correlation between the density and amount of date palm fibers; (**c**) Correlation between the density and amount of polypropylene fibers; (**d**) Correlation between the density and amount of steel fibers.

A linear correlation emerged between the density and amount of date palm, polypropylene, and steel fiber content for the employed of heat–cool cycles and without heat–cool cycles with strong R^2 values, as shown in Figure 7b–d. The density of high-strength concrete comprising date palm, polypropylene, and steel fibers with and without heat–cool cycles might be predicted by the following equations.

$$\gamma_{DF} = -25.54v_f + 2476.07 \; for\ date\ palm\ fibers\ with\ heat - cool\ cycles \tag{14}$$

$$\gamma_{DF} = -22.71v_f + 2520.41 \; for\ date\ palm\ fibers\ without\ heat - cool\ cycles \tag{15}$$

$$\gamma_{PF} = -30.17v_f + 2444.38 \; for\ polypropylene\ fibers\ with\ heat - cool\ cycles \tag{16}$$

$$\gamma_{PF} = -52.25v_f + 2514.18 \; for\ polypropylene\ fibers\ without\ heat - cool\ cycles \tag{17}$$

$$\gamma_{SF} = 132.58v_f + 2382.53 \; for\ steel\ fibers\ with\ heat - cool\ cycles \tag{18}$$

$$\gamma_{SF} = 57.75v_f + 2528.52 \; for\ steel\ fibers\ without\ heat - cool\ cycles \tag{19}$$

where γ_{DF}, γ_{PF}, and γ_{SF} are the density (kg/m^3) of date palm, polypropylene, and steel fiber-reinforced concrete specimens, respectively.

4.4. Water Absorption Capacity

The concrete pore structure is known to play a significant role in the material's durability. The amount of water absorbed by immersion provides an estimate of the concrete's total pore volume [57]. The water absorption capacity of the high-strength date palm, polypropylene, and steel fiber-reinforced concrete was higher than the reference concrete specimens in applied heat–cool cycles and without heat–cool cycles. The water absorption capacity was intensified with the increasing amount of fiber in the concrete in both cases, other than the SF0.2 specimen, as shown in Figure 8a. Since the fibers flow around the mortar and the fibers created a connection with them, thus, those fibers increased the micro-pour inside the concrete. As a consequence, the water absorption capacity of fiber-reinforced was increased significantly compared with the reference specimen. However, the date palm, polypropylene, and steel fiber-reinforced concrete exhibited much lower water absorption capacity since the water absorption capacity in good-grade concrete should have to be lower than 10% by weight [58].

The correlation between the water absorption capacity and amount of date palm, polypropylene, and steel fiber content, respectively for the application of heat–cool cycles and without heat–cool cycles with strong R^2 values as revealed in Figure 8b–d. The water absorption capacity of high-strength concrete encompassing date palm, polypropylene, and steel fibers with and without heat–cool cycles might be projected by the subsequent equations.

$$W_{cDF} = 0.24v_f + 1.70 \; for\ date\ palm\ fibers\ with\ heat - cool\ cycles \tag{20}$$

$$W_{cDF} = 0.14v_f + 1.64 \; for\ date\ palm\ fibers\ without\ heat - cool\ cycles \tag{21}$$

$$W_{cPF} = 0.29v_f + 1.83 \; for\ polypropylene\ fibers\ with\ heat - cool\ cycles \tag{22}$$

$$W_{cPF} = 0.15v_f + 1.90 \; for\ polypropylene\ fibers\ without\ heat - cool\ cycles \tag{23}$$

$$W_{cSF} = 0.25v_f + 1.55 \; for\ steel\ fibers\ with\ heat - cool\ cycles \tag{24}$$

$$W_{cSF} = 0.25v_f + 1.51 \; for\ steel\ fibers\ without\ heat - cool\ cycles \tag{25}$$

where W_{cDF}, W_{cPF}, and W_{cSF} are the water absorption capacity (%) of date palm, polypropylene, and steel fiber-reinforced concrete specimens, respectively.

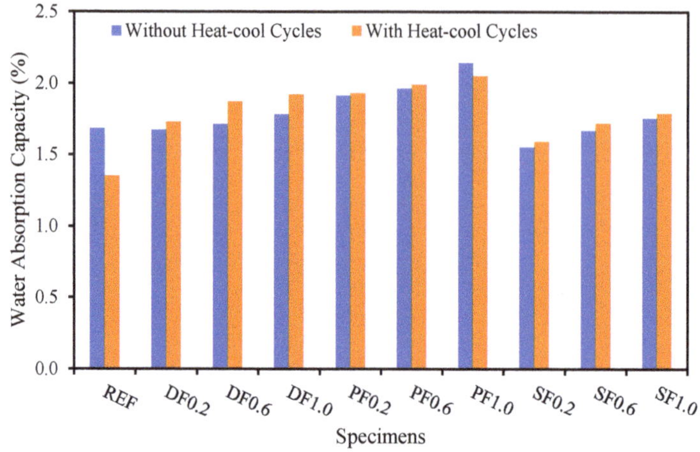

(**a**)

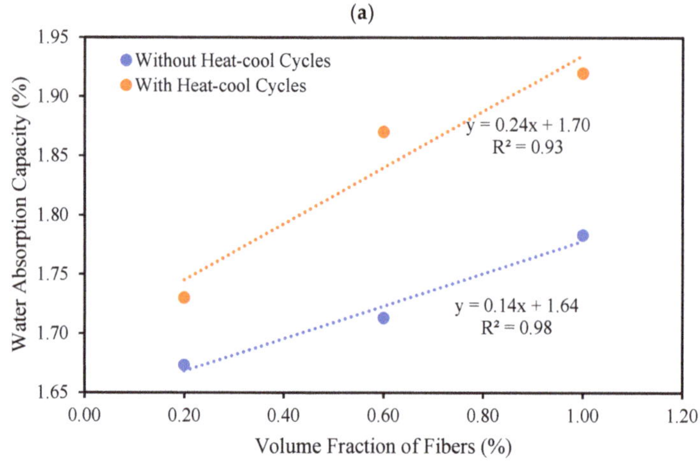

(**b**)

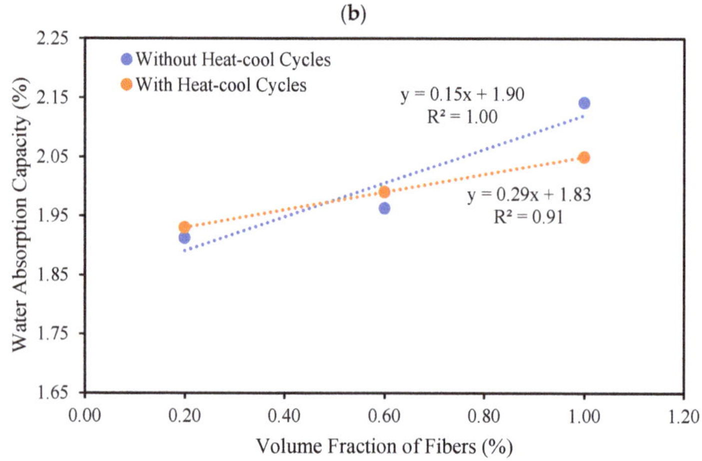

(**c**)

Figure 8. *Cont.*

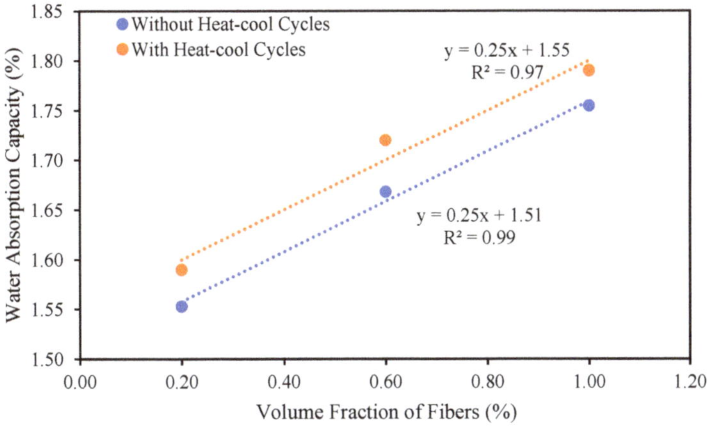

(d)

Figure 8. Variation of water absorption capacity of fibrous concrete due to heat–cool cycles. (**a**) Variation of water absorption capacity due to heat–cool cycles; (**b**) Correlation between the water absorption capacity and amount of date palm fibers; (**c**) Correlation between the water absorption capacity and amount of polypropylene fibers; (**d**) Correlation between the water absorption capacity and amount of steel fibers.

4.5. Ultrasonic Pulse Velocity

One of the most well-known non-destructive methods for evaluating the properties of concrete is UPV [59,60]. "The travel time between the first set and the acceptance of the pulse when the ultrasonic pulse went through the concrete specimen is used to construct the UPV method. The travel route distance between transducers can be used to determine the average wave propagation velocity [61]. Figure 9a depicts the UPV of the high-strength concrete comprising date palm, polypropylene, and steel fibers with and without heat-cool cycles. The date palm, polypropylene, and steel fibers were added to the high-strength concrete with increasing amounts, which helped to reduce ultrasonic wave travel farther and through greater efficiency without heat-cool cycles. There was no influence of heat-cool cycles on the date palm, polypropylene, and steel fibers added to the high-strength concrete with increasing fiber quantities". This was done to ensure the uniformity and homogeneity of this high-strength fiber-reinforced concrete.

The correlation between the UPV and volume fraction of date palm, polypropylene, and steel fiber for the application of heat–cool cycles and without heat–cool cycles as shown in Figure 9b–d. The UPV of high-strength concrete incorporating date palm, polypropylene, and steel fibers with and without heat–cool cycles would be estimated by successive equations.

$$U_{vDF} = 1562.50v_f^2 - 22.50v_f + 4.71\ for\ date\ palm\ fibers\ with\ heat - cool\ cycles \quad (26)$$

$$U_{vDF} = -13.50v_f + 5.07\ for\ date\ palm\ fibers\ without\ heat - cool\ cycles \quad (27)$$

$$U_{vPF} = -15.00v_f + 4.67\ for\ polypropylene\ fibers\ with\ heat - cool\ cycles \quad (28)$$

$$U_{vPF} = -11.49v_f + 4.86\ for\ polypropylene\ fibers\ without\ heat - cool\ cycles \quad (29)$$

$$U_{vSF} = -15937.50v_f^2 + 130v_f + 4.50\ for\ steel\ fibers\ with\ heat - cool\ cycles \quad (30)$$

$$U_{vSF} = -35.74v_f + 5.16\ for\ steel\ fibers\ without\ heat - cool\ cycles \quad (31)$$

where U_{vDF}, U_{vPF}, and U_{vSF} are the UPV (km/s) of date palm, polypropylene, and steel fiber-reinforced high-strength concrete specimens, respectively.

The mathematical equations are very useful for the different mix proportions of fiber-reinforced high-strength concrete, especially when many variables are used. The outcomes of the fiber-reinforced, high-strength concrete might be obtained without the need to conduct experimental investigations or other field studies. It also might be useful to investigate the effects of one or more variables on the performance of high-strength concrete.

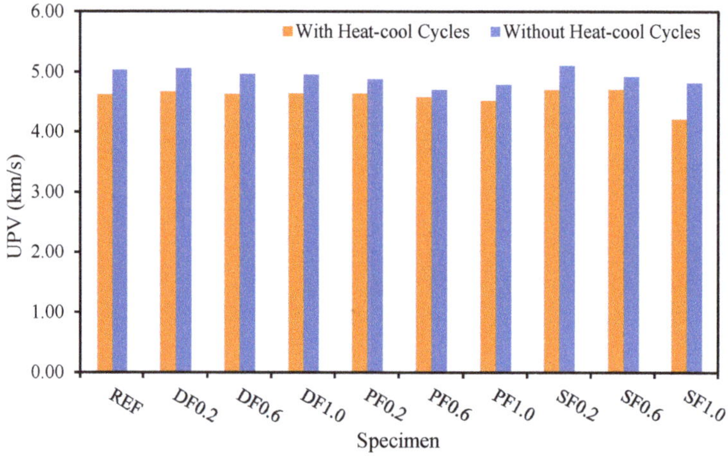

(**a**)

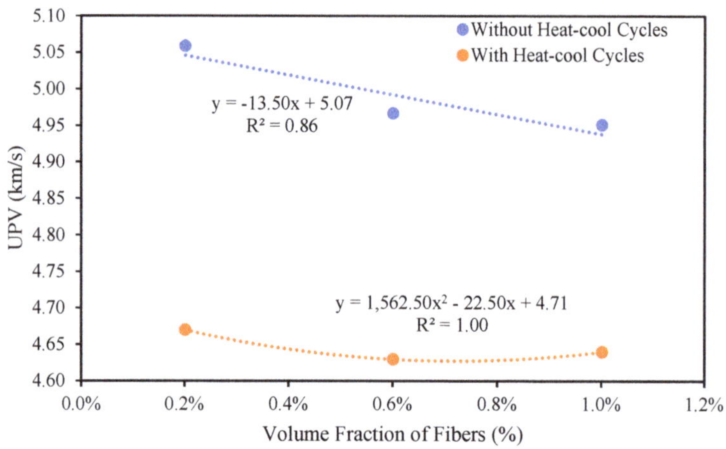

(**b**)

Figure 9. *Cont.*

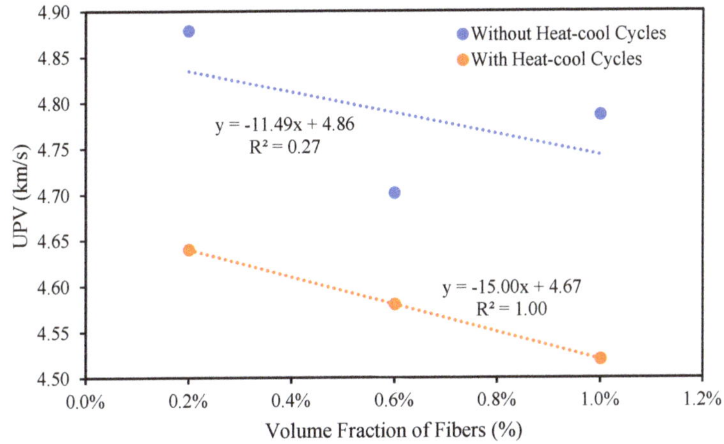

(c)

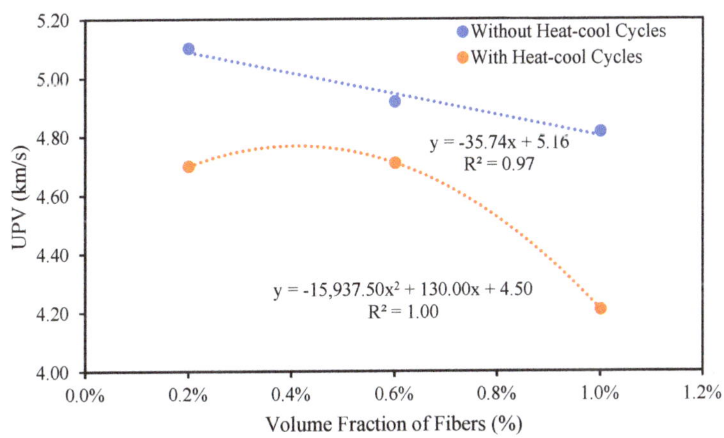

(d)

Figure 9. Deviations of UPV due to heat–cooling for high-strength fiber-reinforced concrete. (**a**) UPV of the fibrous concrete changed due to heat–cool cycles; (**b**) Correlation between the UPV and amount of date palm fibers; (**c**) Correlation between the UPV and amount of polypropylene fibers; (**d**) Correlation between the UPV and amount of steel fibers.

4.6. Energy Absorption Capacity

The energy absorption capacities and their enhancement for high-strength date palm, polypropylene, and steel fiber-reinforced concrete specimens without heat–cool cycles are demonstrated in Figure 10a. The use of date palm, polypropylene, and steel fibers revealed an enhancement in the energy absorption capacity of up to 2%, 2.5%, and 39%, respectively, compared with the reference specimen when no heat–cool cycles were applied. Increasing the amount of fiber reinforcement in the concrete mixes progressively enhanced the load-bearing capacity and energy absorption capacities.

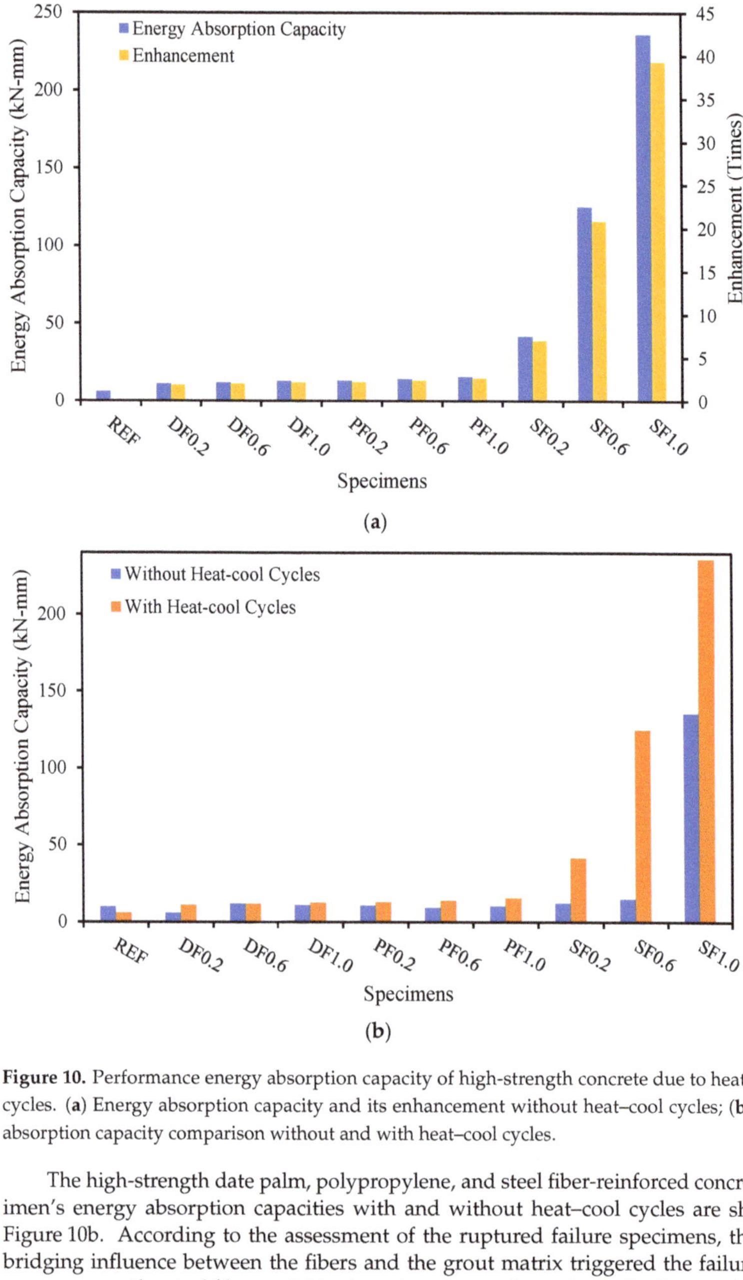

Figure 10. Performance energy absorption capacity of high-strength concrete due to heat–cooling cycles. (**a**) Energy absorption capacity and its enhancement without heat–cool cycles; (**b**) Energy absorption capacity comparison without and with heat–cool cycles.

The high-strength date palm, polypropylene, and steel fiber-reinforced concrete specimen's energy absorption capacities with and without heat–cool cycles are shown in Figure 10b. According to the assessment of the ruptured failure specimens, the crack-bridging influence between the fibers and the grout matrix triggered the failure. As a consequence, the steel fibers exhibited greater energy absorption in both cases (with and without heat–cool cycles) over the date palm and polypropylene fiber-reinforced concrete specimens since the steel fiber had a higher stiffness. Therefore, it should be suggested that the concrete with steel fibers is more applicable in harsh weathering action compared with the date palm and polypropylene fiber concrete specimens.

5. Conclusions

The influence of heat–cool cycles on high-strength concrete reinforced with various percentages (0%, 0.2%, 0.6%, and 1.0%) of date palm, polypropylene, and steel fibers during manufacturing was investigated. The engineering properties were examined, including compressive strength, flexural strength, density, water absorption capacity, ultrasonic pulse velocity, and energy absorption capacity. The following concluding remarks can be illustrated based on the laboratory examination:

The compressive strength of the high-strength concrete comprising date palm, polypropylene, and steel fibers was enhanced significantly with increasing fiber contents without implementing the heat–cool cycles. By contrast, compressive strength was reduced by applying the heat–cool cycles on the specimens containing date palm and polypropylene fibers; however, the steel fibers encompassing specimens expressed a substantial improvement in this case, which was due to the higher compressive load-bearing capacity of the fibers.

The flexural strength substantially improved with increasing the date palm, polypropylene, and steel fibers into the high-strength concrete with and without heat–cool cycles. Increasing the date palm, polypropylene, and steel fibers from 0% to 1.0% enhanced the flexural strength up to 85%, 79%, and 165%, respectively, compared with the reference specimen without the implementation of heat–cool cycles. On the other hand, the flexural strength improved up to 4%, 2%, and 34%, respectively, over the reference specimens with the implementation of heat–cool cycles.

The density was gradually reduced as the date palm and polypropylene fibers increased without applying the heat–cool cycles, whereas the density for steel fibers was noticeably improved by the steel fibers' heavier unit compared to the date palm and polypropylene fibers. In contrast, the densities were decreased by adding the different amounts of fiber for date palm, polypropylene, and steel fibers, except for the SF1.0 specimen in the application of the heat–cool cycles.

The water absorption capacity was increased with the increasing amounts of date palm, polypropylene, and steel fibers into high-strength concrete as fibers increased the micropore in the concrete both with and without the implementation of heat–cool cycles.

With the addition of the date palm, polypropylene, and steel fibers, the high-strength concrete exhibited an improvement in UPV compared with the reference specimen without heat–cool cycles applied. A negligible impact was observed with the implementation of heat–cool cycles on the date palm, polypropylene, and steel fiber-reinforced high-strength concrete specimens.

The addition of date palm, polypropylene, and steel fibers into high-strength concrete showed a substantial improvement in energy absorption capacity compared with the reference specimen in the cases both with and without the implementation of heat–cool cycles.

Therefore, the natural date palm fibers might be used to produce sustainable fibrous high-strength concrete and be applicable in severe weathering conditions.

This study only applied 60 °C for thermal cycles on the specimens and afterward, specimens were cooled for two days at room temperature 25 ± 5 °C (1 cycle) for 180 days.

Author Contributions: Conceptualization, I.H., M.A.H., M.A., S.Q. and Y.Ö.; methodology, I.H., M.A.H., M.A., S.Q. and Y.Ö.; software, I.H., M.A.H., M.A., S.Q. and Y.Ö.; validation, I.H., M.A.H., M.A., S.Q. and Y.Ö.; formal analysis, I.H., M.A.H., M.A., S.Q. and Y.Ö.; investigation, I.H., M.A.H., M.A., S.Q. and Y.Ö.; resources, I.H., M.A.H., M.A., S.Q. and Y.Ö.; data curation, I.H., M.A.H., M.A., S.Q. and Y.Ö.; writing—original draft preparation, I.H., M.A.H., M.A., S.Q. and Y.Ö.; writing—review and editing, I.H., M.A.H., M.A., S.Q. and Y.Ö.; visualization, I.H., M.A.H., M.A., S.Q. and Y.Ö.; supervision, I.H., M.A.H., M.A., S.Q. and Y.Ö.; project administration, I.H., M.A.H., M.A., S.Q. and Y.Ö.; funding acquisition, I.H., M.A.H., M.A., S.Q. and Y.Ö. All authors have read and agreed to the published version of the manuscript.

Funding: This research was funded by the Deanship of Scientific Research at Najran University with grant number (NU/NRP/SERC/11/26).

Institutional Review Board Statement: Not applicable.

Informed Consent Statement: Not applicable.

Data Availability Statement: Not applicable.

Acknowledgments: The authors are thankful to the Deanship of Scientific Research at Najran University for funding this work under the National Research Priorities funding program grant code (NU/NRP/SERC/11/26).

Conflicts of Interest: The authors declare no conflict of interest.

References

1. Ahmmad, R.; Jumaat, M.Z.; Alengaram, U.J.; Bahri, S.; Rehman, M.A.; Bin Hashim, H. Performance evaluation of palm oil clinker as coarse aggregate in high strength lightweight concrete. *J. Clean. Prod.* **2016**, *112*, 566–574. [CrossRef]
2. Hosen, M.A.; Jumaat, M.Z.; Islam, A.S. Side Near Surface Mounted (SNSM) technique for flexural enhancement of RC beams. *Mater. Des.* **2015**, *83*, 587–597. [CrossRef]
3. Hakeem, I.Y.; Amin, M.; Zeyad, A.M.; Tayeh, B.A.; Maglad, A.M.; Agwa, I.S. Effects of nano sized sesame stalk and rice straw ashes on high-strength concrete properties. *J. Clean. Prod.* **2022**, *370*, 133542. [CrossRef]
4. Ma, C.-K.; Apandi, N.M.; Sofrie, C.S.Y.; Ng, J.H.; Lo, W.H.; Awang, A.Z.; Omar, W. Repair and rehabilitation of concrete structures using confinement: A review. *Constr. Build. Mater.* **2017**, *133*, 502–515. [CrossRef]
5. Hakeem, I.Y.; Agwa, I.S.; Tayeh, B.A.; Abd-Elrahman, M.H. Effect of using a combination of rice husk and olive waste ashes on high-strength concrete properties. *Case Stud. Constr. Mater.* **2022**, *17*, e01486. [CrossRef]
6. Hakeem, I.Y.; Amin, M.; Abdelsalam, B.A.; Tayeh, B.A.; Althoey, F.; Agwa, I.S. Effects of nano-silica and micro-steel fiber on the engineering properties of ultra-high performance concrete. *Struct. Eng. Mech.* **2022**, *82*, 295–312.
7. Ali, B.; Qureshi, L.A. Influence of glass fibers on mechanical and durability performance of concrete with recycled aggregates. *Constr. Build. Mater.* **2019**, *228*, 116783. [CrossRef]
8. Koushkbaghi, M.; Kazemi, M.J.; Mosavi, H.; Mohseni, E. Acid resistance and durability properties of steel fiber-reinforced concrete incorporating rice husk ash and recycled aggregate. *Constr. Build. Mater.* **2019**, *202*, 266–275. [CrossRef]
9. Ahmad, S.; Hakeem, I.; Maslehuddin, M. Development of an optimum mixture of ultra-high performance concrete. *Eur. J. Environ. Civ. Eng.* **2016**, *20*, 1106–1126. [CrossRef]
10. Azad, A.K.; Hakeem, I.Y. Flexural behavior of hybrid hollow-core slab built with ultra high performance concrete faces. *Mater. Struct.* **2016**, *49*, 3801–3813. [CrossRef]
11. Richard, P.; Cheyrezy, M. Composition of reactive powder concretes. *Cem. Concr. Res.* **1995**, *25*, 1501–1511. [CrossRef]
12. Amin, M.; Hakeem, I.Y.; Zeyad, A.M.; Tayeh, B.A.; Maglad, A.M.; Agwa, I.S. Influence of recycled aggregates and carbon nanofibres on properties of ultra-high-performance concrete under elevated temperatures. *Case Stud. Constr. Mater.* **2022**, *16*, e01063. [CrossRef]
13. Ahmad, S.; Hakeem, I.; Maslehuddin, M. Development of UHPC mixtures utilizing natural and industrial waste materials as partial replacements of silica fume and sand. *Sci. World J.* **2014**, *2014*, 713531. [CrossRef] [PubMed]
14. Bajaber, M.; Hakeem, I. UHPC evolution, development, and utilization in construction: A review. *J. Mater. Res. Technol.* **2021**, *10*, 1058–1074. [CrossRef]
15. Ahmad, S.; Hakeem, I.; Azad, A.K. Effect of curing, fibre content and exposures on compressive strength and elasticity of UHPC. *Adv. Cem. Res.* **2015**, *27*, 233–239. [CrossRef]
16. Hakeem, I.; Azad, A.K.; Ahmad, S. Effect of steel fibers and thermal cycles on fracture properties of ultra-high-performance concrete. *J. Test. Eval.* **2013**, *41*, 458–464. [CrossRef]
17. Kaur, P.; Talwar, M. Different types of Fibres used in FRC. *Int. J. Adv. Res. Comput. Sci.* **2017**, *8*, 380–383. [CrossRef]
18. Althoey, F.; Zaid, O.; de-Prado-Gil, J.; Palencia, C.; Ali, E.; Hakeem, I.; Martínez-García, R. Impact of sulfate activation of rice husk ash on the performance of high strength steel fiber reinforced recycled aggregate concrete. *J. Build. Eng.* **2022**, *54*, 104610. [CrossRef]
19. Nassiri, S.; AlShareedah, O.; Rodin, H.; Englund, K. Mechanical and durability characteristics of pervious concrete reinforced with mechanically recycled carbon fiber composite materials. *Mater. Struct.* **2021**, *54*, 107. [CrossRef]
20. Mansur de Castro Silva, R.; de Andrade Silva, F. Carbon textile reinforced concrete: Materials and structural analysis. *Mater. Struct.* **2020**, *53*, 17. [CrossRef]
21. Minnaugh, P.L.; Harries, K.A. Fatigue behavior of externally bonded steel fiber reinforced polymer (SFRP) for retrofit of reinforced concrete. *Mater. Struct.* **2009**, *42*, 271–278. [CrossRef]
22. Iqbal, S.; Ali, I.; Room, S.; Khan, S.A.; Ali, A. Enhanced mechanical properties of fiber reinforced concrete using closed steel fibers. *Mater. Struct.* **2019**, *52*, 56. [CrossRef]
23. Peled, A.; Jones, J.; Shah, S.P. Effect of matrix modification on durability of glass fiber reinforced cement composites. *Mater. Struct.* **2005**, *38*, 163–171. [CrossRef]

24. Hussain, I.; Ali, B.; Akhtar, T.; Jameel, M.S.; Raza, S.S. Comparison of mechanical properties of concrete and design thickness of pavement with different types of fiber-reinforcements (steel, glass, and polypropylene). *Case Stud. Constr. Mater.* **2020**, *13*, e00429. [CrossRef]
25. Suksiripattanapong, C.; Phetprapai, T.; Singsang, W.; Phetchuay, C.; Thumrongvut, J.; Tabyang, W. Utilization of Recycled Plastic Waste in Fiber Reinforced Concrete for Eco-Friendly Footpath and Pavement Applications. *Sustainability* **2022**, *14*, 6839. [CrossRef]
26. Ahmad, W.; Farooq, S.H.; Usman, M.; Khan, M.; Ahmad, A.; Aslam, F.; Yousef, R.A.; Abduljabbar, H.A.; Sufian, M. Effect of coconut fiber length and content on properties of high strength concrete. *Materials* **2020**, *13*, 1075. [CrossRef]
27. Chin, S.C.; Tee, K.F.; Tong, F.S.; Doh, S.I.; Gimbun, J. External strengthening of reinforced concrete beam with opening by bamboo fiber reinforced composites. *Mater. Struct.* **2020**, *53*, 141. [CrossRef]
28. Hawreen, A.; Bogas, J.; Kurda, R. Mechanical characterization of concrete reinforced with different types of carbon nanotubes. *Arab. J. Sci. Eng.* **2019**, *44*, 8361–8376. [CrossRef]
29. Geng, K.; Chai, J.; Qin, Y.; Li, X.; Duan, M.; Liang, D. Exploring the brittleness and fractal characteristics of basalt fiber reinforced concrete under impact load based on the principle of energy dissipation. *Mater. Struct.* **2022**, *55*, 78. [CrossRef]
30. Wu, Z.; Shi, C.; He, W.; Wu, L. Effects of steel fiber content and shape on mechanical properties of ultra high performance concrete. *Constr. Build. Mater.* **2016**, *103*, 8–14. [CrossRef]
31. Alabduljabbar, H.; Alyousef, R.; Alrshoudi, F.; Alaskar, A.; Fathi, A.; Mustafa Mohamed, A. Mechanical effect of steel fiber on the cement replacement materials of self-compacting concrete. *Fibers* **2019**, *7*, 36. [CrossRef]
32. Azad, A.K.; Ahmad, S.; Hakeem, I. Effect of cyclic exposure and fibre content on tensile properties of ultra-high-performance concrete. *Adv. Cem. Res.* **2013**, *25*, 273–280. [CrossRef]
33. Lantsoght, E.O. How do steel fibers improve the shear capacity of reinforced concrete beams without stirrups? *Compos. Part B Eng.* **2019**, *175*, 107079. [CrossRef]
34. Torres, J.A.; Lantsoght, E.O. Influence of fiber content on shear capacity of steel fiber-reinforced concrete beams. *Fibers* **2019**, *7*, 102. [CrossRef]
35. Abambres, M.; Lantsoght, E.O. ANN-based shear capacity of steel fiber-reinforced concrete beams without stirrups. *Fibers* **2019**, *7*, 88. [CrossRef]
36. Mishra, S.; Mohanty, A.; Drzal, L.; Misra, M.; Parija, S.; Nayak, S.; Tripathy, S. Studies on mechanical performance of biofibre/glass reinforced polyester hybrid composites. *Compos. Sci. Technol.* **2003**, *63*, 1377–1385. [CrossRef]
37. Mohanty, A.K.; Misra, M.; Drzal, L. Sustainable bio-composites from renewable resources: Opportunities and challenges in the green materials world. *J. Polym. Environ.* **2002**, *10*, 19–26. [CrossRef]
38. Priya, S.P.; Ramakrishna, H.; Rai, S.; Rajulu, A.V. Tensile, flexural, and chemical resistance properties of waste silk fabric-reinforced epoxy laminates. *J. Reinf. Plast. Compos.* **2005**, *24*, 643–648. [CrossRef]
39. Althoey, F.; Hakeem, I.Y.; Hosen, M.; Qaidi, S.; Isleem, H.F.; Hadidi, H.; Shahapurkar, K.; Ahmad, J.; Ali, E. Behavior of Concrete Reinforced with Date Palm Fibers. *Materials* **2022**, *15*, 7923. [CrossRef]
40. Althoey, F.; Hosen, M. Physical and mechanical characteristics of sustainable concrete comprising industrial waste materials as a replacement of conventional aggregate. *Sustainability* **2021**, *13*, 4306. [CrossRef]
41. *ASTM C150/C150M-22*; Standard Specification for Portland Cement. ASTM: West Conshohocken, PA, USA, 2022.
42. Hakeem, I.Y.; Rahman, M.K.; Althoey, F. Experimental Investigation of Hybrid Beams Utilizing Ultra-High Performance Concrete (UHPC) as Tension Reinforcement. *Materials* **2022**, *15*, 5619. [CrossRef] [PubMed]
43. *ASTM C1602/C1602M*; Standard Specification for Mixing Water Used in the Production of Hydraulic Cement Concrete. ASTM International: West Conshohocken, PA, USA, 2018.
44. *C109/C109M-20b*; Standard Test Method for Compressive Strength of Hydraulic Cement Mortars. ASTM: West Conshohocken, PA, USA, 2020.
45. *C293/C293M-16*; Standard Test Method for Flexural Strength of Concrete (Using Simple Beam with Center-Point Loading). ASTM International: West Conshohocken, PA, USA, 2016.
46. *C138/C138M-17a*; Standard Test Method for Density (Unit Weight), Yield, and Air Content (Gravimetric) of Concrete. ASTM International: West Conshohocken, PA, USA, 2017.
47. *BS 1881-122:2011*; Testing Concrete-Part 122: Method for Determination of Water Absorption. British Standards Institution: London, UK, 1998.
48. Hosen, M.A.; Shammas, M.I.; Shill, S.K.; Jumaat, M.Z.; Alengaram, U.J.; Ahmmad, R.; Althoey, F.; Islam, A.S.; Lin, Y. Investigation of structural characteristics of palm oil clinker based high-strength lightweight concrete comprising steel fibers. *J. Mater. Res. Technol.* **2021**, *15*, 6736–6746. [CrossRef]
49. *ASTM C597-16*; Standard Test Method for Pulse Velocity Through Concrete. ASTM International: West Conshohocken, PA, USA, 2016.
50. Hosen, M.A.; Jumaat, M.Z.; Alengaram, U.J.; Sulong, N.R. CFRP strips for enhancing flexural performance of RC beams by SNSM strengthening technique. *Constr. Build. Mater.* **2018**, *165*, 28–44. [CrossRef]
51. Hosen, M.A.; Shammas, M.I.; Shill, S.K.; Al-Deen, S.; Jumaat, M.Z.; Hashim, H. Ductility Enhancement of Sustainable Fibrous-Reinforced High-Strength Lightweight Concrete. *Polymers* **2022**, *14*, 727. [CrossRef] [PubMed]
52. Mishra, R.K.; Behera, B.K.; Chandan, V.; Nazari, S.; Muller, M. Modeling and Simulation of Mechanical Performance in Textile Structural Concrete Composites Reinforced with Basalt Fibers. *Polymers* **2022**, *14*, 4108. [CrossRef]

53. Ralegaonkar, R.; Gavali, H.; Aswath, P.; Abolmaali, S. Application of chopped basalt fibers in reinforced mortar: A review. *Constr. Build. Mater.* **2018**, *164*, 589–602. [CrossRef]
54. Meskhi, B.; Beskopylny, A.N.; Stel'makh, S.A.; Shcherban', E.M.; Mailyan, L.R.; Beskopylny, N.; Chernil'nik, A.; El'shaeva, D. Insulation Foam Concrete Nanomodified with Microsilica and Reinforced with Polypropylene Fiber for the Improvement of Characteristics. *Polymers* **2022**, *14*, 4401. [CrossRef]
55. Kriker, A.; Debicki, G.; Bali, A.; Khenfer, M.; Chabannet, M. Mechanical properties of date palm fibres and concrete reinforced with date palm fibres in hot-dry climate. *Cem. Concr. Compos.* **2005**, *27*, 554–564. [CrossRef]
56. Azad, A.K.; Hakeem, I.Y. Flexural behavior of hybrid concrete beams reinforced with ultra-high performance concrete bars. *Constr. Build. Mater.* **2013**, *49*, 128–133. [CrossRef]
57. De Schutter, G.; Audenaert, K. Evaluation of water absorption of concrete as a measure for resistance against carbonation and chloride migration. *Mater. Struct.* **2004**, *37*, 591–596. [CrossRef]
58. Neville, A.M. *Properties of Concrete*; Longman: London, UK, 1995; Volume 4.
59. Bu, C.; Zhu, D.; Lu, X.; Liu, L.; Sun, Y.; Yu, L.; Xiao, T.; Zhang, W. Modification of Rubberized Concrete: A Review. *Buildings* **2022**, *12*, 999. [CrossRef]
60. Zeyad, A.M.; Hakeem, I.Y.; Amin, M.; Tayeh, B.A.; Agwa, I.S. Effect of aggregate and fibre types on ultra-high-performance concrete designed for radiation shielding. *J. Build. Eng.* **2022**, *58*, 104960. [CrossRef]
61. Althoey, F.; El-Aal, A.K.A.; Shoukry, H.; Hakeem, I. Performance of Cement Mortars Containing Clay Exposed to High Temperature. *Arab. J. Sci. Eng.* **2022**, *47*, 591–599. [CrossRef]

sustainability

MDPI

Article

Influence of Vetiver Root Morphology on Soil–Water Characteristics of Plant-Covered Slope Soil in South Central China

Xuan Wang [1,2], Zhenyu Li [1], Yongjun Chen [1,*] and Yongsheng Yao [1]

1 School of Civil Engineering, Central South University of Forestry and Technology, Changsha 410000, China; xuanwang1996@foxmail.com (X.W.); hdlizhenyu@163.com (Z.L.); yaoyongsheng23@163.com (Y.Y.)
2 China Construction Third Bureau Group Co., Ltd., Wuhan 430000, China
* Correspondence: rxrq007@163.com

Abstract: The soil–water characteristic curve is an important tool to evaluate the water-holding capacity of unsaturated soil. Plant roots can affect the matric suction of soil and the water-holding capacity and permeability of the soil. Therefore, the morphological characteristics of plant roots will lead to the difference in soil–water characteristics between soil slope and plant-covered slope. This study aims to investigate the effect of Vetiver root morphology on soil–water characteristic curves of plant-covered slope soil. The hydrological effect of the root distribution on the root–soil system was also discussed. The results showed that: (1) The root surface area index (*RAI*) and root volume ratio (R_v) of each soil section of the vetiver root system varied with depth in accordance with the Gaussian function distribution; (2) In the process of natural drying, the matric suction generated within the root system is significantly higher than that generated by evaporation of bare soil in the same soil layer. The ability of vegetation soil to enhance soil matrix suction increases with the increase of soil root surface area index; and (3) The α and n values of the SWCC model decreased with the increase of R_v (root volume ratio of soil), while the air entry value increased. Under the same water content, the matric suction corresponding to vegetation soil is significantly greater than bare soil. In addition, the soil–water characteristic curve can be effectively predicted by combining the R_v of vegetated soils.

Keywords: vegetated slope; unsaturated soil; vetiver root system morphology; matrix suction; soil–water characteristic curve

Citation: Wang, X.; Li, Z.; Chen, Y.; Yao, Y. Influence of Vetiver Root Morphology on Soil–Water Characteristics of Plant-Covered Slope Soil in South Central China. *Sustainability* **2023**, *15*, 1365. https://doi.org/10.3390/su15021365

Academic Editor: Ahmed Salih Mohammed

Received: 1 December 2022
Revised: 31 December 2022
Accepted: 9 January 2023
Published: 11 January 2023

1. Introduction

Rigid protection measures are frequently used in engineering to prevent and control the instability of the side slope [1]. With the gradual weathering of the soil and the aging of the concrete, the protection capacity of the side slope is greatly reduced, causing a series of hazards such as landslides [2]. In addition, the traditional protective measures are not conducive to environmental protection. With the strengthening of ecological awareness, plant-based ecological slope protection technology is widely adopted in engineering [3]. This technology mainly relies on the joint action of plant stems and roots to enhance the shear strength and water-holding capacity of the soil, thus improving the safety and stability of the slope [4,5]. However, the hydraulic properties of the root–soil system are inadequately studied because of the relatively complex and uncertain morphological distribution of plant roots.

In the experiment, Vetiver was selected as the slope protection plant, as shown in Figure 1. Vetiver is a perennial herb with a fast growth rate and a well-developed and tough root system. It has the characteristics of high adaptability to the environment with resistance to salinity, drought, and infertility [6]. In North America, vegetation slope protection mainly focuses on erosion control related to road construction. In Malaysia, Thailand and other Southeast Asian countries, vetiver is used to strengthen the stability of ditches and highway subgrade slope protection because of its well-developed root system, deep roots, and greater root tensile strength than ordinary plants [7]. In recent years, ecological slope protection projects have also begun to get a lot of applications in China, and achieved good slope protection effect and vegetation restoration effect [8,9].

Figure 1. Vetiver root system.

With the in-depth research on the mechanism of plant slope protection in recent years, the plant root system has exhibited good mechanical and hydrological effects on slope protection. Cheng et al. [10] conducted a pull-out test on the root system of the vetiver plant, and found that the tensile strength of the root system is positively related to the root diameter. Because of the reinforcement of the root system, the shear strength of the shallow soil on the slope is increased by about 35–100% compared with the plain soil. Xiao et al. [11] comprehensively considered root length and root diameter of the vetiver plant, further improved the original formula for calculating root tensile strength, and obtained the relevant calculation model parameters for vetiver plant roots. Deljouei's research [12] shows that the main factor affecting the change in fine root tension is the DBH of tree species and sampling trees, In the follow-up study of Deljouei [13] it was found that the increase in RAR in different species also had an impact on the enhancement of root reinforcement. In addition, after considering the effect of DBH the elevation significantly affected the root reinforcement of *C. betulus*.

In recent years, there are still many controversies about the contribution of plant roots to soil fixation hydraulic effect. The previous study shows that 90% of road collapses are caused by rainwater, suggesting that rainwater is the primary cause of side slope instability [14]. The water absorption of plant roots has an enhanced effect on slope stability. Under continuous precipitation, the maximum runoff coefficient for the plant-protected slopes is 36.96% higher than for the slopes without plant protection [15]. By investigating *Setaria viridis* and *Artemisia*, scholars have found that plants with cut leave still provide good soil consolidation under rainfall conditions [16]. Yao [17] showed that the soil–water content was the highest in the middle part of the slope. The soil at the foot of the slope should have the highest water content affected by runoff. However, the development of plant roots results in more water absorption and transpiration, leading to less water content at the foot than at the middle part of the slope. The permeability coefficient of soils with plant cover is greater than bare soils [18]. When there are more roots in the soil the rainfall infiltration in the soil is higher. At sufficient rainfall, the rainwater infiltration paths of root-rich soils differ significantly from those of bare soils, with the root system directing the lateral preferential flow of water [19]. Ng et al. [8,20] studied the interactions between the atmosphere, vegetation, and soil. The influence pattern of different plant root morphological characteristics on the stability of shallow slopes was obtained by studying the relationship between characteristic parameters, including leaf area, root area index

(*RAI*), vegetation biomass, root volume ratio, and root cross-sectional area ratio of plants and soil suction.

The soil–water characteristic curve of unsaturated soils is influenced by many factors, such as soil structure, compaction, dry density, initial porosity ratio, and stress history [21]. Li [22] believed that the mineral composition and pore structure of the soil are the most critical factors influencing the soil–water characteristic curve. If the main mineral components of the soil are all hydrophilic minerals, the soil usually has more suction and less dehumidification. The characteristic curve of the soil is also flatter, but the residual moisture is higher. Sun [23] found that different pore distribution curves of soils can cause significant differences in soil–water characteristic curves. Hou [24] compared the pore structure of original loess in different soil layers and their soil–water characteristic curves. The results showed that under the same matrix suction, different soil layers have different volumetric water contents due to the differences in pore characteristics. Cai [25] showed that the influence of pore structure on the soil–water characteristic curve was significant, and the soil–water characteristic curves of compact and loose soils showed the trends of "double-drop" and "single-drop", respectively. Yao et al. [26,27] discovered that the soil–water characteristic curve of the soil varied significantly in the range of matric suction from 0 to 200 kPa after the application of vertical stress. This result suggested that the compaction and stress state of the soil are closely related to the water-holding capacity.

In order to understand the influence of the root system on the water-holding capacity of the soil, the characteristic parameters of the root morphology of the Vetiver were measured by in situ full excavation method. Moreover, the variation in water content and matric suction of soils with different root contents in the natural environment were measured by the tensiometer method. Combined with the relevant theory of unsaturated soil mechanics, the effect of Vetiver root morphology on the soil–water characteristic curve model of side slope soils was investigated.

2. Materials and Methods

2.1. Materials

2.1.1. Experimental Site

The test site was located at Central South University of Forestry and Technology in Changsha, Hunan Province, China (28°8′ N, 112°59′ E, 90 m a.s.l.). It belongs to the subtropical monsoon climate. During the project construction, the excavated cutting or manually filled embankment slopes are mostly barren earth rock slopes. The test site was filled with barren, strongly weathered red sandy soil. During the filling process, in order to keep the soil uniformity, void ratio and other parameters consistent, the method of layered filling and compaction was adopted. After standing for a period of time, the soil was gradually compacted under its own weight, and then with the same method, so as to finally complete the filling in the test slope. The physical and mechanical properties of the red soil on the test side slope are shown in Table 1. The particle size distribution curve is shown in Figure 2. The size of the test side slope is 20 m × 2 m × 2 m. The slopes were divided into two areas: one with Vetiver planted slopes; the other with bare soil slopes as the control group. The environmental conditions such as rainfall, temperature and light, were kept consistent for all slopes.

Table 1. Properties of soils samples.

Proportion	Natural Density ($g{\cdot}cm^3$)	Maximum Density ($g{\cdot}cm^3$)	Optimum Moisture Content (%)	Liquid Limit (%)	Plastic Limit (%)	Saturated Penetration Coefficient ($mm{\cdot}h^{-1}$)
2.72	1.82	1.62	21.5	41.5	25.4	1.426

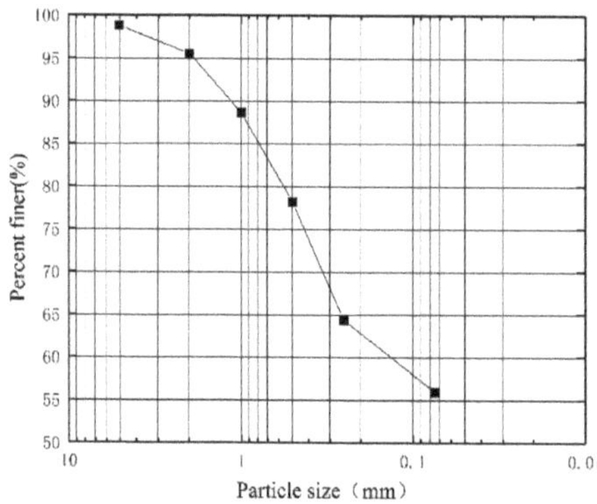

Figure 2. Gradation curve of soil sample.

The test started on 15 February 2021, when the seeds of Vetiver were sown. After planting the vetiver grass, in order to ensure the water demand for its early development and growth the vetiver grass should be watered manually at regular intervals in the first two weeks after planting until it can grow and reproduce normally. Then, according to climate conditions or test needs, decide whether to supplement water to plants to ensure that vetiver will not stop growing due to excessive water shortage, and observe and record the growth of vetiver in the test site every day.

2.1.2. Monitoring System and Period

The monitoring system consisted of two types of sensors (Figure 3a) and a data collection system (Figure 3b). In order to measure the volumetric soil moisture content of each measurement point, WaterScout SM100 with a range of 0%~saturation was adopted, and its precision is 0.1% of the range. To measure matric suction Watermark Soil Moisture Sensors were used. The measuring range of this type of tensiometer is 0–200 kPa. In order to ensure that the sensor is effective, it should be saturated before installation and the sensor should be preconditioned by repeated drying and wetting cycles (soaking for 30 min, and then drying for several hours). Then use a small shovel to drill and take soil at the slope site and dig a hole slightly larger than the size of the sensor to prevent wear of the sensor film. Fill the hole with water and carefully push the sensor down into the hole. When the sensor is completely pushed into the hole, backfill and tamp the hole in time to eliminate the hole cavitation of tensiometer.

During the test, soil moisture sensors and tensiometers pre-buried in the specimen soil were used to determine the change in volumetric soil moisture content and the matric suction in each soil layer under the natural environment. The specific location of the instrument and the measurement method are shown in Figure 3a, respectively. Each measurement point was located in the vertical direction of the planting location, and soil moisture sensors and tensiometers were pre-built at the measurement points. According to the previous research [28] on the root system of vetiver by the research group and in order to make the test results have better discrimination, the first measurement point was located at a depth of 150 mm, the second at a depth of 350 mm, and the third at a depth of 550 mm; the measurement point of bare slope is the same as above (Figure 3a). The monitoring of the volumetric soil moisture content and the matric suction began on 7 July 2021, when the root system of the tested vetiver grows for about six months. Within 47 days during the test data were recorded every 24 h at each measurement point using a data collection system. After 47 days of monitoring, the monitored vetiver was sampled and analyzed to describe the effect of plant root morphological characteristics on the soil matric suction.

During the test it was rainy season in Hunan Province. The total precipitation was 225.46 mm, and the temperature was within the range of 16~32 °C. The specific weather temperature and precipitation in the test site are shown in Figure 4.

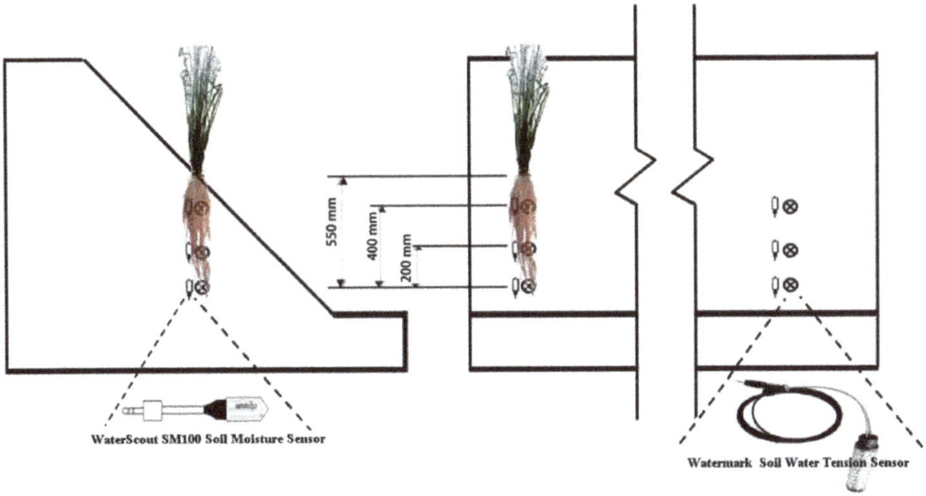

(a)

(b)

Figure 3. (**a**) Installation position of soil moisture content and matrix suction measuring equipment; (**b**) Soil sensor reader.

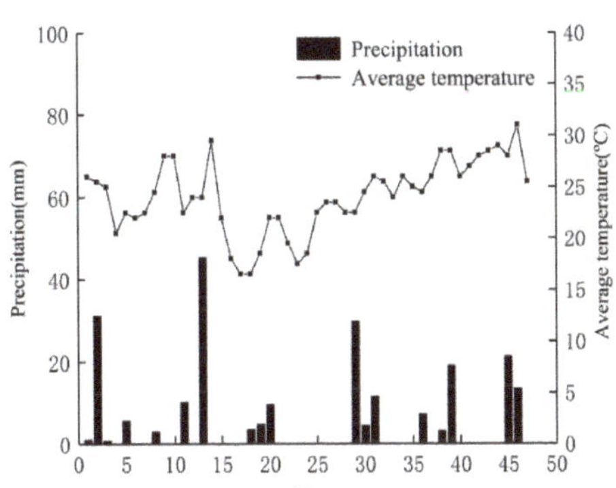

Figure 4. Temperature and precipitation during the test.

2.2. Methods

2.2.1. Vetiver Root Morphological Characteristics

In order to quantitatively describe the effect of plant root morphological characteristics on the soil matric suction, the roots of the experimental Vetiver with a growth period of six months were collected. The main parameters measured in this experiment included the number of roots, root diameter, and root length. With the root branching order as the ordering standard [29], the root ordering method was used to classify the coarse and fine roots. The classification of this method is based on structure. For example, the root system located at the extreme end without branches is defined as the order 3 root, while the parent root of the order 3 root is defined as the order 2 root. Based on the planar structure of the Vetiver root system (Figure 5), it can be classified into three root orders. The root tips of Vetiver are mostly order 2 and order 3 roots, which are the main water-absorbing areas of the root system. The root trichomes of order 1 roots are relatively few and have a weak water absorption capacity.

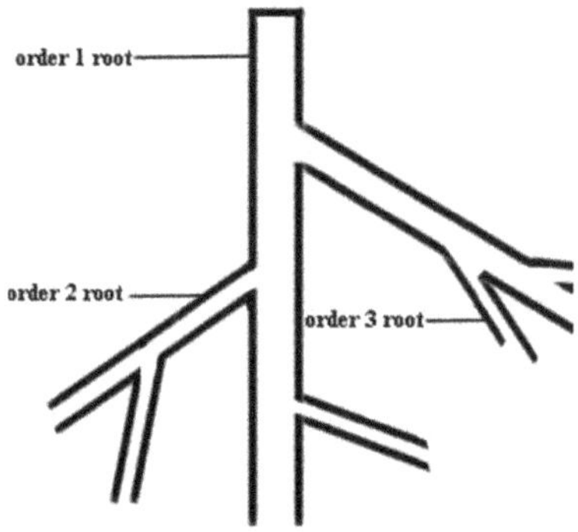

Figure 5. Plane configuration of vetiver root system.

Plant roots are in direct contact with the soil and are the main organ to absorb water and other nutrients. Due to the different distributions of roots in different soil sections, the soil suction is also significantly affected by the root distribution. The *RAI* is defined as the ratio of the root surface area of order 2 and 3 roots in the longitudinal section at a given depth to the area of the root extension area in the horizontal direction, as shown in Equation (1):

$$RAI = \frac{\sum_{i=1}^{n} \pi d_i \Delta h}{\frac{\pi D_r^2}{4}} \tag{1}$$

where Δh is the calculated depth range, usually 10 mm; D_r is the maximum extension diameter of the root system in a horizontal direction; d_i is the diameter of the *i*th root system; the total number of roots is n.

The calculation of *RAI* requires first determining the number of order 2 and order 3 roots and their sizes. Before measurement the roots are excavated, and the soil attached to the roots is washed with a small stream of water. This process reduces the damage to the root system and increases the accuracy and efficiency of the measurement.

The root volume ratio R_v is defined as the total volume of all plant roots at a certain depth and for the proportion of the total soil volume in that soil layer, see Equation (2):

$$R_v = \frac{\sum_{i=1}^{n} \frac{\pi d_i^2}{4} \Delta h}{\frac{\pi D_r^2}{4} \Delta h} \tag{2}$$

The measurement of R_v is approximately the same as that of *RAI*. After excavating the overall root system of the plant the number of all roots and the diameter of the roots were measured for all sections at each depth. The plant roots are then grouped by 0.5 mm diameter, and the number of roots in each diameter range is counted.

2.2.2. Soil–Water Characteristic Curve

In order to establish the mathematical model of the soil–water characteristic curve, the effective saturation S_e defined by the saturated water content and residual water content was introduced, which can be expressed by Equation (3) [30]:

$$S_e = \frac{\theta - \theta_r}{\theta_s - \theta_r} \tag{3}$$

where S_e is effective saturation; θ_s is saturated water content; θ_r is residual moisture content. When the residual water content is 0, the effective saturation S_e equals the soil saturation.

The Van Genuchten model [31] is the most commonly used mathematical model for describing soil-water characteristic curves, as shown in Equation (4). This model has high fitting accuracy for different types of soils and can describe the soil–water characteristic curve in the range of full negative pore water pressure:

$$S_e = \frac{1}{\left[1 + (\alpha\varphi)^n\right]^m} \tag{4}$$

where α, m and n are fitting parameters; ψ is matric suction.

Combined with Equation (3), the model can be expressed as:

$$\theta = \theta_r + \frac{\theta_s - \theta_r}{\left[1 + (\alpha\varphi)^n\right]^m} \tag{5}$$

The fitting constants of the soil–water characteristic curve model depend on the physical properties of the soil such as pore size distribution and air entry pressure values.

The reason for the change in the hydraulic properties of the vegetated soil is the decrease in pore space and increase in water-holding capacity due to the root system occupying the pore space in the soil. Considering the influence of plant roots on the soil–water characteristic curve, the prediction order of suction change in vegetated soil can be improved, providing the basis for analyzing the stability and erosion resistance of shallow vegetated soil. Leung et al. [32] proposed a predicted model for the pore ratio of rooted soils based on the concept of a root system occupying the volume of the soil pore, as shown in Equation (6):

$$e = \frac{e_0 - R_v(1 + e_0)}{1 + R_v(1 + e_0)} \tag{6}$$

where e_0 is Void ratio of bare soil; R_v is the volume ratio of root system, which varies with the depth of soil. See Section 2.2.1 for the volume ratio depth distribution characteristics of vetiver root system.

To simulate the effect of the root system on the water-holding capacity of the soil, Gallipoli et al. [33] proposed an equation with soil porosity ratio as a function; see Equation (7):

$$S_r = \left[1 + \left(\frac{se^{m_4}}{m_3}\right)^{m_2}\right]^{-m_1} \tag{7}$$

where S_r is the saturation of soil mass, s is the matrix suction of soil mass, e is the void ratio, m_1, m_2, m_3 and m_4 are all fitting parameters of the model, dimensionless.

3. Results

3.1. Vetiver Root Morphology Characteristics

The Vetiver root system in this experiment is shown in Figure 6. The length, root diameter, and root depth were recorded by manual measurement.

Figure 6. The experimental Vetiver root system.

The number of order 1 roots obtained from the experiment is 223, and the number of order 2 and 3 roots is 3158. The root diameters of order 1, order 2, and order 3 roots in each depth section were summed up at a unit depth of 10 mm to obtain the total root diameter of the root system, as shown in Figure 7.

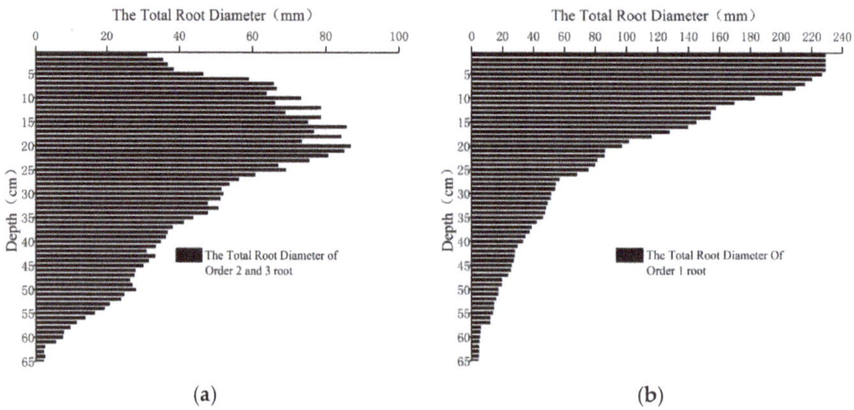

Figure 7. Characteristics of depth distribution of (**a**) order 2 and 3 roots of Vetiver; (**b**) order 1 root of Vetiver.

The results show that order 2 and 3 roots are mainly concentrated in the surface layer from 0 to 20 cm, accounting for 73% of the total root system in the soil section. Then, the number of roots decreases with increasing depth. In soils below 20 cm, order 2 and 3 roots are also relatively enriched in some areas but the number is still lower than at the surface layer.

The RAI and R_v of each soil section of the Vetiver root system were fitted. The results showed that the variation in RAI and R_v with depth is consistent with the Gaussian function (Equation (8)), as shown in Figure 8.

$$f(z) = ae^{-(z-b)^2/2c^2} \tag{8}$$

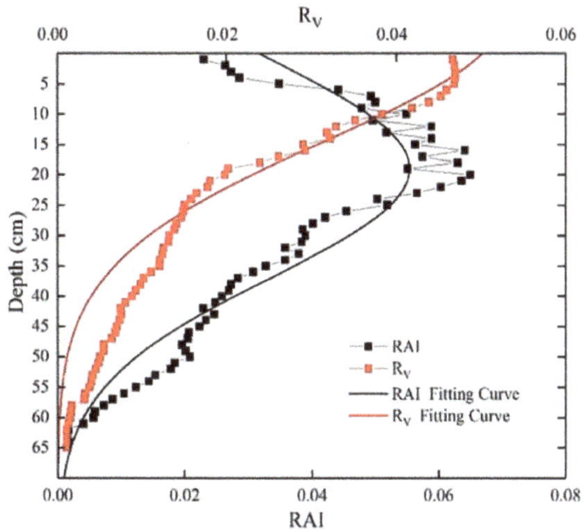

Figure 8. Root area index (*RAI*) and Root volume ratio (R_v) of each soil layer section.

The fitting results are shown in Table 2. The variation in *RAI* with depth is roughly similar to the variation in root diameter with soil depth, which resembles an upside-down clock in shape. The root system of the test vetiver has a maximum *RAI* of 0.05541 at a depth of 18.97 cm. In the surface soil with a depth of 15 cm, the *RAI* increases with depth. The R_v of vetiver shows a slight increase in the surface soil with a depth of 6 cm, and then it began to decrease with depth in the form of a Gaussian function.

Table 2. Fitting parameters of distribution characteristic curve of vetiver root system.

	Fitting Parameters			Correlation Coefficient (R^2)
	a	*b*	*c*	
RAI	0.05541	18.97	25.39	0.9149
R_v	0.05241	−5.9241	28.59	0.9560

3.2. Influence of Vetiver Root Morphology on Matrix Suction

In this study, the effect of Vetiver transpiration and soil evaporation on side slope suction was studied by comparing and analyzing the changes in matric suction of side slopes with and without plant cover. The suction variation at the slope soil depths of 150 mm, 350 mm, and 550 mm was monitored by a tensiometer buried in the slope soil, and the results are shown in Figure 9.

The magnitude and variation in matric suction of vegetated soil were larger than those for bare soil in the same layer, indicating that the water transpiration effect of plant roots is significant. Compared to bare soil, the root system in the vegetated soil increased the matric suction by absorbing water from the soil, significantly reducing the water content. The root system near the surface soil was more densely distributed and strongly influenced by the atmosphere. Therefore, the soil closer to the slope surface showed a greater variation in matric suction, while the soil in the deeper layer had a smaller variation in matric suction.

Combined with the root morphology, the weather from day 20 to day 28 of the test period has been sunny with zero rainfall, and the slope soil is in a naturally dry state. The effect of transpiration and evaporation of vegetated soil and evaporation of bare soil on the soil matric suction during this period is shown in Figure 10. Table 3 shows the matric suction Δs generated in bare soil and vegetated soil in each period, i.e., $\Delta s = s_1 - s_0$, where s_0 is the initial matric suction before soil drying, s_1 is the matric suction after 48 h of drying.

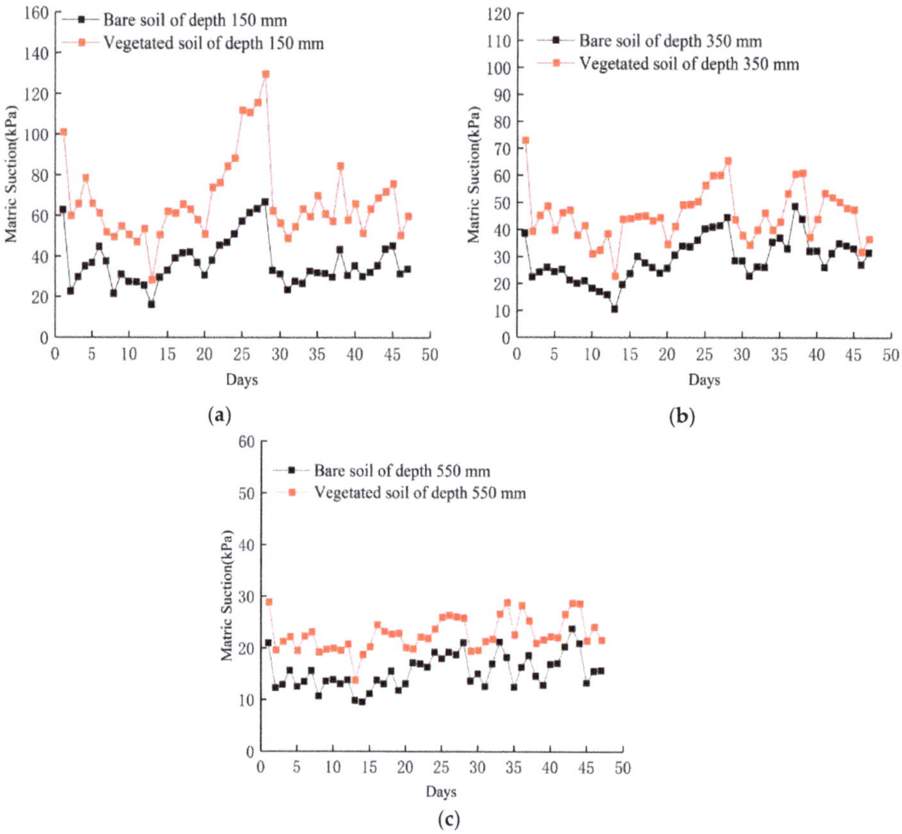

Figure 9. Change in matrix suction of (**a**) 150 mm; (**b**) 350 mm; (**c**) 550 mm soil layer.

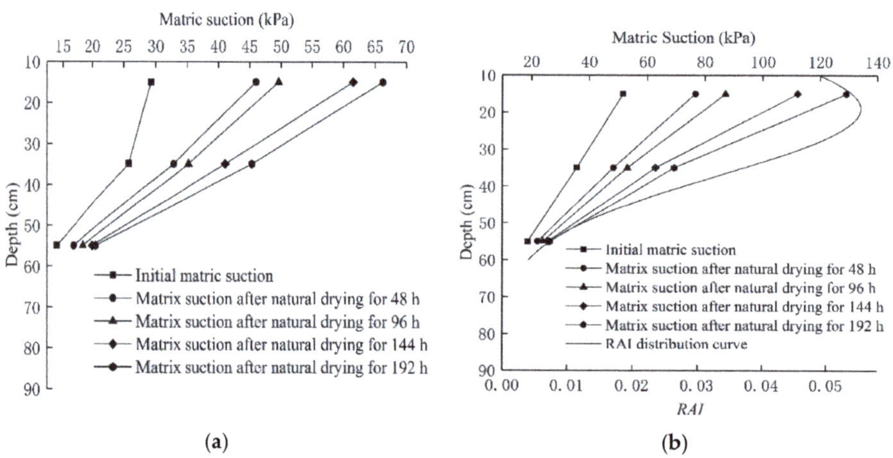

Figure 10. Change in matrix suction of (**a**) bare soil and (**b**) vegetated soil at each depth under natural dry environment.

Table 3. Matrix suction Δs (kPa) generated by soil in each period.

Period	Soil Types	150 mm	350 mm	550 mm
0–48 h	Bare Soil	16.7	7	2.8
	Vegetated Soil	25	12.6	3.4
48–96 h	Bare Soil	3.6	2.4	1.6
	Vegetated Soil	10.5	4.7	1.8
96–144 h	Bare Soil	9.9	5.9	1.3
	Vegetated Soil	23.1	9.8	1.5
144–192 h	Bare Soil	4.7	4.2	0.6
	Vegetated Soil	17.1	6.4	0.7

From Figure 10a, it can be seen that the matric suction of the bare slope soil varies within 8 days of natural drying, with the greatest variation at a depth of 150 mm, increasing from 29.3 kPa to 66.2 kPa. The suction of the soil grows from 25.8 kPa to 45.3 kPa at a soil depth of 350 mm. The soil located at a depth of 550 mm has the smallest suction increase, growing from 14 kPa to 20.6 kPa. These results indicate a close relationship between the evaporation rate of the soil and the soil depth. The surface soils are strongly influenced by the atmosphere, with stronger evaporation and more significant suction changes than the deeper soils.

Comparing the relationship between *RAI* and soil matric suction in each soil layer, it can be found that the matric suction generated in the root system is significantly higher than that generated in the bare soil at the same layer. The matric suction of both vegetated soil and bare soil significantly increases at the beginning of the natural drying process (0–48 h). The matric suction of the soil covered by Vetiver reaches 76.6 kPa at a depth of 150 mm, and the matric suction generated by transpiration and evaporation reaches 25 kPa during that time, which is 8.3 kPa higher than that generated by the bare soil. At a soil depth of 350 mm the matric suction of the Vetiver-covered soil reaches 48.3 kPa, and the matric suction generated by transpiration and evaporation is 12.6 kPa, which is 5.6 kPa higher than that generated by the bare soil. At a soil depth of 550 mm, the substrate suction produced by both Vetiver-covered soil and bare soil is at a low order, with the former only 0.6 kPa greater than the bare soil. At 96 h of natural soil drying, the matric suction generated by both vegetated soil and bare soil is lower than that generated in other periods. The matric suction generated by each layer of vegetated soil at this period is 10.5 kPa, 4.7 kPa, and 1.8 kPa, which decreases by 58%, 62.7%, and 47%, respectively. These differences may be caused by the relatively weak evaporation rate of the soil, photosynthesis, and transpiration of the plants in cloudy or low-temperature environments. The enhancement of soil matric suction by plant transpiration is most significant during 96–144 h of the soil drying phase. The enhancement of matric suction is 23.1 kPa in the vegetated soil at a depth of 150 mm, 13.2 kPa higher than that of bare soil.

During the natural drying of the soil, the variation in soil suction in each soil layer was significantly influenced by the *RAI* of Vetiver roots, as shown in Figure 11.

It can be seen from Figure 11a that due to soil evaporation and plant transpiration, the surface soil is strongly influenced by the atmosphere and is rich in the root system. Therefore, the matric suction generated during natural drying is significantly greater in the surface soil than in the deep soil, where the vegetated soil generates a total of 77.6 kPa of matric suction, which is increased by 40.7 kPa compared to the bare soil. As the root volume in the deep soil decreases, the matric suction generated by the vegetated soil also decreases. At a depth of 550 mm, the matric suction generated by the vegetated soil during drying is only 1.1 kPa higher than that of bare soil.

As shown in Figure 11b, the *RAI* of the soil layer at a depth of 150 mm is 0.0563. Due to the environmental factors at each time, the matric suction generated by the vegetated soil is 6.9 kPa to 13.1 kPa higher than the bare soil. The *RAI* of the soil layer at a depth of 350 mm is 0.0328, and the increase in matrix suction generated by the vegetated soil at this depth ranges from 2.2 kPa to 5.6 kPa. For the sparse root system at 550 mm soil layer with an *RAI* of only 0.0124, the increase in matric suction in the vegetated soil reaches 0.6 kPa at the early drying stage and is less than or equal to 0.2 kPa at all other times. Based on the above results it can be concluded that plant roots can effectively enhance the matric suction during soil drying, and the enhancement increases with the *RAI* of the root-bearing soil and is affected by environmental factors.

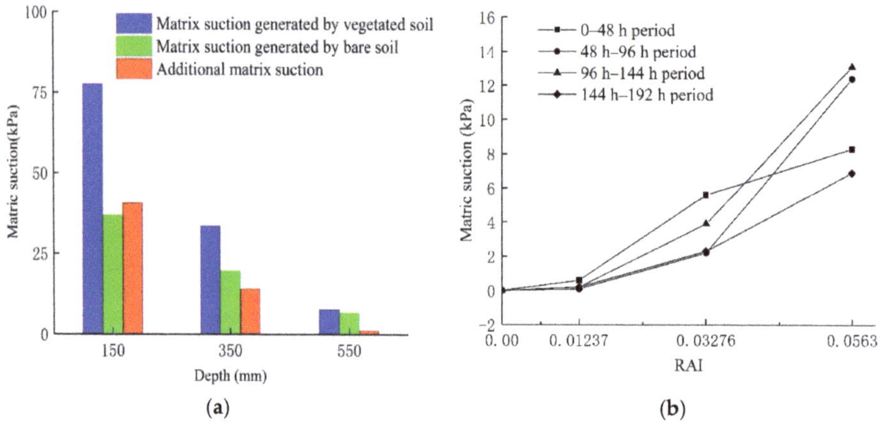

Figure 11. (**a**) Matrix suction generated by natural drying and dehumidification of soil mass; and (**b**) Relevance between matrix suction and root surface area coefficient.

3.3. The Soil–Water Characteristic Curve Model of Vetiver-Covered Slope Soil

The above results suggest that the matric suction of each soil layer varies significantly with the decrease in the volumetric water content during the test period. On this basis, it can be assumed that the soil is in a four-phase state (solid–liquid–gas–shrinkage film) during the whole field test, which can be reflected by the transition of the soil–water characteristic curve.

Because of the limited variation range of water content and matric suction by the natural environment in the field test, it is often difficult to obtain the soil–water characteristic curve in the full suction range. Before fitting the test data to obtain the parameters α, m, and n, the saturated water content θ_s and the residual water content θ_r of the soil must be determined first. Because there is no uniform test standard the residual water content θ_r is generally obtained by fitting. The saturated water content θ_s can be calculated from the pore ratio e, see Equation (9) [32]:

$$\theta_s = \frac{e}{1+e} \tag{9}$$

For vegetated soils, the growth of roots occupies part of the soil pore space and changes the pore structure, resulting in a change in pore size. Therefore, the pore ratio of root-penetrated soil can be expressed by Equation (6).

The parameters of the soil–water characteristic curve for each soil sample obtained by fitting the V-G model are shown in Table 4. Soil–water characteristic curves of soil with and without root systems are shown in Figure 12.

It can be seen that the fitted soil–water characteristic curves have three stages, including the boundary effect stage, the transition stage, and the unsaturated residual stage. In the boundary effect stage, the suction of the soil slightly increases and the water content maintains the saturation value. In the transition stage, the suction increases to the air entry suction value of the soil. The gas starts to enter the large pores in the soil, the soil starts to drain, and the water content starts to decrease significantly. In the unsaturated residual stage, the water content is gradually weakened by suction. The water content no longer decreases with the increase in suction at a certain critical point, and the water content at this point is the residual water content.

Table 4. Fitting parameter values of V-G model for different soil layer.

Soil Types	θ_s	Fitting Parameters				Correlation Coefficient (R^2)
		θ_r	α	n	m (1 − 1/n)	
Bare soil	41.6735	6.461	0.13295	1.916	0.4781	0.90483
Vetiver covered slope 150 mm soil layer	38.7736	2.885	0.07981	1.7231	0.4197	0.92725
Vetiver covered slope 350 mm soil layer	40.474	4.134	0.09389	1.797	0.4435	0.93672
Vetiver covered slope 550 mm soil layer	41.1739	5.193	0.10868	1.8541	0.4607	0.92746

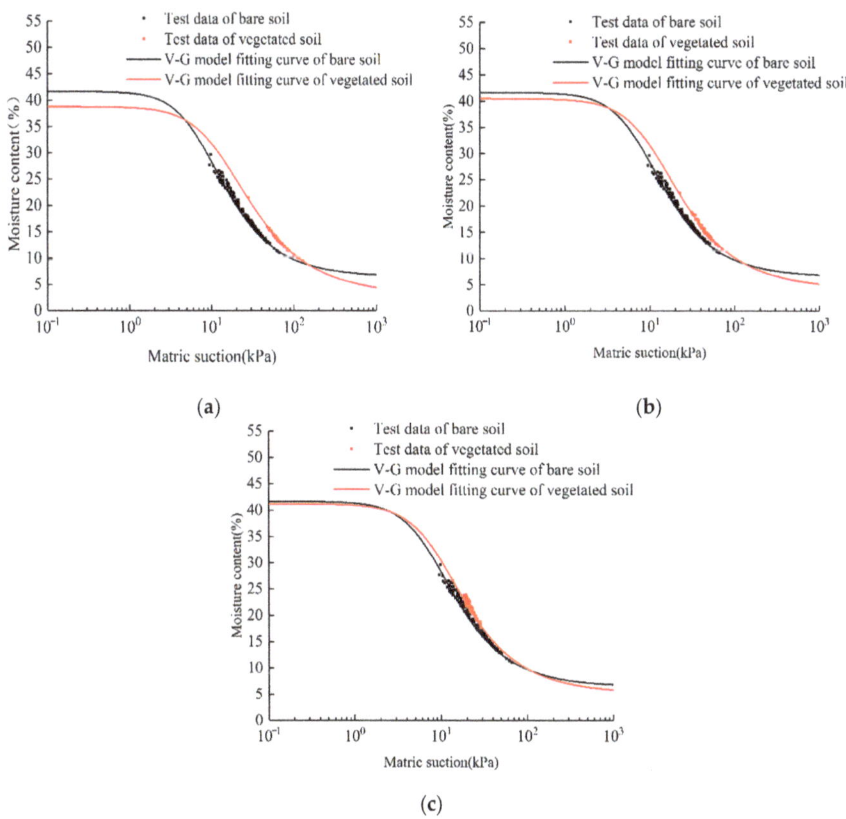

Figure 12. Fitting soil–water characteristic curve of (**a**) 150 mm soil layer; (**b**) 350 mm soil layer; (**c**) 550 mm soil layer.

The soil sample data match the Van Genuchten model, and the fitted coefficients of determination R^2 are all greater than 0.9. With the increase in matric suction the soil–water characteristic curves of each soil layer cross near the matric suction of 10 kPa. With further increase in the substrate suction the curves converge again. The saturated volumetric water content of the soil decreases with the increasing R_v of the soil layer, e.g., the saturated water content of the soil at 150 mm decreases by about 2.89% compared with that of the bare soil. The parameter α associated with the air entry value also varies with the R_v of the soil layer, with a larger R_v indicating the smaller the fitting parameter α. At the soil layer with a depth of 550 mm the fitted parameter α is 0.10868 for the vegetated soil, smaller than that for the bare soil (0.13295), and the curve is slightly smoother for the vegetated soil than for the bare soil. The most significant change in the soil–water characteristic curve of the bare soil is found at a depth of 150 mm. For the same water content at a gentler characteristic curve, the matric suction corresponding to the vegetated soil is significantly greater than that of bare soil. These results are of great significance for controlling surface soil erosion and shallow stability.

Determination of the air entry value of the soil–water characteristic curve is necessary for studying unsaturated soils. To obtain the air entry value, it is crucial to determine the inversion point on the soil–water–soil characteristic curve. The calculation process is shown below.

The coordinate of the inversion point is:

$$\begin{cases} \varphi_i = \frac{1}{\alpha m^{1/n}} \\ \theta_i = \theta_r + \frac{\theta_s - \theta_r}{(1+1/m)^m} \end{cases} \tag{10}$$

After substituting the coordinate of the inversion point into the slope equation, the slope of the inversion point is obtained:

$$K_i = \theta'(\varphi) = \frac{d\theta(\varphi)}{dlg\varphi} = -\frac{ln(10)(\theta_s - \theta_r)n}{[1+1/m]^{m+1}} \tag{11}$$

After obtaining the coordinate of a point and the slope of the tangent line, the equation of the tangent line at that point can be expressed as:

$$\theta - \frac{1}{\left(1+\frac{1}{m}\right)^m} = -\frac{(\theta_s - \theta_r)n}{[1+1/m]^{m+1}}\left[lg\varphi - lg\left(1/\alpha m^{\frac{1}{n}}\right)\right] \tag{12}$$

Finally, a horizontal tangent line tangent to the soil–water characteristic curve is created at the point (θ_s, 0), and the coordinate of its intersection with the tangent line of the inversion point is calculated. The obtained horizontal coordinate of the intersection point is the air entry value ψ_{aev} of the soil, as shown in Table 5. The results show that the air entry value of the soil increases with the increasing R_v of the soil layer. The air entry value of the bare soil is 3.2337 kPa. The R_v of the vegetated soil at a depth of 150 mm reaches 0.029, significantly increasing the air entry value which is 2.652 kPa higher than that of the bare soil. The presence of the root system also affects the location of the inversion point. With a larger R_v of the soil layer, the horizontal coordinate of the inversion point on the soil–water characteristic curve is larger. This result suggests that the period before the residual stage is prolonged, enhancing the water-holding capacity of the soil. It can be found through Table 5 that the slope of bare soil reaches -0.827 at the inversion point, while the slope of vegetated soil decreases with the increase in R_v. Therefore, the soil–water characteristic curve of vegetated soil can be smoother with increasing R_v compared with that of bare soil. Under the same water content, the matric suction of vegetated soil is greater than that of bare soil.

The presence of the root system has a significant and regular effect on the soil–water characteristic curve of the soil. The relationship between the soil–water characteristic curve parameters and R_v based on the V-G model is shown in Figure 13.

It can be seen from Figure 13a that when R_v is 0 (the soil does not contain roots), the value of α is 0.13295. When R_v is 0.029, the value of α decreases to 0.07981. Consistent with the reference [34], the value of parameter α in the V-G model tends to decrease with increasing R_v, with lower α indicating a higher air entry value. When R_v is 0, the α value reaches its maximum, corresponding to the minimum air entry value of 3.2337 kPa. When R_v is 0.029, the α value reaches its minimum and its corresponding air entry value reaches the maximum of 5.8857 kPa. Therefore, the air entry value of the soil increases with the increase in R_v.

As can be seen in Figure 13b, the values of the fitted parameters n of the V-G model for the test soil range from 1.7231 to 1.916, which increases with R_v.

Table 5. Basic parameters of Soil–Water Characteristic Curve of each soil layer.

Soil Types	Basic Parameters of Soil–Water Characteristic Curve		
	Air Entry Value (ψ_{aev})	Inversion Point (ψ_i, θ_i)	Slope (k_i)
Bare soil	3.2337	(11.06, 0.2699)	−0.827
Vetiver covered slope 150 mm soil layer	5.8857	(20.74, 0.24)	−0.688
Vetiver covered slope 350 mm soil layer	5.0486	(16.74, 0.2567)	−0.742
Vetiver covered slope 550 mm soil layer	3.9824	(13.98, 0.2634)	−0.789

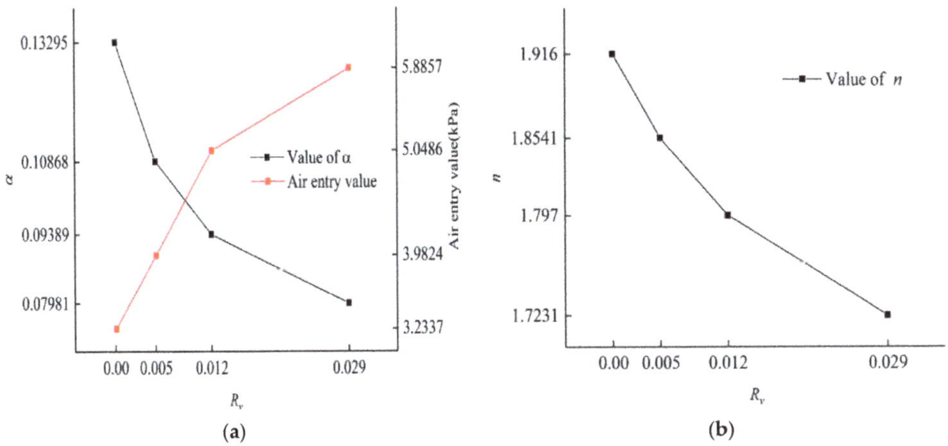

Figure 13. Variation law of (**a**) parameters α and air entry value; (**b**) n value with R_v.

3.4. Prediction of the Soil–Water Characteristic Curve by R_v

The soil–water characteristic curves of both bare soil and vegetated soil are obtained using the V-G model, better characterizing the relationship between volumetric water content and matric suction of the soil. To better visualize the effect of plant roots on the soil–water characteristic curve, the equation related to the porosity ratio proposed by Gallipoli et al. [33] is adopted to investigate the effect of R_v on the water-holding capacity of the soil. Four fitting parameters (m_1, m_2, m_3, and m_4) are obtained by fitting the Gallipoli model to the test data points of bare soil, as shown in Table 6. Finally, the soil–water characteristic curve of the vegetated soil is obtained by combining the parameter R_v of the root system based on the soil–water characteristic curve, as shown in Figure 14.

From Figure 14, it can be found that the prediction curves obtained from the parametric root volume ratio R_v of the root system are in good agreement with the measured values. By analyzing the soil–water characteristic curves of the three sets of soil samples with roots, it can be seen that the presence of roots increases the air entry value of the soil.

Table 6. Parameter values of Gallipoli model.

Soil Types	Parameters of Gallipoli Model					R_v
	m_1	m_2	m_3	m_4	e	
Bare soil	0.4695	1.807	3.657	2.932	0.7161	0
Vetiver covered slope 150 mm soil layer	0.4695	1.807	3.657	2.932	0.6332	0.029
Vetiver covered slope 350 mm soil layer	0.4695	1.807	3.657	2.932	0.6799	0.012
Vetiver covered slope 550 mm soil layer	0.4695	1.807	3.657	2.932	0.6961	0.005

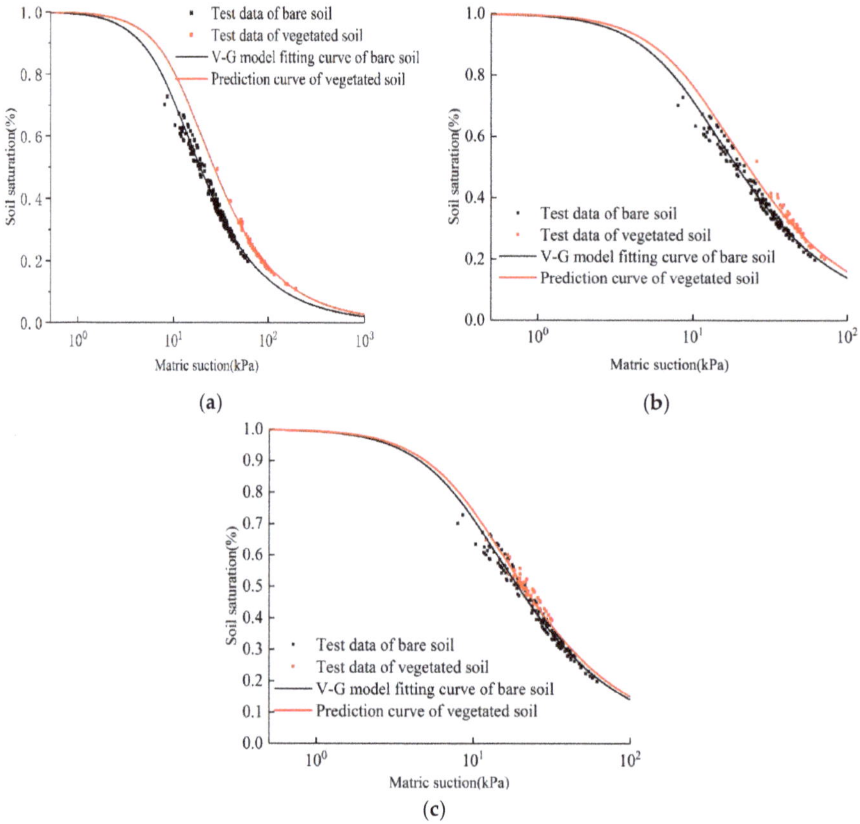

Figure 14. Comparison between soil–water characteristic curve predicted by plant root parameter R_v and measured value (**a**) vegetation soil at 150 mm soil layer; (**b**) vegetation soil at 350 mm; (**c**) vegetation soil at 550 mm.

4. Discussion

Compared with Li's research [28] on the morphology of biennial vetiver root system, the results show that the distribution of Vetiver roots in depth direction exhibits similar patterns in different growth periods. The root system of vetiver is very dense within 0~20 cm below the ground surface. With the increase in soil depth, the biomass of grass root system decreases regularly [35]. It is more economical for plants to grow their roots closer to the soil surface where they can exploit most of the available nutrients and water from natural rainfall [36]. Vetiver grass belongs to the vertical root type, and its root group diffusion range is about 20~25 cm in diameter. The diffusion range of the Vetiver root group sampled in this study is 21.5 cm in diameter. The growth range of vetiver root group first increased and then decreased with the depth of soil layer, and its distribution conforms to the Gauss curve [28]. According to the change pattern of *RAI* with depth (Figure 7), it can be inferred that during the growth period the vetiver root system mainly produces more order 2 and 3 roots through order 1 roots, so as to meet the demand of expanding around the soil and absorbing more water and nutrients.

In this study the transpiration of plant roots is obvious. Compared with bare soil, where there is only soil evaporation, the roots in vegetation soil absorb water from the soil, thereby greatly reducing the soil moisture content and increasing the soil matrix suction. Ng [8] found that after 20 days of plant transpiration test, the maximum suction of Bermuda grass vegetation soil can reach 1.3 times that of bare soil, which is similar to the test results in this paper: the vegetation soil with the largest root surface area index *RAI* produces 40.7 kPa more than that of bare soil in the 8-day natural drying process. Moein et al. [37] also found that soil–water content was negatively correlated with very fine root density (length and biomass) by studying the relationship between root distribution and soil–water content of *C. japonica* and *C. obtusa*.

Rahardjo et al. [38] found that roots increased the air entry value of silty soils by about 4 kPa. Yan [39] also showed that silty soils were affected by the root system, with the air entry value increasing by about 3 kPa. The conclusions in the present study are similar to those of the above studies. In the study of the parameter n of the V-G model, Chen et al. [40] found that the flatness of the soil–water characteristic curve was affected by the value of parameter n, with a smaller n value indicating a higher flatness degree. There is a correlation between the n value and the pore size distribution of the soil. When the pore size distribution of the soil is uniform (the filling effect of the pore space between each soil group is poor), the n value is larger and the curve is steeper. Thus, it can be assumed that the presence of roots has a similar effect to the increased grain size of the soil for improving soil gradation.

Leung et al. also found similar changes caused by roots in soil–water characteristic curves [32]. Both the measured results and the Gallipoli model prediction curves show that the roots of plants can enhance the air entry value and the water-holding capacity of the soil. Romero et al. [41] and Ng et al. [42] also demonstrated that a lower porosity ratio indicated a higher air entry value of the soil.

5. Conclusions

In this study, Vetiver, a common slope protection plant, was selected as the test object, the root morphology of Vetiver was analyzed, and the characteristic parameters of root morphology were obtained. Through outdoor tests, the matric suction and moisture content of soil at different soil depths with or without plant coverage was measured, the variation rules of soil moisture content and matric suction under the action of plant roots were studied, and the relationship between root morphology and matric suction variation was analyzed. Through fitting the test data, the soil–water characteristic curves of soil layers at different depths with or without vegetation coverage were obtained. The soil–water characteristic curve model of slope soil considering the root form of Vetiver grass was established. The effects of root volume ratio on air entry value, saturated water content, and parameters of the V-G model were studied. The main conclusions are as follows:

The root surface area index *RAI* and root volume ratio R_v of each soil section of the Vetiver root system varied with depth in accordance with the Gaussian function distribution.

In the process of natural drying, the matric suction generated within the root system is significantly higher than that generated by the evaporation of bare soil in the same soil layer. The ability of vegetation soil to enhance soil matrix suction increases with the increase in soil root surface area index.

The V-G model fits the discrete data points of matrix suction and water content of bare soil and Vetiver vegetation soil obtained from the test well. Through the analysis of the soil–water characteristic curve of soil with or without vegetation coverage, it is found that a larger root volume ratio of soil leads to smaller α and n. As a result, the air entry value is larger, and the soil–water characteristic curve is gentler. Under the same water content, the matric suction corresponding to vegetation soil is obviously greater than that corresponding to bare soil.

According to the soil–water characteristic curve model related to the void ratio proposed by Gallipoli and the root volume ratio R_v of soil, the soil–water characteristic curve of vegetation soil is predicted. The results show that the method in this study can effectively predict the soil–water characteristic curve of vegetation soil mass. Furthermore, the steps are simplified by this method, thus obtaining the soil–water characteristic curve of vegetation soil mass when the plant root morphology characteristic parameters are available.

Author Contributions: Conceptualization, X.W. and Z.L.; methodology, X.W. and Z.L.; software, X.W.; validation, Y.C., Y.Y. and Z.L.; formal analysis, Y.C.; investigation, X.W. and Z.L.; resources, Z.L.; data curation, X.W.; writing—original draft preparation, X.W. and Y.Y.; writing—review and editing, X.W. and Y.C.; visualization, Y.C.; supervision, Z.L.; project administration, Z.L.; funding acquisition, Y.Y. All authors have read and agreed to the published version of the manuscript.

Funding: This research was funded by the Key Research and Development Program of Hunan Province, grant number 2021SK2050; the National Natural Science Foundation of China, grant number 51908562; the Science and Technology Project of the Department of Transportation of Jiangxi Province, grant number 2022H0024; the Research Foundation of Education Bureau of Hunan Province, China, grant number 19B581; the Standardization Project of Hunan Province; the Natural Science Foundation of Hunan Province, grant number 2020JJ5987.

Institutional Review Board Statement: Not applicable.

Informed Consent Statement: Not applicable.

Data Availability Statement: Not applicable.

Conflicts of Interest: The authors declare no conflict of interest.

References

1. Zeng, L.X.; Xiao, L.; Zhang, J.-H.; Gao, Q.-F. Effect of the characteristics of surface cracks on the transient saturated zones in colluvial soil slopes during rainfall. *Bull. Eng. Geol. Environ.* **2020**, *79*, 11. [CrossRef]
2. An, R.; Cai, J.R.; Qin, Y. Analysis on effect of vegetation root-system morphology on slope stability. *Water Resour. Hydropower Eng.* **2018**, *49*, 7.
3. Zhang, J.-H.; Li, F.; Zeng, L.; Zheng, J.-L.; Zhang, A.-S.; Zhang, Y.-Q. Effect of cushion and cover on moisture distribution in clay embankments in southern China. *J. Cent. South Univ.* **2020**, *27*, 14. [CrossRef]
4. Zhou, Y.Y.; Chen, J.P.; Wang, X.M. Progress of study on soil reinforcement mechanisms by root and its expectation. *Ecol. Environ.* **2012**, *21*, 7.
5. Li, X.-W.; Kong, L.-W.; Guo, A.-G. Permeability and mechanical characteristics of expansive soil and cut slope protection mechanism under vegetation action. *Rock Soil Mech.* **2013**, *34*, 7.
6. Zhang, D.; Wang, B. Mechanics Research on Vetiver Grass in Railway Side Slope Protection. *Bull. Soil Water Conserv.* **2006**, *26*, 3.
7. LI, S.; Sun, H.; Yang, Z.; He, L.; Cui, B. Mechanical Characteristics of Interaction between Root System of Plants and Rock for Rock Slope Protection. *Chin. J. Rock Mech. Eng.* **2006**, *25*, 2051–2057.
8. Ng, C.W.W. Atmosphere-plant-soil interactions: Theories and mechanisms. *Chin. J. Geotech. Eng.* **2017**, *39*, 47.
9. Jie, H.; Jun, G. Geosynthetics used to stabilize vegetated surfaces for environmental sustainability in civil engineering. *Front. Struct. Civ. Eng.* **2017**, *11*, 10.
10. Cheng, H.; Zhang, X.Q. An Experimental Study on Herb Plant Root System for Strength Principle of Soil-fixation. *Bull. Soil Water Conserv.* **2002**, *22*, 4.
11. Hongbin, X.I.A.O.; Liang, Z.H.A.O.; Zhenyu, L.I.; Weidong, L.; Wen, Y.; Qingqing, T.; Wei, L. Experimental study on *Vetiveria zizanioides* root system distribution and tensile strength. *J. Cent. South Univ. For. Technol.* **2014**, *3*, 5.
12. Deljouei, A.; Abdi, E.; Schwarz, M.; Majnounian, B.; Sohrabi, H.; Dumroese, R.K. Mechanical Characteristics of the Fine Roots of Two Broadleaved Tree Species from the Temperate Caspian Hyrcanian Ecoregion. *Forests* **2020**, *11*, 345. [CrossRef]
13. Deljouei, A.; Cislaghi, A.; Abdi, E.; Borz, S.A.; Majnounian, B.; Hales, T.C. Implications of hornbeam and beech root systems on slope stability: From field and laboratory measurements to modelling methods. *Plant Soil* **2022**. [CrossRef]
14. Qiao, J. Present situation of landslide disaster and prevention countermeasures in China. *J. Nat. Disasters* **2007**, *16*, 7.
15. Liu, W. *Study on Hydrological Effect of Plant Slope Protection and Its Influence on Slope Stability*; Central South University of Forestry and Technology: Changsha, China, 2017.

16. Yang, F.; Cheng, J.; Zhang, H.; Zhou, Z. Effect of Herb Plants on Soil Detachment and Erosion Dynamics. *Trans. Chin. Soc. Agric. Mach.* **2016**, *47*, 9.
17. Shan, Y.; Xie, J.; Lei, N. Spatial Distribution Characteristics and Influencing Factors of Soil Moisture on Loess Cutting Slope. *Bull. Soil Water Conserv.* **2014**, *34*, 5.
18. Chen, J..-L.; Li, J.-H.; Cheng, P.; Song, L.; Zhou, T. Field test on seepage performance of soil cover with different types of vegetation. *Rock Soil Mech.* **2018**, *39*, 7.
19. Ide, J.I.; Shinohara, Y.; Higashi, N.; Komatsu, K.; Otsuki, K. A preliminary investigation of surface runoff and soil properties in a moso-bamboo (*Phyllostachys pubescens*) forest in western Japan. *Hydrol. Res. Lett.* **2010**, *4*, 5. [CrossRef]
20. Ng, C.W.W.; Leung, A.K.; Woon, K.X. Effects of soil density on grass-induced suction distributions in compacted soil subjected to rainfall. *Can. Geotech. J.* **2014**, *51*, 11. [CrossRef]
21. Wang, D.L.; Luan, M.T.; Yang, Q. Experimental study of soil-water characteristic curve of remolded unsaturated clay. *Rock Soil Mech.* **2009**, *30*, 6.
22. Li, Z.Q.; Hu, R.L.; Wang, L.C.; Li, Z.X. Study on SWCC of unsaturated expansive soil. *Rock Soil Mech.* **2006**, *27*, 15.
23. Sun, D.A.; You, G.; Annan, Z.; Daichao, S. Soil–water retention curves and microstructures of undisturbed and compacted Guilin lateritic clay. *Bull. Eng. Geol. Environ.* **2016**, *75*, 11. [CrossRef]
24. Xiao-kun, H.O.U.; Tong-lu, L.I.; Xiao, X.I.E. Effect of undisturbed Q3 loess's microstructure on its SWCC. *J. Hydraul. Eng.* **2016**, *47*, 8.
25. Cai, G.Q.; Liu, W.; Xu, R.Z.; Li, J.; Zhao, C.G. Experimental investigation for soil-water characteristic curve of red clay in full suction range. *Chin. J. Geotech. Eng.* **2019**, *41*, 4.
26. Yao, Y.; Ni, J.; Li, J. Stress-dependent water retention of granite residual soil and its implications for ground settlement. *Comput. Geotech.* **2021**, *129*, 103835. [CrossRef]
27. Yao, Y.; Li, J.; Xiao, Z.; Xiao, H. Soil-Water Characteristics and Creep Deformation of Unsaturated Expansive Subgrade Soil: Experimental Test and Simulation. *Front. Earth Sci.* **2021**, *9*, 1141. [CrossRef]
28. Zhenyu, L.I.; Lifeng WA, N.G.; Hongbin, X.I.A.O. Distribution Characteristics of Vetiver's Roots in Highway Slope. *J. Basic Sci. Eng.* **2017**, *25*, 11.
29. Xi, B. Morphology, distribution, dynamic characteristics of poplar roots and its water uptake habits. *J. Beijing For. Univ.* **2019**, *41*, 13.
30. Zhang, J.; Peng, J.; Zheng, J.; Yao, Y. Characterisation of stress and moisture-dependent resilient behaviour for compacted clays in South China. *Road Mater. Pavement Des.* **2020**, *21*, 14. [CrossRef]
31. Van Genuchten, M.T. A closed form equation for predicting the hydraulic conductivity of unsaturated soils. *Soil Sci. Soc. Am. J.* **1980**, *44*, 7. [CrossRef]
32. Leung, A.; Garg, A.; Ng, C. Effects of plant roots onsoil–water retention and induced suction in vegetated soil. *Eng. Geol.* **2015**, *193*, 15. [CrossRef]
33. Gallipoli, D.; Wheeler, S.J.; Karstunen, M. Modelling the variation of degree of saturation in a deformable unsaturated soil. *Géotechnique* **2003**, *53*, 8. [CrossRef]
34. Bao, C. Behavior of unsaturated soil and stability of expansive soil slope. *Chin. J. Geotech. Eng.* **2004**, *26*, 15.
35. Karimi, Z.; Abdi, E.; Deljouei, A.; Cislaghi, A.; Shirvany, A.; Schwarz, M.; Hales, T.C. Vegetation-induced soil stabilization in coastal area: An example from a natural mangrove forest. *Catena* **2022**, *216*, 106410. [CrossRef]
36. Mickovski, S.B.; Van Beek, L.P.H. Root morphology and effects on soil reinforcement and slope stability of young vetiver (*Vetiveria zizanioides*) plants grown in semi-arid climate. *Plant Soil* **2009**, *324*, 43–56. [CrossRef]
37. Farahnak, M.; Mitsuyasu, K.; Hishi, T.; Katayama, A.; Chiwa, M.; Jeong, S.; Otsuki, K.; Sadeghi, S.M.M.; Kume, A. Relationship between Very Fine Root Distribution and Soil Water Content in Pre- and Post-Harvest Areas of Two Coniferous Tree Species. *Forests* **2020**, *11*, 1227. [CrossRef]
38. Himmelbauer, M.L. Estimating length, average diameter and surface area of roots using two different image analysis systems. *Plant Soil* **2004**, *1*, 110.
39. Ng, C.W.W.; Garg, A.; Leung, A.K.; Hau, B.C.H. Soil-water characteristics of Relationships between leaf and root area indices and soil suction induced during drying-wetting cycles. *Ecol. Eng.* **2016**, *91*, 6. [CrossRef]
40. Chen, W.J.; Cheng, D.H.; Tao, W. Physical significance of the parameters in the van Genuchten model. *Hydrogeol. Eng. Geol.* **2017**, *44*, 7.
41. Romero, E.; Gens, A.; Lloret, A. Water permeability, water retention and microstructure of unsaturated compacted boom clay. *Eng. Geol.* **1999**, *54*, 11. [CrossRef]
42. Ng, C.W.; Pang, Y.W. Experimental investigations of the soil-water characteristics of a volcanic soil. *Can. Geotech. J.* **2000**, *37*, 1252–1264. [CrossRef]

sustainability

MDPI

Article

Meso-Mechanical Simulation of the Mechanical Behavior of Different Types of Steel Fibers Reinforced Concretes

Haifeng Wang [1], Yicheng Jiang [2,*] and Ling Liu [2]

[1] China Railway 12th Bureau Group Co., Ltd., Nantong Railway Construction Component Co., Ltd., Nantong 226000, China
[2] School of Aerospace Engineering and Applied Mechanics, Tongji University, Shanghai 200092, China
* Correspondence: 1830929@tongji.edu.cn

Abstract: Introducing steel fibers into traditional concrete can improve its mechanical properties and crack resistance, but few studies have considered how the steel fiber shape and the bond-slip effect between fibers and matrix affect the mechanical behavior of concrete. This paper establishes a three-dimensional representative volume element (3D RVE) of steel fiber-reinforced concrete (SFRC) with random distribution, different shapes, and different interfacial strengths of steel fibers using Python, Abaqus and Hypermesh. Uniaxial tensile behaviors and failure modes of the SFRC are systematically simulated and analyzed. The results show that when the interfacial strength of steel fiber/concrete is changed from 1 to 3 MPa, the tensile strength of the SFRC increases accordingly. When the interfacial strength is greater than 3 MPa, it has no effect on tensile strength. Additionally, if the interfacial strength is 1 MPa, the tensile strength of the SFRC with end-hook steel fibers is increased by 7% when compared to the SFRC with straight steel fibers, whereas if the interfacial strength reaches 2.64 MPa (strength of pure concrete), the fiber shape has little effect on the tensile strength of the SFRC. Moreover, the simulation results also show that interfacial damage dominates when the interfacial strength is less than 1 MPa, and the crack propagation rate in the end-hook steel fiber-modified SFRC is lower than that in a straight steel fiber-modified SFRC. Therefore, this research reveals that using end-hook steel fibers can improve the strength of the SFRC under low interfacial strength, but the ideal strength of the SFRC can be achieved only by using straight fibers when the interfacial strength between steel fibers and concrete is relatively high.

Keywords: steel fiber; concrete; representative volume element (RVE); meso-mechanical

Citation: Wang, H.; Jiang, Y.; Liu, L. Meso-Mechanical Simulation of the Mechanical Behavior of Different Types of Steel Fibers Reinforced Concretes. *Sustainability* **2022**, *14*, 15803. https://doi.org/10.3390/su142315803

Academic Editor: Ahmed Salih Mohammed

Received: 26 October 2022
Accepted: 18 November 2022
Published: 28 November 2022

Publisher's Note: MDPI stays neutral with regard to jurisdictional claims in published maps and institutional affiliations.

1. Introduction

Concrete is a typical building engineering material. During the manufacturing, transportation and construction of concrete parts, it is easy to produce micro-cracks and cause damage because of its low tensile properties, which creates a potential safety hazard in the subsequent service [1–6]. The strength of concrete can be improved by introducing an appropriate amount of short steel fibers [7,8]. The added volume fraction of steel fiber (V_{sf}) is generally about 0–2%. If the V_{sf} is too high, it is easy to agglomerate during stirring and mixing and cause initial defects, while the reinforcing and crack resistance of the concrete with low V_{sf} are not obvious [9–13]. Previous studies have reported that the splitting strength, compressive strength, and flexural strength of short steel fibers-reinforced concrete (SFRC) could be maximally enhanced by about 50% [14–16], 86% [3], and 20% [15], respectively, when compared to the unmodified concrete with a V_{sf} of 0–2%. Therefore, the SFRC shows better mechanical properties when compared with the traditional concrete.

The mechanical properties of the SFRC are synergistically determined by the mechanical properties of steel fibers, concrete and doping characteristics. In order to explore how to obtain high performance in SFRC, the meso-mechanical analysis method of composite materials can be used for predicting and thereby reducing the experimental cost [16–22].

Zhao et al. used LS-DYNA finite element software to establish a three-dimensional cylindrical representative volume element (3D RVE) for the SFRC. The mechanical properties and crack evolution under axial compression were numerically simulated [23]. Wu et al. used LS-DYNA to build an RVE model composed of mortar and steel fibers for the Hopkinson compression bar test, and how the V_{sf} affected the dynamic performance of the concrete was studied comprehensively [24]. Tian et al. proposed a cylindrical RVE model for the SFRC using LS-DYNA, and studied the uniaxial compression performance of the cylinder [25]. Kim et al. reported the effect of the aggregate–mortar interface on the concrete properties and crack orientation through the pre-inserted cohesive element model (CZM) [26]. Zhou et al. studied the pull-out behavior of steel fibers via the cohesive element and determined the basic interfacial parameters of the SFRC [27]. The afore-mentioned studies reveal that by establishing RVE to perform the mesoscopic simulation analysis, the influence trend of doped steel fibers on the macroscopic mechanical properties of SFRC can be obtained. When the V_{sf} is 0–2%, the tearing strength and compressive strength can both be increased by 0–30%, and the crack direction and failure modes are obviously different from those of pure concrete, that is, the steel fibers show good strengthening and toughening effects on the concrete. Additionally, the CZM can well fit the bond-slip behavior between steel fibers and concrete [28], in which the hook-shaped steel fibers greatly influence the interfacial bonding performance. The longer the embedded length, the greater the pull-out load. Abbas et al. [29] and Gao et al. [30] estimated the relationship between strength and aspect ratio/V_{sf} via experimental methods, with the results suggesting that the aspect ratio has a more obvious effect on the strength of the concrete than the V_{sf}.

However, due to the complexity of the numerical model and the mesh, the current studies still have some obvious limitations. Firstly, the steel fibers are usually considered as line elements, and the concrete and the fibers are coupled by an embedded method. Secondly, the mechanical properties or damage cracks of SFRC are mainly analyzed using two-dimensional RVE in many studies, and the involved calculation method can preliminarily calculate the influence of V_{sf} on the concrete strength, but cannot take the bridging effect of steel fibers into account. Thirdly, in terms of the influence mechanism of the doped steel fibers on the mechanical properties and failure modes, few studies have considered the bond-slip effect between steel fibers and concrete as well as establishing three-dimensional (3D) RVE for the SFRC. At last, the combination of Abaqus–Python–Hypermesh can realize the automation and parameterization of RVE generation, greatly reducing the generation time. Based on this, this paper combines Python–Abaqus–Hypermesh to establish a 3D RVE for the SFRC, which can greatly reduce the construction complexity of the traditional 3D model. Based on the established model, the influence trend of steel fiber shapes as well as the interfacial strength between steel fibers and concrete on the tensile mechanical properties, crack initiation, and failure modes was systematically analyzed.

2. Analytical Method

2.1. Materials and Constitutive Model

SFRC composites are inhomogeneous at the mesoscale. The establishment of RVE corresponds to the stress–strain distribution of homogeneous materials, so that the calculation of macroscopic SFRC composites is decoupled from the microscopic ones. Based on this, the mechanical response of RVE can reflect the mechanical properties of macroscopic SFRC composites. The V_{sf} is an important index that affects the performance of SFRC. Based on literature reports [26], two types of steel fibers, the straight type and end-hook type with a diameter of 0.7 mm and a length of 30 mm, were used in this paper. The interfacial strength was set as 1, 2, 3, 4 and 5 MPa and the concrete chosen was model C50. The performance parameters of steel fibers and concrete are shown in Table 1. Because of the tensile and compressive anisotropic hardening and damage of the concrete, the built-in CDP (Concrete Damaged Plastic) model in Abaqus was adopted for the concrete [31], and the plastic parameters of the plastic damage model of GB50010-2010 specification was used as reference [32], as shown in Table 2.

Table 1. Material parameters.

Material	Density/t/mm³	Modulus/GPa	Poisson's Ratio
Steel fiber	7.8×10^{-9}	200	0.25
Concrete	2.5×10^{-9}	36.6	0.2

Table 2. Plastic parameters of CDP damage.

Dilation Angle/°	Eccentricity	Stress Ratio	Viscosity Parameter
30.5	0.1	0.666	0.001
Yield stress /MPa	Inelastic strain	Yield stress /MPa	Cracking strain
12.96	0	2.64	0
16.51	5.23×10^{-5}	2.54	5.13×10^{-5}
20.73	1.05×10^{-4}	2.36	6.72×10^{-5}
24.30	1.75×10^{-4}	2.17	8.34×10^{-5}
27.23	2.63×10^{-4}	1.99	9.93×10^{-5}

For the steel fiber/concrete interface, the CZM was used to simulate the interfacial bond-slip effect, as shown in Figure 1, showing the CZM damage evolution constitutive relationship. Here, t_n^0 is the interface normal strength, t_s^0 and t_t^0 are the interface shear strengths, and the green dotted line represents the secondary stress criterion controlled by the above three parameters, that is, the initial damage criterion of the Cohesive element [27] can be given as:

$$\left(\frac{\langle t_n \rangle}{t_n^0}\right)^2 + \left(\frac{\langle t_s \rangle}{t_s^0}\right)^2 + \left(\frac{\langle t_t \rangle}{t_t^0}\right)^2 = 1 \tag{1}$$

where t_n is the normal stress of the interface; t_s and t_t are the two shear stresses of the interface, respectively; <> is the Macaulay bracket that is used to indicate that the pure compressive deformation (or stress) state will not cause damage; $<t_n>$ indicates that when the item is negative (i.e., compressive stress), its value is zero. In addition, the blue and red triangles in Figure 1 are the tensile–separation curves of shear and normal modes, respectively; the areas enclosed by the red and blue solid lines represent modes I (G_{IC}) and II/III (G_{IIC}/G_{IIIC}) critical strain energy release rate, respectively; the areas shaded in red and blue represent the strain energy release rates in the normal and two tangential directions of the cohesive element, namely, G_I and G_{II} (or G_{III}); the shaded area in the black triangle represents the total critical strain energy release rate G_C. Thus, the B-K criterion for judging the interface damage extension is:

$$G^C = G_{IC} + (G_{IIC} - G_{IC})\left(\frac{G_{II} + G_{III}}{G_I + G_{II} + G_{III}}\right)^{\eta} \tag{2}$$

In Equation (2), η is the material parameter (usually taken as 1.35). When using Equations (1) and (2) to determine the initial damage and the crack growth, the performance parameters used for the cohesive element are presented in Table 3 [33].

Table 3. Interface parameters of steel fiber/concrete.

Modulus/MPa	t_n^0/MPa	t_s^0/t_t^0/MPa	G_{IC}/N/mm¹	$G_{IIC} = G_{IIIC}$/N/mm¹
1×10^7	1–5	1–5	0.025	0.625

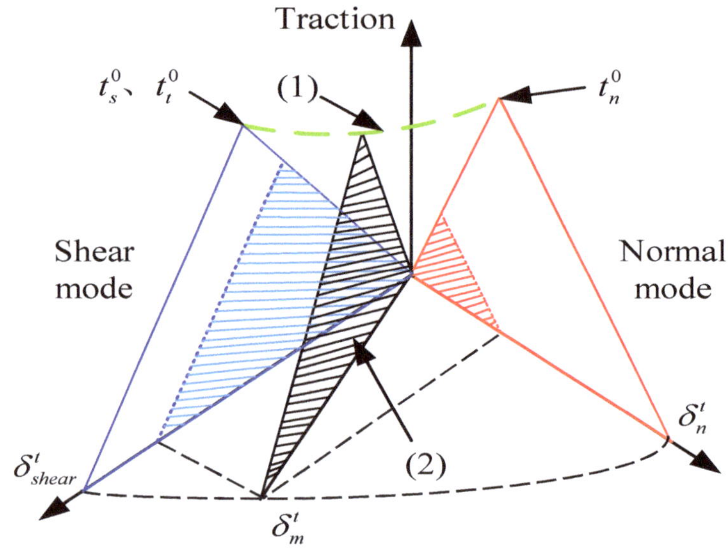

Figure 1. Constitutive model of cohesive element.

2.2. 3D RVE Tensile Model

Here, a secondary development of Python–Abaqus was used to establish the RVE model of the random steel fibers/concrete [33]. Figure 2a presents the models of end-hook and straight steel fibers in Abaqus, and the corresponding physical fibers are given in Figure 2b. Since this paper seeks to explore the effects of different shapes of steel fibers on the strength of the SFRC, the number and random positions of these two steel fibers should be the same. The Python algorithm was used to determine the node collection point whereat two fibers would not cross each other, as shown in Figure 2c. Firstly, the random position was controlled by a set of random coordinates and two corner values to construct the distance function between any two-line segments in space and the function of any line rotating around any point in space. Secondly, the end-hook fiber was divided into four segments from the broken line. We then calculated the shortest distance between the newly generated random end-hook fiber and the generated fibers. If all the distances were less than the steel fiber diameter, then the newly generated fiber was stored in the collection, otherwise another random fiber was regenerated. The two kinds of fibers generated by the algorithm have the same endpoints, which can ensure that the random position is exactly the same. By comparing the actual number of fibers with the predesigned fiber volume fraction, we judged whether to continue inserting new fibers, and repeat the process until the actual fibers reached the target value, at which point the generation process was completed. The final obtained RVE model is shown in Figure 2d.

The above method was used to generate two kinds of RVE with different shapes of steel fibers. Due to the relatively large aspect ratio of steel fibers in the RVE model, Abaqus has limitations when used in processing meshes, so Hypermesh was used to divide the model into tetrahedral meshes. We read the surface elements on the interface between steel fibers and concrete in batches to create a collection, and output the Abaqus inp file. By modifying the keyword, the grid set was transferred to the grid component, and so the edit mesh would generate the interface unit, as shown in Figure 3a.

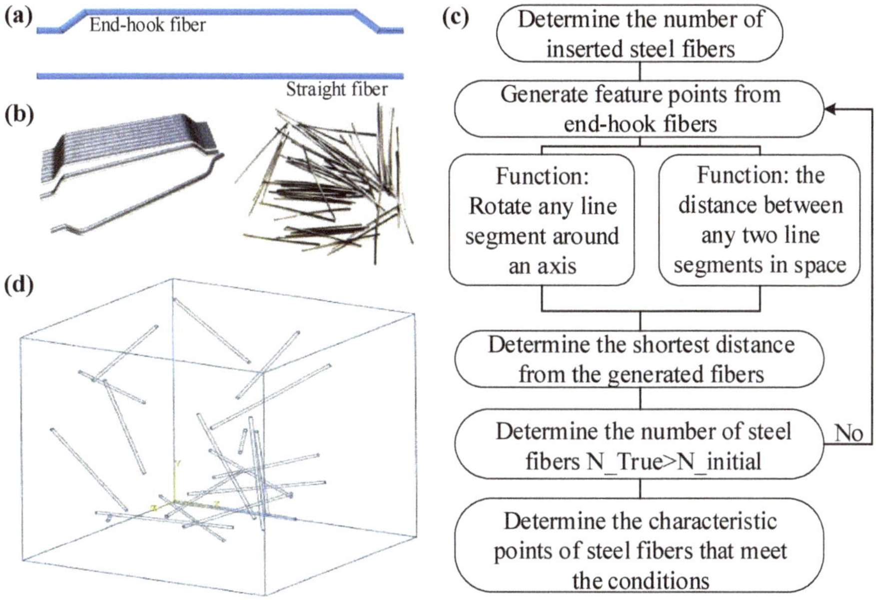

Figure 2. (**a**) 3D models of end-hook and straight steel fiber, (**b**) real end-hook and straight steel fibers, (**c**) generation process of RVE model, (**d**) 3D RVE geometric model.

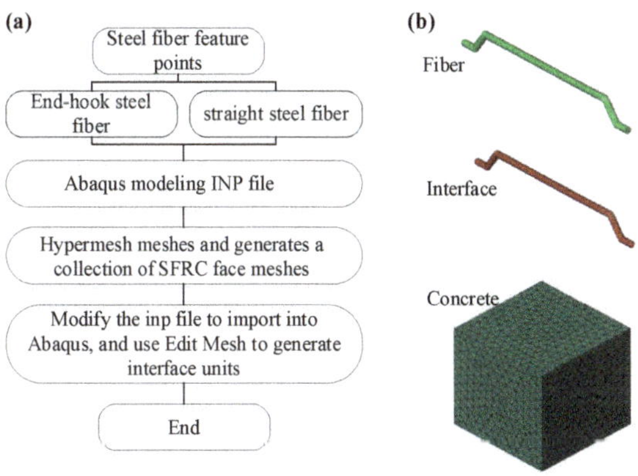

Figure 3. (**a**) Mesh part generation process, (**b**) meshes of steel fiber, interface and concrete.

Secondly, to qualitatively analyze the influence of steel fibers on the mechanical properties of the SFRC from the point of view of the bond-slip effect, the bilinear elastic–plastic model was adopted for the steel fibers, and the plastic damage model was used for the concrete. Therefore, in the 3D RVE model, C3D6 tetrahedral elements were used for both the concrete and the steel fibers, and the COH3D6 cohesive element was used to simulate the bond-slip effect between steel fibers and the concrete. The meshing was given in Figure 3b. In order to visualize the damage phenomenon of the concrete, the quasi-static display analysis step was used, and the reaction force and displacement of the reference point were selected from the field variable output.

Based on the above proposed RVE models and taking interfacial strength parameters and fiber shapes into consideration, ten RVE models were established. The first series of RVE contained straight steel fibers (SFs) with interfacial strengths of 1, 2, 3, 4, and 5 MPa, respectively. The second series of RVE contained end-hook fibers (EFs) with interfacial strengths of 1, 2, 3, 4 and 5 MPa, respectively. Table 4 presents these ten groups of RVE models, from which the influence of steel fiber types and interfacial strengths on the tensile properties of SFRC could be obtained.

Table 4. SFRC simulation groups.

Group	Interface Strength/MPa	Steel Fiber Type
SFs (1–5)	1\2\3\4\5	Straight fibers
EFs (1–5)	1\2\3\4\5	End-hook fibers

3. Results and Discussion

3.1. *Effect on the Tensile Strength*

According to the pre-analysis results, the fracture toughness shown in Table 3 has little influence on the ultimate strength of the SFRC. Therefore, the fracture toughness remains unchanged in the calculation. Additionally, when the interfacial strength is greater than 3 MPa, the yielded stress–strain curves almost overlap each other. Therefore, Figure 4a shows the tensile stress–strain curves of SFRC with interfacial strengths of 1–3 MPa. It can be seen that the steel fiber shapes and interfacial strengths have little effect on the stiffness of the SFRC. The obtained tensile strength is given in Figure 4b. It can be seen from Figure 4b that when the interfacial strength is lower than 2 MPa, the reinforcing effect of EFs is better than that of SFs. Additionally, if the interfacial strength is 1 MPa, the tensile strength of the SFRC with end-hook steel fibers is increased by 7% when compared to the SFRC with straight steel fibers. Additionally, when the interfacial strength approaches or exceeds 2.64 MPa (strength of concrete), these two steel fibers show similar reinforcing effects on the SFRC. Moreover, when the interfacial strength reaches 3 MPa (about 14% higher than the strength of concrete), these two steel fibers no longer show a reinforcing ability in relation to the strength of the SFRC. Therefore, Figure 4 proves that when the interfacial strength between steel fibers and concrete is relatively weak (less than 2 MPa), using the end-hook fibers as reinforcing is a better choice, while if the interfacial strength is high enough (close to or greater than the strength of concrete), choosing the straight fibers for reinforcing is relatively simple and effective.

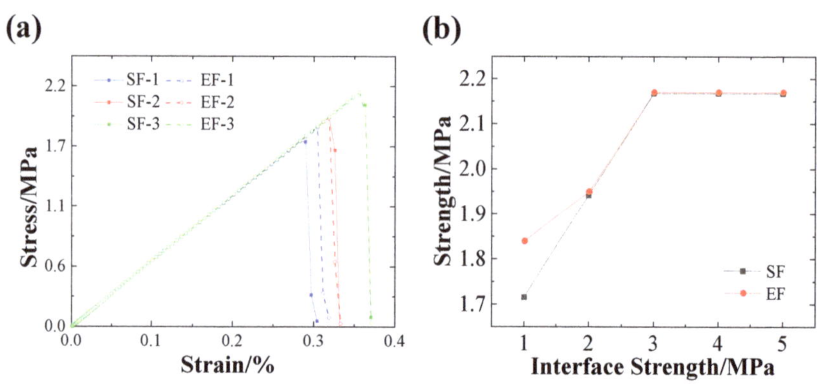

Figure 4. Tensile simulation results of SFRC with various shapes of fibers and different interfacial strengths, (**a**) stress–strain curves, (**b**) tensile strength.

3.2. *Effects on the Interfacial Damage and Failure Modes*

3.2.1. With Low Interfacial Strength

Figure 5 gives the mesoscale failure process of the SFRC with SFs at a low interfacial strength of 1 MPa. Figure 5a indicates the initial failure of fibers/concrete interfaces in the

RVE, which shows that when the external load reaches 3.23 kN, cracks begin to initiate at the end interface because of stress concentration, and cracks propagate along the interface due to the fact that the interfacial strength is lower than the strength of concrete. Figure 5b suggests that when the external load achieves 4.13 kN, crack propagation occurs at the interface perpendicular to the loading direction. This is because the interface perpendicular to the loading direction bears a large normal load, while the interface parallel to the loading direction mainly bears shear load; therefore, the damage mode is dominated by a mode-I crack. As shown in Figure 5c,d, when the crack propagates from the end interface to the concrete matrix, the crack extends quickly with the increase in load. When the crack reaches the SFRC surface, a penetrating crack is formed, thus the SFRC fails macroscopically, and the ultimate load is 6.18 kN.

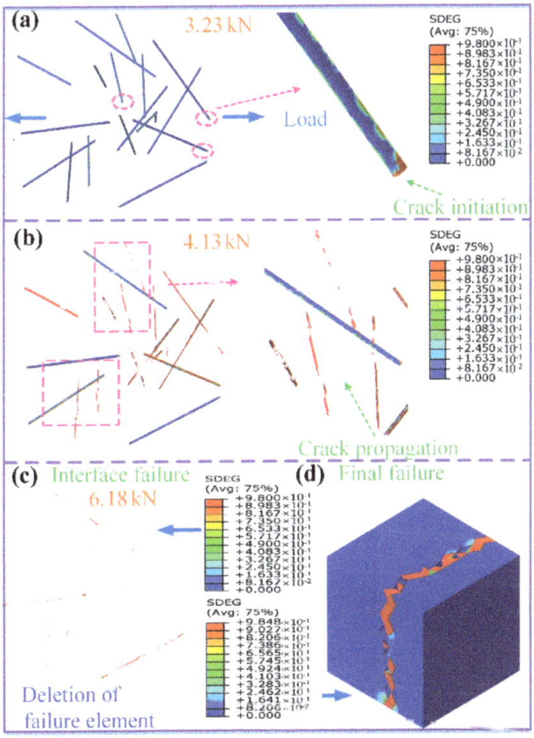

Figure 5. Mesoscale failure process of SFRC with SFs at low interfacial strength 1 MPa, (**a**) initial interface damage, (**b**) propagation of interface crack, (**c**) interface failure, (**d**) SFRC failure.

Figure 6 shows the mesoscale failure process of SFRC with EFs at a low interfacial strength of 1 MPa. Figure 6a gives the initial interface damage of the SFRC RVE, which indicates that when the external load reaches 1.64 kN, cracks begin to occur at the end interfaces and at the fiber corners due to stress concentration. It can be seen from Figure 6b that when the external load reaches 4.13 kN, cracks begin to propagate along the interface, which propagate more significantly at these interfaces nearly perpendicular to the loading direction. This is because these interfaces bear large normal loads, and cracks are easy to occur. Moreover, Figure 6c,d suggest that the SFRC with EFs is still dominated by interface damage, where cracks initiate at the fiber ends, propagate into the concrete matrix, and finally fail macroscopically. The obtained ultimate load is 6.60 kN.

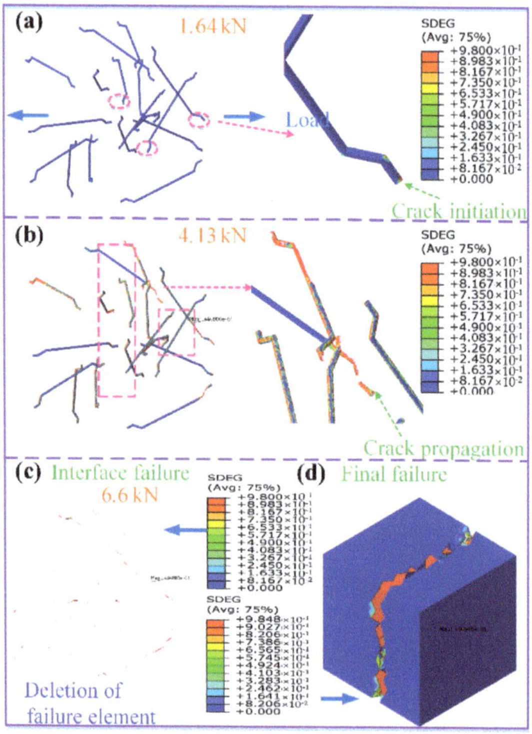

Figure 6. Mesoscale failure process of SFRC with EFs at low interfacial strength 1 MPa, (**a**) initial interface damage, (**b**) propagation of interface crack, (**c**) interface failure, (**d**) SFRC failure.

It can be seen from Figures 5 and 6 that the failure modes of the SFRC with SFs or EFs are similar at low interfacial strengths. However, due to the additional corners of the EFs, the external load induced by the initial interface damage of the SFRC with EFs is lower than that of the SFRC with SFs. Alternatively, it can be seen from Figures 5 and 6 that the SFs-modified SFRC has more damaged areas than the EFs-modified SFRC (SDEG > 0.98, shown in red), which means the initial crack onset rate at the interface of the SFRC with EFs is lower than that of the SFs-modified SFRC. Additionally, the external load corresponding to interfacial crack propagation for these two SFRCs is the same, obtained as 4.13 kN, where the propagation rate of interfacial cracks of the EFs-modified SFRC is obviously slower than that of the SFs-modified SFRC. More importantly, the tensile strength of the EFs-modified SFRC is 7% higher than that of the SFs-modified SFRC when the interfacial strength is 1 MPa.

3.2.2. With High Interfacial Strength

Figure 7 gives the mesoscale damage process of the SFs-modified SFRC at the relatively high interfacial strength of 5 MPa. As shown in Figure 7a, when the external load reaches 6.9 kN, a small number of initial cracks occur at the end interfaces. With the external load increasing to 7.32 kN (Figure 7b), the cracks propagate a short distance along the interfaces and immediately extend into the concrete matrix because the interfacial strength is higher than that of the concrete matrix. Figure 7c,d present the damage diagrams of the interfaces and the concrete matrix at the final failure moment, respectively. It can be seen that with high interfacial strength, the damage percentage of interfaces at the final failure moment is much smaller than that of the concrete matrix, and the concrete damage is the dominating failure mode of the SFRC. The obtained ultimate load is 7.80 kN. Once again, Figure 8 presents the mesoscale damage process of EFs-modified SFRC at the high interfacial strength of 5 MPa. When the external load reaches 4.5 kN (Figure 8a), tiny initial cracks appear at the end interfaces. As the external load continues to increase to 7.32 kN (Figure 8b), the cracks extend from both ends of the fibers to their middle parts, and then quickly extend into the concrete matrix. Figure 8c,d give the damage diagrams of the interfaces and the concrete matrix at the final failure moment, respectively. Here, the cracks have propagated to the surface of the SFRC, accompanied by a large damage area in the concrete (Figure 8d), which eventually leads to macroscopic damage to the SFRC, with an ultimate load of 7.81 kN.

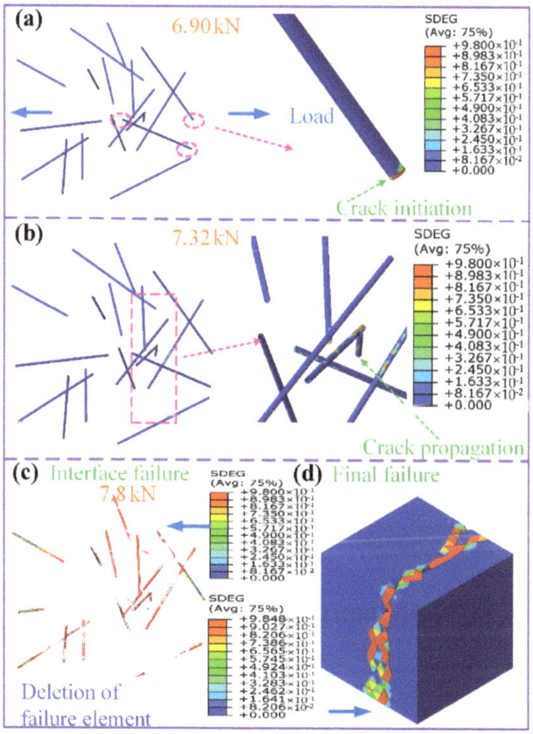

Figure 7. Mesoscale failure process of SFRC with SFs at high interfacial strength of 5 MPa, (**a**) initial interface damage, (**b**) propagation of interface crack, (**c**) interface failure, (**d**) SFRC failure.

It can be seen from Figures 7 and 8 that due to the additional corners of the EFs, cracks easily initiate in the EFs-modified SFRC. Under high interfacial strength, both interfacial strengths and fiber shapes reach their reinforcing threshold, and the failure mode is dominated by the concrete matrix. That is, when the interfacial strength is 5 MPa, the EFs-modified SFRC and the SFs-modified SFRC show the same damage modes and ultimate strengths.

It can be concluded from Figures 4–8 that the external load corresponding to the initial damage is relatively high in the case of high interfacial strength (Figures 7 and 8), and otherwise is reverse (Figures 5 and 6). The tensile strength of the SFRC increases with the increase in the interfacial strength. The addition of low-fiber volume content steel fibers into concrete does not show a reinforcing effect, and the induced interface is very easy to crack.

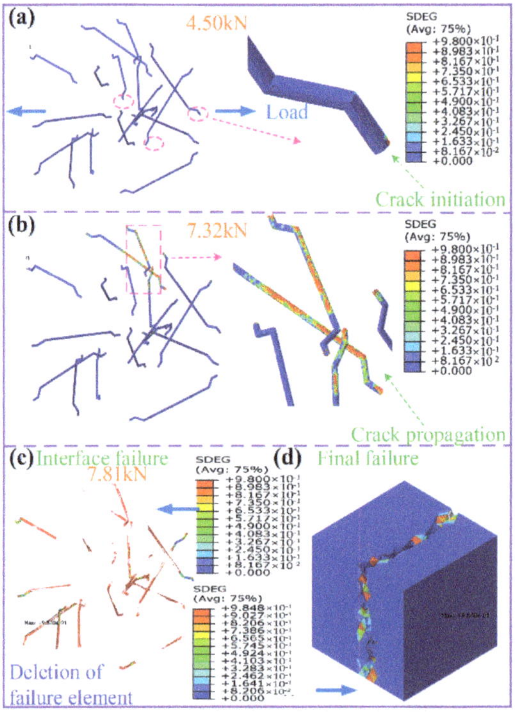

Figure 8. Mesoscale failure process of SFRC with EFs at high interfacial strength of 5 MPa, (**a**) initial interface damage, (**b**) propagation of interface crack, (**c**) interface failure, (**d**) SFRC failure.

4. Conclusions

This paper uses Abaqus–Python–Hypermesh to establish a 3D RVE for the SFRC with different types of steel fibers and different interfacial strengths. The results show that when the interfacial strength is less than 2 MPa, end-hook steel fibers have a better reinforcing effect than straight steel fibers, with a maximum enhancement of about 7% in the concrete compared to the straight steel fibers. Additionally, when the interfacial strength approaches the concrete's strength, both fibers show similar reinforcing effects. Moreover, interfacial damage is the dominating failure mode in cases of low interfacial strength, and the main failure mode is the concrete if the interfacial strength is high. Furthermore, the induced initial damage load in the end-hook steel fibers-modified SFRC is lower than that in the straight steel fibers-modified SFRC, because more ends and corners present in the former would cause more stress concentration points. However, the crack propagation rate in the end-hook steel fibers-modified SFRC is slower than that in the straight steel fibers-modified SFRC. Summarily, interfacial strength and fiber type are the main parameters that determine the effect of the addition of fibers into concrete. This research provides a better design guideline for the modification of concrete. Real interface strength testing, SFRC tensile experimental verification, and more finite element groups of high-fiber volume content will be considered in future research.

Author Contributions: Conceptualization, H.W.; methodology, H.W.; software, H.W. and Y.J.; formal analysis, H.W.; data curation, Y.J.; writing—original draft, Y.J.; writing—review and editing, L.L. All authors have read and agreed to the published version of the manuscript.

Funding: This research received no external funding.

Institutional Review Board Statement: Not applicable.

Informed Consent Statement: Not applicable.

Data Availability Statement: The data for the study are not available publicly.

Acknowledgments: This research was financially supported by the Nantong Railway Construction Component Co., Ltd. The simulation analysis support from engineer Changtao Wang and Zhanji Wang is appreciated.

Conflicts of Interest: The authors declare no conflict of interest.

References

1. Abdallah, S.; Fan, M.; Rees, D.W.A. Bonding mechanisms and strength of steel fiber–reinforced cementitious composites: Overview. *J. Mater. Civ. Eng.* **2018**, *30*, 04018001. [CrossRef]
2. Marcos-Meson, V.; Michel, A.; Solgaard, A.; Fischer, G.; Edvardsen, C.; Skovhus, T.L. Corrosion resistance of steel fibre reinforced concrete-A literature review. *Cem. Concr. Res.* **2018**, *103*, 1–20. [CrossRef]
3. Shah, A.A.; Ribakov, Y. Recent trends in steel fibered high-strength concrete. *Mater. Des.* **2011**, *32*, 4122–4151. [CrossRef]
4. Divyah, N.; Prakash, R.; Srividhya, S.; Sivakumar, A. Parametric study on lightweight concrete-encased short columns under axial compression-Comparison of design codes. *Struct. Eng. Mech.* **2022**, *83*, 387–400.
5. Divyah, N.; Thenmozhi, R.; Neelamegam, M.; Prakash, R. Characterization and behavior of basalt fiber-reinforced lightweight concrete. *Struct. Concr.* **2021**, *22*, 422–430. [CrossRef]
6. Prakash, R.; Thenmozhi, R.; Raman, S.N.; Subramanian, C.; Divyah, N. Mechanical characterisation of sustainable fibre-reinforced lightweight concrete incorporating waste coconut shell as coarse aggregate and sisal fibre. *Int. J. Environ. Sci. Technol.* **2021**, *18*, 1579–1590. [CrossRef]
7. Prakash, R.; Divyah, N.; Srividhya, S.; Avudaiappan, S.; Amran, M.; Naidu Raman, S.; Guindos, P.; Vatin, N.I.; Fediuk, R. Effect of Steel Fiber on the Strength and Flexural Characteristics of Coconut Shell Concrete Partially Blended with Fly Ash. *Materials* **2022**, *15*, 4272. [CrossRef]
8. Prakash, R.; Thenmozhi, R.; Raman, S.N.; Subramanian, C. Characterization of eco-friendly steel fiber-reinforced concrete containing waste coconut shell as coarse aggregates and fly ash as partial cement replacement. *Struct. Concr.* **2020**, *21*, 437–447. [CrossRef]
9. Liang, X.; Wu, C. Meso-scale modelling of steel fibre reinforced concrete with high strength. *Constr. Build. Mater.* **2018**, *165*, 187–198. [CrossRef]
10. Wang, W.T.; Wei, Y.Y.; Han, J.; Hong, J.; Zeng, Y. Mechanical Properties of Ultrahigh Performance Steel Fiber Reinforced Concrete: Experimental Study. *Appl. Math. Mech.* **2014**, *4*, 295–298.
11. Nihat, K.; Bahadur, A. Glass Fiber–Reinforced Sprayed Concrete: Physical, Mechanical, and Durability Properties. *J. Mater. Civ. Eng.* **2021**, *33*, 04020396.
12. Zhang, R.; Jin, L.; Du, X. Three-dimensional meso-scale modelling of failure of steel fiber reinforced concrete at room and elevated temperatures. *Constr. Build. Mater.* **2021**, *278*, 122368. [CrossRef]
13. Banthia, N.; Nandakumar, N. Crack growth resistance of hybrid fiber reinforced cement composites. *Cem. Concr. Compos.* **2003**, *25*, 3–9. [CrossRef]
14. Liu, R.; Li, H.; Jiang, Q.; Meng, X. Experimental investigation on flexural properties of directional steel fiber reinforced rubberized concrete. In *Structures*; Elsevier: Amsterdam, The Netherlands, 2020; Volume 27, pp. 1660–1669.
15. Ramírez, J.A.L.; Juan, L.M.; Carrillo, J. Material Damage Evolution for Plain and Steel-Fiber-Reinforced Concrete under Unconfined Compression Loading by Dynamic Ultrasonic Tests. *Arab. J. Sci. Eng.* **2018**, *43*, 5667–5675. [CrossRef]
16. Wang, C. Experimental investigation on behavior of steel fiber reinforced concrete (SFRC). *UC Res. Repos.* **2006**. [CrossRef]
17. Xu, M.; Wille, K. Calibration of K & C concrete model for UHPC in LS-DYNA. *Adv. Mater. Res.* **2015**, *141*, 04015051.
18. Wu, Z.; Shi, C.; He, W.; Wu, L. Effects of steel fiber content and shape on mechanical properties of ultra high performance concrete. *Constr. Build. Mater.* **2016**, *103*, 8–14. [CrossRef]
19. Ferrotto, M.F.; Fischer, O.; Cavaleri, L. A strategy for the finite element modeling of FRP-confined concrete columns subjected to preload. *Eng. Struct.* **2018**, *173*, 1054–1067. [CrossRef]
20. Youssf, O.; ElGawady, M.A.; Mills, J.E.; Ma, X. Finite element modelling and dilation of FRP-confined concrete columns. *Eng. Struct.* **2014**, *79*, 70–85. [CrossRef]
21. Pan, J.; Zhong, W.; Wang, J.; Zhang, C. Size effect on dynamic splitting tensile strength of concrete: Mesoscale modeling. *Cem. Concr. Compos.* **2022**, *128*, 104435. [CrossRef]
22. Xu, S.C.; Wu, C.Q.; Liu, Z.Z.; Su, Y. Effects of the Nano Materials and Steel Fibre on Early-Age Properties of Ultra-High Performance Concrete. *Bull. Chin. Ceram. Soc.* **2014**, *33*, 542–546.
23. Zhao, Q.S.; Xu, S.C.; Liu, Z.C. Microscopic numerical simulation of the uniaxial compression of steel fiber reinforced ultra-high performance concrete. *Acta Mater. Compos. Sin.* **2018**, *35*, 13.
24. Wu, P.T.; Wu, C.Q.; Liu, Z.X.; Xu, S.C. Numerical simulation of SHPB test of ultra-high performance fiber reinforced concrete with meso-scale mode. *Sci. China Phys. Mech. Astron.* **2020**, *50*, 13.

25. Tian, H.W.; Zhou, Z.; Lu, J.P.; Peng, Z. Meso-scale numerical simulation of axial compression performance of fiber reinforced polymer composite-confined ultra-high performance concrete. *Acta Mater. Compos. Sin.* **2020**, *37*, 10.
26. Kim, S.M.; Al-Rub, R.K.A. Meso-scale computational modeling of the plastic-damage response of cementitious composites. *Cem. Concr. Res.* **2011**, *41*, 339–358. [CrossRef]
27. Zhou, R.; Lu, Y. A mesoscale interface approach to modelling fractures in concrete for material investigation. *Constr. Build. Mater.* **2018**, *165*, 608–620. [CrossRef]
28. Trawinski, W.; Tejchman, J.; Bobinski, J. A three-dimensional meso-scale modelling of concrete fracture based on cohesive elements and X-ray CT images. *Eng. Fract. Mech.* **2018**, *189*, 27–50. [CrossRef]
29. Abbas, S.; Soliman, A.M.; Nehdi, M.L. Exploring mechanical and durability properties of ultra-high performance concrete incorporating various steel fiber lengths and dosages. *Constr. Build. Mater.* **2015**, *75*, 429–441. [CrossRef]
30. Zhou, R.; Chen, H.M.; Lu, Y. Mesoscale modelling of concrete under high strain rate tension with a rate-dependent cohesive interface approach. *Int. J. Impact Eng.* **2020**, *139*, 103500. [CrossRef]
31. Abaqus Inc. *Abaqus Analysis User's Manual v.2016*; Dassault Systemes: Paris, France, 2016.
32. China Association for Engineering Construction Standardization. *CECS13:19, Test Methods for Steel Fiber Reinforced Concrete*; China Association for Engineering Construction Standardization: Beijing, China, 1989.
33. Thilakarathna, P.S.M.; Baduge, K.K.; Mendis, P.; Vimonsatit, V.; Lee, H. Mesoscale modelling of concrete–a review of geometry generation, placing algorithms, constitutive relations and applications. *Eng. Fract. Mech.* **2020**, *231*, 106974. [CrossRef]

sustainability

MDPI

Article

Performance of Self-Compacted Geopolymer Concrete Containing Fly Ash and Slag as Binders

Aryan Far H. Sherwani [1,2,*], Khaleel H. Younis [3,4], Ralf W. Arndt [2,*] and Kypros Pilakoutas [5]

1 Department of Civil Engineering, Faculty of Engineering, Soran University, Soran 44008, Kurdistan Region, Iraq
2 Department of Civil Engineering, Fachhochschule Erfurt-University of Applied Sciences, 99084 Erfurt, Germany
3 Department of Surveying and Road Construction, Erbil Technology College, Erbil Polytechnic University, Erbil 44001, Kurdistan Region, Iraq
4 Civil Engineering Department, Tishk International University, Erbil 44001, Kurdistan Region, Iraq
5 Civil and Structural Engineering Department, University of Sheffield, Sheffield S1 3JD, UK
* Correspondence: aryanfar.abd@soran.edu.iq (A.F.H.S.); ralf.arndt@fh-erfurt.de (R.W.A.); Tel.: +964-7504522250 (A.F.H.S.)

Abstract: Geopolymers can replace cement and help reduce the environmental impact of concrete construction, but research is needed to ensure their mechanical properties, durability and practicability. The aim of this investigation is to examine the influence of ground granulated blast furnace slag (slag) content on the performance, at the fresh and hardened states, of fly ash (FA) based self-compacted geopolymer concrete (SCGC). For this purpose, four SCGC mixtures containing 450 kg/m^3 of total binder were examined. The alkaline-to-binder ratio was 0.5 for all mixes. FA was substituted with slag at 0%, 30%, 50%, and 100% of the total binder content. The fresh properties in terms of flowability, passing ability, viscosity, and segregation resistance, as well as the mechanical properties in terms of compressive strength and splitting tensile strength, were quantified. The durability behavior of SCGC was also studied to determine sorptivity and long-term free drying shrinkage. The results confirm that slag adversely affects the workability of SCGC mixtures except for the resistance to sieve segregation. Performance of SCGC in hardened states is in general enhanced with slag inclusion but at increased shrinkage strain. Predictions of splitting tensile strength were made using the ACI 318, ACI 363, Eurocode CEB-FIB, and Lee and Lee models. The ACI 363 and Eurocode CEB-FIB models were found to be inaccurate, except for the 30% slag mix. Predicted values obtained from the Lee and Lee model were very close to the actual values of the FA-based SCGC mix. The results of this work could lead to more sustainable concretes using geopolymers instead of OPC.

Keywords: self-compacted geopolymer concrete (SCGC); slag/fly ash; fresh properties; mechanical properties; empirical equation; sorptivity; long-term free drying shrinkage

Citation: Sherwani, A.F.H.; Younis, K.H.; Arndt, R.W.; Pilakoutas, K. Performance of Self-Compacted Geopolymer Concrete Containing Fly Ash and Slag as Binders. *Sustainability* **2022**, *14*, 15063. https://doi.org/10.3390/su142215063

Academic Editor: Ahmed Salih Mohammed

Received: 2 August 2022
Accepted: 1 September 2022
Published: 14 November 2022

1. Introduction

Industrialization, urbanization, and population growth are the main drivers of environmental pollution and climate change and the construction industry is a major contributor. To accommodate infrastructure development, the current worldwide concrete consumption per person exceeds one cubic meter per annum [1]. Increasing demand for concrete using ordinary Portland cement (OPC) is responsible for environmental pollution, depletion of natural resources and the emission of substantial amounts of carbon dioxide (6–7% of worldwide emissions) [2,3]. Geopolymer concrete that can be made using other industrial wastes may help the concrete industry reduce its environmental impact and carbon emissions in an economical manner [4].

The term "geopolymer," was first coined in 1978 by Davidovits [5]. The polymerization process involves an aluminosilicate source material including silica (Si) and alumina (Al) and an alkaline liquid, leading to an amorphous structure. Due to its environmental credentials, geopolymer concrete has gained popularity and attracts the interest of many scientists. Geopolymer raw materials do not require a large energy input because they are not calcined at high temperatures, a prerequisite for producing cement raw materials. It has been established that the production of geopolymers emits five to six times less carbon dioxide than Portland cement [6]. In addition, geopolymer concrete offers similar or superior fresh and mechanical properties compared to conventional concrete [7]. This type of concrete is affected by many mixed proportion parameters as well as curing conditions [8–10]. Furthermore, this type of concrete reduces energy consumption, waste disposal, and construction costs [10]. To synthesize geopolymers, it is desirable to combine fly ash (FA) with some other high silica sources, such as slag, rice husk ash, silica fume, etc. [11]. Li et al. [12] observed that slag, a by-product of iron production from blast furnace, containing calcium, magnesium silicates, and aluminosilicates, is another candidate waste product available worldwide in huge quantities. The production of one ton of slag emits just 70 kg of carbon dioxide (CO_2), which is only 7% of the CO_2 emitted during cement production. However, the use of slag in FA-based geopolymers decreases setting time and workability. Nonetheless, the calcium oxide (CaO) of slag forms calcium-silicate-hydrates (C-S-H) with alumino-silicate gel, which boosts the mechanical characteristics of concrete [13–17]. The substitution of slag with FA thus improves the initial setting time of geopolymer paste [18]. It was reported that the inclusion of slag leads to the reduction of the fresh state properties of SCGC, whilst improving the hardened state properties [19].

Self-compacted concrete (SCC) was introduced to provide optimum compaction and to help place concrete in confined spaces [20–22]. The fundamental aspects of SCC are flow, filling, and passage ability that can resist segregation [23]. Self-compacted geopolymer concrete (SCGC) is an innovation that would provide both environmental and practical benefits [24,25]. The fresh SCGC must comply with specifications recommended by the European Federation of National Associations Representing for Concrete (EFNARC) [26]. Currently, few investigations were conducted into SCGC, hence there is a need for further research to verify the efficacy of SCGC in both its fresh and hardened form and to develop materials for practical applications. To reduce costs and enhance the workability and mechanical properties of concrete, supplementary cementitious materials such as FA and slag are currently widely used in concrete [27].

Drying shrinkage is another important characteristic of concrete that needs to be examined, as it is critical for the durability and long-term serviceability of concrete structures [28]. Restrained drying shrinkage can cause cracking, and while it may not necessarily compromise the structural integrity of reinforced concrete structures, it may cause serious durability issues [29]. Research shows that drying shrinkage of oven-cured geopolymer concrete is often relatively low compared to conventional concrete. As reported by Wallah and Rangan [30], the majority of the water generated during the chemical reaction in FA-based geopolymer concrete may evaporate during the curing phase. As a result, the excess water in the hardened concrete's micropores is minimal, and thus the drying shrinkage is very limited. Other researchers also confirm that conventionally cured geopolymer concrete made from FA has good engineering performance and low drying shrinkage [31,32]. However, Wang et al. [33] showed that slag-based geopolymer concretes activated with sodium silicate might result in higher shrinkage strains than OPC concrete due to the development of silica-rich gel. For this reason, the long-term free drying shrinkage of FA-based SCGC made with/without slag needs to be examined.

Details of the binder type used and investigated properties of previous research on plain SCGC made with/without FA and slag are summarized in Table 1. Details of the current investigation are also displayed at the bottom of Table 1.

Table 1. Details of binder type used and examined properties of SCGC in the previous research and current study.

Refs	Binder Type	Fresh Properties	Mechanical Properties	Durability Properties	Others
[34]	FA (100,75,50,25,0%) slag (0,25,50,75,100%)	Slump, T_{500} and V-funnel flow time, L-box height	Compressive, splitting, fracture parameters	-	Statistical Evaluation, correlation
[35]	FA (50%), slag (50%), NS [1] (5–10 kg/m^3)	Slump, T_{500} and V-funnel flow time, L-box height	Compressive, fracture parameters, bond strength,	-	Statistical Evaluation, correlation
[36]	FA (25%), slag (75%)	Slump, T_{500} and V-funnel flow time, L-box and J-ring height, U-box	Compressive	Rapid chloride and water permeability, Sorptivity, Abrasion, Acid and sulphate attack, shrinkage	
[37]	slag,NS (9–10 kg/m^3)	Slump, T_{500} and V-funnel flow time, and J-ring height	Compressive, flexural, bond strength	Sorptivity	XRD
[38]	FA (Class F & C)	Slump, T_{500} flow time, L-box height, U-box	Compressive	Rapid chloride permeability, Sorptivity, Acid and sulphate attack, corrosion	SEM
[4]	FA (100%), slag (100,95,85, 75%), RHA [2] (5, 15, 25%)	Slump, T_{500} and V-funnel flow time, L-box and J-ring height	Compressive, splitting, flexural	Sorptivity	SEM
[39]	FA (100%), slag (100,95,85, 75%), RHA (5, 15, 25%)	Slump, T_{500} and V-funnel flow time, L-box and J-ring height	Compressive, splitting, flexural	-	SEM
[40]	FA	Slump, V-funnel flow time, L-box height	Compressive, splitting	Water absorption	
[24]	FA	Slump, T_{500} and V-funnel flow time, L-box and J-ring height	Compressive	-	-
[41]	FA	Slump, T_{500} and V-funnel flow time, L-box and J-ring height	Compressive	-	-
[42]	FA	Slump, T_{500} and V-funnel flow time, L-box and J-ring height	Compressive	-	SEM
[43]	FA (100,80,60,40,20,0%) , slag (100,80,60,40,20,0%)	Slump, T_{500} and V-funnel flow time, L-box height	Compressive, splitting, flexural	-	-
[25]	FA	Slump, T_{500} and V-funnel flow time, L-box and J-ring height, sieve segregation	Compressive	-	-
[44]	FA (100,95,90,85,80%) MK [3] (5,10,15,20%) GSA [4] (5,10,15,20%)	Slump, T_{500} and V-funnel flow time, L-box and J-ring height	Compressive, splitting, flexural	Water permeability	-
[45]	MK	-	Flexural strength	Water absorption	SEM
[46]	slag	Slump, T_{500} and V-funnel flow time, L-box height	Compressive, splitting, flexural	Carbonation depth, drying shrinkage, acid resistance, water absorption,	SEM
[47]	FA	Slump, T_{500} and V-funnel flow time, J-ring height	Compressive, splitting, flexural strength	-	-
[48]	FA (50%), slag (50%)	Slump, T_{500} and V-funnel flow time, L-box height	Compressive	-	-
[49]	FA (100–70%), slag (10,20,30%), SF [5] (5,10,15%)	Slump, T_{500} and V-funnel flow time, L-box height, U-box	Compressive, splitting, flexural strength	-	-
[50]	Slag (100, 98%), NS (2%)	Slump, T_{500} and V-funnel flow time, L-box height	Compressive, splitting, fracture parameters, modulus of elasticity	-	-

Table 1. *Cont.*

Refs	Binder Type	Fresh Properties	Mechanical Properties	Durability Properties	Others
[51]	FA	Slump, T_{500} and V-funnel flow time, L-box and J-ring height	Compressive	-	-
[52]	FA (100,90,80,70%), MK (10,20,30%)	Slump, T_{500} and V-funnel flow time, L-box height	Compressive, splitting, flexural	-	-
[32]	FA (100,90%), SF (10%)	-	-	Shrinkage	-
[53]	Slag (100,70,60,50,40,30%)FA (30,40,50,60,70%)	Slump, T_{500} and V-funnel flow time, L-box and J-ring height	Compressive, splitting, flexural	-	XRD, SEM, correlation, modeling
The current study	FA (100,70,50,0%) slag (0,30,50,100%)	Slump and J-ring flow, T_{500}, TJ_{500}, and V-funnel flow time, L-box and J-ring height sieve segregation	Compressive, splitting	Sorptivity, long-term free drying shrinkage, long-term mass loss	Correlation, an empirical equation

Where [1] NS is nano-silica, [2] RHA is rice husk ash, [3] MK is metakaolin, [4] GSA is groundnut shell ash, [5] SF is silica fume.

Studies on the influence of slag content on fresh, mechanical, and durability behaviors, as well as mechanical property predictive models, are limited. The study presented here aims at filling these gaps by examining the impact of slag inclusion on the various fresh, mechanical, and durability properties of FA-based SCGC specimens. Slump flow, T_{500}, J-ring flow, TJ_{500}, J-ring height, L-box, V-funnel, and sieve segregation resistance are the fresh property tests to be used in this work. In terms of mechanical properties, both the compressive and splitting tensile strength will be determined. In terms of durability, sorptivity and long-term free drying shrinkage will be investigated. Correlations between different properties will be attempted and comparisons with existing predictive models made.

2. Materials

Class F fly ash (FA) and ground granulated blast furnace slag (slag) were utilized as a binder. Their chemical and physical features are presented in Table 2.

Table 2. Chemical and physical features of FA and slag in the presented study.

Component %	CaO	SiO_2	Al_2O_3	Fe_2O_3	MgO	Various	Specific Gravity	Loss on Ignition	Blain Fineness (cm^2/g)
FA	4.0	56	24	7.0	2.0	7.0	2.28	3.0	3098
Slag	40	36	11	0.4	7.6	5.0	2.80	2.3	4250

The alkaline activator utilized in this study was a blend of sodium silicate (Na_2SiO_3) and sodium hydroxide (NaOH). The alkaline/binder ratio was kept at 0.5 in this study, the molarity (M) constant at 12, and the Na_2SiO_3/NaOH ratio at 2.5 [34]. The mass of Na_2SiO_3 used in the production of SCGC was 45% dry (of which 15% was Na_2O and 30% SiO_2) and 55% water content. The NaOH was 99% pure. The NaOH solution was made by dissolving solid NaOH pellets in water, targeting at 12 M molarity.

Coarse and fine natural aggregates used in this study were gravel and sand, respectively. The physical features of the used aggregates are shown in Table 3.

Table 3. Physical features of gravel and sand in the presented study.

Type of FA	Size (mm)	Specific Gravity	Water Absorption (%)
Gravel	4–16	2.70	0.5
Sand	0–4	2.66	0.8
Standard	BS EN 933-1+ A1 2005 [54]	BS EN 1097-6:2013 [55]	BS EN 1097-6:2013 [55]

The superplasticizer (SP) used was MC Power flow evo 502, a SP based on polycarboxylate ether that meets the standards of EN 934-2: T3.1/3.2. It is a fifth generation MC-superplasticizer with improved rheological properties. It is yellow in color and has a density of 1.03 kg/m^3. Tap water content was maintained at 40 kg/m^3 throughout the research.

3. Mixing Procedure

Four SCGC mixtures were prepared with 450 kg/m^3 total binder content [35]. Different amounts of slag were used to make these combinations, while the other parts remained constant. Slag was used in this study to replace the FA at 0%, 30%, 50%, and 100% replacement levels. Table 4 shows the detailed mix proportions. In the following mix codes, G% denotes the slag percentage used.

Table 4. Mix proportions of SCGC in the presented study.

Mix Code	Binder (kg/m^3)	FA (%)	Slag (%)	Gravel (kg/m^3)	Sand (kg/m^3)	Alkaline/Binder	Molarity (M)	SP (%)	Water (kg/m^3)
G0	450	100	0	800	825	0.5	12	7	40
G30		70	30						
G50		50	50						
G100		100	100						

3.1. Mixing and Casting

A comparable mixing process was employed to attain consistency and homogeneity in each mixture. In the first phase, the dry ingredients, including aggregates (gravel and sand), and binder (FA and slag), were blended for approximately 2 min in an electric concrete mixer with a 75-L capacity. After properly blending the dry ingredients, the blended liquids of alkaline activator and water were fed to the mixer and wet mixed for two more minutes. The superplasticizer was added to the wet batch, and mixing proceeded for an extra two minutes. After the mixing process was completed, SCGC mixtures were tested in their fresh state, and then samples for hardened state characteristics were cast. Before placing the concrete into the molds, the fresh concrete was re-mixed in the mixer for thirty seconds to ensure the homogeneity of the mixture.

3.2. Curing Method

The samples were kept for 24 h in the laboratory [56] prior to being oven-cured at 85 °C for 24 h, except for the shrinkage tests. It should be noted that the SCGC specimens made with 100% FA (G0) did not harden enough after 24 h due to the low content of CaO, which is considered the main compound that affects the initial setting time of concrete. Slag can be added to the FA-based SCGC to enhance the setting time [34]. Therefore, in this study, the geopolymer concrete specimens were subjected to heat-curing after one day without being removed from the molds. After oven curing, the hardened samples were stored at ambient temperature until the testing date.

4. Testing Procedure

4.1. Fresh Properties

The flowability of the SCGC mixtures were evaluated using slump flow, J-ring flow, and V-funnel tests, passing-ability was assessed through the use of the L-box and J-ring height tests, and viscosity was evaluated via V-funnel, $T_{500\ mm}$, and $TJ_{500\ mm}$, according to the EFNARC recommendations [57]. Segregation resistance was measured via a sieve segregation test [58]. The J-ring test is a combination test to assess the filling-and-passing-ability of the freshly mixed SCGC through a restricted and congested area without blocking. During this test, J-ring flow diameter, $TJ_{500\ mm,}$ and J-ring height were calculated.

Figure 1 shows photos of the fresh tests conducted in this study. Flow diameters in the x and y directions were monitored in the slump flow and J-ring flow tests, and the $T_{500\,mm}$ and $TJ_{500\,mm}$ flow times at which the flow diameter reached 500 mm were also recorded. The J-ring height, or passing ability (*PJ*), which refers to the blocking of fresh concrete through congested reinforcement bars, was measured once the concrete flow had stopped moving. Then, in the center of the J-ring, the straight rod was placed with its flat side down and to measure the distance (in mm) between the lower edge of the rod and the concrete surface (*ho*). Following that, the height (mm) of four spots outside the J-ring, two along the x-axis (*hx1*, *hx2*) and two along the y-axis (*hy1*, *hy2*) were measured. The following equation was used to determine the PJ value in mm:

$$PJ = \frac{hx1 + hx2 + hy1 + hy2}{4} - ho \tag{1}$$

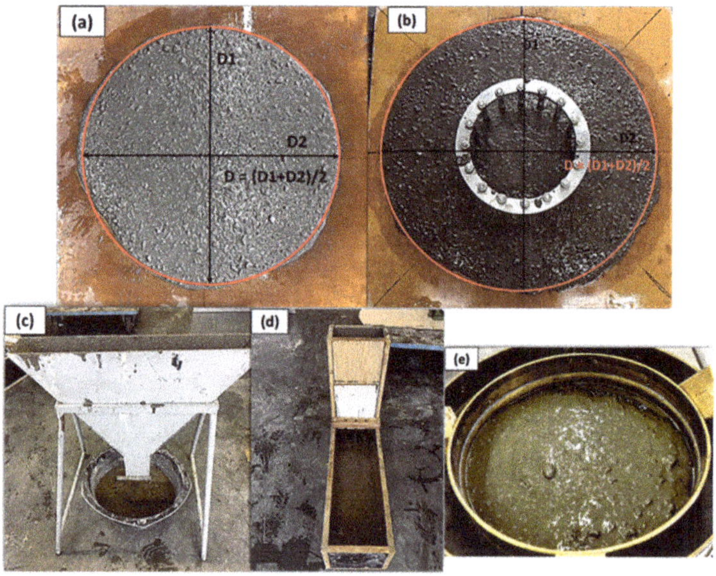

Figure 1. Fresh test apparatus: (**a**) slump test, (**b**) J-ring test, (**c**) V-funnel test, (**d**) L-box test, and (**e**) sieve segregation test.

The V-funnel test involves completely filling the V-shaped section with concrete, allowing the concrete to discharge, and recording the discharge time. The concrete viscosity in its fresh condition can be assessed indirectly using the V-funnel flow time and $T_{500\,mm}$ test measurements, as well as the V-funnel and $TJ_{500\,mm}$ test records. The SCGC's passing-ability (PA) between tiny gaps of bars was determined by means of the L-box test by dividing the horizontal section's concrete height by the vertical section's height after the concrete flow has stopped. The sieve segregation test was conducted using a sieve having 5 mm holes, a 300 mm diameter, and a 30 mm height. After finishing the mixing process, approximately 5 kg of the fresh concrete was placed in a plastic container and set aside for 15 min. The pan was then weighed dry (m_p). After that, a sieve and pan were placed on the scale and approximately 4.8 kg of fresh SCGC was carefully placed at half a meter height, and the weight was recorded (m_s). After two minutes, the sieve was carefully lifted, and the amount of concrete that had passed through the 5 mm holes and remained in the pan were weighed (m_{ps}). The sieve segregation index (*SI*) was calculated using the following equation:

$$SI,\ \% = \frac{(m_{ps} - m_p) * 100}{m_s} \tag{2}$$

Table 5 Displays the upper and lower bounds that EFNARC has established for the performance of SCC mixtures in their fresh state [57].

Table 5. Fresh test assessment following the EFNARC guidelines.

Flowability Classes			
Classes	**Slump Flow and J-Ring Flow Diameter (mm)**		
Slump flow (SF) 1	550–650		
Slump flow (SF) 2	660–750		
Slump flow (SF) 3	760–850		
Viscosity Classes			
Class	**T_{500} (sec)**	**TJ_{500} (sec)**	**V-Funnel time (sec)**
Viscosity (VS1/VF1)	$\leq$2	$\leq$2	$\leq$8
Viscosity (VS2/VF2)	>2 and $\leq$5	>2 and $\leq$6	9 to 25
Passing ability			
Classes	**J-Ring passing ability**		
Passing-ability (PJ) 1	$\leq$10 mm		
Passing-ability (PJ) 2	$\geq$10, 20 mm		
Classes	**L-Box height ratio**		
Passing-ability (PA) 1	$\geq$0.8 with two rebar		
Passing-ability (PA) 2	$\geq$0.8 with three rebar		
Segregation resistance [58]			
Classes	**Sieve segregation resistance (%)**		
SI 1	$\leq$20%		
SI 2	$\leq$15%		

4.2. Mechanical Properties

The compressive strength of hardened SCGC specimens was obtained at 7 and 28 days of age from three 100 mm cube according to BS EN 12390 [59]. The splitting tensile strength was determined according to BS EN 12390-6 [60] using the average of three cylindrical samples of Ø150 × 300 mm at 28 days of age.

4.3. Correlation between Splitting Tensile and Compressive Strength

Normally, there is a good correlation between the mechanical properties of concrete (compressive strength, elastic modulus and tensile/flexural/splitting strength). Hence, the 28 day splitting tensile strength can be predicted from the compressive strength using empirical equations proposed by ACI 318-11 [61], ACI 363R-10 [62], CEB-FIB [63], and Lee and Lee [64], as tabulated in Table 6. The empirical equations provided by codes and literature are based on the cylindrical compressive strength values $f_{\acute{c}}$ in MPa. As cubes were tested in this study the following conversion equation was used [65]:

$$f_{\acute{c}} = 0.8 \times f_{cu} \quad (3)$$

where 0.8 is the conversion factor.

Table 6. ACI predictive equations for the splitting tensile strength of concrete.

Property	Standard	Equation	
Splitting Tensile Strength, f_t (MPa)	ACI 318 [61]	$f_t = 0.56\sqrt{f}$	(4)
	ACI 363 [62]	$f_t = 0.59\sqrt{f}$ for 21 MPa < $f\acute{c}$ < 83 MPa	(5)
	CEB-FIB [63]	$f_t = 0.3f^{\frac{2}{3}}$	(6)
	Lee and Lee [64]	$f_t = 0.45\sqrt{f}$	(7)

4.4. Durability Properties

4.4.1. Capillary Water Absorption (Sorptivity)

The rate of water absorption per unit area was measured from the capillary water absorption of three 100 mm cubes. After 28 days, the specimens' sides were taped to prevent water absorption from the sides. The specimens were then placed on a tray and rested on knife-edge supports to obtain a depth of water immersion of 2.0 mm, in accordance with BS EN 13057 [66] (see Figure 2). The mass resulting from water absorption from the bottom surface was determined over time (0 min, 12 min, 30 min, 60 min, 120 min, 240 min, and 1440 min). The water uptake was determined by plotting the weight gain per unit area against the square root of time, and the slope of the best fit line was used to calculate the Sorptivity coefficient (index) [67,68].

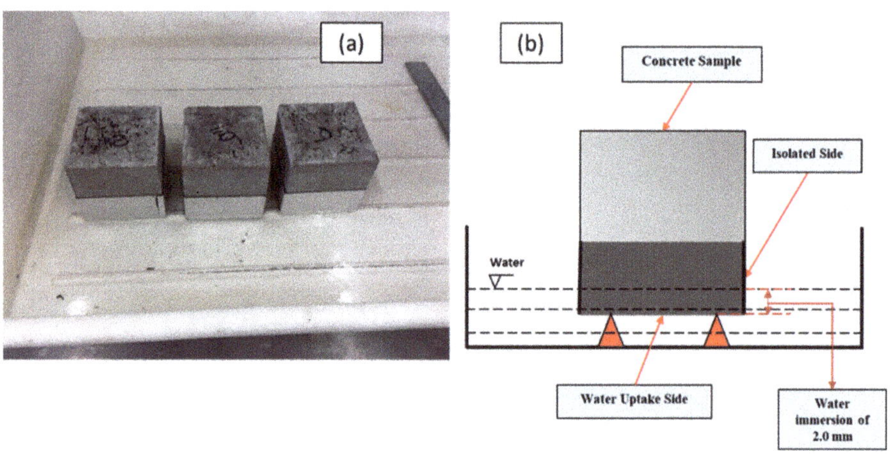

Figure 2. Capillary absorption test: (**a**) specimen set up, (**b**) schematic representation of specimens.

4.4.2. Free Drying Shrinkage

To quantify drying shrinkage and mass loss of SCGC, two 100 × 100 × 300 mm prisms were used for the G0 and G50 mixes according to ASTM C 157 [69]. After casting, the prisms were left at ambient conditions for 24 h, then exposed to oven curing at 40 °C for three days [70]. Following curing, the prisms were demolded, and demec points were glued to the specimens at a gauge length of 200 mm mid-height on two sides (see Figure 3). The change in length was determined by a dial gauge extensometer, with a strain accuracy of 5 με (see Figure 4). Initial measurements of length and weight were carefully recorded. The prisms were then placed in a controlled climate chamber and subjected to a drying condition (22 ± 2 °C and 50 ± 2% relative humidity (RH)). Measurements of length and weight were taken over a period of a year; every 24 h for the first week, four times a week for the later three weeks, once per week for the second month, once per two weeks for the third month, and then once per month for the long-term periods (six, nine, and 12 months). The values were averaged for the two samples.

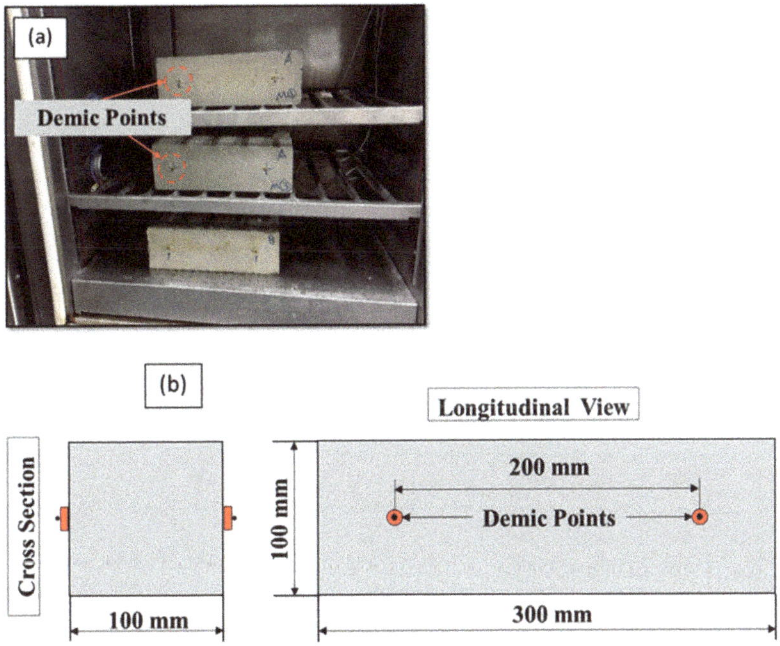

Figure 3. Free drying shrinkage test: (**a**) prisms are placed in a controlled climate chamber, (**b**) geometry of the specimens.

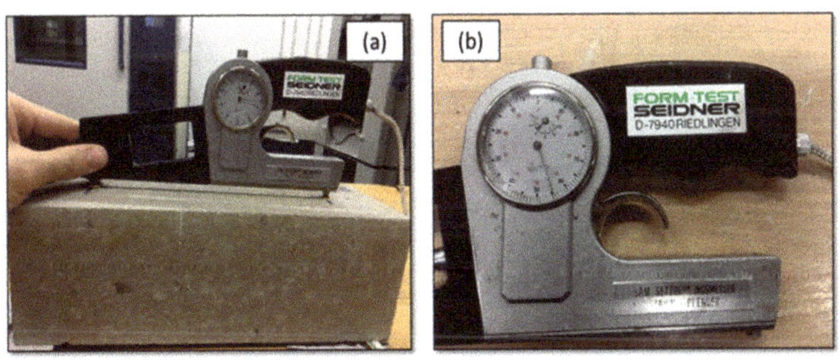

Figure 4. Free drying shrinkage measurement: (**a**) specimen measurement, (**b**) strain gauge measurement device developed by BAM.

5. Result and Discussion

5.1. Fresh Properties

The impact of utilizing slag as a binder on the performance of FA based SCGC mixes was investigated, and the results were compared with the limits of EFNARC [57]. The results of the fresh properties of SCGC mixtures with various slag contents are presented in Table 7 and Figures 5–18.

Table 7. Fresh test results of the SCGC in the presented study.

Mix Code	Slag (%)	Slump (mm)	J-Ring (mm)	T_{500} (s)	TJ_{500} (sec)	PJ (mm)	V-Funnel (s)	L-Box	Segregation Index (%)
G0	0	788	780	2.12	2.42	0.0	7.5	1.00	28.7
G30	30	775	760	2.31	2.71	1.0	10.3	0.99	22.4
G50	50	750	738	2.94	3.31	2.7	12.0	0.96	18.4
G100	100	680	665	4.10	4.43	7.1	17.1	0.84	11.4

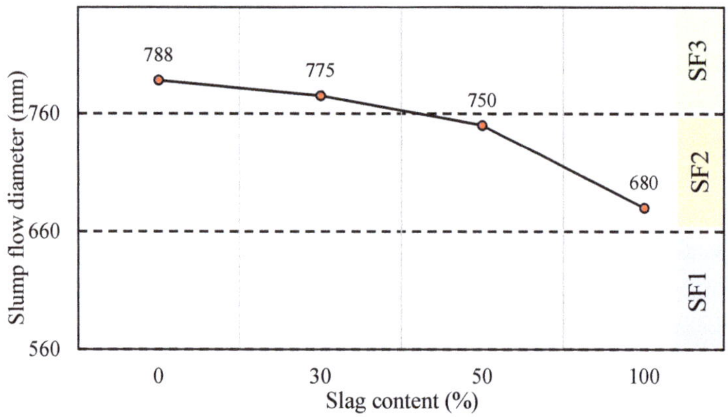

Figure 5. Influence of slag content on the slump results.

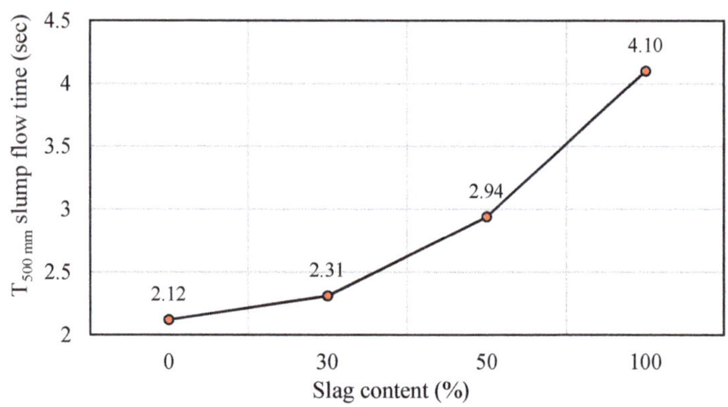

Figure 6. Influence of slag content on the T500 mm flow time.

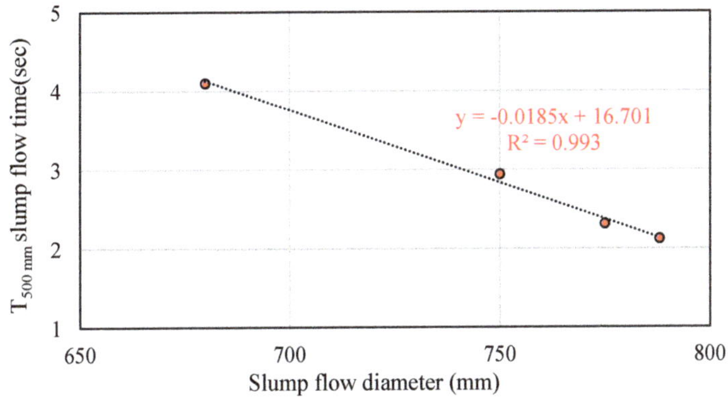

Figure 7. Correlation between $T_{500\ mm}$ slump flow time and slump flow diameter.

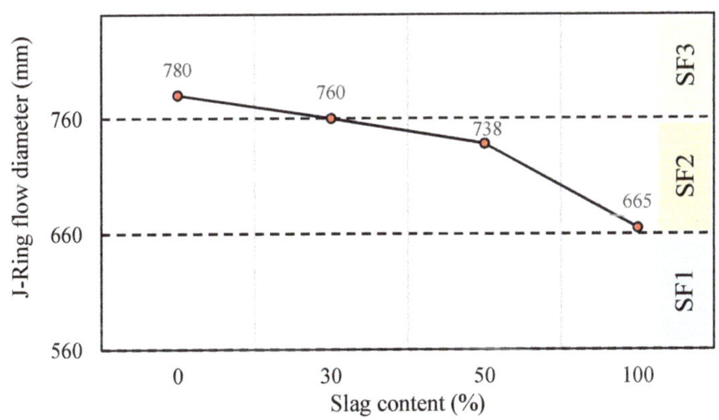

Figure 8. Influence of slag content on the J-ring flow diameter.

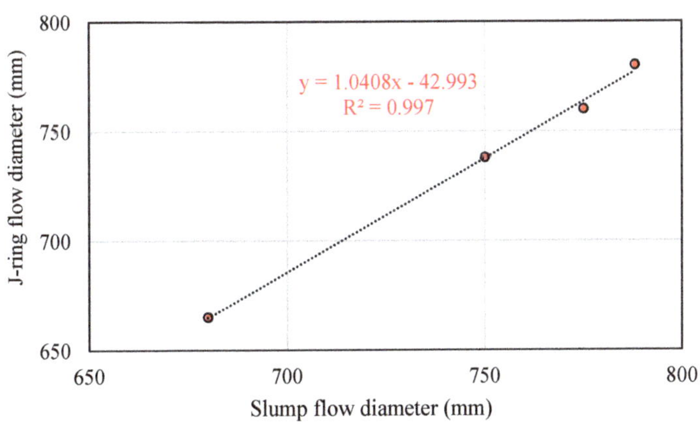

Figure 9. Correlation between J-ring and slump flow diameter.

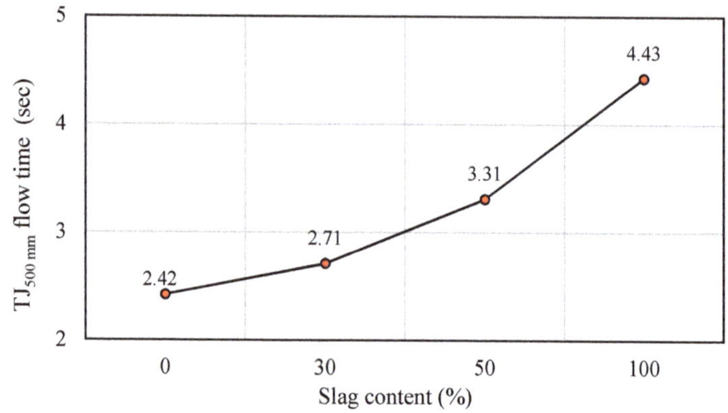

Figure 10. Influence of slag content on the $TJ_{500\ mm}$ flow time.

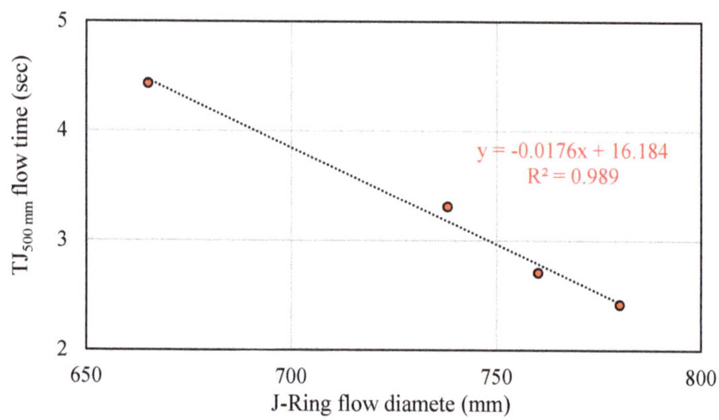

Figure 11. Correlation between $TJ_{500\ mm}$ slump flow time and J-ring flow diameter.

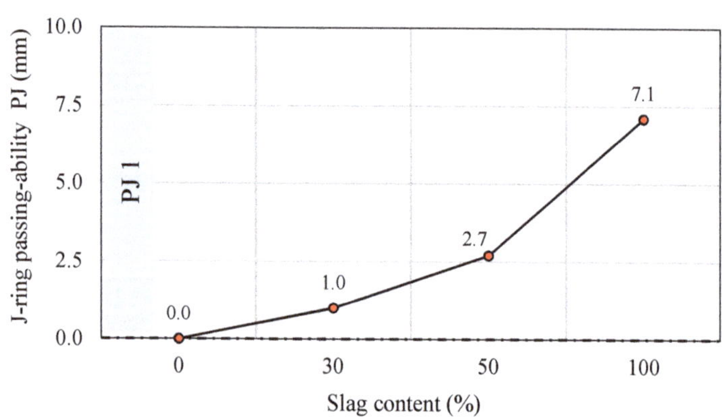

Figure 12. Influence of slag content on the J-ring passing-ability.

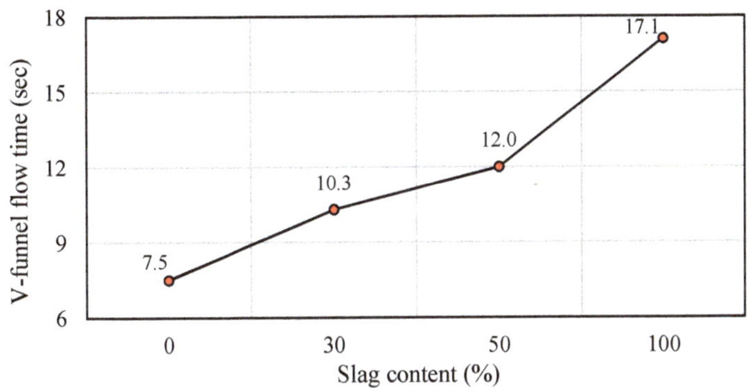

Figure 13. Influence of slag content on the V-funnel flow time.

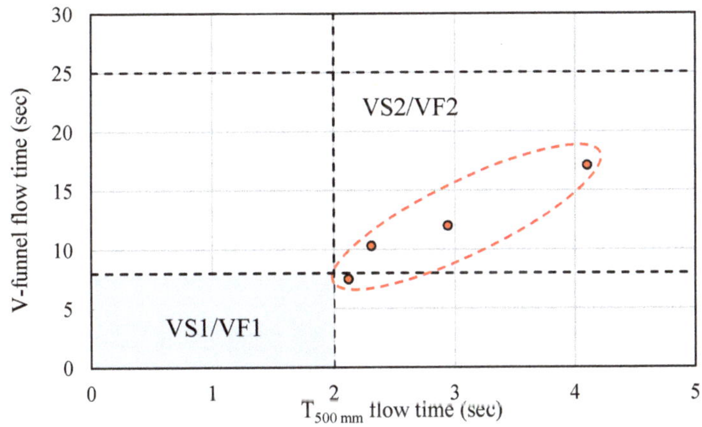

Figure 14. Viscosity class variation with V-funnel and $T_{500\ mm}$.

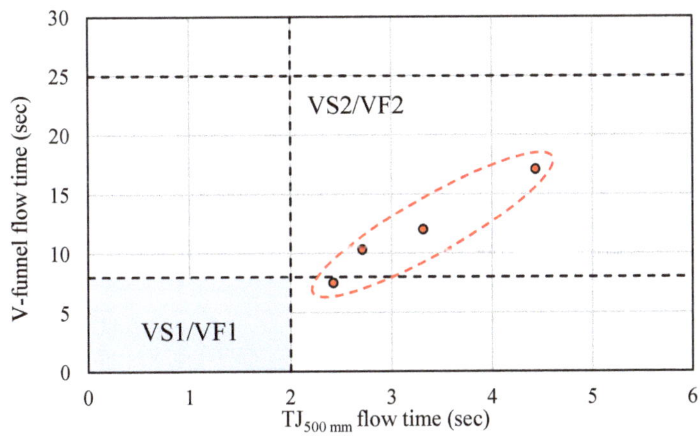

Figure 15. Viscosity class variation with V-funnel and $TJ_{500\ mm}$.

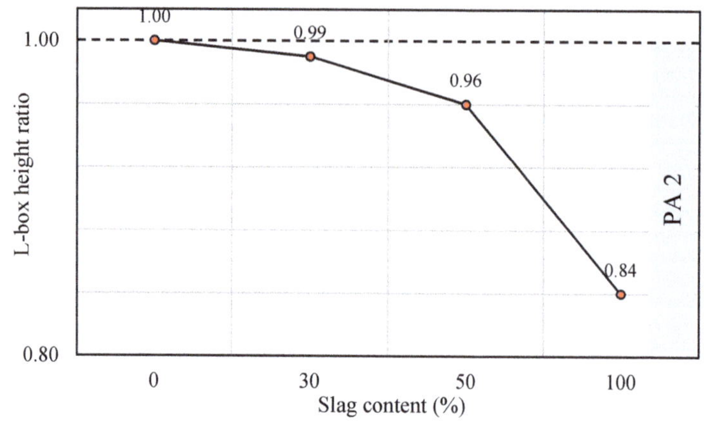

Figure 16. Influence of slag content on the L-box height ratio.

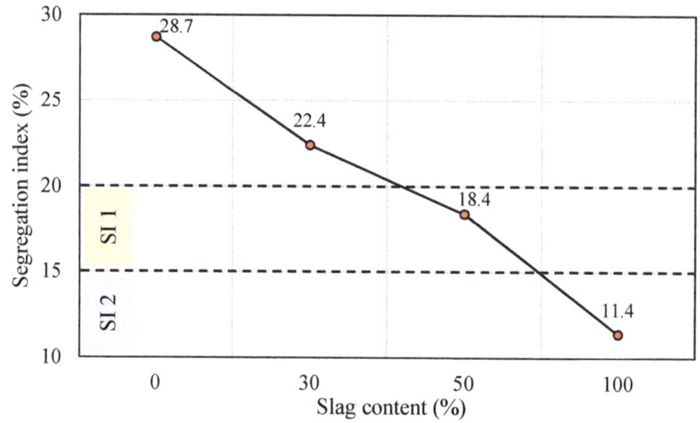

Figure 17. Influence of slag content on the segregation index.

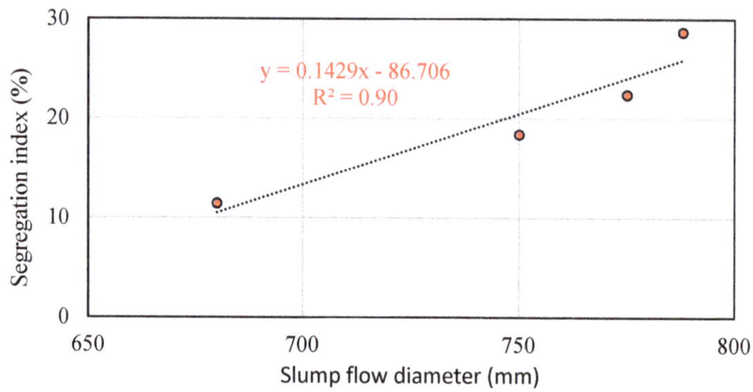

Figure 18. Correlation between segregation index and slump flow diameter.

5.1.1. Slump Flow Diameter

The slump flow diameter values of the produced geopolymer concrete in mm with respect to the slag content are shown in Figure 5. The slump values range from 788 to

680 mm, showing a systematic decrease with the increase in slag content. This is because slag binders have a larger specific surface area than FA binders, which demands a higher quantity of mixing water, thereby diminishing the workability of the freshly mixed concrete.

The slump flow values for the produced SCGCs of the mixes, including 0% and 30% slag content are in the range of the SF3 class, and the mixes containing 50% and 100% slag content in the range of the SF2 class of EFNARC limits. Based on the EFNARC limitations [57], the SF2 class can be used for constructions with complex geometries, vertical applications, and filling beneath the formwork. However, SCGC in the SF3 class provides a better surface finish than SCGC in the SF2 class. On the other hand, controlling segregation resistance is more difficult in the SF3 class. Previous studies recorded values between 755 and 650 mm for the blended fly ash and slag blended SCGC [34]. Moreover, the results align with the BS EN 12350-8 standard [71] which specifies that slump values should have an average diameter greater than 600 mm.

5.1.2. $T_{500\ mm}$ Slump Flow Time

Figure 6 shows the influence of slag on the $T_{500\ mm}$ slump flow time, in which time is measured up to the point at which the flow diameter of the freshly mixed concrete reaches 500 mm. In the results of T_{500}, it can be observed that by increasing the slag content, the slump flow time is increased. The recorded $T_{500\ mm}$ slump flow in this study was between 2.12 and 4.10 s. $T_{500\ mm}$ meet the requirements of the EFNARC specifications [57] and the BS EN 12350-8 standard (less than 6 s) [71].

To examine the correlation between these two tests, the inverse linear relationship between flow time and flow diameter data is depicted in Figure 7. The R-square value of 0.993 shows a strong correlation between the T_{500} and Slump flow diameter values of the mixes conducted in this study.

5.1.3. J-Ring Flow Diameter

The J-ring flow diameter test is used to determine the flow spread of the fresh mixed SCC as it flows through the J-ring. In this test, the J-ring flow diameter in mm for the various slag content is evaluated, as shown in Figure 8. It can be seen that the J-ring flow diameter decreases with the increase in slag content. The J-ring results ranged between 780 and 665 mm for the slag contents of 0% and 100%, respectively. The mixes of the SCGC can be classified into the SF3 class when the slag content ratio ranges between (0 to 30%) and the SF2 class when the slag content ratio exceeds 30% up to 100%. The test values are within the EFNARC limits [57]. Furthermore, the linear correlation between J-ring and slump flow diameter is found in Figure 9. From the figure, the R-square is 0.997, indicating a strong relationship between J-ring-and-slump flow diameter outcomes. This means that both results are proportional to the slag content. Safiuddin et al. [72], who studied the fresh properties of SCC and achieved a 0.998 correlation coefficient between J-ring and slump flow diameter, also confirmed this result

5.1.4. $TJ_{500\ mm}$ Flow Time

The $TJ_{500\ mm}$ flow time outcomes of the SCGC with different slag content are shown in Table 7 and Figure 10. The flow time measured for the $TJ_{500\ mm}$ test varies between 2.42 and 4.43 s. According to EFNARC specifications and guidelines [57], the $TJ_{500\ mm}$ of SCC typically takes between 2 and 6 sec. Hence, the $TJ_{500\ mm}$ values are within a satisfactory range. The influence of slag content on the $TJ_{500\ mm}$ of SCGC is obvious from Figure 10, and the $TJ_{500\ mm}$ results decreased with the increase in slag content. Moreover, the correlation coefficient (R-square) for the $TJ_{500\ mm}$ slump flow time and J-ring flow diameter is depicted in Figure 11. The R^2 of 0.989 shows a very strong relationship between $TJ_{500\ mm}$ flow time via J-ring flow diameter. As expected, there is an inverse linear relationship between the two measurements.

5.1.5. J-Ring Passing-Ability

The J-ring passing ability values with respect to slag content are illustrated in Figure 12. It can be seen that the passing-ability increases with the inclusion of slag. The test values were located between 0.0 and 7.1 mm for the slag content of 0% and 100%, respectively. According to the EFNARC specifications [57], the J-ring passing-ability for values smaller than 10 mm is classed as PJ 1. Therefore, the PJ values are typically in the range of PJ 1 class for the mixes in the study, with values less than 10 mm, which indicates good passing ability.

5.1.6. V-Funnel Flow Time

The discharge time from the V-funnel flow time test is comparable to the slump flow test results, as depicted in Figure 13. The test values increase with increasing slag content. The EFNARC requirements [57] specify viscosity classes based on V-funnel, $T_{500\ mm}$ slump, and $TJ_{500\ mm}$ slump flow time test results, as can be seen in Table 7. In addition, the viscosity class variations between V-funnel vs. $T_{500\ mm}$ are illustrated in Figure 14. According to the test findings, all mixtures belong to the VS2/VF2 viscosity category, except for the mix created without slag (G0), which is within the VS1/VF1 viscosity class based on the V-funnel results, whereas the G0 is classified as VS2/VF2 referring to the $T_{500\ mm}$ outcomes. According to EFNARC, the VS1/VF1 viscosity category has a high filling capacity, even with dense reinforcements, with a tendency to bleeding and segregation. The viscosity class VS2/VF2 offers excellent resistance to segregation. However, the VS2/VF2 class has an inadequate smooth surface texture and may be susceptible to SCGC mix flow stoppage. The results show that increasing the slag level increases the flow duration, which is advantageous for avoiding segregation and bleeding but may result in insufficient filling ability. The EN 12350-9 [73] standard also specifies that the V-funnel outcome should be less than 15 sec to ensure adequate filling capacity. Likewise, as illustrated in Figure 15, a similar trend can be observed for the viscosity class variances between V-funnel and $TJ_{500\ mm}$ flow time data.

5.1.7. L-Box Height Ratio

The L-box test determines the mixes' capacity to pass through three-bar restricted openings. The test values of SCGC mixtures should be between 0.8 and 1.0 to confirm certain passing abilities according to the EFNARC guidelines [57] and the EN12350-10 standard [71]. From the test results presented in Figure 16, it is feasible to ascertain that all mixtures possess an appropriate passing capacity (passing-ability > 0.8), and the L-box values are classified in the class of passing ability 2. The highest passing ability of 1.00 is for the mixes without slag content. Then, the L-box values decrease as the slag content increases.

5.1.8. Sieve Segregation

The segregation index (%) is depicted in Figure 17. With increased slag content, the segregation index begins to diminish. This is due to the fact that slag has a larger specific surface area than FA. The test values achieved in the mixes mentioned above range from 11.4% to 28.7%. Based on BS EN12350-11 [58], the segregation index should be lower than 20%. This means that the mixes with 0% and 30% slag content are out of the allowable range, whereas the mixes with 50% and 100% slag content satisfy the SI 1 and SI 2 classes, respectively. Figure 18 depicts the relationship with both the segregation index and the slump flow diameter. It can be seen that the segregation index and slump flow of the SCC mixtures have a significant linear relationship. The correlation coefficient R-square value is 0.90, indicating a significant correlation.

5.2. Mechanical Properties

Table 8 shows the results for the different mechanical properties. The SCGC specimens were cured at 85 °C for 24 h and then kept at room temperature until they reached 28 days of age. As expected, increasing the slag content of the geopolymer concrete was found to enhance the compressive strength [34].

Table 8. Hardened properties of SCGC in the presented study.

Mix Code	Slag Content (%)	Compressive (MPa)		Splitting Tensile (MPa)
		7 Days	28 Days	
G0	0	38.70	44.21	2.51
G30	30	58.17	65.72	4.31
G50	50	76.63	81.67	5.35
G100	100	80.92	85.10	5.60

5.2.1. Compressive Strength (f_{cu})

Figure 19 shows the impact of the slag content on the compressive strength of SCGC. The FA-based geopolymer concrete mixes yield the lowest strength values, whereas the inclusion of slag significantly increases the compressive strength of SCGC. The compressive strength of the seven-day-old specimens were between 38.7–80.9 MPa for the G0 and G100 mixtures, respectively. These findings are line with a study by Al-Rawi and Tayşi [34]. The compressive strength values at 28 days are very close to those at seven days of age. This may be due to the high early age development rate of geopolymer concrete due to heat curing compared to normal concrete, as has been reported by previous studies [74].

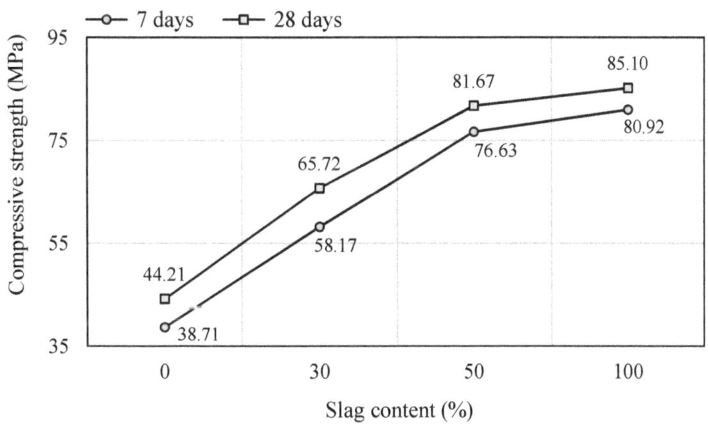

Figure 19. Influence of slag content on the compressive strength at seven and 28 days.

Figure 20 shows the percentage relative increase in compressive strength for the various slag content compared to the mix without slag. At 7 days, the improvement in strength of SCGC was 98.0% and 109.0%, for the slag contents 50%, and 100%, respectively. The results confirm that the biggest increase in strength comes with the replacement of 50% slag.

The effect of slag, FA, and blended FA and slag binders on the compressive strength of OPC and geopolymer mortars was also examined by Chi and Huang [75]. The results showed that the compressive strength of geopolymer mortars (excluding FA-based geopolymer mortars) was higher than that of ordinary Portland cement mortars. During X-Ray Diffraction tests, specimens containing 100% FA reveal a low rate of reactive calcium, resulting in a low level of C-S-H. As a result, specimens made of FA-based geopolymer have lower mechanical properties [75]. Previous research on FA and slag mixtures also found that strength increased with increasing slag content [76–79]. This is due to an increase in calcium content in the mix with an increase in slag concentration. Ismail and Bernal [80] observed that calcium-rich pastes typically form a C-S-H gel, which forms a dense structure and increases strength. Nevertheless, as Si increases and calcium decreases, the Sodium-Alumino-Silicate-Hydrate (N-A-S-H) gel forms, resulting in a loss of strength. The compressive strength increases with slag content after seven and 28 days of curing.

Furthermore, the presence of CaO in slag encourages the hydration process, so a 100% slag content mixture has more compressive strength [81].

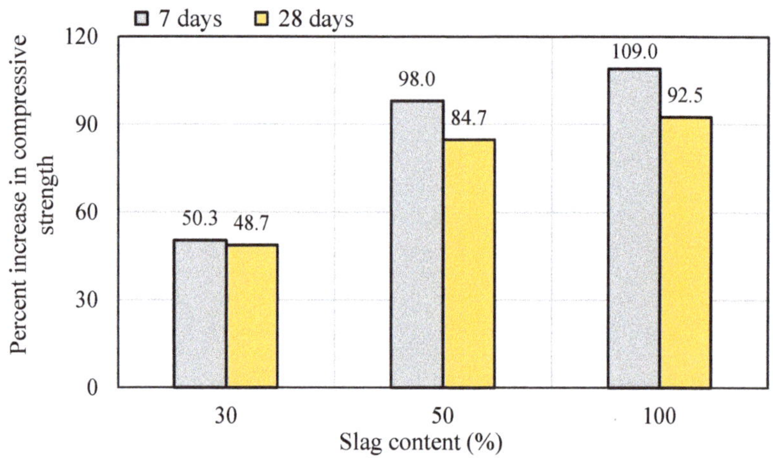

Figure 20. Percent increase in compressive strength vs. slag content.

5.2.2. Splitting Tensile Strength (f_t)

Figure 21 demonstrates the test values of splitting tensile strength (f_t), while Figure 22 shows the relative increase. It can be observed that increasing the slag content increases the splitting tensile. The maximum value of splitting achieved in this study was 5.60 MPa for the slag-based geopolymer concrete. The increase was 71.7%, 113.1%, and 123.1% for the G30, G50, and G100 mixes. It was also found that the increase in splitting values was 123% for the SCGC mixes having 100% slag content.

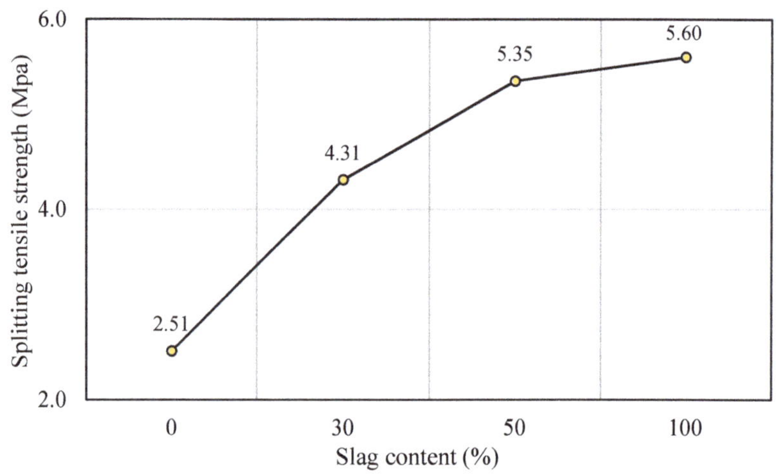

Figure 21. Influence of slag content on the compressive strength at 28 days.

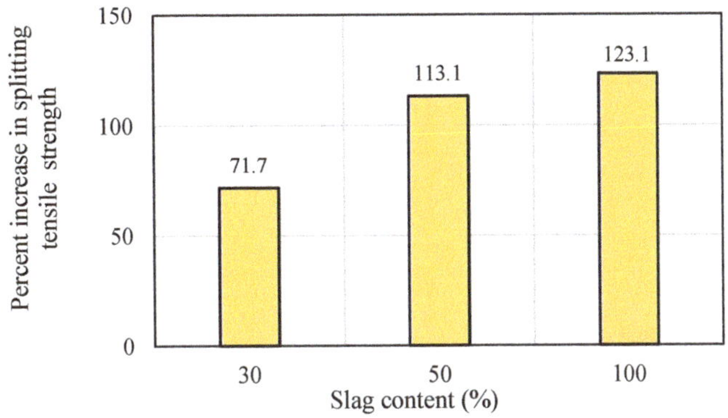

Figure 22. Percent increase in splitting tensile strength.

The 28-day failure surface of the SCGC specimens is shown in Figure 23. Coarse aggregate is equally distributed all across the specimens and does not segregate, except for the mix made with 100% FA, where a sign of segregation can be seen near the top casting side of the specimens. It is clear from Figure 23b,c, that for the mixes made with slag, the failure surface of SCGC specimens mostly went through the geopolymer paste matrix and the aggregates. However, in the FA-based geopolymer concrete mix (See Figure 23a), the failure path propagated through the interfacial transition zone and geopolymer paste matrix and rarely through the aggregates. This highlights the disparity in strength of geopolymer pastes of FA and slag-based mixes. According to Shen et al. [82], the surface's roughness and broken aggregates directly correlate with the splitting tensile strength. When the roughness is increased and more aggregates were cracked, the tensile splitting strength is higher [82]. The correlation between the splitting tensile and compressive strength in FA-based SCGC with various slag content is illustrated in Figure 24. It is obvious that an increase in compressive strength typically results in a corresponding subsequent increase in tensile strength [34]. Despite the usage of slag binder, the R-Square value of 0.994 indicates that the splitting and compressive strength values are strongly correlated with a direct linear relationship. For the results presented in Figure 24, the following correlation between splitting tensile strength and compressive strength was found:

$$f_t = 0.025 f_{cu}{}^{1.225} \quad (8)$$

where f_t is the splitting tensile strength (MPa) and f_{cu} is the cube compressive strength (MPa).

Figure 23. Failure path of splitting tensile strength test; (**a**) G0 (0% slag content), (**b**) G30 (30% slag content), and (**c**) G50 (50% slag content).

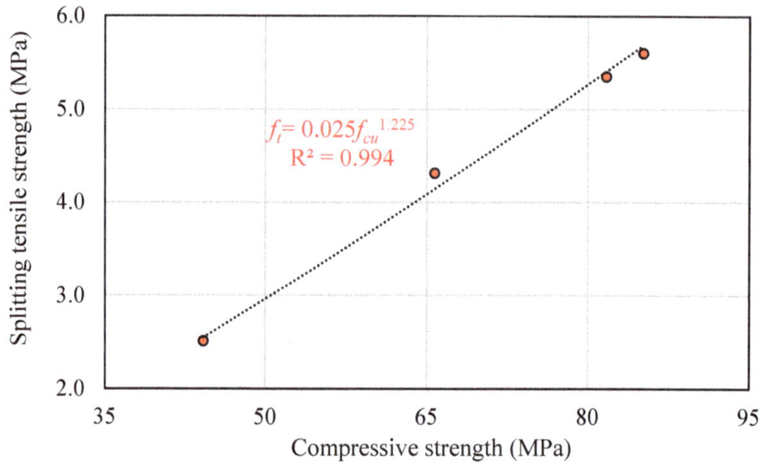

Figure 24. Correlation between splitting tensile and compressive strength.

5.3. Prediction of Splitting Tensile Strength

Though predictive equations for tensile strength exist for normal concrete, there is a lack of equations for geopolymer and self-compacted geopolymer concrete. The results from this study are compared with the results calculated from proposed equations by ACI 318-11 [61], ACI 363R-10 [62], Eurocode CEB-FIB [63], and Lee and Lee [64]. All the empirical models presented above were developed for normal concrete, except for the model proposed by Lee and Lee, which proposed for FA and slag blended geopolymer concrete. Figure 25 and Table 9 show the actual and predicted 28-day mechanical values using Equations (4)–(7) in Table 6 and a proposed model (Equation (8)).

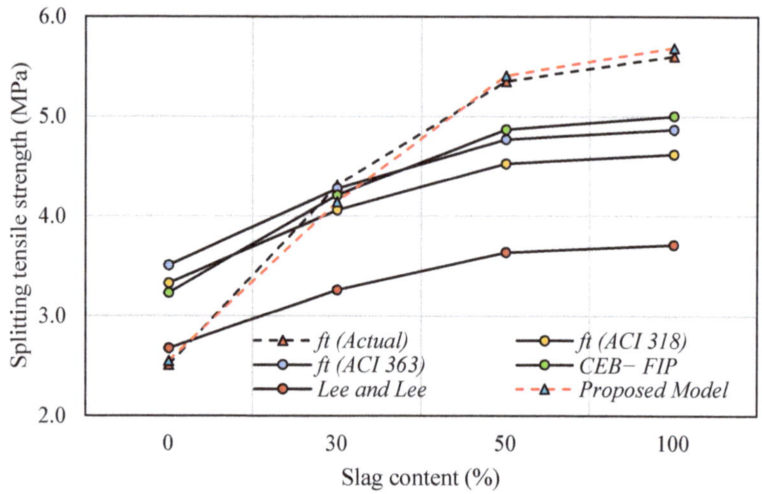

Figure 25. Actual and predicted splitting tensile strength values.

Table 9. Actual and predicted mechanical properties of SCGC recommended by ACI 318-11 [61], ACI 363R-10 [62], European code CEB-FIB [63], Lee and Lee [64], and self-developed model.

Mix Code	Slag Content (%)	Splitting Tensile Strength (MPa)					
		Actual Result	ACI 318	ACI 363	CEB−FIP	Lee and Lee	Proposed Model (Equation (8))
G0	0	2.51	3.33	3.51	3.23	2.68	2.55
G30	30	4.31	4.06	4.28	4.21	3.26	4.14
G50	50	5.35	4.53	4.77	4.87	3.64	5.41
G100	100	5.60	4.62	4.87	5.00	3.71	5.68

Splitting Tensile Strength (f_t)

Figure 25 compares the splitting tensile results obtained from this investigation versus predicted outcomes by ACI 318-11 [61], ACI 363R-10 [62], Eurocode CEB-FIB [63], Lee and Lee [64], and the proposed model (Equation (8)). For the reference mix G0, the tensile strength is close to the model that Lee and Lee proposed, while the predicted values by ACI models and CEB-FIB provide overestimates. However, surprisingly, the tensile strength predicted by Lee and Lee does not increase much with the compressive strength. The same can be said for the other predictions as well. This is because the splitting tensile strength in this study grows faster than the compressive strength. This is reflected in Equation 8 by the exponent of the compressive strength being higher than 1. This result is surprising, and indicates a more fundamental property of slag that enhances the tensile strength, possibly via the penetration of hydration products into the ITZ and aggregates themselves. This clearly requires further investigation.

The actual splitting tensile results of 70% FA and 30% slag mix are very similar to those predicted by ACI 318, ACI 363, and Eurocode CEB-FIB, but the proposed equations by Lee and Lee underestimate them. For instance, the experimental values were 5.8%, 0.7%, 2.3%, and 24.3% higher than the predicted values proposed by ACI 318, ACI 363, Eurocode CEB-FIB, and Lee and Lee, respectively, for the mix of 30% slag content. Using 50% and 100% slag content, the above variations between actual and predicted models become higher. However, ACI 363 and Eurocode CEB-FIB are the closest to the real values for G50 and G100 mixtures, which range between 9.0 and 13% lower than what was predicted. Faridmehr et al. [53] investigated SCGC made with the combined use of FA and slag. They also found that the proposed relationship by ACI 318 between compressive and splitting tensile strength does not correctly estimate the splitting tensile values of SCGC. Consequently, the above findings might help researchers predict the splitting tensile values of FA-based SCGC mixes made with or without slag with respect to the experimental compressive strength values, but to reach more decisive conclusions, more investigations and analysis are needed.

5.4. Durability Properties

Table 10 shows the capillary water absorption, free drying shrinkage, and mass loss results. The results indicate that increasing slag content reduces mass loss and capillary absorption but increases drying shrinkage.

Table 10. Results of durability properties of SCGC.

Mix Code	Slag Content (%)	Capillary Absorption(mm/min$^{0.5}$)	Free Drying Shrinkage after 365 Days	
			Max. Shrinkage (Microstrain)	Max. Mass Loss (g)
G0	0	0.0513	104	201
G30	30	0.0377	——	——
G50	50	0.0291	175	115
G100	100	0.0242	——	——

5.4.1. Capillary Water Absorption (Sorptivity)

Capillary water absorption tests determine SCGC's ability to absorb water from a single surface [83]. This index properly assesses the quality of the concrete surface layer, which controls reinforcement corrosion [83]. The lower the sorptivity, the more durable the concrete is and the better it performs in a harsh environment. The 28-day capillary water absorption coefficient for various slag contents is plotted in Figure 26. The relative decrease in capillary absorption coefficient is presented in Figure 27.

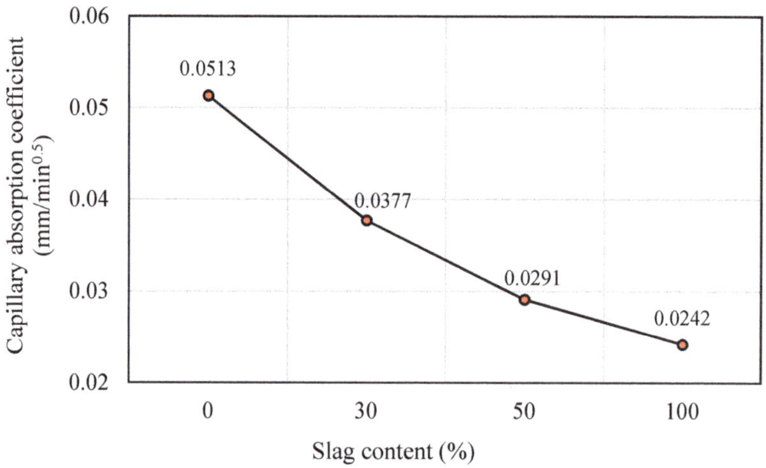

Figure 26. Influence of slag content on the capillary water absorption after 28 days.

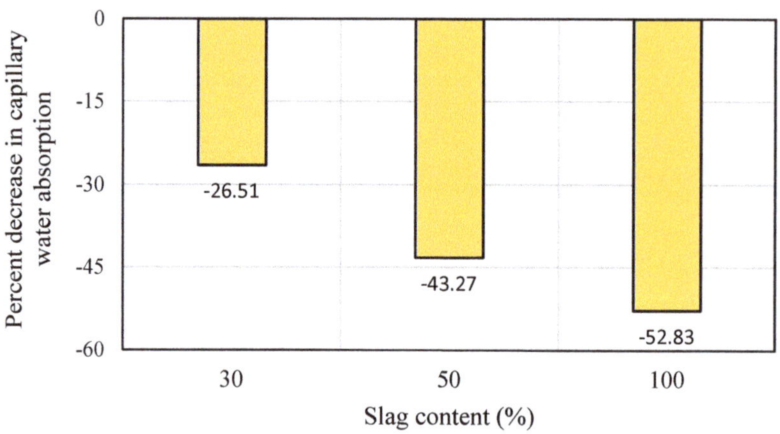

Figure 27. Percent decrease in capillary water absorption.

From the test results, a systematic decrease can be detected due to increased slag content. With the increasing of the slag content from 0 to 100%, a reduction in sorptivity of 52.83% is recorded. The trends observed in the results are in line with results by other researchers such as Shaikh [84], who found that geopolymer concrete displays lower sorptivity than normal concrete. Patel and Shah [4] studied blended FA and slag-based SCGC incorporating rice husk ash cured in ambient conditions. They reported that higher sorptivity is achieved by FA-based geopolymer concrete. The values of capillary suction for all SCGC specimens were in a range of 0.069–0.136 mm/min$^{0.5}$. According to recent research on the effects of mineral admixtures on the characteristics of geopolymer concrete by Jindal et al. [85], raising the slag replacement ratio reduces the percentage of water absorption. Furthermore, permeability is determined by pore size distribution and structure. The pores' continuity influences water absorption. Geopolymer gel formation fills pores in the microstructure of the geopolymer concrete, changing the pore configuration and densifying the microstructure. As the age develops from 1 day to 28 days, the pores are gradually filled up by the production of geopolymer and C-S-H gels [4]. In geopolymer concrete, the slag binder provides the C-S-H gel formation, which holds CaO oxide in the chemical composition, resulting in enhanced strength.

5.4.2. Drying Shrinkage and Mass Loss

Curing is a key factor in evaluating the free drying shrinkage of concrete. It was previously reported that the shrinkage of geopolymers is highly dependent on the curing regime and liquid/binder ratio [86]. Additionally, the shrinkage of geopolymer mortar at a low temperature (40 °C) is very high compared to those cured at a higher temperature (60 °C) [86]. Furthermore, the compressive strength of geopolymer concrete improves with increasing curing temperature. The fib MC 2010 and B4 models both predict shrinkage based on the compressive strength [87]. For the above reasons, the effect of the curing condition and compressive strength on the shrinkage of geopolymer concrete is of great importance. Setting time and strength development of low-calcium FA-based geopolymer concrete is very low compared to blended FA and slag-based geopolymer concrete. Therefore, apart from the other parameters, different curing conditions were investigated in this study: (1) curing at an elevated temperature (85 °C), and (2) the prisms are cast for a free drying shrinkage test cured in an oven at a low temperature (40 °C) for three days. Figure 28 shows the compressive strength of 100 mm cubical specimens of the SCGC mixes cured at 40 °C for three days. The compressive strengths were enhanced by increasing slag content from 0% to 50%. A similar trend was reported in the previous sections of this paper.

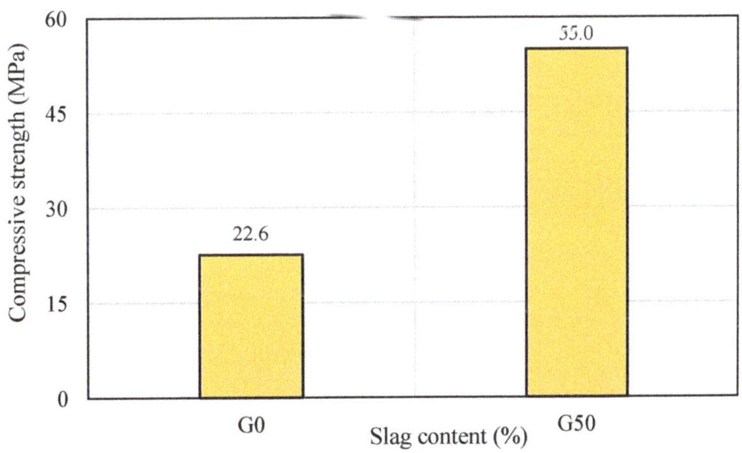

Figure 28. Compressive strength of SCGC specimens cured at 40 °C for three days.

Free Drying Shrinkage

Concrete drying shrinkage is characterized as a volumetric change as a result of concrete drying. Initially, free water is lost, resulting in minimal to no shrinkage. As the concrete continues to dry, the absorbed water is removed. Hydrostatic tension holds this adsorbed water in tiny capillaries. Tensile stresses are created when this water is lost, causing the concrete to shrink. The shrinkage caused by such a water loss is substantially more than the shrinkage caused by free water loss [88]. Drying shrinkage is a long-term process that depends on the water-cement ratio, hydration process, curing condition, moisture content, drying interval, aggregate properties, additives, and chemical properties of the cement [89–91]. Typical curves of free drying shrinkage vs. drying time for the SCGC mixes with and without the inclusion of slag are given in Figure 29. Regardless of the slag content, a sharp increase in the free drying shrinkage test measurements is noticed up to 90 days of drying time; then, the curve becomes flat. The inclusion of slag in SCGC results in higher shrinkage values. The values of drying shrinkage at 365 days were 104 and 175 microstrain for the mix codes of G0 and G50, respectively. These values are lower than for OPC concrete. However, in a different study [70], it was reported that the drying shrinkage of FA-based geopolymer concrete cured in ambient conditions reduces with the addition of slag, and the six-month shrinkage of geopolymer concrete mixes ranged between 482 and 722 microstrain. Moreover, geopolymer concrete attained lower values of shrinkage compared to ordinary concrete for comparable compressive strength values. Additionally, it has been reported that the drying shrinkage strains of geopolymer concrete cured via oven are generally less than those concrete cured at ambient conditions [92]. Furthermore, it was reported that low-calcium binders are less reactive than high-calcium binders. In alkali-activated mixtures, class F fly ash reduced the shrinkage of concrete compared to class C fly ash and slag binders [93,94].

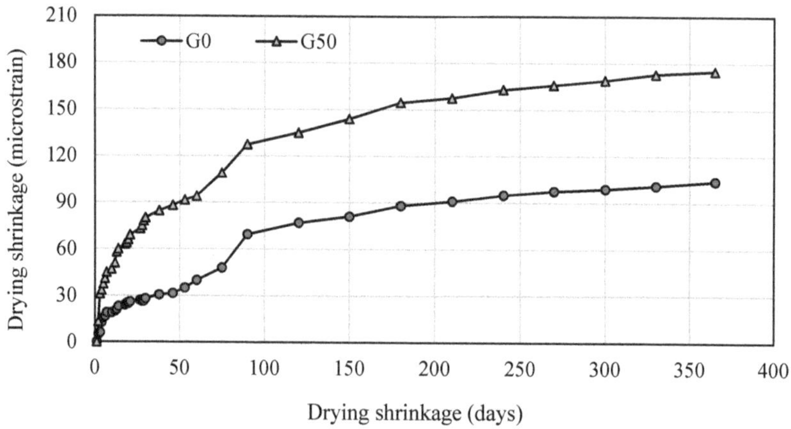

Figure 29. Drying shrinkage strain development of SCGC.

Mass Loss

The results of mass loss versus drying time are plotted in Figure 30. SCGC incorporating slag displayed a lower mass loss than the mix made with 100% FA. At 365 days, the mass losses of the mixes made of 0% and 50% slag were 201 g and 115 g, respectively. According to previous studies, there might not be a direct relationship between drying shrinkage and SCC's mass loss measurements. The mass loss indicator alone cannot provide substantial information on the variation in the drying shrinkage of concrete [95], as a number of other variables influence drying shrinkage in addition to mass loss [90,91]. The obtained behavior is contrary to expectations and there are no available studies regarding this property, hence more work is required in this direction.

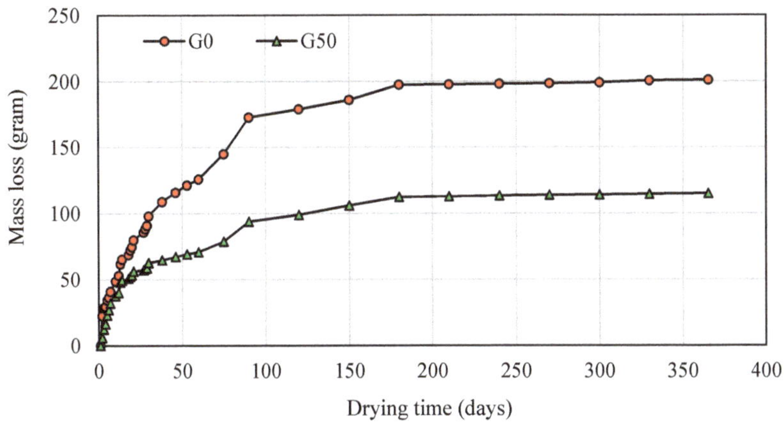

Figure 30. Mass loss variations of SCGC.

6. Conclusions

Based on the findings, the following conclusions can be drawn on the impact of slag content on the fresh and hardened qualities of SCGC containing FA:

1. Slag inclusion has a negative impact on the flowability of SCGC in terms of slump flow and J-ring flow diameter. The slump and J-ring flow values decrease by increasing the slag in the mixes, but this is still within EFNARC's acceptable limits. This may be due to the morphology, lower specific surface area, and lesser reactivity of the FA binder compared to slag.
2. T_{500}, TJ_{500}, and V-funnel flow times increase with the increase in slag replacement levels. The relationship of V-funnel via T500 and V-funnel via TJ500 also indicates that slag-rich SCGCs mixtures belong to the viscosity class (VS2/VF2) as defined by EFNARC. It can also be concluded that this kind of concrete might help improve resistance to segregation or reduce pressure on the formwork.
3. Slag leads to a reduction in the passing-ability of fresh concrete.
4. Slag has a significant impact on the segregation resistance of SCGC mixes. The allowable range of segregation index (%) is attained when the usage of slag exceeds 30% of the total binder content.
5. The low-calcium content of FA binders delays hardening, and using slag in geopolymer concrete mixes can help accelerate strength gain.
6. The utilization of slag results in a considerable increase in compressive and splitting tensile strength. The highest relative benefit is observed at 50% replacement.
7. The splitting tensile strength was found to increase faster than the compressive strength, indicating a potential enhancement of the ITZ. A new predictive equation is proposed for these mixes.
8. Based on the sorptivity test, a higher slag content enhances the durability of SCGC. By utilizing slag, SCGA specimens can achieve a higher rate of resistance to absorption. As the percentage of slag climbed from 0% to 100%, the sorptivity value declined to 48.5%.
9. FA-based SCGC made with or without slag exhibits low free drying shrinkage. SCGC made with 100% FA displays lower shrinkage strains than the mix made with 50% FA and 50% slag due to the lower reactivity of the FA binder compared to slag. After one year of exposing samples to drying conditions, the free drying shrinkage values of SCGC specimens cured at 40 °C were 104 and 175 microstrain for the G0 and G50 mixes, respectively.
10. Fully FA-based SCGC specimens show greater mass loss than the mix with 50% slag content. Therefore, it may be inferred that the mass loss parameter does not provide sufficient information about the variety in free drying shrinkage results.

7. Recommendations

It is recommended to investigate the influence of slag content on the durability performance of FA-based SCGC; specifically, gas permeability and rapid chloride permeability. In addition, further studies are needed on the long-term free drying shrinkage and mass loss. Furthermore, restrained shrinkage could be a topic of interest to the researchers. Investigating the microstructure of the FA-based SCGC with various slag content is also important to justify the behavior of the mechanical strength and durability of such mixes.

Author Contributions: Conceptualization, A.F.H.S., K.H.Y., R.W.A. and K.P.; methodology, A.F.H.S., K.H.Y., R.W.A. and K.P; software, A.F.H.S., K.H.Y. and R.W.A.; validation, A.F.H.S., K.H.Y., R.W.A. and K.P.; formal analysis, A.F.H.S., K.H.Y. and R.W.A.; investigation, A.F.H.S., K.H.Y. and R.W.A.; resources, A.F.H.S., K.H.Y. and R.W.A.; data curation, A.F.H.S. and K.H.Y.; writing—original draft preparation, A.F.H.S., K.H.Y. and R.W.A.; writing—review and editing, A.F.H.S., K.H.Y., R.W.A. and K.P.; visualization, A.F.H.S., K.H.Y., R.W.A. and K.P.; supervision, K.H.Y., R.W.A. and K.P.; project administration, A.F.H.S., K.H.Y. and R.W.A.; funding acquisition, A.F.H.S., K.H.Y. and R.W.A. All authors have read and agreed to the published version of the manuscript.

Funding: This research received no external funding.

Institutional Review Board Statement: Not applicable.

Informed Consent Statement: Not applicable.

Acknowledgments: The authors would like to thank Soran University and Erbil Polytechnical University especially Bashdar A. Abdulla for supporting this research and the University of Applied Sciences Erfurt for their hospitality as well as administrative and practical support, especially Cornelia Witter from FHE International Office and Wolfgang Hezel from FHE Concrete laboratory.

Conflicts of Interest: The authors declare that they have no conflicts of interest.

Abbreviations

FA, Fly Ash; Slag, Ground Granulated Blast Furnace Slag; SCC, Self-Compacted Concrete; SCGC, Self-Compacted Geopolymer Concrete; ACI, American Concrete Institute; CEB-FIB, Constructions Électriques de Beaucourt-International Federation for Structural Concrete; OPC, Ordinary Portland Cement; Si, Silica; AL, Alumina; CaO, Calcium Oxide; CO_2, Carbon Dioxide; C-S-H, Calcium-Silicate-Hydrates; N-A-S-H, Sodium-Alumino-Silicate-Hydrate; NS, Nano-Silica; RHA, Rice Husk Ash; MK, Metakaolin; GSA, Groundnut Shell Ash; SF, Silica Fume; Na_2SiO_3, Sodium Silicate; NaOH, Sodium Hydroxide; Na_2O, Sodium Oxide; SiO_2, Silicon Dioxide; BS EN, British and European Standard Specifications; SP, Superplasticizer; EFNARC, European Federation of National Associations Representing for Concrete; PJ, J-ring Passing-ability; PA, L-box Passing-ability; SI, Sieve Segregation Index; f_{cu}, Cubical Compressive Strength; f_c', Cylindrical Compressive Strength; RH, Relative Humidity.

References

1. Antiohos, S.K.; Tapali, J.G.; Zervaki, M.; Sousa-Coutinho, J.; Tsimas, S.; Papadakis, V.G. Low Embodied Energy Cement Containing Untreated RHA: A Strength Development and Durability Study. *Constr. Build. Mater.* **2013**, *49*, 455–463. [CrossRef]
2. Prasittisopin, L.; Trejo, D. Hydration and Phase Formation of Blended Cementitious Systems Incorporating Chemically Transformed Rice Husk Ash. *Cem. Concr. Compos.* **2015**, *59*, 100–106. [CrossRef]
3. Aprianti, E.; Shafigh, P.; Bahri, S.; Farahani, J.N. Supplementary Cementitious Materials Origin from Agricultural Wastes—A Review. *Constr. Build. Mater.* **2015**, *74*, 176–187. [CrossRef]
4. Patel, Y.J.; Shah, N. Enhancement of the Properties of Ground Granulated Blast Furnace Slag Based Self Compacting Geopolymer Concrete by Incorporating Rice Husk Ash. *Constr. Build. Mater.* **2018**, *171*, 654–662. [CrossRef]
5. Rangan, B.V. *Fly Ash-Based Geopolymer Concrete*; Curtin University of Technology: Perth, Australia, 2010.
6. Davidovits, J. *Geopolymer CHemistry and Applications*, 3rd ed.; Geopolymer Institute: Saint-Quentin, France, 2011.
7. Ahmed, H.U.; Mahmood, L.J.; Muhammad, M.A.; Faraj, R.H.; Qaidi, S.M.; Sor, N.H.; Mohammed, A.S.; Mohammed, A.A. Geopolymer Concrete as a Cleaner Construction Material: An Overview on Materials and Structural Performances. *Clean. Mater.* **2022**, *5*, 100111. [CrossRef]
8. Ahmed, H.U.; Mohammed, A.S.; Faraj, R.H.; Qaidi, S.M.A.; Mohammed, A.A. Compressive Strength of Geopolymer Concrete Modified with Nano-Silica: Experimental and Modeling Investigations. *Case Stud. Constr. Mater.* **2022**, *16*, e01036. [CrossRef]

9. Ahmed, H.U.; Mohammed, A.S.; Qaidi, S.M.A.; Faraj, R.H.; Hamah Sor, N.; Mohammed, A.A. Compressive Strength of Geopolymer Concrete Composites: A Systematic Comprehensive Review, Analysis and Modeling. *Eur. J. Environ. Civ. Eng.* **2022**, *26*, 1–46. [CrossRef]
10. Ahmed, H.U.; Mohammed, A.A.S.; Rafiq, S.; Mohammed, A.A.S.; Mosavi, A.; Sor, N.H.; Qaidi, S.M.A.A. Compressive Strength of Sustainable Geopolymer Concrete Composites: A State-of-the-Art Review. *Sustainability* **2021**, *13*, 13502. [CrossRef]
11. Wongpa, J.; Kiattikomol, K.; Jaturapitakkul, C.; Chindaprasirt, P. Compressive Strength, Modulus of Elasticity, and Water Permeability of Inorganic Polymer Concrete. *Mater. Des.* **2010**, *31*, 4748–4754. [CrossRef]
12. Li, C.; Sun, H.; Li, L. A Review: The Comparison between Alkali-Activated Slag (Si + Ca) and Metakaolin (Si + Al) Cements. *Cem. Concr. Res.* **2010**, *40*, 1341–1349. [CrossRef]
13. Chindaprasirt, P.; De Silva, P.; Sagoe-Crentsil, K.; Hanjitsuwan, S. Effect of SiO_2 and Al_2O_3 on the Setting and Hardening of High Calcium Fly Ash-Based Geopolymer Systems. *J. Mater. Sci.* **2012**, *47*, 4876–4883. [CrossRef]
14. Mehta, A.; Williams, V.; Parajuli, B. Child with Dysuria and/or Hematuria. *Indian J. Pediatr.* **2017**, *84*, 792–798. [CrossRef] [PubMed]
15. Saha, S.; Rajasekaran, C. Enhancement of the Properties of Fly Ash Based Geopolymer Paste by Incorporating Ground Granulated Blast Furnace Slag. *Constr. Build. Mater.* **2017**, *146*, 615–620. [CrossRef]
16. Nath, P.; Sarker, P.K.; Rangan, V.B. Early Age Properties of Low-Calcium Fly Ash Geopolymer Concrete Suitable for Ambient Curing. *Procedia Eng.* **2015**, *125*, 601–607. [CrossRef]
17. Nath, P.; Sarker, P.K. Effect of GGBFS on Setting, Workability and Early Strength Properties of Fly Ash Geopolymer Concrete Cured in Ambient Condition. *Constr. Build. Mater.* **2014**, *66*, 163–171. [CrossRef]
18. Hadi, M.N.S.; Farhan, N.A.; Sheikh, M.N. Design of Geopolymer Concrete with GGBFS at Ambient Curing Condition Using Taguchi Method. *Constr. Build. Mater.* **2017**, *140*, 424–431. [CrossRef]
19. Sherwani, A.F.H.; Younis, K.H.; Arndt, R.W. Fresh, Mechanical, and Durability Behavior of Fly Ash-Based Self Compacted Geopolymer Concrete: Effect of Slag Content and Various Curing Conditions. *Polymers* **2022**, *14*, 3209. [CrossRef]
20. Okamura, H.; Ouchi, M. Self Compacting Concrete—Research Paper. *J. Adv. Concr. Technol.* **2003**, *1*, 5–15. [CrossRef]
21. Faraj, R.H.; Ali, H.F.H.; Sherwani, A.F.H.; Hassan, B.R.; Karim, H. Use of Recycled Plastic in Self-Compacting Concrete: A Comprehensive Review on Fresh and Mechanical Properties. *J. Build. Eng.* **2020**, *30*, 101283. [CrossRef]
22. Faraj, R.H.; Sherwani, A.F.H.; Daraei, A. Mechanical, Fracture and Durability Properties of Self-Compacting High Strength Concrete Containing Recycled Polypropylene Plastic Particles. *J. Build. Eng.* **2019**, *25*, 100808. [CrossRef]
23. Faraj, R.H.; Sherwani, A.F.H.; Jafer, L.H.; Ibrahim, D.F. Rheological Behavior and Fresh Properties of Self-Compacting High Strength Concrete Containing Recycled PP Particles with Fly Ash and Silica Fume Blended. *J. Build. Eng.* **2020**, *34*, 101667. [CrossRef]
24. Memon, F.A.; Nuruddin, F.; Shafiq, N.; Fareed Ahmed, M.; Nuruddin, M.F.; Shafiq, N. Compressive Strength and Workability Characteristics of Low-Calcium Fly Ash-Based Self-Compacting Geopolymer Concrete. *World Acad. Sci. Eng. Technol.* **2011**, *74*, 8–14.
25. Younis, K.H.; Salihi, K.; Mohammedameen, A.; Sherwani, A.F.H.; Alzeebaree, R. Factors Affecting the Characteristics of Self-Compacting Geopolymer Concrete. *IOP Conf. Ser. Earth Environ. Sci.* **2021**, *856*, 012028. [CrossRef]
26. Astuti, P.; Afriansya, R.; Anisa, E.A.; Randisyah, J. Mechanical Properties of Self-Compacting Geopolymer Concrete Utilizing Fly Ash. *AIP Conf. Proc.* **2022**, *2453*, 020028. [CrossRef]
27. Nazari, A.; Torgal, F.; Cevik, A.; Sanjayan, J. Compressive Strength of Tungsten Mine Waste- and Metakaolin-Based Geopolymers. *Ceram. Int.* **2014**, *40*, 6053–6062. [CrossRef]
28. Gilbert, R.I. Creep and Shrinkage Models for High Strength Concrete—Proposals for Inclusion in AS3600. *Aust. J. Struct. Eng.* **2002**, *4*, 95–106. [CrossRef]
29. Srishaila, J.M.; Ahamed, P.U.; Vishwanath, K.N.; Prakash, P. Experimental Study on Workability and Strength Characteristics of Fly Ash and GGBS Based Self-Compacting Geo Polymer Concrete. *Int. J. Eng. Res. Dev.* **2014**, *10*, 68–77.
30. Wallah, S.E.; Rangan, B.V. *Low-Calcium Fly Ash-Based Geopolymer Concrete: Long-Term Properties*; Curtin University of Technology: Perth, Australia, 2006.
31. Olivia, M.; Nikraz, H. Properties of Fly Ash Geopolymer Concrete Designed by Taguchi Method. *Mater. Des.* **2012**, *36*, 191–198. [CrossRef]
32. Nuruddin, M.F.; Memon, F.A.; Shafiq, N.; Demie, S. Drying Shrinkage of Fly Ash-Based Self-Compacting Geopolymer Concrete. *Appl. Mech. Mater.* **2014**, *567*, 362–368. [CrossRef]
33. Wang, S.-D.; Pu, X.-C.; Scrivener, K.L.; Pratt, P.L. Alkali-Activated Slag Cement and Concrete: A Review of Properties and Problems. *Adv. Cem. Res.* **1995**, *7*, 93–102. [CrossRef]
34. Al-Rawi, S.; Tayşi, N. Performance of Self-Compacting Geopolymer Concrete with and without GGBFS and Steel Fiber. *Adv. Concr. Constr.* **2018**, *6*, 323–344. [CrossRef]
35. Gülşan, M.E.; Alzeebaree, R.; Rasheed, A.A.; Niş, A.; Kurtoğlu, A.E. Development of Fly Ash/Slag Based Self-Compacting Geopolymer Concrete Using Nano-Silica and Steel Fiber. *Constr. Build. Mater.* **2019**, *211*, 271–283. [CrossRef]
36. Nagaraj, V.K.; Babu, D.L.V. Assessing the Performance of Molarity and Alkaline Activator Ratio on Engineering Properties of Self-Compacting Alkaline Activated Concrete at Ambient Temperature. *J. Build. Eng.* **2018**, *20*, 137–155. [CrossRef]

37. Saini, G.; Vattipalli, U. Assessing Properties of Alkali Activated GGBS Based Self-Compacting Geopolymer Concrete Using Nano-Silica. *Case Stud. Constr. Mater.* **2020**, *12*, e00352. [CrossRef]
38. Ganeshan, M.; Venkataraman, S. Durability and Microstructural Studies on Fly Ash Blended Self-Compacting Geopolymer Concrete. *Eur. J. Environ. Civ. Eng.* **2019**, *25*, 2074–2088. [CrossRef]
39. Patel, Y.J.; Shah, N. Development of Self-Compacting Geopolymer Concrete as a Sustainable Construction Material. *Sustain. Environ. Res.* **2018**, *28*, 412–421. [CrossRef]
40. Manjunath, R.; Ranganath, R.V. Performance Evaluation of Fly-Ash Based Self-Compacting Geopolymer Concrete Mixes. *IOP Conf. Ser. Mater. Sci. Eng.* **2019**, *561*, 012006. [CrossRef]
41. Nuruddin, M.F.; Demie, S.; Shafiq, N. Effect of Mix Composition on Workability and Compressive Strength of Self-Compacting Geopolymer Concrete. *Can. J. Civ. Eng.* **2011**, *38*, 1196–1203. [CrossRef]
42. Demie, S.; Nuruddin, M.F.; Shafiq, N. Effects of Micro-Structure Characteristics of Interfacial Transition Zone on the Compressive Strength of Self-Compacting Geopolymer Concrete. *Constr. Build. Mater.* **2013**, *41*, 91–98. [CrossRef]
43. Reddy, K.M.; Kumar, G.N. Experimental Study on Self Compacting Geopolymer Concrete. *Int. J. Eng. Res.* **2017**, *4*, 953–957.
44. Bheel, N.; Awoyera, P.; Tafsirojjaman, T.; Hamah Sor, N.; sohu, S. Synergic Effect of Metakaolin and Groundnut Shell Ash on the Behavior of Fly Ash-Based Self-Compacting Geopolymer Concrete. *Constr. Build. Mater.* **2021**, *311*, 125327. [CrossRef]
45. Kamseu, E.; Ponzoni, C.; Tippayasam, C.; Taurino, R.; Chaysuwan, D.; Sglavo, V.M.; Thavorniti, P.; Leonelli, C. Self-Compacting Geopolymer Concretes: Effects of Addition of Aluminosilicate-Rich Fines. *J. Build. Eng.* **2016**, *5*, 211–221. [CrossRef]
46. Muttashar, H.L.; Ariffin, M.A.M.; Hussein, M.N.; Hussin, M.W.; Ishaq, S. Bin Self-Compacting Geopolymer Concrete with Spend Garnet as Sand Replacement. *J. Build. Eng.* **2018**, *15*, 85–94. [CrossRef]
47. Nurudin, M.F.; Memon, F.A.; Nuruddin, M.F.; Memon, F.A.; Nurudin, M.F.; Memon, F.A. Properties of Self-Compacting Geopolymer Concrete. *Mater. Sci. Forum* **2015**, *803*, 99–109. [CrossRef]
48. Sashidhar, C.; Guru Jawahar, J.; Neelima, C.; Pavan Kumar, D. Preliminary Studies on Self Compacting Geopolymer Concrete Using Manufactured Sand. *Asian J. Civ. Eng.* **2016**, *17*, 277–288.
49. Ushaa, T.G.; Anuradha, R.; Venkatasubramani, G.S. *Performance of Self-Compacting Geopolymer Concrete Containing Different Mineral Admixtures*; NISCAIR-CSIR: New Delhi, India, 2015; Volume 22.
50. Eren, N.A.; Alzeebaree, R.; Çevik, A.; Niş, A.; Mohammedameen, A.; Gülşan, M.E. Fresh and Hardened State Performance of Self-Compacting Slag Based Alkali Activated Concrete Using Nanosilica and Steel Fiber. *J. Compos. Mater.* **2021**, *55*, 4125–4139. [CrossRef]
51. Memon, F.A.; Nuruddin, M.F.; Demie, S.; Shafiq, N. Effect of Superplasticizer and Extra Water on Workability and Compressive Strength of Self-Compacting Geopolymer Concrete. *Res. J. Appl. Sci. Eng. Technol.* **2012**, *4*, 407–414.
52. Arun, B.R.; Nagaraja, P.S.; Srishaila, J.M. *An Effect of NaOH Molarity on Fly Ash—Metakaolin-Based Self-Compacting Geopolymer Concrete*; Springer: Singapore, 2019; Volume 25, ISBN 9789811333170.
53. Faridmehr, I.; Nehdi, M.L.; Huseien, G.F.; Baghban, M.H.; Sam, A.R.M.; Algaifi, H.A. Experimental and Informational Modeling Study of Sustainable Self-Compacting Geopolymer Concrete. *Sustainability* **2021**, *13*, 7444. [CrossRef]
54. *BS EN 933-1+A1 2005*; Tests for Geometrical Properties of Aggregates—Part 1: Determination of Particle Size Distribution—Sieving Method. BSI Standards Publication: London, UK, 2008.
55. *BS EN 1097-6*; Tests for Mechanical and Physical Properties of Aggregates—Part 6: Determination of Particle Density and Water Absorption. BSI Standards Publication: London, UK, 2013.
56. Sata, V.; Wongsa, A.; Chindaprasirt, P. Properties of Pervious Geopolymer Concrete Using Recycled Aggregates. *Constr. Build. Mater.* **2013**, *42*, 33–39. [CrossRef]
57. EFNARC. The European Guidelines for Self-Compacting Concrete. *Eur. Guidel. Self Compact. Concr.* **2005**, *22*, 63.
58. *BS EN 12350-11*; Testing Fresh Concrete—Part 11: Self-Compacting Concrete—Sieve Segregation Test. BSI Standards Publication: London, UK, 2010.
59. *BS EN 12390-3*; Testing Hardened Concrete—Part 3: Compressive Strength of Test Specimens. BSI Standards Publication: London, UK, 2009.
60. *BS EN 12390-6:2000*; Testing Hardened Concrete—Part 6: Tensile Splitting Strength of Test Specimens. BSI Standards Publication: London, UK, 1993.
61. *ACI 318M-11*; Building Code Requirements for Structural Concrete and Commentary. American Concrete Institute: Farmington Hills, MI, USA, 2011.
62. *ACI 363 R-10*; Report on High-Strength Concrete. American Concrete Institute: Farmington Hills, MI, USA,, 2010.
63. Comité Euro-International du Béton. *CEB-FIP Model Code 1990: Design Code*; Thomas Telford Publishing: London, UK, 1993; ISBN 978-0-7277-3944-5.
64. Lee, N.K.; Lee, H.K. Setting and Mechanical Properties of Alkali-Activated Fly Ash/Slag Concrete Manufactured at Room Temperature. *Constr. Build. Mater.* **2013**, *47*, 1201–1209. [CrossRef]
65. İpek, S.; Ayodele, O.A.; Mermerdaş, K. Influence of Artificial Aggregate on Mechanical Properties, Fracture Parameters and Bond Strength of Concretes. *Constr. Build. Mater.* **2020**, *238*, 117756. [CrossRef]
66. *BS EN 13057*; Products and Systems for the Protection and Repair of Concrete Structures—Test Methods—Determination of Resistance of Capillary Absorption. BSI Standards Publication: London, UK, 2002.

67. Khatib, J.M.; Mangat, P.S. Absorption Characteristics of Concrete as a Function of Location Relative to Casting Position. *Cem. Concr. Res.* **1995**, *25*, 999–1010. [CrossRef]
68. Khatib, J.M.; Clay, R.M. Absorption Characteristics of Metakaolin Concrete. *Cem. Concr. Res.* **2004**, *34*, 19–29. [CrossRef]
69. *ASTM:C157/C157M-08*; Test Method for Length Change of Hardened Hydraulic-Cement Mortar and Concrete. ASTM International: West Conshohocken, PA, USA, 2008; Volume 8, pp. 1–7. [CrossRef]
70. Deb, P.S.; Nath, P.; Sarker, P.K. Drying Shrinkage of Slag Blended Fly Ash Geopolymer Concrete Cured at Room Temperature. *Procedia Eng.* **2015**, *125*, 594–600. [CrossRef]
71. *BS EN 12350-8*; Testing Self Compacting Concrete: Slump Flow Test. British Standards Institution: London, UK, 2010; pp. 5–8.
72. Mohammed, N.; Sarsam, K.; Hussien, M. The Influence of Recycled Concrete Aggregate on the Properties of Concrete. *MATEC Web Conf.* **2018**, *162*, 02020. [CrossRef]
73. *BS EN 12350-9*; Testing Fresh Concrete Self-Compacting Concrete. V-Funnel Test. British Standards Institution: London, UK, 2010.
74. Akgaoglu, T.; Qubukguoglu, B.; Awad, A. A Critical Review of Slag and Fly-Ash Based Geopolymer Concrete. *Comput. Concr.* **2019**, *24*, 453–458. [CrossRef]
75. Chi, M.; Huang, R. Binding Mechanism and Properties of Alkali-Activated Fly Ash/Slag Mortars. *Constr. Build. Mater.* **2013**, *40*, 291–298. [CrossRef]
76. Puertas, F.; Martínez-Ramírez, S.; Alonso, S.; Vázquez, T.; Martínez-Ramírez, S.; Alonso, S.; Vázquez, T.; Martínez-Ramírez, S.; Alonso, S.; Vázquez, T. Alkali-Activated Fly Ash/Slag Cements: Strength Behaviour and Hydration Products. *Cem. Concr. Res.* **2000**, *30*, 1625–1632. [CrossRef]
77. Qiu, J.; Zhao, Y.; Xing, J.; Sun, X. Fly Ash/Blast Furnace Slag-Based Geopolymer as a Potential Binder for Mine Backfilling: Effect of Binder Type and Activator Concentration. *Adv. Mater. Sci. Eng.* **2019**, *2019*, 2028109. [CrossRef]
78. Wang, W.C.; Wang, H.Y.; Lo, M.H. The Fresh and Engineering Properties of Alkali Activated Slag as a Function of Fly Ash Replacement and Alkali Concentration. *Constr. Build. Mater.* **2015**, *84*, 224–229. [CrossRef]
79. Sasui, S.; Kim, G.; Nam, J.; Koyama, T.; Chansomsak, S. Strength and Microstructure of Class-C Fly Ash and GGBS Blend Geopolymer Activated in NaOH & NaOH + Na_2SiO_3. *Materials* **2020**, *13*, 59. [CrossRef]
80. Ismail, I.; Bernal, S.A.; Provis, J.L.; San Nicolas, R.; Hamdan, S.; Van Deventer, J.S.J. Modification of Phase Evolution in Alkali-Activated Blast Furnace Slag by the Incorporation of Fly Ash. *Cem. Concr. Compos.* **2014**, *45*, 125–135. [CrossRef]
81. Samantasinghar, S.; Singh, S.P. Effect of Synthesis Parameters on Compressive Strength of Fly Ash-Slag Blended Geopolymer. *Constr. Build. Mater.* **2018**, *170*, 225–234. [CrossRef]
82. Shen, Q.; Chen, W.; Liu, C.; Zou, W.; Pan, L. The Tensile Strength and Damage Characteristic of Two Types of Concrete and Their Interface. *Materials* **2020**, *13*, 16. [CrossRef]
83. Atewi, Y.R.; Hasan, M.F.; Güneyisi, E. Fracture and Permeability Properties of Glass Fiber Reinforced Self-Compacting Concrete with and without Nanosilica. *Constr. Build. Mater.* **2019**, *226*, 993–1005. [CrossRef]
84. Shaikh, F.U.A. Effects of Alkali Solutions on Corrosion Durability of Geopolymer Concrete. *Adv. Concr. Constr.* **2014**, *2*, 109–123. [CrossRef]
85. Jindal, B.B.; Jangra, P.; Garg, A. Effects of Ultra Fine Slag as Mineral Admixture on the Compressive Strength, Water Absorption and Permeability of Rice Husk Ash Based Geopolymer Concrete. *Mater. Today Proc.* **2020**, *32*, 871–877. [CrossRef]
86. Ridtirud, C.; Chindaprasirt, P.; Pimraksa, K. Factors Affecting the Shrinkage of Fly Ash Geopolymers. *Int. J. Miner. Metall. Mater.* **2011**, *18*, 100–104. [CrossRef]
87. Caron, R.; Patel, R.A.; Dehn, F. Extension of the Fib MC 2010 for Basic and Drying Shrinkage of Alkali-Activated Slag Concretes. *Struct. Concr.* **2022**, 1–14. [CrossRef]
88. Neville, A.M. *Concrete: Neville's Insights and Issues;* Thomas Telford Ltd.: London, England, 2006; ISBN 0727734687.
89. Shh, S.P.; Krguller, M.E.; Sarigaphuti, M. Effects of Shrinkage-Reducing Admixtures on Restrained Shrinkage Cracking of Concrete. *ACI Mater. J.* **1992**, *89*, 289–295. [CrossRef]
90. Wiegrink, K.; Surendra, P.; Shah, S.M. Shrinkage Cracking of High-Strength Concrete. *ACI Mater. J.* **1996**, *93*, 409–415. [CrossRef]
91. Gesoğlu, M.; Ozturan, T.; Güneyisi, E.; Gesoglu, M.; Özturan, T.; Güneyisi, E.; Gesoğlu, M.; Ozturan, T.; Güneyisi, E. Shrinkage Cracking of Lightweight Concrete Made with Cold-Bonded Fly Ash Aggregates. *Cem. Concr. Res.* **2004**, *34*, 1121–1130. [CrossRef]
92. Brito, M.E.; Case, E.; Kriven, W.M.; Salem, J.; Zhu, D. (Eds.) Developments in Porous, Biological and Geopolymer Ceramics. In Proceedings of the 31st International Conference on Advanced Ceramics and Composites, Daytona Beach, FL, USA, 21–26 January 2007; Volume 28.
93. Hossain, D.S.; Strength, K.M.A. Shrinkage and Early Age Characteristics of One-Part Alkali-Activated Binders with High-Calcium Industrial Wastes, Solid Reagents and Fibers. *J. Compos. Sci.* **2021**, *5*, 315.
94. Adesina, A.; Das, S. Drying Shrinkage and Permeability Properties of Fibre Reinforced Alkali-Activated Composites. *Constr. Build. Mater.* **2020**, *251*, 119076. [CrossRef]
95. Öz, H.Ö.; Gesoglu, M.; Güneyisi, E.; Sor, N.H. Self-Consolidating Concretes Made with Cold-Bonded Fly Ash Lightweight Aggregates. *ACI Mater. J.* **2017**, *114*, 385–395. [CrossRef]

sustainability

MDPI

Article

Enhancing the Performance of Recycled Aggregate Concrete Using Micro-Carbon Fiber and Secondary Binding Material

Syed Safdar Raza [1,*], Muhammad Fahad [2], Babar Ali [3], Muhammad Talha Amir [4,5], Yasser Alashker [6,7] and Ahmed Babekar Elhag [6]

1 Faculty of Civil Engineering, Bahauddin Zakariya University, Multan 66000, Pakistan
2 Department of Civil Engineering, University of Engineering & Technology (UET), Peshawar 25000, Pakistan
3 Department of Civil Engineering, COMSATS University Islamabad, Sahiwal Campus, Sahiwal 57000, Pakistan
4 Department of Civil Engineering, University of Engineering and Technology, Taxila 47050, Pakistan
5 House of Property and Entrepreneurship (HOPE), Wallayat Complex, Plaza 74, Phase 7 Bahria Town, Islamabad 46000, Pakistan
6 Department of Civil Engineering, College of Engineering, King Khalid University, Abha 61413, Saudi Arabia
7 Structural Engineering Department, Faculty of Engineering, Zagazig University, Zagazig 44519, Egypt
* Correspondence: safdarshah91@bzu.edu.pk

Abstract: In this study, the effect of micro-carbon fiber on the properties of concrete incorporating recycled coarse aggregate at three different levels, i.e., 0%, 50%, and 100% by volume replacement of natural coarse aggregate, was studied. Carbon fiber was incorporated at a dosage of 0.5% by volume fraction. The effect of silica fume or micro-silica on the efficacy of fiber reinforcement was also investigated. Studied parameters include important mechanical properties, such as compressive strength, splitting tensile strength, and flexural strength, and physical/quality parameters such as water absorption capacity and ultrasonic pulse velocity. The results showed that the mechanical and durability performance deteriorates with the increasing percentage of recycled coarse aggregate. Carbon fiber can significantly improve the tensile properties of recycled aggregate concrete. The combination of carbon fiber and silica fume proved to be highly useful in addressing both mechanical and durability concerns simultaneously. Concrete made with 50% recycled coarse aggregate, 8% silica fume, and 0.5% carbon fiber yielded 20% greater tensile and flexural strength compared to the control mix. Likewise, concrete containing 100% recycled coarse aggregate with silica fume and carbon fiber yielded higher tensile strength compared to the control mix. Silica fume ameliorated the bonding between fibers and matrix and improved the overall efficacy of fiber reinforcement.

Keywords: silica fume; micro-fibers; recycling; tensile properties; non-destructive properties

Citation: Raza, S.S.; Fahad, M.; Ali, B.; Amir, M.T.; Alashker, Y.; Elhag, A.B. Enhancing the Performance of Recycled Aggregate Concrete Using Micro-Carbon Fiber and Secondary Binding Material. *Sustainability* **2022**, *14*, 14613. https://doi.org/10.3390/su142114613

Academic Editor: José Ignacio Alvarez

Received: 8 October 2022
Accepted: 4 November 2022
Published: 7 November 2022

1. Introduction

As a result of the urbanization and modernization of cities, a large number of existing infrastructures will be renovated, transformed, and/or demolished, causing the generation of a large amount of concrete waste (CW). Re-utilization or eco-friendly consumption of CW is a crucial problem to be addressed to save precious landfilling space and conserve natural resources. Currently, the use of CW is very limited in several countries and commonly used disposal methods include stacking and landfilling, consequently causing hazardous effects on the environment [1]. The effective mitigation of the negative impacts of CW on the environment requires future planning. The use of recycled coarse aggregate (RCA) as construction aggregate can resolve the environmental issues associated with the disposal of CW.

The use of RCA in concrete as a replacement for natural aggregate leads us towards preservation of natural resources, a circular economy, and environmental safety [2]. Various studies have investigated the effect of RCA incorporation levels on the mechanical and durability performance of concrete. Replacement levels of 30–50% of RCA do not noticeably affect the performance of concrete [3]. However, at the 100% replacement level, RCA-incorporating concrete shows a noticeable decline in mechanical and durability performance compared to 100% natural aggregate concrete (NAC) [4,5]. The degree of reduction in the mechanical performance of concrete due to the incorporation of RCA is also the dependent quality of CW. RCA-sourced, high-quality CW (high-strength, high-performance concretes) yields properties comparable to those of natural coarse aggregate (NCA), while low-quality CW (lean concretes, normal-strength concretes, lightweight concretes) yields RCA samples of inferior quality [6].

Mainly, the presence of adhered mortar makes RCA less dense and weaker compared to NCA. Normally, RCA samples comprise 30–35% old cement–sand mortar and 65–70% old natural aggregate. The former is more porous and weaker than the latter; subsequently, the performance of recycled aggregate concrete (RAC) is inferior compared to that of NAC. The drawbacks of RCA use in concrete can be controlled by using additional materials such as secondary binders, chemical admixtures, and fibers [4,7–9]. Mineral binders have proven more effective and eco-efficient by far in advancing the mechanical and long-term mechanical performance of RAC [10–12]. Silica fume (SF) was found to be more effective compared to other alternative binders (fly ash, ground steel slag, and rice husk ash) in successfully enhancing the compressive strength and splitting tensile strength of RAC [12,13]. Furthermore, with the superior filler effect and pozzolanic react-ability, SF contributes more to the mechanical performance and imperviousness of RAC than other secondary binders [12]. The bonding of porous RCA with the binder matrix is strengthened due to the filler action of SF and chemical reactions between portlandite and SF on the interface of the RCA and the binder [14,15]. SF is a waste or by-product of ferrosilicon alloys and possesses very fine and highly reactive silica; its consumption in concrete would also benefit sustainable development.

Mineral admixtures or secondary binders are used to advance the mechanical strength and durability parameters of concrete [11,16], but they do not contribute a significant change in the flexural or tensile behavior of concrete. The brittleness of plain cement concrete is a pressing issue to be resolved. Therefore, the use of fibers is being encouraged to supplement the tensile strength and post-cracking toughness of plain concrete. With the addition of a small volume fraction of fiber, RAC can yield superior tensile and flexural ductility compared to plain NAC [6,15,17]. Fibers impart special characteristics such as impact resistance, fracture toughness, and fire and freeze–thaw resistance [18–21]. They improve the tenacity, ductility, and deformability of brittle materials [22]. The efficiency of fiber is dependent on the dosage, shape, and material properties of the fibers. Among all fibers, steel fiber is most commonly used due to its wider availability and high efficiency in plain concrete. However, the use of other non-metallic (basalt, carbon, glass, etc.) and synthetic fibers (polypropylene, polyvinyl, nylon, etc.) is also being encouraged due to their high tensile strength, superior durability, and light weight.

Carbon fibers (CF) have several benefits, e.g., high tensile strength, toughness, high strength-to-weight ratio, good chemical resistance, and low thermal expansion. CF-reinforced concrete can be used for corrosion-resistant construction. The inclusion of CF improves the flexural and splitting tensile strength of high-performance concrete and delays the rupture of plain concrete [23]. The inclusion of CF also provides small improvements in the absorption resistance of concrete [23]. A companion study revealed that incorporation can improve ductility and fire resistance in reactive powder concrete [24]. The available literature implies that CF can be used as a fiber reinforcement to supplement the strength of RAC. However, systematic investigation is still needed to evaluate the engineering performance of CF-reinforced RAC.

The durability benefits of SF and ductility benefits of CF in RAC can be combined by the composite addition of SF and CF. Until now, most research has focused on the use of macro-steel fibers with secondary binders [13,25–27]; however, research dealing with the mechanical and durability issues of RAC by combined incorporation of micro-fibers and secondary binders is rare [15]. Thus, this research is devoted to investigating high-strength concrete incorporating RCA, CF, and SF. The aim is to develop sustainable, ductile, durable, and high-performance cementitious composite. For this purpose, high-strength concrete was produced as a control mix. RCA was incorporated at 0%, 50%, and 100% by volume replacement of NCA. With the replacement level of RCA, CF and SF were incorporated at optimum doses to modify the properties of RAC. The mechanical parameters studied in this research include compressive strength—f_{cm}, splitting tensile strength—f_{ct}, and flexural strength—f_b. To estimate the permeability-related durability, the water absorption (WA) capacity of modified mixes was evaluated. A non-destructive ultrasonic pulse velocity (UPV) test was performed to assess the quality of the concrete. The experimental results of this investigation will benefit the development of ductile, high-strength concrete using sustainable materials in the construction industry.

2. Materials and Methods

2.1. Characteristics of Constituent Materials

2.1.1. Portland Cement and Silica Fume

In the preparation of all mixes, Type I Portland cement was used as the primary binder. The characteristics of this cement meet the standard requirements of ASTM C150 [28]. It is commercially available as 53 Grade Maple Leaf cement in Pakistan.

Silica fume (SF) was used as the secondary binder (by partial replacement of the primary binder). It is highly reactive and possesses a superior filling effect due to its specific surface area of around 27,000 m^2/kg. SF particles are 70 times finer than Portland cement particles. Particle size distribution of both primary and secondary binders is illustrated in Figure 1. SF is almost entirely composed of silicon dioxide (SiO_2). The X-ray diffraction (XRD) results revealed that ultra-fine SF shows an amorphous state, as shown in Figure 2. The XRD diffraction spectrum of the SF sample with the peak at around $2\theta = 23°$ indicates the presence of micro-crystalline silica (a form of porous silicon) [29]. Scanning electron microscopic (SEM) images revealed that most particles in the SF sample had diameters of less than 1 micron, as shown in Figure 3. This qualified it as a 'micro-binder' suitable for filling the gaps between the particles of the main binder.

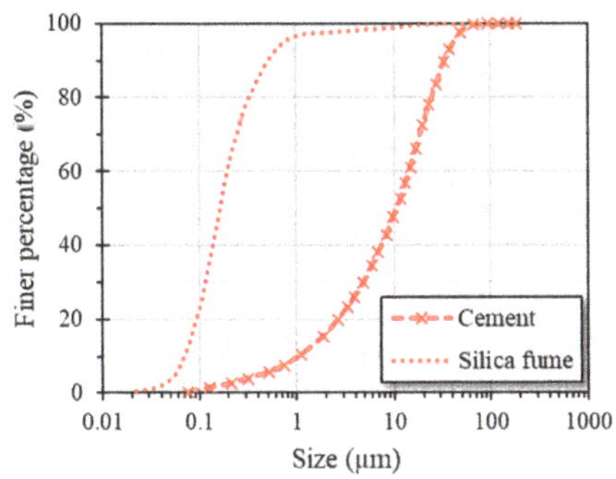

Figure 1. Gradation of cement and SF.

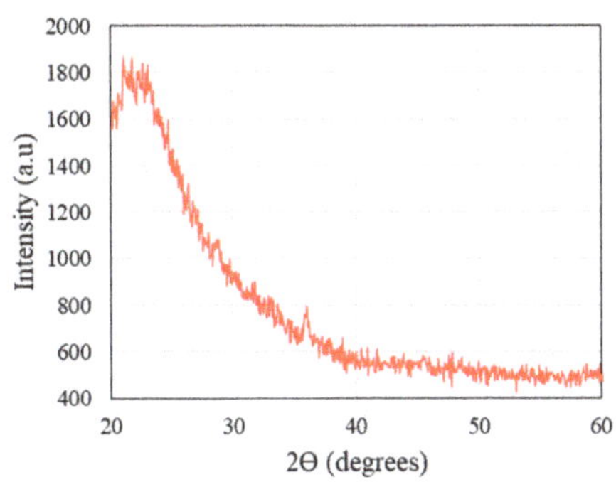

Figure 2. XRD analysis of SF.

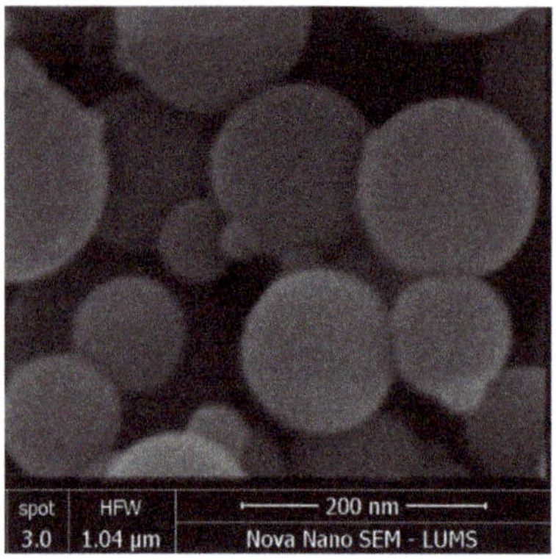

Figure 3. SEM image of SF at 200 nm resolution.

2.1.2. Natural and Recycled Aggregate

Fine aggregate was sourced from the Lawrancepur quarry in Pakistan. The fineness modulus of this siliceous sand is around 2.9. For the manufacture of fiber-reinforced concrete, the maximum size of coarse aggregate was chosen to be 12.5 mm. Dolomitic sandstone was used as NCA. Due to the non-availability of the CW recycling plant, the old concrete samples were processed by hand. The gradation or distribution of particles was kept almost the same in the samples of both RCA and NCA. Important engineering characteristics of fine and coarse aggregates are presented in Table 1. The aggregates' gradation charts are shown in Figure 4.

Table 1. Characteristics of aggregates.

Characteristic	Aggregate		
Class	Fine	NCA	RCA
Material	Siliceous	Dolomitic sandstone	Laboratory concrete waste
Quarry	Lawrancepur	Kirana Hills	-
Max. size (mm)	4.75	12.5	12.5
Min. size (mm)	0.075	2.36	2.36
Relative density	2.66	2.68	2.46
Water absorption (%)	1.09	1.17	3.35

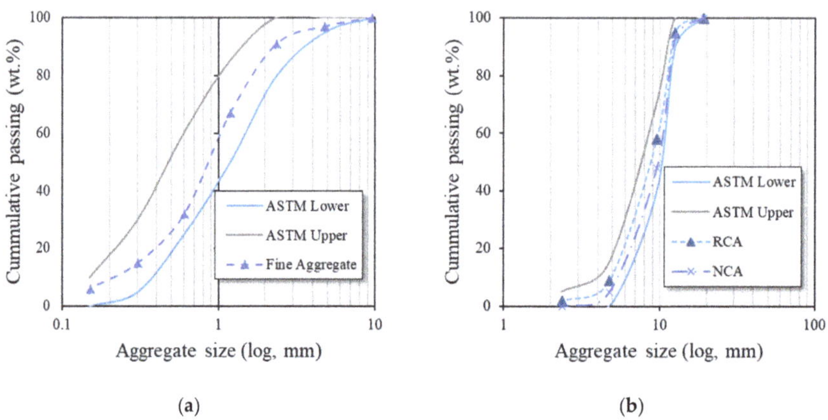

Figure 4. Gradation charts: (**a**) fine aggregate; (**b**) NCA and RCA.

2.1.3. Carbon Fiber

To enhance the tensile properties of RAC, CF was used as fiber reinforcement. The length and diameter of the CF were 15 mm and 7 microns, respectively. The length of the CF was decided considering the workability issues associated with longer filament lengths and the fact that shorter lengths yield insignificant effects on the ductility response of concrete [23]. Thus, a medium length of 15 mm was chosen for the CF. The tensile strength value of CF is above 3000 MPa, and it has an elastic modulus of 230 GPa. The material density of CF is 1800 kg/m^3. The macro–micro overview of CF is shown in Figure 5.

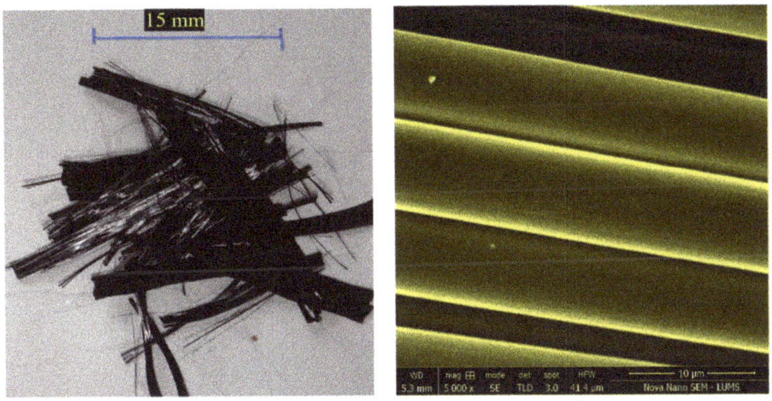

(a) (b)

Figure 5. CF sample: (**a**) direct observation; (**b**) SEM image.

2.1.4. Tap Water and Water Reducer

The workability of all mixes in the desired range was attained using 'Sika Viscocrete 3110'. It is a type G admixture and is designed to meet the specifications of ASTM C494 [30]. The preparation/manufacture and curing of all mixes were performed with tap water.

2.2. *Characteristics of Constituent Materials*

Three concrete families were designed using RCA as 0%, 50%, and 100% volumetric replacement of NCA. The first mix containing 0% RCA was designed as the control mix. The target f_{cm} of the control mix was 65–70 MPa at 28 days, while the workability of fresh concrete was chosen for a highly flowable high-performance mix with an Abram's cone slump value of 150–230 mm. RCA was then incorporated as 50% and a full replacement of NCA in the control mix. RCA mixes were produced with and without SF. The substitution level of SF was considered as 8% by volume replacement of the cement. Based on the literature [31–33], the optimum substitution level of SF lies between 5% and 10% to achieve maximum strength (Yunchao et al. [31] recommended 6% SF out of 3% SF, 6% SF, and 9% SF, Xie et al. [32] recommended 8–12% SF out of 4% SF, 8% SF, and 12% SF; Ali et al. [33] recommended 5% SF out of 5% SF and 10% SF). Thus, an intermediate percentage of 8% was chosen for SF incorporation. Similar to the selection process of SF percentage, the dosage of 0.5% CF was also selected based on maximum mechanical and durability performance according to the findings of a companion study [23] (out of 0.15% CF, 0.25% CF, 0.5% CF, 0.75% CF, and 1% CF). In the RCA incorporating mixes, CF was added as 0.5% by volume fraction with and without SF. Eventually, a total of 12 concrete mixes were produced. Complete details about the nomenclature and composition of the mixes are illustrated in Table 2. The inclusion of both CF and SF is damaging to the workability of concrete; thus, a superplasticizer or water reducer was used to maintain the workability of the fresh concrete. All 12 concrete mixes were prepared in the laboratory, as detailed by a companion study [34].

Table 2. Proportions of concrete mixes.

Batch ID	OPC (kg/m^3)	SF (kg/m^3)	Fine Aggregate (kg/m^3)	Coarse Aggregate (kg/m^3)		Water (kg/m^3)	WR (kg/m^3)	CF (kg/m^3)
				NCA	RCA			
R0	550	0	650	1075	0	181.5	2.15	0.00
R0/SF	506	31	650	1075	0	181.5	2.25	0.00
R0/CF	550	0	644	1069	0	181.5	3.65	9.25
R0/SF/CF	506	31	644	1069	0	181.5	3.98	9.25
R50	550	0	650	538	489	181.5	2.21	0.00
R50/SF	506	31	650	538	489	181.5	2.46	0.00
R50/CF	550	0	644	534	486	181.5	3.94	9.25
R50/SF/CF	506	31	644	534	486	181.5	4.15	9.25
R100	550	0	650	0	978	181.5	2.31	0.00
R100/SF	506	31	650	0	978	181.5	2.58	0.00
R100/CF	550	0	644	0	972	181.5	3.84	9.25
R100/SF/CF	506	31	644	0	972	181.5	4.35	9.25

2.3. Testing Techniques

For all tests, three replicate samples of mixes were prepared, cured, and tested under the same conditions. Then, their average value is presented in this research paper with standard deviation values using error bars. All tests were conducted on samples after curing for 28 days. To evaluate the f_{cm} of the concrete mixes, 100 mm cubic samples of the concrete were tested according to BS: EN 12390-3 [35]. The compression testing setup is shown in Figure 6a. The tensile strength of concrete is indirectly assessed by measuring the f_{ct} of concrete. For this purpose, 100 mm diameter × 200 mm height samples of concrete were subjected to splitting tensile load according to ASTM C496 [36]. The splitting tensile test setup is illustrated in Figure 6b. A third-point bending test was performed to estimate the modulus of rupture or bending strength (f_{cb}) of the concrete samples. Prismatic specimens of 100 × 100 × 350 mm^3 were tested under third-point loading according to ASTM C1609 [37]. The clear span between simple supports was 300 mm. The bending test setup is shown in Figure 6c.

The durability of concrete structures is highly dependent on the voids connected to the surface of the concrete. The measure of the permeable pore volume of concrete is related to the durability assessment. Therefore, the WA capacity of all mixes was evaluated. Concrete discs of 100 mm diameter × 50 mm thickness were tested according to the ASTM C938 [38]. The percentage difference between the dried and saturated concrete samples is regarded as the WA capacity of concrete. The UPV test is a non-destructive field test to predict the strength and durability of concrete. The change in the speed of the pulse wave through a sample reveals the change in the porosity or density of concrete. The UPV test was conducted on cubic samples, as shown in Figure 6d.

(**a**) (**b**)

Figure 6. *Cont.*

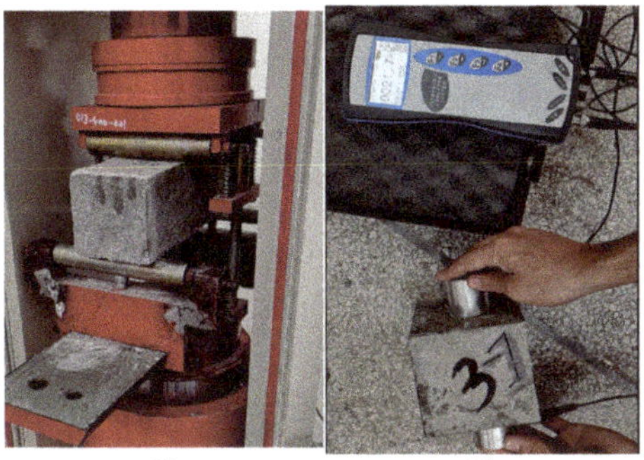

(**c**) (**d**)

Figure 6. The overview of experimental testing: (**a**) compression test; (**b**) splitting tensile test; (**c**) bending test; and (**d**) pulse velocity test.

3. Results and Discussion

3.1. Compressive Strength

The effect of SF and CF addition on the f_{cm} of concrete with and without RCA is shown in Figure 7. The f_{cm} values of concrete mixes relative to the control mix are illustrated in Figure 8. It can be seen that the f_{cm} was reduced noticeably with an increasing percentage of RCA. At 50% and full replacement of NCA with RCA, the f_{cm} was decreased by 14% and 25%, respectively. Since RCA is more porous and weaker than NCA, it reduces the strength of the concrete. The attached mortar in RCA absorbs a high amount of water, and, thus, RAC has more voids than the NAC. The strength reduction due to RCA incorporation has been linked to the increase in porosity and voids [39].

The addition of SF as a partial substitution for cement caused a noticeable improvement in the f_{cm}. For R0, the f_{cm} was increased by 5% due to SF addition, while R50 and R100 experienced improvements of around 9% compared to the control mix. Thus, SF played an effective role in overcoming the strength deficit of RAC. It was also noted that SF addition is more useful in RCA-incorporating families. This is because the overall portlandite (CH) content in RAC is greater than in the R0 family, and the presence of RCA offers a high potential for pozzolanic reactions. This finding is in line with Dilbas et al. [7]. At the interfacial transition zones, reactions between the silica and free CH strengthen the aggregate–matrix bond [14].

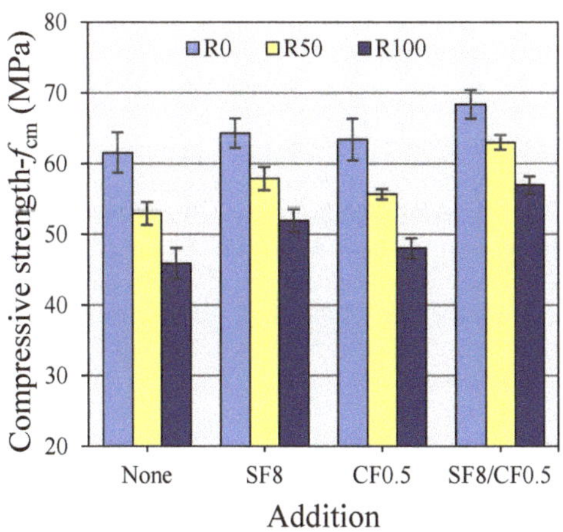

Figure 7. Effect of CF and SF addition on 28-day f_{cm} of concrete with different percentages of RCA.

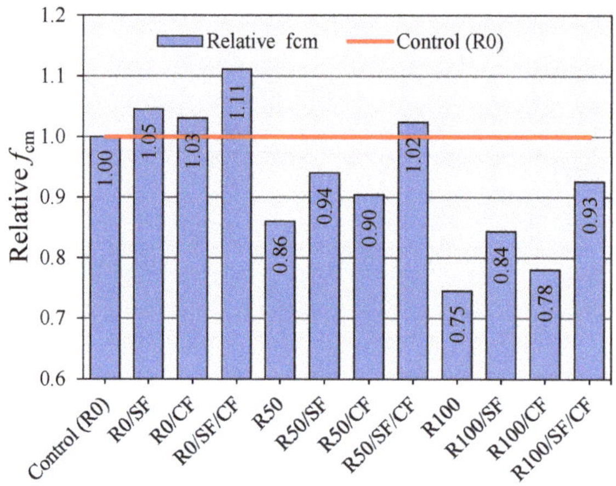

Figure 8. Relative f_{cm} of concrete mixes with SF, CF, and RCA.

A singular CF addition did not cause a noteworthy change in the f_{cm}. A nominal improvement of 3% was observed in the f_{cm} of all concrete families. It can be credited to the improvement in the axial stiffness of concrete due to advanced crack resistance caused by the fibers [40]. Previous studies [41,42] also reported that micro-fibers cause a minimal change in the f_{cm}, since they are only majorly valuable for boosting the tensile strength and fracture toughness.

The combined use of CF and SF augmented the f_{cm} of concrete by significant margins. As is shown in Figure 8, around an 11% improvement in f_{cm} of R0 can be achieved by using 8% SF and 0.5% CF. Due to the combined use of CF and SF, R0, R50, and R100 concrete experienced 11%, 16%, and 18% improvements, respectively. It was clear that SF and CF have synergistic benefits since the benefits of their combined incorporation were marginally greater than the sum of the benefits due to their singular incorporations. R50/SF/CF showed an f_{cm} comparable to that of the control mix.

3.2. Splitting Tensile Strength

In practical engineering, f_{ct} is measured instead of direct tensile strength since it provides a simpler and easier assessment of the tensile strength of concrete. Figure 9 shows the effect of SF and CF on the f_{ct} with and without incorporation of RCA, while the f_{ct} value of all mixes relative to the control mix is shown in Figure 10. A downward trend was noticed in f_{ct}, increasing the percentage of RCA. The f_{ct} decreased by 7% and 16% at 50% and 100% RCA, respectively. These reductions in f_{ct} were anticipated due to the inherent weakness of RCA. The incorporation of SF provided minor improvements in the f_{ct}. The f_{ct} of R0, R50, and R100 was improved by 4%, 9%, and 10%, respectively, due to the inclusion of SF as an 8% replacement of cement. The pozzolanic reactions and filler effect of micro-silica particles strengthen the binder matrix. In the case of mixes incorporating RCA, the chemical reactions may also occur across the bond between silica-modified matrix and aggregates. Kurda et al. [43] systematically showed that the utilization ratio of pozzolanic binders is greater in the case of RAC than in NAC.

The incorporation of CF had an upward effect on the f_{ct}. The tensile strength was increased significantly by 18–20% with the addition of 0.5% CF. Therefore, the tensile strength deficit of RAC families was completely overwhelmed by the CF addition. Both R50 and R100 concretes attained higher tensile strengths than the control mix. This is because the use of CF increases the bonding force of concrete and enhances the f_{ct} [44]. Furthermore, the efficacy of fibers under pulling action or tension is more than under compressive forces. Fibers activate earlier under tension loads and supplement the bonding force of concrete.

Raza and Qureshi [24] also found that the effect of fibers on f_{ct} is more promising than their effect on f_{cm}.

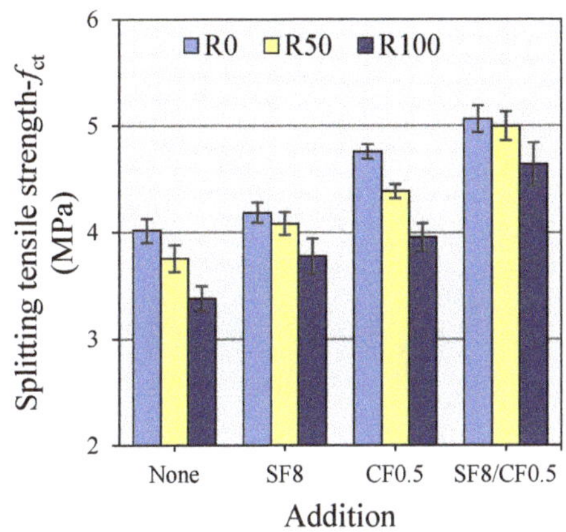

Figure 9. Effect of CF and SF addition on 28-day f_{ct} of concrete with different percentages of RCA.

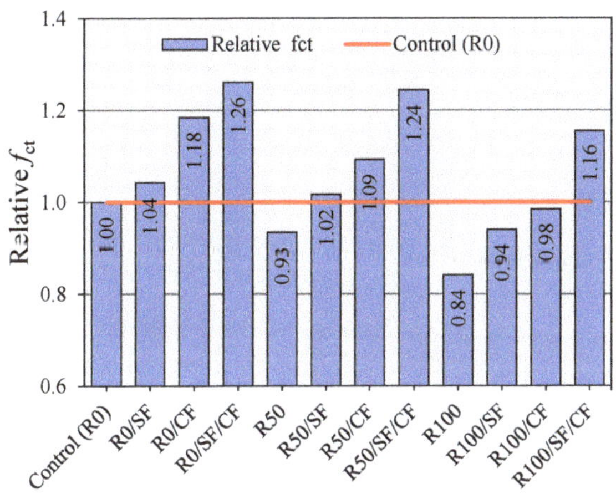

Figure 10. Relative f_{ct} of concrete mixes with SF, CF, and RCA.

The combined incorporation of CF and SF showed maximum improvement in f_{ct}. R0, R50, and R100 concrete experienced 26%, 31%, and 32% improvements due to the combined use of CF and SF. These results confirmed that the use of CF and SF has a synergistic effect on the f_{ct}. The surplus net gain due to the simultaneous addition of CF and SF is credited to the improvement in the bond strength of fibers. The pore refinement and growth of the extra calcium silicate hydrate (CSH) gels improve the pulling force of the cement matrix. Thus, the strengthening of the cement matrix and refinement of the pore structure improves the interfacial bond properties and causes further enhancement or synergistic effect on the macro performance.

3.3. Bending Strength

Similar to f_{ct}, f_{cb} is an indirect measure of the true tensile strength of plain and fibrous concretes. However, unlike f_{ct}, f_{cb} can be directly used in the design of concrete elements such as tunnels, slabs, pavements, etc. Unlike f_{cm} and f_{ct}, f_{cb} is not a simplistic measure as it requires high-quality control and is strongly influenced by the support conditions, fiber orientations, and sample size. The effect of RCA, CF, and SF addition on f_{cb} is illustrated in Figure 11. The f_{cb} values of all mixes relative to the control mix are presented in Figure 12.

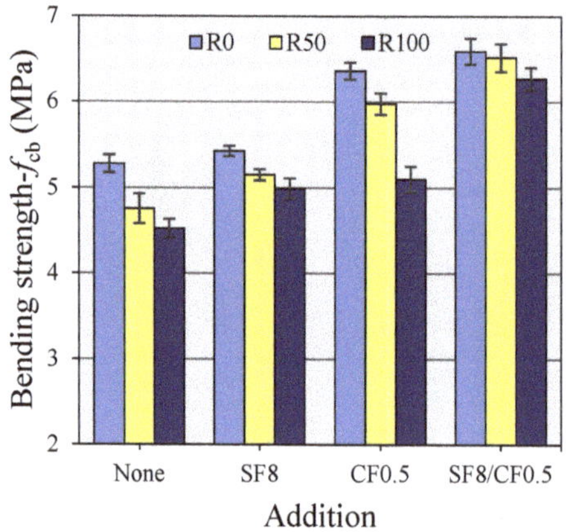

Figure 11. Effect of CF and SF addition on 28-day f_{cb} of concrete with different percentages of RCA.

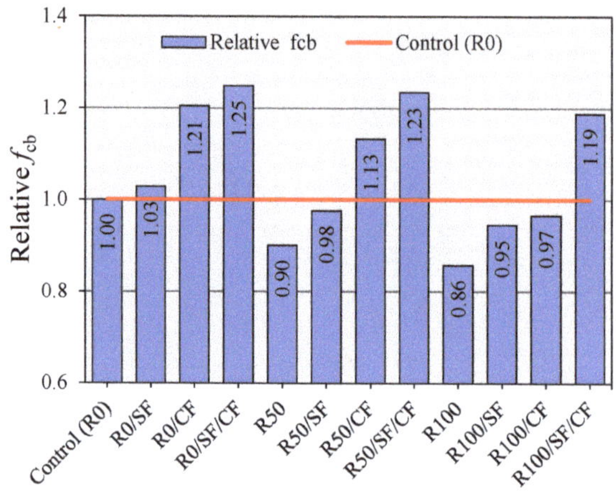

Figure 12. Relative f_{cb} of concrete mixes with SF, CF, and RCA.

The addition of SF had a minimal effect on the f_{cb} of R0. However, it showed phenomenal improvements of around 10% in the f_{cb} of both R50 and R100 concretes. The net efficiency of SF was high in mixes containing RCA. This behavior was observed in the results of f_{ct} as well. The inclusion of fibers caused a remarkable increment in the f_{cb} of all RCA families. For instance, R0, R50, and R100 experienced net improvements of 21%, 23%, and 11% due to the addition of CF, respectively. Fiber inclusion in the matrix of concrete effectively overcame the f_{cb} deficit of R50 and R100 against the control mix.

The concurrent inclusion of SF and CF was demonstrated to be highly beneficial in boosting the f_{cb} of all RCA families. Not only were the f_{cb} increments of SF and CF combined by their conjunctive use, but they also caused a synergistic effect similar to that observed in other mechanical results. For instance, in R50, the singular addition of SF and CF caused net increments of 8% and 23%, respectively but their joint incorporation led to an improvement of 33%. Similarly, for the R100 concrete, the addition of SF and CF showed improvements of 9% and 11%, but their joint inclusion caused f_{cb} to increase by 32%. The results of tensile testing (i.e., f_{ct} and f_{cb}) showed that the individual addition of fiber in the matrix of RAC is enough to overcome the strength gap or deficit compared to plain NAC.

However, the further enhancement effect on the performance of fibers can be achieved by the addition of SF.

The variation in f_{ct} and f_{cb} with the incorporation of RCA, SF, and CF almost showed a similar trend. The crack-bridging effect of fiber is highly useful for both tensile properties of concrete; therefore, the f_{ct} and f_{cb} correlated with high accuracy, as shown in Figure 13. As we know that f_{cb} measurement is difficult and sensitive, it can be assessed accurately from the f_{ct}.

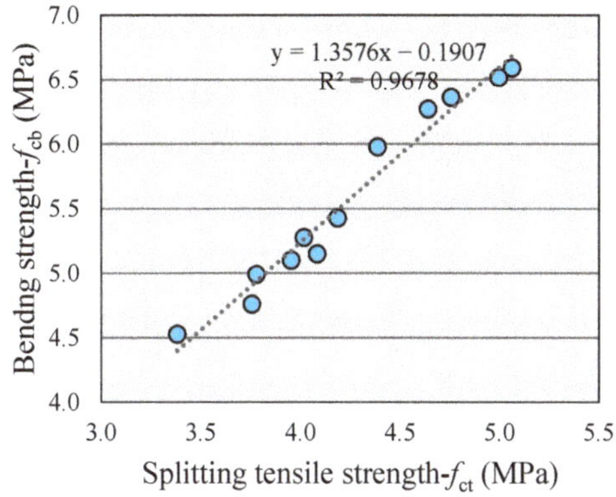

Figure 13. Correlation between f_{cb} and f_{ct}.

3.4. Water Absorption

The change in the permeable porosity of concrete can be used as a durability indicator because the ingress rate of harmful chemicals is entirely dependent on the porosity of the concrete. The results of WA testing are illustrated in Figure 14. RCA incorporation is detrimental to the porosity of concrete due to the attached mortar. The pores present inside RCA increase the connectivity of micro-channels, leading to an increased value of WA. At full replacement of NCA with RCA, a 40% increase in the WA capacity of concrete was noticed. This drawback of high porosity and loosely attached mortar associated with RCA also had a detrimental effect on the mechanical properties of concrete. A secondary binder, such as SF, due to its micro size, effectively closes the spaces in the binder matrix and on the surface of RCA. The meandering effect of fine particles also slows or reduces the penetration of water inside the concrete [10,13]. The improvement in the interfacial properties of concrete results in a tremendous WA reduction.

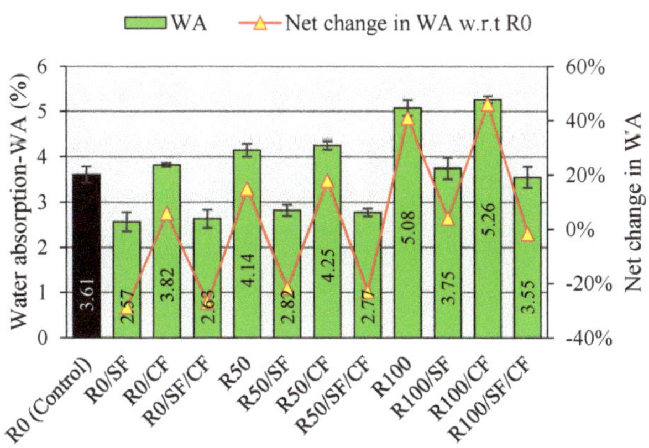

Figure 14. Effect of CF and SF addition on 28-day WA capacity of concrete with different percentages of RCA.

The incorporation of CF aggravates the imperviousness of concrete. The tangling effect of fiber filaments and improper compaction can lead to an increase in the size of pores and connectivity between them. The CF filaments have a high surface area; due to this reason, they can be very difficult to disperse. Thus, CF introduces voids or pores inside the cementitious matrix. At 0.5% volume of CF, the WA capacity of R0, R50, and R100 was increased by 6%, 3%, and 5%, respectively.

SF plays a vital role in managing the negative effects of both RCA and fibers. The increased CSH gel growth and pore refinement lead to better interfacial properties for both RCA and fibers. The increase in the imperviousness of the binder matrix causes a balancing effect on the pores created due to the 'balling' or 'tangling' effect of the fibers. It can be concluded that the role of SF is very crucial for suppressing the harmful effects of RCA and micro-fibers on the imperviousness of concrete. Owing to the superior filling effect of SF, R100/SF/CF yielded a WA value around 2% lower than that of the control mix, while R50/SF/CF yielded 23% less WA than the control.

3.5. Ultrasonic Pulse Velocity

To assess the quality of concrete in the field, UPV is widely used as a non-destructive testing method. Table 3 provides the interpretation of UPV value in terms of concrete quality as per BIS: 13311 [45]. A UPV value between 3.5 and 4.5 km/s usually corresponds to normal-strength and medium-strength concretes, whereas UPV values above 4.5 km/s correspond to high-performance or high-strength concrete grades. The results of UPV testing of all mixes are illustrated in Figure 15. The control mix showed a UPV value above 4.5 km/s, which shows its excellent quality. The inclusion of SF refines the micro-structure and improves the imperviousness of concrete, which leads to further improvement in the UPV value of concrete. Unlike SF, fiber addition has a downward effect on the UPV of concrete. The increase in porosity due to fiber addition may reduce the speed of the pulse [46]. The reduction in UPV value can also be linked to the increase in heterogeneity on fiber addition. A mix containing both SF and CF attained a UPV value similar to that of the control or unmodified mix. SF balances the negative effect of micro-fibers on UPV.

Table 3. Interpretation of UPV value.

UPV (km/s)	Concrete Grading (Quality)
Above 4.5	Excellent
3.5 to 4.5	Good
3.0 to 3.5	Medium
Below 3.0	Doubtful or inferior

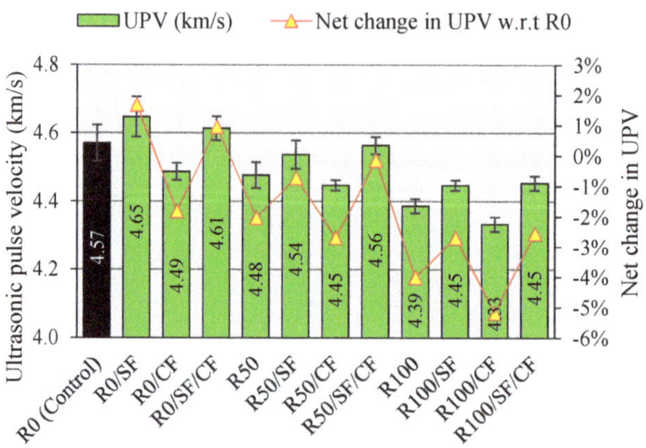

Figure 15. Effect of CF and SF addition on 28-day UPV of concrete with different percentages of RCA.

The decrease in UPV was observed due to an increase in the RCA level. The increase in the porosity and voids due to RCA addition is responsible for slowing the speed of the pulse wave. All mixes of the R100 family fell under the good quality category. SF caused a noticeable improvement in the UPV value of RAC, while CF had a declining effect on the UPV values of R0, R50, and R100 concrete. In RCA families, only R50/SF and R50/SF/CF yielded UPV values above 4.5 km/s, which falls under excellent quality. Since UPV, WA, and f_{cm} are exclusively dependent on the density and micro-structural growth of concrete, these parameters can be correlated to UPV's high accuracy, as shown in Figure 16. The increase in UPV due to the modification of concrete indicates an improvement in the imperviousness and f_{cm}. For example, the increase in UPV due to SF addition can be used as an indicator of an f_{cm} increment and a decline in WA.

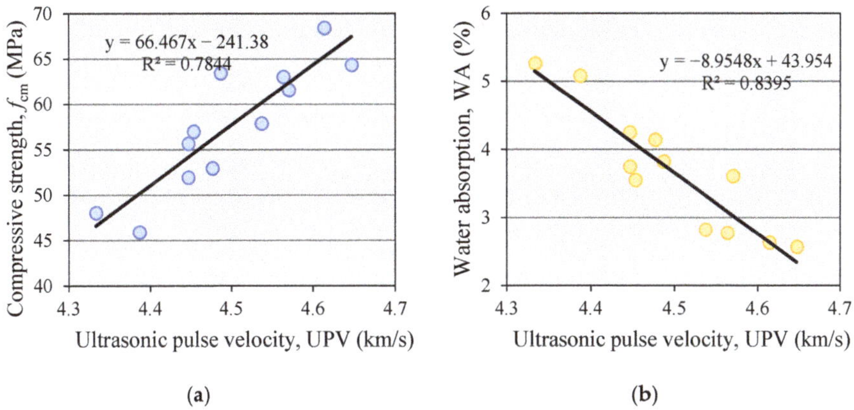

Figure 16. Correlations between (**a**) f_{cm} and UPV and (**b**) WA and UPV.

4. Conclusions

In this research, the effect on the properties of 'high-strength concrete' of singular and combined incorporation of silica fume and carbon fiber was studied with three different levels of recycled coarse aggregates as a replacement for natural coarse aggregates. The following are the important findings of this research:

- The modification method using silica fume is highly effective in improving the compressive strength of recycled aggregate concrete. Concretes containing 50% and 100% recycled coarse aggregates experienced strength increments of around 9% at the addition of 8% silica fume. Carbon fiber caused nominal upgradation of 2–4% in compressive strength. At the combined use of 0.5% carbon fiber and 8% silica fume, the compressive strength of 100% recycled aggregate concrete was improved by 18%;
- The use of carbon fiber is effective in the upgradation of tensile strength. At the addition of 0.5% carbon fiber, concrete mixes made with 50% and 100% recycled aggregate experienced 16% and 14% improvements in splitting tensile strength, respectively. The combined use of silica fume and carbon fiber caused a maximum improvement of 32% in the splitting tensile strength of recycled aggregate concrete. The singular use of fiber or combined use of fiber and silica fume can overcome the tensile strength deficit of 100% recycled aggregate concrete;
- The singular addition of fiber is more useful than silica fume in upgrading the bending strength. Concretes with 50% and 100% recycled aggregates gained net improvements of 23% and 11%, respectively, due to the inclusion of 0.5% carbon fiber;
- The use of silica fume enhances the interfacial properties of aggregates and fibers; both fibers and silica cause a synergistic improvement in the tensile properties. The net effect of fiber on mechanical performance also improves with the addition of silica fume;

- Due to the addition of 0.5% carbon fiber and 8% silica fume, 100% recycled aggregate concrete achieved 19% greater bending strength than the control mix;
- Both recycled aggregate and fiber have a downward effect on the imperviousness of concrete. Thus, they can lead to a decline in the durability of concrete. The use of silica fume is highly effective in managing the negative effects of both fibers and inferior aggregates on the water absorption capacity and durability of concrete;
- The use of recycled aggregate noticeably reduces the pulse velocity; hence, the quality of concrete degrades. Fiber inclusion further has a minor but declining effect on the ultrasonic pulse velocity due to a possible increase in the porosity of concrete. Silica fume proved advantageous in refining the quality of concrete with and without recycled aggregate. Concrete made with 50% recycled coarse aggregate achieved excellent quality, with a pulse velocity value well above 4.5 km/s, owing to the filling and pozzolanic action of micro-silica.

Author Contributions: Conceptualization, S.S.R. and M.F.; methodology, B.A.; software, B.A.; validation, M.F., B.A. and M.T.A.; formal analysis, S.S.R.; investigation, Y.A.; resources, A.B.E. and S.S.R.; data curation, S.S.R.; writing—original draft preparation, S.S.R.; writing—review and editing, B.A. and M.T.A.; visualization, S.S.R.; supervision, S.S.R.; project administration, B.A. and A.B.E.; funding acquisition, Y.A. and A.B.E. All authors have read and agreed to the published version of the manuscript.

Funding: The authors extend their appreciation to the Deanship of Scientific Research at King Khalid University for funding this work through the Large Groups Project under grant number **R.G.P. 2/152/43.**

Institutional Review Board Statement: Not applicable.

Informed Consent Statement: Not applicable.

Data Availability Statement: Not applicable.

Acknowledgments: The authors extend their appreciation to the Deanship of Scientific Research at King Khalid University for funding this work through the Large Groups Project under grant number **R.G.P. 2/152/43.**

Conflicts of Interest: The authors declare no conflict of interest.

References

1. Revilla-Cuesta, V.; Skaf, M.; Faleschini, F.; Manso, J.M.; Ortega-López, V. Self-compacting concrete manufactured with recycled concrete aggregate: An overview. *J. Clean. Prod.* **2020**, *262*, 121362. [CrossRef]
2. Junaid, M.F.; ur Rehman, Z.; Kuruc, M.; Medved', I.; Bačinskas, D.; Čurpek, J.; Čekon, M.; Ijaz, N.; Ansari, W.S. Lightweight concrete from a perspective of sustainable reuse of waste byproducts. *Constr. Build. Mater.* **2022**, *319*, 126061. [CrossRef]
3. Kurda, R.; Silvestre, J.D.; de Brito, J. Life cycle assessment of concrete made with high volume of recycled concrete aggregates and fly ash. *Resour. Conserv. Recycl.* **2018**, *139*, 407–417. [CrossRef]
4. Ahmadi, M.; Farzin, S.; Hassani, A.; Motamedi, M. Mechanical properties of the concrete containing recycled fibers and aggregates. *Constr. Build. Mater.* **2017**, *144*, 392–398. [CrossRef]
5. Kaplan, G.; Gulcan, A.; Cagdas, B.; Bayraktar, O.Y. The impact of recycled coarse aggregates obtained from waste concretes on lightweight pervious concrete properties. *Environ. Sci. Pollut. Res.* **2021**, *28*, 17369–17394. [CrossRef]
6. Afroughsabet, V.; Biolzi, L.; Ozbakkaloglu, T. Influence of double hooked-end steel fibers and slag on mechanical and durability properties of high performance recycled aggregate concrete. *Compos. Struct.* **2017**, *181*, 273–284. [CrossRef]
7. Dilbas, H.; Şimşek, M.; Çakır, Ö. An investigation on mechanical and physical properties of recycled aggregate concrete (RAC) with and without silica fume. *Constr. Build. Mater.* **2014**, *61*, 50–59. [CrossRef]
8. Barbudo, A.; de Brito, J.; Evangelista, L.; Bravo, M.; Agrela, F. Influence of water-reducing admixtures on the mechanical performance of recycled concrete. *J. Clean. Prod.* **2013**, *59*, 93–98. [CrossRef]
9. Ali, B.; Ahmed, H.; Ali Qureshi, L.; Kurda, R.; Hafez, H.; Mohammed, H.; Raza, A. Enhancing the Hardened Properties of Recycled Concrete (RC) through Synergistic Incorporation of Fiber Reinforcement and Silica Fume. *Materials* **2020**, *13*, 4112. [CrossRef]
10. Kurda, R.; Silvestre, J.D.; de Brito, J.; Ahmed, H. Optimizing recycled concrete containing high volume of fly ash in terms of the embodied energy and chloride ion resistance. *J. Clean. Prod.* **2018**, *194*, 735–750. [CrossRef]

11. Kurda, R.; de Brito, J.; Silvestre, J.D. Water absorption and electrical resistivity of concrete with recycled concrete aggregates and fly ash. *Cem. Concr. Compos.* **2019**, *95*, 169–182. [CrossRef]
12. Kou, S.; Poon, C.; Agrela, F. Comparisons of natural and recycled aggregate concretes prepared with the addition of different mineral admixtures. *Cem. Concr. Compos.* **2011**, *33*, 788–795. [CrossRef]
13. Qureshi, L.A.; Ali, B.; Ali, A. Combined effects of supplementary cementitious materials (silica fume, GGBS, fly ash and rice husk ash) and steel fiber on the hardened properties of recycled aggregate concrete. *Constr. Build. Mater.* **2020**, *263*, 120636. [CrossRef]
14. Sasanipour, H.; Aslani, F.; Taherinezhad, J. Chloride ion permeability improvement of recycled aggregate concrete using pretreated recycled aggregates by silica fume slurry. *Constr. Build. Mater.* **2021**, *270*, 121498. [CrossRef]
15. Alyousef, R.; Ali, B.; Mohammed, A.; Kurda, R.; Alabduljabbar, H.; Riaz, S. Evaluation of Mechanical and Permeability Characteristics of Microfiber-Reinforced Recycled Aggregate Concrete with Different Potential Waste Mineral Admixtures. *Materials* **2021**, *14*, 5933. [CrossRef] [PubMed]
16. Mangi, S.A.; Raza, M.S.; Khahro, S.H.; Qureshi, A.S.; Kumar, R. Recycling of ceramic tiles waste and marble waste in sustainable production of concrete: A review. *Environ. Sci. Pollut. Res.* **2022**, *29*, 18311–18332. [CrossRef]
17. Ali, B.; Qureshi, L.A. Influence of glass fibers on mechanical and durability performance of concrete with recycled aggregates. *Constr. Build. Mater.* **2019**, *228*, 116783. [CrossRef]
18. Guler, S.; Öker, B.; Akbulut, Z.F. Workability, strength and toughness properties of different types of fiber-reinforced wet-mix shotcrete. In *Structures*; Elsevier: Amsterdam, The Netherlands, 2021; Volume 31, pp. 781–791.
19. Köksal, F.; Altun, F.; Yiğit, I.; Şahin, Y. Combined effect of silica fume and steel fiber on the mechanical properties of high strength concretes. *Constr. Build. Mater.* **2008**, *22*, 1874–1880. [CrossRef]
20. Guler, S.; Akbulut, Z.F.; Siad, H.; Lachemi, M. Effect of macro polypropylene, polyamide and steel fibers on the residual properties of SCC at ambient and elevated temperatures. *Constr. Build. Mater.* **2021**, *289*, 123154. [CrossRef]
21. Guler, S.; Akbulut, Z.F. Effect of high-temperature on the behavior of single and hybrid glass and basalt fiber added geopolymer cement mortars. *J. Build. Eng.* **2022**, *57*, 104809. [CrossRef]
22. Ur Rehman, Z.; Khalid, U. Optimization of COVID-19 face mask waste fibers and silica fume as a balanced mechanical ameliorator of fat clay using response surface methodology. *Environ. Sci. Pollut. Res.* **2022**, *29*, 17001–17016. [CrossRef] [PubMed]
23. Raza, S.S.; Amir, M.T.; Azab, M.; Ali, B.; Abdallah, M.; El Ouni, M.H.; Elhag, A.B. Effect of micro-silica on the physical, tensile, and load-deflection characteristics of micro fiber-reinforced high-performance concrete (HPC). *Case Stud. Constr. Mater.* **2022**, *17*, e01380. [CrossRef]
24. Raza, S.S.; Qureshi, L.A. Effect of carbon fiber on mechanical properties of reactive powder concrete exposed to elevated temperatures. *J. Build. Eng.* **2021**, *42*, 102503. [CrossRef]
25. Mastali, M.; Dalvand, A. Use of silica fume and recycled steel fibers in self-compacting concrete (SCC). *Constr. Build. Mater.* **2016**, *125*, 196–209. [CrossRef]
26. Xie, J.; Fang, C.; Lu, Z.; Li, Z.; Li, L. Effects of the addition of silica fume and rubber particles on the compressive behaviour of recycled aggregate concrete with steel fibres. *J. Clean. Prod.* **2018**, *197*, 656–667. [CrossRef]
27. Nazarimofrad, E.; Shaikh, F.U.A.; Nili, M. Effects of steel fibre and silica fume on impact behaviour of recycled aggregate concrete. *J. Sustain. Cem. Mater.* **2017**, *6*, 54–68. [CrossRef]
28. *ASTM-C150*; Standard Specification for Portland Cement. ASTM International: West Conshohocken, PA, USA, 2018.
29. Wang, Y.; Xu, Z.; Wang, J.; Zhou, Z.; Du, P.; Cheng, X. Synergistic effect of nano-silica and silica fume on hydration properties of cement-based materials. *J. Therm. Anal. Calorim.* **2020**, *140*, 2225–2235. [CrossRef]
30. *ASTM-C494*; Standard Specification for Chemical Admixtures for Concrete. AASHTO: Washington, DC, USA, 2015.
31. Yunchao, T.; Zheng, C.; Wanhui, F.; Yumei, N.; Cong, L.; Jieming, C. Combined effects of nano-silica and silica fume on the mechanical behavior of recycled aggregate concrete. *Nanotechnol. Rev.* **2021**, *10*, 819–838. [CrossRef]
32. Xie, J.; Zhang, Z.; Lu, Z.; Sun, M. Coupling effects of silica fume and steel-fiber on the compressive behaviour of recycled aggregate concrete after exposure to elevated temperature. *Constr. Build. Mater.* **2018**, *184*, 752–764. [CrossRef]
33. Ali, B.; Kurda, R.; Herki, B.; Alyousef, R.; Mustafa, R.; Mohammed, A.; Raza, A.; Ahmed, H.; Fayyaz Ul-Haq, M. Effect of Varying Steel Fiber Content on Strength and Permeability Characteristics of High Strength Concrete with Micro Silica. *Materials* **2020**, *13*, 5739. [CrossRef]
34. Farooq, M.A.; Fahad, M.; Ali, B.; ullah, S.; El Ouni, M.H.; Elhag, A.B. Influence of nylon fibers recycled from the scrap brushes on the properties of concrete: Valorization of plastic waste in concrete. *Case Stud. Constr. Mater.* **2022**, *16*, e01089. [CrossRef]
35. *EN-12390-3*; Testing of Hardened Concrete, Part 3: Compressive Strength of Test Specimens. British Standards Institution: London, UK, 2002.
36. *ASTM-C496*; Standard Test Method for Splitting Tensile Strength of Cylindrical Concrete Specimens. ASTM International: West Conshohocken, PA, USA, 2017.
37. *ASTM-C1609*; Standard Test Method for Flexural Performance of Fiber-Reinforced Concrete (Using Beam With Third-Point Loading). ASTM International: West Conshohocken, PA, USA, 2019.
38. *ASTM-C948 Standard*; Test Method for Dry and Wet Bulk Density, Water Absorption, and Apparent Porosity of Thin Sections of Glass-Fiber Reinforced Concrete. ASTM International: West Conshohocken, PA, USA, 2016.
39. Kurda, R.; de Brito, J.; Silvestre, J.D. Combined influence of recycled concrete aggregates and high contents of fly ash on concrete properties. *Constr. Build. Mater.* **2017**, *157*, 554–572. [CrossRef]

40. Kazmi, S.M.S.; Munir, M.J.; Wu, Y.-F.; Patnaikuni, I.; Zhou, Y.; Xing, F. Axial stress-strain behavior of macro-synthetic fiber reinforced recycled aggregate concrete. *Cem. Concr. Compos.* **2019**, *97*, 341–356. [CrossRef]
41. Khan, M.; Cao, M.; Ali, M. Effect of basalt fibers on mechanical properties of calcium carbonate whisker-steel fiber reinforced concrete. *Constr. Build. Mater.* **2018**, *192*, 742–753. [CrossRef]
42. Khan, U.A.; Jahanzaib, H.M.; Khan, M.; Ali, M. Improving the tensile energy absorption of high strength natural fiber reinforced concrete with fly-ash for bridge girders. In *Key Engineering Materials*; Trans Tech Publications Ltd.: Wollerau, Switzerland, 2018; Volume 765, pp. 335–342.
43. Kurda, R.; de Brito, J.; Silvestre, J.D. Indirect evaluation of the compressive strength of recycled aggregate concrete with high fly ash ratios. *Mag. Concr. Res* **2018**, *70*, 204–216. [CrossRef]
44. Zheng, Y.; Zhuo, J.; Zhang, P.; Ma, M. Mechanical properties and meso-microscopic mechanism of basalt fiber-reinforced recycled aggregate concrete. *J. Clean. Prod.* **2022**, *370*, 133555. [CrossRef]
45. *BIS 13311-1*; Method of Non-Destructive Testing of Concret, Part 1. Bureau of Indian Standards: Delhi, India, 1992.
46. Das, C.S.; Dey, T.; Dandapat, R.; Mukharjee, B.B.; Kumar, J. Performance evaluation of polypropylene fibre reinforced recycled aggregate concrete. *Constr. Build. Mater.* **2018**, *189*, 649–659. [CrossRef]

sustainability

MDPI

Article

Acid Resistance of Alkali-Activated Natural Pozzolan and Limestone Powder Mortar

Khaled A. Alawi Al-Sodani [1,*], Adeshina A. Adewumi [1,*], Mohd Azreen Mohd Ariffin [2], Babatunde Abiodun Salami [3], Moruf O. Yusuf [1], Mohammed Ibrahim [4], Ali H. AlAteah [1], Mohammed M. H. Al-Tholaia [1], Sami M. Ibn Shamsah [5] and Mohammad Ismail [2]

1 Department of Civil Engineering, College of Engineering, University of Hafr Al Batin, P.O. Box 1803, Hafr Al Batin 39524, Saudi Arabia
2 Faculty of Engineering, School of Civil Engineering, Universiti Teknologi Malaysia, Johor Bahru 81310, Malaysia
3 Interdisciplinary Research Center for Construction and Building Materials, Research Institute, King Fahd University of Petroleum and Minerals, Dhahran 31261, Saudi Arabia
4 Applied Research Center for Metrology, Standards and Testing, Research Institute, King Fahd University of Petroleum and Minerals, Dhahran 31261, Saudi Arabia
5 Department of Mechanical Engineering, College of Engineering, University of Hafr Al Batin, P.O. Box 1803, Hafr Al Batin 39524, Saudi Arabia
* Correspondence: kalsodani@uhb.edu.sa (K.A.A.A.-S.); adeshina@uhb.edu.sa (A.A.A.); Tel.: +966-1372-05171 (K.A.A.A.-S.)

Abstract: The development of sustainable, environmentally friendly alkali-activated binder has emerged as an alternative to ordinary Portland cement. The engineering and durability properties of alkali-activated binder using various precursor combinations have been investigated; however, no study has focused on the impact of high-volume natural pozzolan (NP) on the acid resistance of alkali-activated NP and limestone powder. Therefore, the current study assesses the impact of high-volume natural pozzolan (volcanic ash) on the durability properties of alkali-activated natural pozzolan (NP) and limestone powder (LSP) mortar by immersion in 6% H_2SO_4 for 365 days. The samples were prepared with different binder ratios using alkaline activators (10 M $NaOH_{(aq)}$ and Na_2SO_4) combined in a 1:1 ratio and cured at 75 °C. NP was combined with the LSP at three different combinations: NP:LSP = 40:60 ($AAN_{40}L_{60}$), 50:50 ($AAN_{50}L_{50}$), and 60:40 ($AAN_{60}L_{40}$), representing low-volume, balanced, and high-volume binder combinations. Water absorption, weight change, and compressive strength were examined. The microstructural changes were also investigated using FTIR, XRD, and SEM/EDS characterization tools. Visual examination showed insignificant deterioration in the sample with excess natural pozzolan ($AAN_{60}L_{40}$) after 1 year of acid exposure, and the maximum residual strengths were 20.8 MPa and 6.68 MPa in $AAN_{60}L_{40}$ and $AAN_{40}L_{60}$ with mass gain (1.37%) and loss (10.64%), respectively. The high sulfuric acid resistance of $AAN_{60}L_{40}$ mortar was attributed to the high Ca/Si = 10 within the C-A-S-H and N-A-S-H formed. The low residual strength recorded in $AAN_{40}L_{60}$ was a result of gypsum formation from an acid attack of calcium-dominated limestone powder. The controlling factor for the resistance of the binder to acid corrosion was the NP/LSP ratio, whose factor below 0.6 caused significant debilitating effects.

Keywords: natural pozzolan; limestone powder; acid resistance; sulfuric acid; alkali activation

Citation: Al-Sodani, K.A.A.; Adewumi, A.A.; Mohd Ariffin, M.A.; Salami, B.A.; Yusuf, M.O.; Ibrahim, M.; AlAteah, A.H.; Al-Tholaia, M.M.H.; Shamsah, S.M.I.; Ismail, M. Acid Resistance of Alkali-Activated Natural Pozzolan and Limestone Powder Mortar. *Sustainability* **2022**, *14*, 14451. https://doi.org/10.3390/su142114451

Academic Editor: Ahmed Salih Mohammed

Received: 24 September 2022
Accepted: 1 November 2022
Published: 3 November 2022

Publisher's Note: MDPI stays neutral with regard to jurisdictional claims in published maps and institutional affiliations.

1. Introduction

Civil infrastructure deterioration due to industrial and biogenic acid attacks has been a cause of concern owing to the high cost of rehabilitation and retrofitting [1]. The resistance to acid of concrete structures is critical for long-term durability assessment and economic viability. In this context, cementitious materials with high acid resistance are required for various acid-exposed critical infrastructures to enhance their durability. Although the ordinary Portland cement (OPC) binder is known to be resistant to chemical attacks, it is highly susceptible to acid attack through either the decomposition and leaching of hydration products or the negative effect of newly formed compounds on the binder's internal microstructure [2]. To improve the durability performance of OPC binders, researchers are constantly striving to produce better-performing binders through the modification of OPC binders via novel chemical or mineral admixtures, among other approaches. Mineral admixtures such as fly ash, silica-fume, ground granulated blast furnace slag, rice husk ash, silicomanganese, etc., are commonly used to increase the long-term durability of concrete through the formation of a uniform and consolidated internal matrix [3–11].

The development of alkali-activated materials (AAMs), which have been studied and shown to have better acid resistance than OPC, is another strategy adopted by researchers to develop durable materials [12,13]. The choice of source material, which has been shown to affect an AAM's internal structure, has a substantial impact on the durability performance of the AAM [14]. Relatively new [15–17] and old [12,13] studies have also confirmed that AAM offers superior durability performance in comparison to OPC binders. Although it is reported that AAMs outperform OPC binder in terms of durability, the quality of binder products required to resist acid attack depends on the calcium level in its source material [13]. The difference in the quality of the produced binder is responsible for significant variance in the durability test outcomes. The mechanism of deterioration, for instance, with varying exposure concentrations of acid, is significantly different [12,13,15]. As a result, there has been considerable skepticism about alkali-activated materials' superior durability performance in surviving the harshness of acids during exposure. However, new [16–18] and old [12,13] studies have confirmed AAMs' superior durability performance compared to OPC materials.

In a recent study, Reu et al. [16] investigated the degradation of slag/fly ash mortar (mixed in varying proportions) in an acidic medium comprising phosphoric and sulfuric acids at pH of 2.5 ± 0.5 for 150 days of exposure. The findings indicated variation in the aggressiveness of acid attack, with phosphoric acid being the most aggressive, and that the degradation level of AAM was far less than that of OPC-based mortar. Chen et al. [15] studied the durability performance of pavement repair mortar materials developed from alkali-activated metakaolin–GGBFS (MK-GGBFS) and OPC upon exposure to a strong sulfuric acid solution (pH = 1). The superior durability of the binary blend of alkali-activated MK-GGBFS over OPC was due to the void-filling effect and the formation of alkali-activated binder gel (N-A-S-H and C-S-H). The degradation of the mortar in the sulfuric acid environment was attributed to the leaching of metal cations, the breakdown of aluminosilicate bonds, and the formation of microcracks forming gypsum within the matrix [19]. In another comparative study [20], the resistance to acid solution of alkali-activated glass cullet (AAGC) synthesized from three types of recycled glasses (flat, hollow, and windshield) was reportedly better than that of OPC.

Many new materials have crept into the durability research space of alkali-activated materials; some of these materials include waste ceramic tile powder (WCP) [21,22], light-burnt dolomite powder [23], waste glass powder [24–26], volcanic ash [26], iron-rich laterite soil [27], silico-manganese fumes [28], glass cullet [20], palm oil fuel ash (POFA) [18], etc. Despite the different precursors, varied acid concentrations, and exposure durations, AAM's better performances were consistent in the reviewed comparative durability studies. This has encouraged researchers to explore other types of aluminosilicate precursors. Furthermore, 75 wt % of the Earth's crust is made up of silica and alumina, allowing for a significant number of raw materials, which could either be natural materials or waste by-products, available to be explored for alkaline activation [28].

Natural pozzolan (NP) is a siliceous or siliceous and aluminous material with a polymeric silicon–oxygen–aluminum framework. The alkali activation of NP involves the use of alkali activators ($NaOH_{(aq)}$ and $Na_2SiO_{3(aq)}$) for the hydrolysis and dissolution of complex aluminosilicate compounds present in NP. The dissolution process leads to the release of monomers such as Ca-O, Si-O-Si, Al-O-Al, and Al-O-Si. This process is followed by the orientation of the oligomers, followed by the polycondensation of chains to form coagulated structures. Alkali activation of NP results in the formation of phillipsite ($(Na,K,Ca)_{1-2}(Si,Al)_8O_{16}.6(H_2O)$), CSH, calcium aluminum silicate hydrate ($(Ca_2Al_4Si_8O_{24})12H_2O$), C-A-S-H), and sodium aluminosilicate hydroxide hydrate ($Na_8(AlSiO_4)_6(OH)_2$-$4H_2O$), N-A-S-H); these binder gels have been reported to enhance the strength development of the AAM [29,30].

Natural pozzolan (NP) is abundantly available in volcanic regions and is rich in silica content [31]. However, NP is deficient in CaO, which could cause a delay in the setting time of the binder. This has necessitated the binary blending of volcanic ash natural pozzolan and other CaO-containing waste, such as slag and limestone waste powder (LSP) [29,32]. LSP is generated during the tile production process; about 30 million tons are generated per year in Turkey, while the U.K. generates 22.2 million tons of LSP annually [29]. The synergy between NP and LSP as binary precursors in the development of the alkali-activated binder has enhanced the mechanical properties by improving microstructures [33].

There has been successful deployment of natural pozzolan (pumice, perlite, and volcanic ash) as a base material in alkali-activated binder, where its effect (either solely or supplemented with other materials) on mechanical and durability performance improvement has been reported [34–41]. However, these studies, especially on the durability performance of alkali-activated concrete AAC, either with NP alone or in combination with supplementary materials, are still limited. There are several studies on AAC durability against sulfate, acid attacks, diffusion, permeability, depth of oxygen and chloride ion penetration in NP-based binders [42–45]. In one of the studies on chloride ion transport in NP-based AAC, a high amount of chloride ions was recorded in the solution after the rapid chloride permeability test irrespective of the voltage applied [37,44]. However, improved or enhanced durability performance was recorded in the durability studies where NP was blended with other materials. For instance, in a recent work by Aguirre-Guerrero et al. [43], the corrosion behavior of blended NP/GBFS-based reinforced AAC exposed to a chloride-bearing environment was studied. Higher resistance to chloride ion penetration was recorded compared to ordinary Portland cement-reinforced concrete used as a control. This better performance was attributed to improved gel composition, microstructural characteristics, and the matrix's capacity to bind the chloride ions before reaching the rebar surface.

In another NP blended study, Ibrahim et al. assessed the acid resistance of NP blended with nano-silica AAC subjected to 5% sulfuric acid. A reduction in specimen strength and weight was recorded with the blended NP in comparison with NP-based AAC and OPC-based concrete. The role played by the nano-silica, according to the authors, in achieving better microstructure leading to enhanced performance was due to hydration products such as C-A-S-H with increased absorption of aluminum ions into the binder matrix. The exposure to sulfuric acid resulted in the disintegration of the paste between the aggregates. Aside from sulfuric acid exposure, the curing temperature was also identified as a dominant factor in the performance of NP-based AAC. NP-based AAC cured at lower temperatures performed better during exposure to 5% sulfuric acid [39].

There have been studies conducted mostly on the fresh properties and strength development of AAC using NP and other emerging materials; an example is NP and limestone powder [33,45]. However, little is understood about its performance in terms of exposure to acid attacks. Thus, this research investigated the impact of high-volume volcanic ash natural pozzolan on the acid resistance of alkali-activated natural pozzolan and limestone powder. To explore the possibility of synthesizing a sustainable and durable blend of alkali-activated mortar using volcanic natural pozzolan and limestone powder, the hardened mortar was exposed to 6% sulfuric acid resistance for 3, 6, 9, and 12 months. The binder's morphology was examined through scanning electron microscopy–energy-dispersive spectroscopy (SEM/EDS), the difference in mineral phases was observed through X-ray diffraction (XRD), and changes in the frequency of vibration of different functional groups and their bond characteristics were studied using Fourier transform infrared (FTIR) spectroscopy. The findings of this study can reduce the waste that leads to environmental and health hazards.

2. Materials and Methods

2.1. Materials

2.1.1. Natural Pozzolan and Limestone Powder

The natural pozzolan (NP) used for this research was volcanic ash from the coast of the Red Sea, Saudi Arabia, and was commercially acquired from a local company. The limestone rock was mined in large sizes from a limestone quarry site in Riyadh (Saudi Arabia) and was transported to a tile-making factory in Hafr Al-Batin, Saudi Arabia. The dust generated during the cutting process of producing tiles was used for this research work. The two base materials were initially oven-dried at 105 ± 5 °C for 24 h, after which they were sieved through a No. 200 sieve. A particle size analyzer (HELOS (H3533) and QUIXEL brand) was used to determine the particle size distributions of the precursor materials. The analyzer worked using laser scattering principles. The precursors were dispersed in distilled water using the wetting method; then, the system was vigorously agitated to avoid sedimentation. The characteristics of the raw materials (NP and LSP) are given in Table 1. The specific surface areas (BET) of precursors were determined with Micromeritics ASAP2020 using nitrogen gas adsorption. The surface of the powdered sample was cleared of all adsorbed gasses (outgassing process). Then, the sample was cryogenically cooled using liquid nitrogen, followed by dosing of nitrogen gas into the system at reduced pressure. X-ray fluorescence (XRF) spectrometry was used to determine the elemental compositions of precursors. An XRF machine works by sending radiation waves to the samples in an intense X-ray beam from a radioisotope source. The primary source of rays excites the sample by detaching the tightly bound inner shell electrons from the excited atoms of the samples. When the excited atoms are relaxed to the original state, a fluorescent X-ray is emitted. The energies of these emitted rays are detected using an energy-dispersive detector that identifies the elemental traces in the sample, while the intensity of the X-rays is used to determine the quantity of the elements. X-ray diffraction (XRD) is a rapid and simple technique for non-destructive characterizations of crystalline materials. The results were analyzed by MATCH XRD software using the COD database. The chemical compositions of NP and LSP from X-ray fluorescence (XRF) spectrometry are shown in Table 2. The test provides information especially on structures, phases, preferred crystal orientations, and structural parameters. The mineralogical composition and phase nature of PMs were explored using an XRD Bruker instrument, model d2-Phaser, with Cu Ka radiation (40 kV, 40 mA) by continuous scanning within an angle 2-theta range of 4–80° and at a scan speed of 2.5°/min.

Table 1. Physical characteristics of NP and LSP.

Materials	Specific Gravity	Volume Mean Diameter (μm)	Specific Surface Area (cm^2/g)	d_{90} (μm)	d_{50} (μm)	d_{10} (μm)
LSP	2.7	12.1	0.6	31.0	6.4	1.20
NP	2.3	5.8	3.1	11.6	4.8	1.4

Table 2. Chemical constituents of NP and LSP.

Oxide Components (%)	CaO	SiO_2	Al_2O_3	Fe_2O_3	MgO	Na_2O	K_2O	L.O.I
NP	2.0	74.0	13.0	1.5	0.5	4.0	5.0	5.0
LSP	94.1	2.5	0.8	1.2	0.6	-	0.3	44.0

2.1.2. Alkaline Activators

The alkali activator used for this study is a combination of 10 M $NaOH_{(aq)}$ (NS) and $NaSiO_{3(aq)}$ (SS), the SS contains H_2O: 62.11%, SiO_2: 29.13% and Na_2O: 8.76% with silica modulus, and SiO_2/Na_2O of 3.3. It should be noted that sodium hydroxide was made a day in advance of mixing in order to allow the solution to cool.

2.1.3. Aggregates

The fine aggregate used for this study was desert sand from dunes that met ASTM C33's gradation size requirements, as shown in Table 3. The fine aggregates' specific gravity in saturated surface dry conditions was 2.63, and its fineness modulus was 2.6.

Table 3. Physical characteristics of NP and LSP.

S/N	Sieve Size (mm)	Cumulative Passing (%)
1	9.6	100
2	4.75	95
3	2.63	81
4	1.18	50.5
5	0.6	25.4
6	0.3	6.2
7	0.15	1.9

2.1.4. Acid Solution

The acidic solution used for the durability test was concentrated 6% $H_2SO_{4(aq)}$ solution supplied by Ajax Finechem Pty Ltd. with a molar mass of 98.08 g and specific gravity of 1.84. The purity level was 96.5%.

2.2. Experimental Program

2.2.1. Mix Design for the Mortar Preparation

The mortar was produced with three different precursor combinations (NP:LSP = 40:60, 50:50, and 60:40). The mortar was developed using a mass ratio of $Na_2SiO_{3(aq)}/10$ M $NaOH_{(aq)}$ (SS/NH) = 1.0. The mixes were designated as AAN_xL_{100-x}, where x represents the percentage of NP present in the mix. The mix was designed to study the effect of a high volume of LSP ($AAN_{40}L_{60}$), an equal volume of NP and LSP ($AAN_{50}L_{50}$), and a high volume of NP ($AAN_{60}L_{40}$). A fine aggregate to binder ratio of 2.0 was maintained for all mixes. The ($NaOH_{(aq)}$ + $Na_2SiO_{3(aq)}$)/(NP + LSP) ratio was maintained at 0.5, while free water was 10%. Table 4 depicts the mix proportions used for this research.

Table 4. Mix proportions in kg/m^3 of the developed mortar.

Mix No.	Mix ID	NP	LSP	NH Molarity	SS/NH	SS	SH	H_2O	Fine Aggregate
M1	$AAN_{60}L_{40}$	363.0	242.0	10	1.0	151.5	151.5	60.5	1210
M2	$AAN_{50}L_{50}$	302.5	302.5	10	1.0	151.5	151.5	60.5	1210
M3	$AAN_{40}L_{60}$	242.0	363.0	10	1.0	151.5	151.5	60.5	1210

2.2.2. Mortar Preparation

The required dry quantities of materials, shown in Table 4, were mixed with the Liya 5.0 L planetary bench mixer. The precursor was mixed with the fine aggregate for 3 min. The alkaline activators (10 $NaOH_{(aq)}$ + $Na_2SiO_{3(aq)}$) and water were mixed to achieve a homogeneous mixture. The mix was then placed in a cube mold of 50 × 50 × 50 mm^3, conforming to ASTM C109, followed by consolidation using a Liya vibrating table. The surface was then meticulously smoothed with a trowel, covered with a polythene sheet to stop moisture evaporation, and finally oven-dried for 24 h at 75 °C. The demolded samples were subsequently cured over the course of 28 days by curing at a laboratory temperature of 20 ± 5 °C for strength development. The samples were then immersed in 6% $H_2SO_{4(aq)}$ for 12 months and tested for 3-, 6-, 9-, and 12-month residual strength and change in mass. The sample morphology, bond characterization, and mineral phases after immersion were examined using SEM/EDX, FTIR, and XRD analysis, respectively.

2.2.3. Evaluation of the Resistance to Acid

To evaluate the resistance of the developed alkaline-activated mortar to acid attack, a cubic sample of 50 × 50 × 50 mm^3 was used. The impact of acid on the mortar under study was investigated using 6% H_2SO_4, which was prepared using distilled water. Before being submerged in the solution, the 28-day-old specimens were weighed. The samples were immersed for 365 days and the solution was changed every month to maintain a pH of 1 throughout the test. The performance of geopolymer mortar specimens was evaluated using physical observation and by measuring the loss in weight and the residual strength.

2.3. *Evaluation Methods*

2.3.1. Water Absorption

Absorption is a measure of pore spaces present in the matrix of a specimen. The determination of water absorption is usually performed by controlling the drying of specimens in an oven until a constant mass is achieved, which signifies the complete removal of existing moisture. Water absorption of binder specimens was determined following ASTM C-642, with the exception that the samples were dried in an oven at 75 °C (similar to curing temperature) for 24 h and their masses were recorded. The samples were then placed in water for 24 h and the masses of the samples were measured in surface-dried conditions. The water absorption was evaluated using Equation (1).

$$\text{Water absorption } (\%) = \frac{m_2 - m_1}{m_1} \times 100 \tag{1}$$

where m_1 is the average mass of the oven-dried specimen in air and m_2 is the average mass of the surface-dried specimen after immersion.

2.3.2. Weight Loss

To study the resistance to acid attack, the cubic sample of 50 × 50 × 50 mm^3 of known original weight was fully immersed in 6% $H_2SO_{4(aq)}$ after 28 days of air-curing for 12 months. The solutions were changed every 30 days to keep the acidic solution's concentration at a pH level of 1. The loss in weight after 3, 6, 9, and 12 months of immersion was measured by sensitive weighing balance upon drying the sample surface water using a dry towel to ensure dry surface conditions. Three replicates were used for each measurement, and the average was determined. The results are recorded as shown in Equation (2).

$$\text{Percentage weight change } (\%) = \frac{W_2 - W_1}{W_2} \times 100 \tag{2}$$

where W_1 is the average surface-dried mass after immersion and W_2 is the average air-dried specimen before immersion.

2.3.3. Compressive Strength

The mortar compressive strength was determined according to ASTM C150 using a Liya compression testing machine with a uniform loading rate of 0.9 kN/s. The testing on three replicates was determined after 28 days of curing and 3, 6, 9, and 12 months after immersion in acidic solution. The loss in the compressive strength was determined using Equation (3).

$$\text{Loss in compressive strength (MPa)} = \frac{\sigma_2 - \sigma_1}{\sigma_2} \times 100 \tag{3}$$

where σ_1 is the average loss in compressive strength of the specimen before immersion and σ_2 is the average loss in compressive strength of the specimen after immersion.

2.3.4. Binder Products Characterization

The morphology of the binders before and after immersion in the acidic solution was examined with the JSM-5800LV SEM device using an accelerating voltage of 20 kV, while the elemental composition of each spectrum was determined with an energy-dispersive X-ray spectroscopic analyzer. The bonds and functional groups of the binder products were characterized using Perking Elmer 880 (KBr pellet technique) Fourier transform infrared (FTIR) spectroscopy.

3. Results and Discussion

3.1. Characterization of Precursors (NP and LSP)

Figure 1 shows the particle size distribution (PSD) curves for NP and LSP. LSP (Figure 1, LSP) had a mean particle size of 12.1 μm, whereas NP powder had a mean particle size of 5.8 μm (Figure 1, NP). Critical observation of the SEM shown in Figure 2 reveals that the geometry of the particles of the materials differs significantly. LSP is spherical in shape while NP is flaky. This points to the fact that the preponderance of NP over LSP will enhance the permeability due to the presence of more interstitial pores within the matrix of the binder, as shown in Figure 2. The excessive absorption in NP-dominated binder is equally evident in Figure 2. The NP contains a diffuse hump at $2\theta \approx 23°$, which indicates the amorphous nature of NP. NP consists of plagioclase (Ca, Na) $Al_2Si_2O_8$, microcline ($KAl_2Si_2O_8$), and quartz (SiO_3), as revealed in Figure 3. LSP is crystalline in nature and mostly contains quartz (SiO_2) with calcite ($CaCO_3$). The activation of NP and LSP resulted in the formation of N-A-S-H and C-A-S-H products, while the unreactive silica acts as a filler, thereby reducing the porosity of the resulting binder [29].

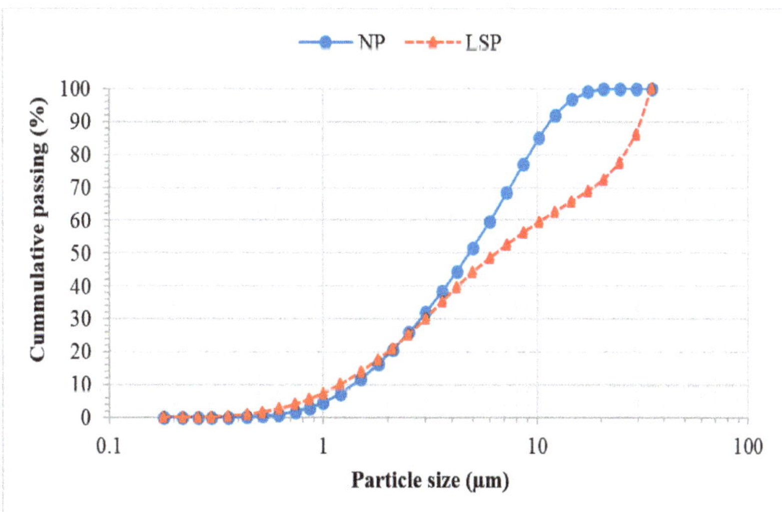

Figure 1. NP and LSP particle size distribution.

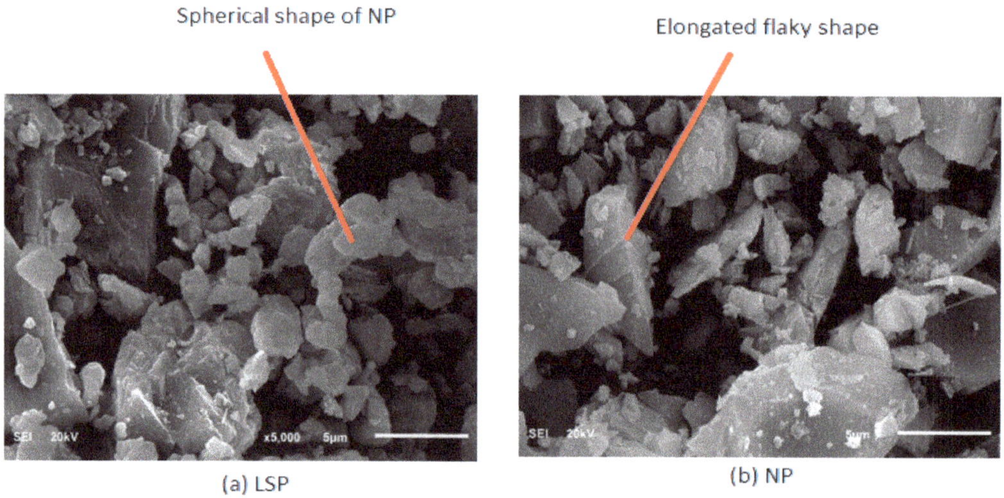

Figure 2. Micrograph images.

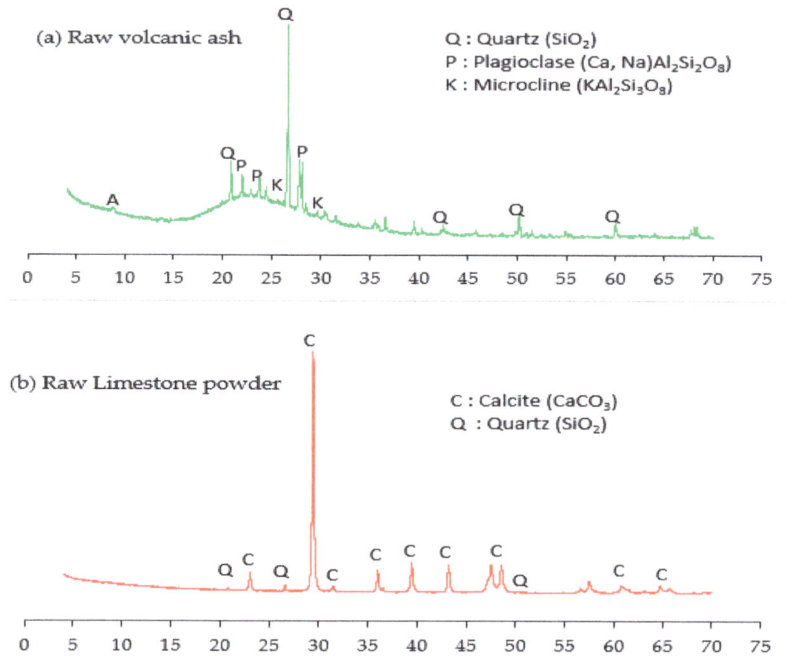

Figure 3. XRD: (**a**) NP and (**b**) LSP.

3.2. Effect of Binder Combination on Absorption

Figure 4 shows the absorption rate of the alkaline-activated mortar with different NP/LSP binder ratios ($AAN_{40}L_{60}$, $AAN_{50}L_{50}$, $AAN_{60}L_{40}$). The absorption rate increased as the immersion duration increased but decreased as the NP/LSP ratio increased such that the increment in NP/LSP from 0.67 to 1.0 and 1.5 led to 3.16% and 6.8% moisture contents, respectively. The sample with the high-volume NP ($AAN_{60}L_{40}$) had the highest absorption value; this can be attributed to the higher specific surface area of NP (3.1 cm^2/g) (Table 1) and the angular flaky shape of LSP, as revealed in the SEM results shown in Figure 2. The preponderance of NP over LSP ($AAN_{60}L_{40}$) makes the mortar more porous, thereby increasing its absorption capacity.

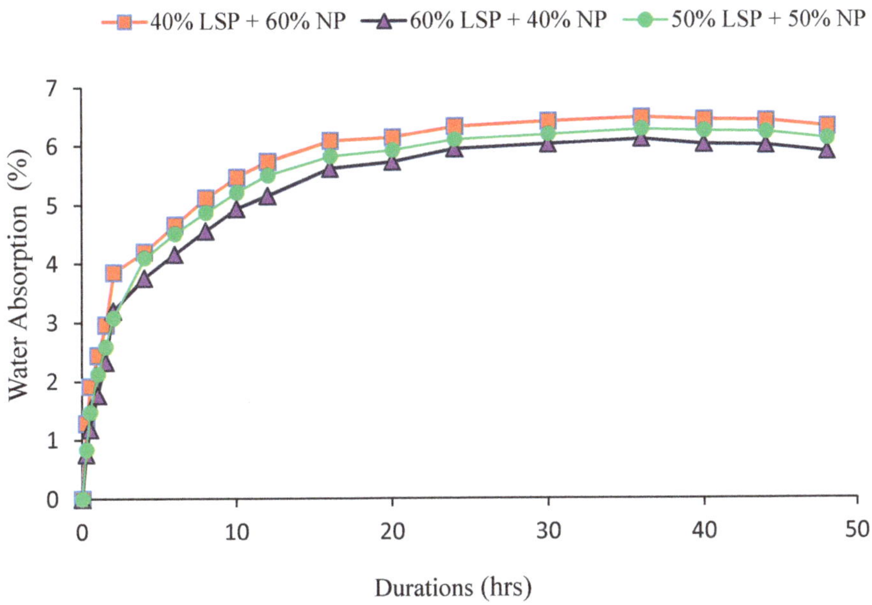

Figure 4. Effect of binder combination on absorption.

3.3. Visual Characterization in Acidic Exposure

The visual examination of the AANL mortar mixtures after one year of exposure to an acid solution (6% H_2SO_4) is depicted in Figure 5. It was observed that $AAN_{60}L_{40}$ exhibited little surface deterioration after 12 months in acid solution, unlike $AAN_{50}L_{50}$, whose surface became gritty after 12 months of exposure with minor surface distortion. When the NP content was further reduced to 40% ($AAN_{40}L_{60}$), the sample exhibited major surface deterioration and cracks with edge delamination. This implies that the sample with a high volume of NP showed better resistance to acid attack than the sample with a high volume of limestone powder.

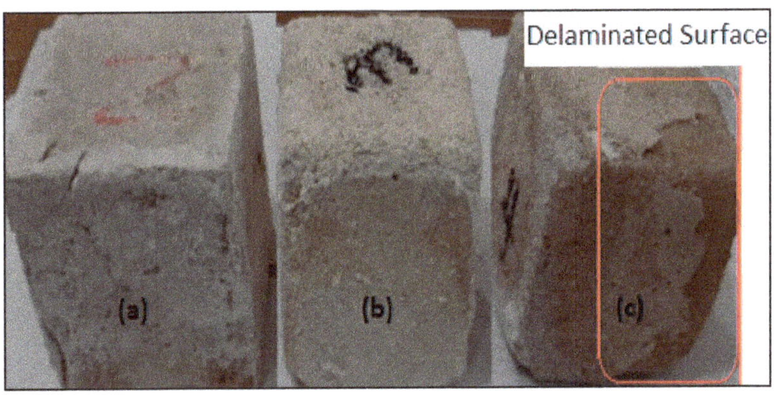

Figure 5. Visual examination of samples exposed to 6% H_2SO_4 attack: (**a**) $AAN_{60}L_{40}$, (**b**) $AAN_{50}L_{50}$, and (**c**) $AAN_{40}L_{60}$.

3.4. Effect of Binder Combination on Change in Weight of acid Exposed Specimen

Figure 6 shows the samples exposed to 6% H_2SO_4 for 3 months; all the specimens, i.e., $AAN_{40}L_{60}$, $AAN_{50}L_{50}$, and $AAN_{60}L_{40}$, experienced weight gain. However, after 6 months of exposure, the samples experienced weight reduction by 23%, 31.12%, and 12.7%, respectively. After 9 months of exposure, $AAN_{40}L_{60}$ and $AAN_{50}L_{50}$ experienced substantial losses in weight, while $AAN_{60}L_{40}$ exhibited marginal weight loss. $AAN_{40}L_{60}$ and $AAN_{50}L_{50}$ deteriorated further after 12 months, with weight loss of 10.59% and 7.62%, respectively, while $AAN_{60}L_{40}$ maintained its stability. This shows that high-volume natural pozzolan with high silica content enhanced the corrosion resistance against acid attack.

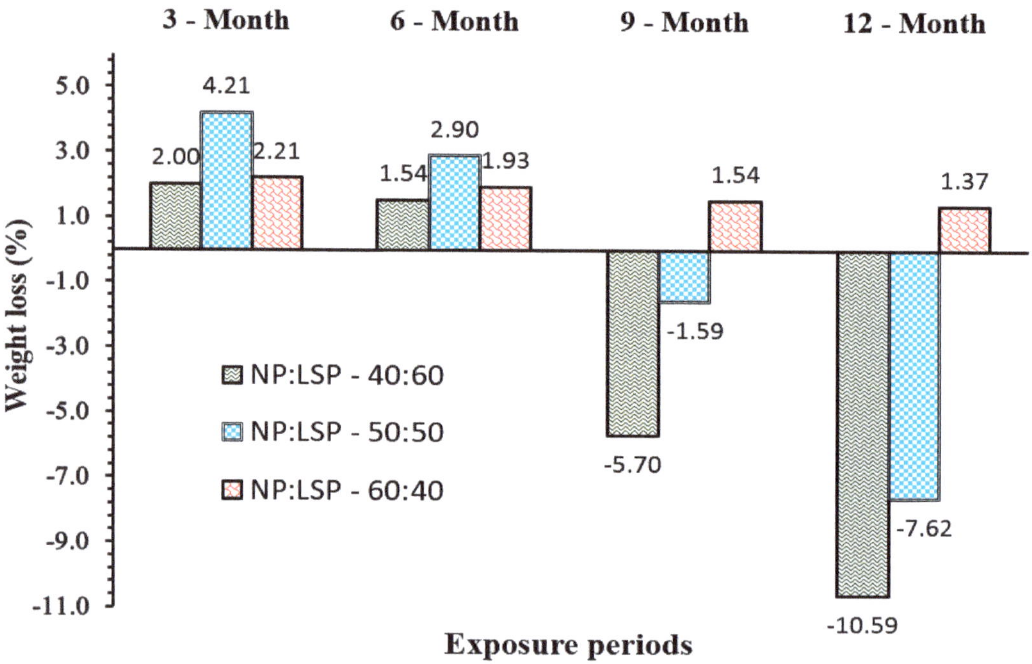

Figure 6. Change in weight of samples exposed to 6% H_2SO_4.

3.5. Residual Compressive Strength after Acid Attack

The residual compressive strengths of mortar mixtures exposed to 6% H_2SO_4 for 3, 6, 9, and 12 months are shown in Figure 7. Comparing the strength values of 28 days to various months of exposure, it was observed that for all samples exposed to acid attack (Figure 7), there was a sharp reduction in the compressive strength recorded throughout the exposure period, except for $AAN_{60}L_{40}$. The $AAN_{40}L_{60}$ sample experienced the highest residual strength loss of 75.56%, as shown in Figure 8, while $AAN_{60}L_{40}$ had the lowest loss of 16.80% after one year of exposure. At the balanced proportion of NP and LSP, the strength loss was 34%, 42%, 53%, and 69% for 3, 6, 9, and 12 months, respectively (Figure 8). There was no significant difference between the strength loss at 6 and 9 months in $AAN_{60}L_{40}$. Generally, the sample synthesized with a high volume of limestone (LSP) powder showed less resistance to acid attack, whereas $AAN_{60}L_{40}$ with a high volume of natural pozzolan (NP) exhibited the highest resistance to acid attack.

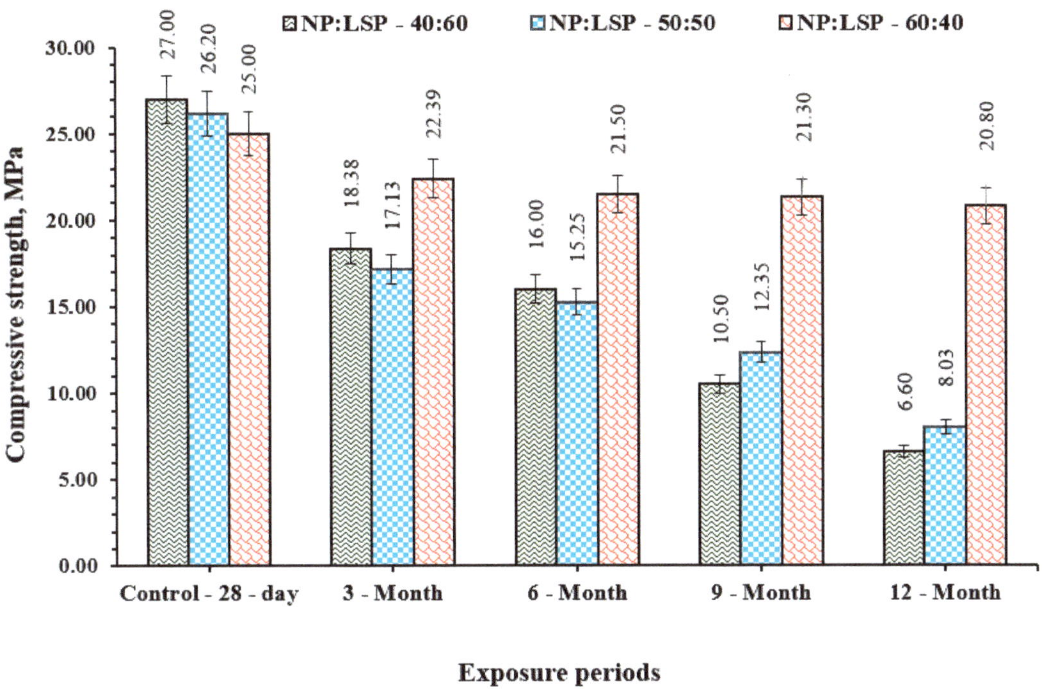

Figure 7. Compressive strength of samples exposed to 6% H_2SO_4.

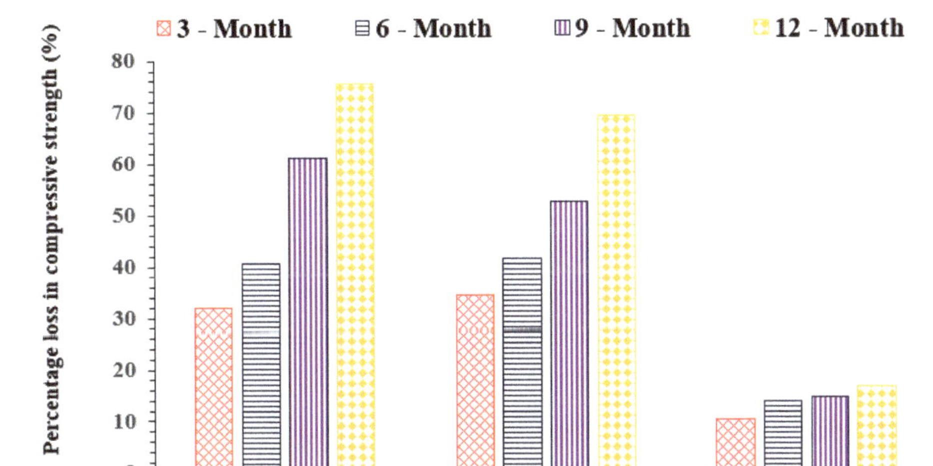

	AAN40L60	AAN50L50	AAN60L40
3 - Month	31.9	34.62	10.44
6 - Month	40.74	41.79	14
9 - Month	61.11	52.86	14.8
12 - Month	75.56	69.35	16.8

Figure 8. Percentage loss in compressive strength with 1 year of exposure to acid attack.

3.6. Mineralogy, Morphology, and Bond Characteristics of Mortar after Exposure to Acid Environment

3.6.1. XRD Characterization of AANL Binder

The XRD spectra of the exposed and unexposed binder are depicted in Figure 9. $AAN_{60}L_{40}$, $AAN_{50}L_{50}$, and $AAN_{40}L_{60}$ have almost similar microstructural characteristics before exposure to a severe acidic environment, as reflected in the compressive strength results (Figure 7); thus, $AAN_{40}L_{60}$ is used as the control for the unexposed samples. The alkaline activators can dissolve complex aluminosilicate present in the base materials to release monomers that come together to form the product through condensation, as shown in Equations (4)–(6). The alkaline-activated products formed are gehlenite ($Ca0.Al_2O_3.SiO_2$), anorthite ($CaAl_2Si_2O_8$), and albite ($NaAlSi_3O_8$) (Figure 9a).

$$CaCO_3 \rightarrow CaO + CO_2 \quad (4)$$

$$Na_2SiO_{3(aq)} \rightarrow Na_2O + SiO_2 + H_2O \quad (5)$$

$$NaOH_{(aq)} \rightarrow Na_2O + SiO_2 + H_2O \quad (6)$$

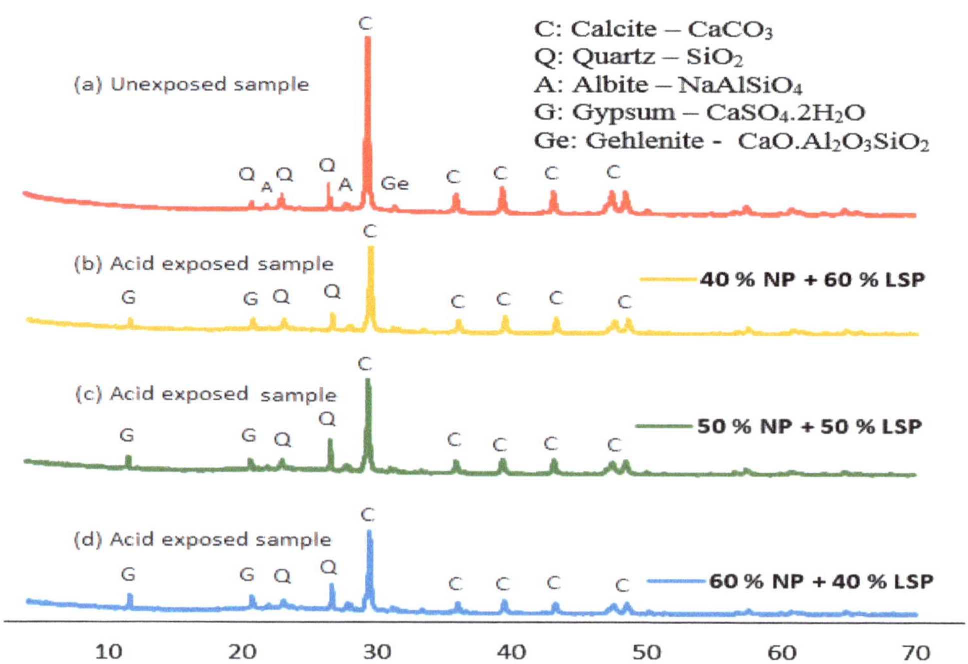

Figure 9. X-ray diffractograms of the exposed and unexposed samples.

These products enhanced the strength development, but exposure to sulfuric acid destroyed phases such as gehlenite and albite in the LSP-dominated binder, as shown in Figure 9. The destruction of gehlenite and albite could be due to ion exchange between hydroxonium ion of an acid and calcium in LSP to form gypsum, as shown in Equation (7).

$$CaCO_3 + H_2SO_4 + H_2O \rightarrow CaSO_4.2H_2O + CO_2 \quad (7)$$

The formation of gypsum on all the tested binders after acidic exposure is evidenced in Figure 9b,c. The gypsum formed caused leaching of the alkali-activated product, subsequently leading to mass and strength reduction in the mortar; however, this effect is not prominent in the binder developed with a high content of NP ($AAN_{60}L_{40}$).

3.6.2. FTIR Analysis of Samples after Acid Attack

Figure 10 shows the FTIR image of the mortar before and after exposure to sulfuric acid attack. The major functional groups affected were the hydroxyl group, binding water present within the pores, and Si-O-Si reorganization. The wavenumbers at 3500 cm^{-1} and 2375 cm^{-1} showing the presence of hydroxyl group (O-H) stretching and water molecule (H-O-H) bending at the wavenumbers of 1645 cm^{-1} were observed in the unexposed samples (Figure 10a). A stretching vibration of C-O-O (CO_3^{2-}) was observed at a wavenumber of 1418 cm^{-1} in the unexposed sample. However, upon exposure to the acidic environment, the broad peak of the hydroxyl group was decomposed due to acid attack, as revealed in Figure 10b–d. Furthermore, the H-O-H bending at wavenumber 2335 cm^{-1} in the unexposed sample changed to a deeper peak due to the precipitation of more water molecules entrapped in the binder matrix due to $H_2SO_{4(aq)}$. The carbonyl group around 1427 cm^{-1} in the unexposed sample was observed to be depleted after acid attack.

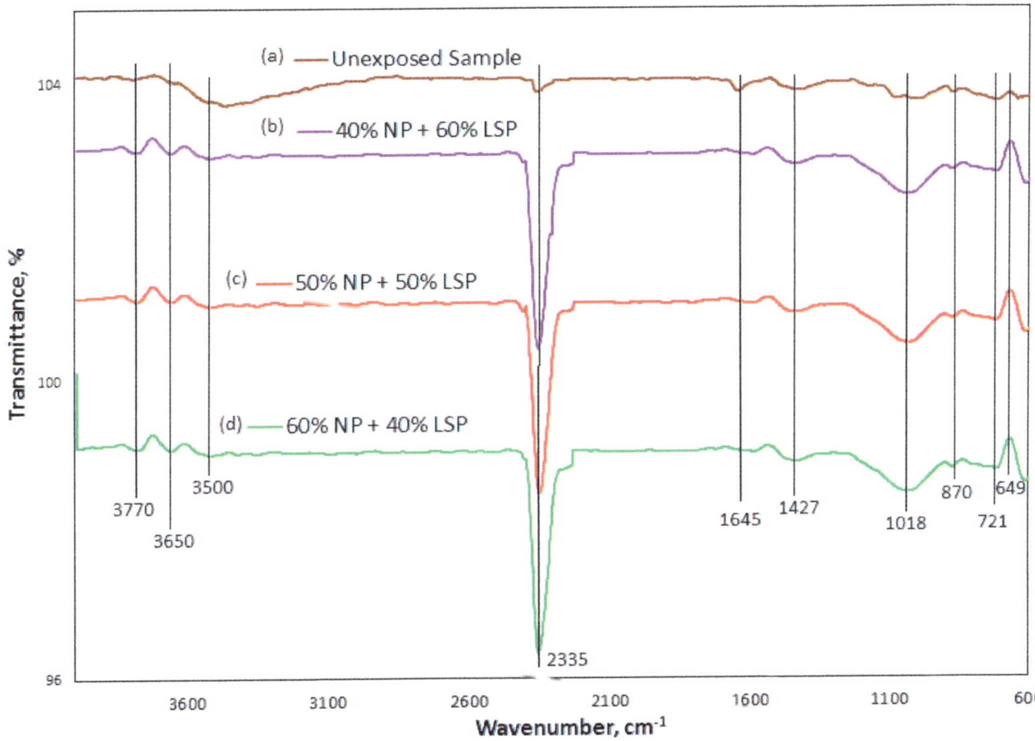

Figure 10. FTIR spectra of specimen unexposed and exposed to 6% H_2SO_4.

Furthermore, the intensity band of Si-O-T from the aluminosilicate framework at 1018 cm^{-1} due to asymmetric vibration, which corresponds to binder gel, becomes broadened after acid attack. This implies that the depolymerization of C-A-S-H and N-A-S-H products occurred due to the elimination of Al and Si atoms from the binder. A weak absorption band of dihydrate gypsum (S-O) at a wavenumber of 649 cm^{-1} was also found in the exposed samples, whereas it was absent in the unexposed samples. For the unexposed sample, in-plane bending vibration of C-O was found in a weak absorption peak of 721 cm^{-1}, and this was shifted to a lower band of 711 cm^{-1}. The high strength loss recorded in $AAN_{60}L_{40}$ and $AAN_{50}L_{50}$ exposed samples is caused by the formation of gypsum after decalcification of the binder matrix, while the marginal strength loss in

$AAN_{60}L_{40}$ is due to the presence of high silica content that hindered high degradation of the aluminosilicate framework.

3.6.3. SEM and EDX Analysis of Samples after Acid Attack

Figures 11–14 depict the SEM and EDS findings for the AANL binder under acidic and non-acidic conditions. The control sample's SEM image showed a uniform and denser microstructure as a result of more alkaline-activated product production (Figure 11). From the EDS (unexposed sample) results, the Si/Na values of 0.96–1.4, Si/Ca = 0.84–2.33 and Si/Al = 4.84–7.11 were noted in spectra 5 and 6 (Figure 11). This indicates the formation of aluminosilicate products such as C-A-S-H and N-A-S-H, thereby enhancing strength development.

However, after exposure to 6% H_2SO_4 for 1 year, spectrum 4 of $AAN_{60}L_{40}$ (Figure 14) exposed to acid attack showed the presence of gypsum, C-A-S-H, and N-A-S-H products with a dense microstructure. The exposed sample EDS results show that the Si/Al ratio decreases with an increased in the percentage of NP. The highest value of Si/Na = 8.7, was observed on spectrum 17 for $AAN_{40}L_{60}$ (Figure 13); Si/Na = 7.2 for $AAN_{50}L_{50}$, as shown in spectrum 10 (Figure 14) and Si/Na = 6.6 for $AAN_{60}L_{40}$, as depicted in spectrum 3 (Figure 14). Furthermore, there was a higher Ca/Si ratio of 10.5 in $AAN_{60}L_{40}$, as revealed in spectrum 3, than in $AAN_{50}L_5$ (spectrum 10), with Ca/Si = 0.5. For $AAN_{40}L_{60}$, as shown in spectrum 18, Ca/Si = 2.1. $AAN_{60}L_{40}$ with the highest value of Ca/Si and the lowest value of Si/Al had the highest resistance to H_2SO_4, with residual compressive strength of 20.8 MPa after 1 year of exposure to acid attack. $AAN_{40}L_{60}$ and $AAN_{50}L_{50}$ have a porous microstructure, as shown in the magnified micrographs (Figures 13b and 14b), which accounted for the loss in strength recorded.

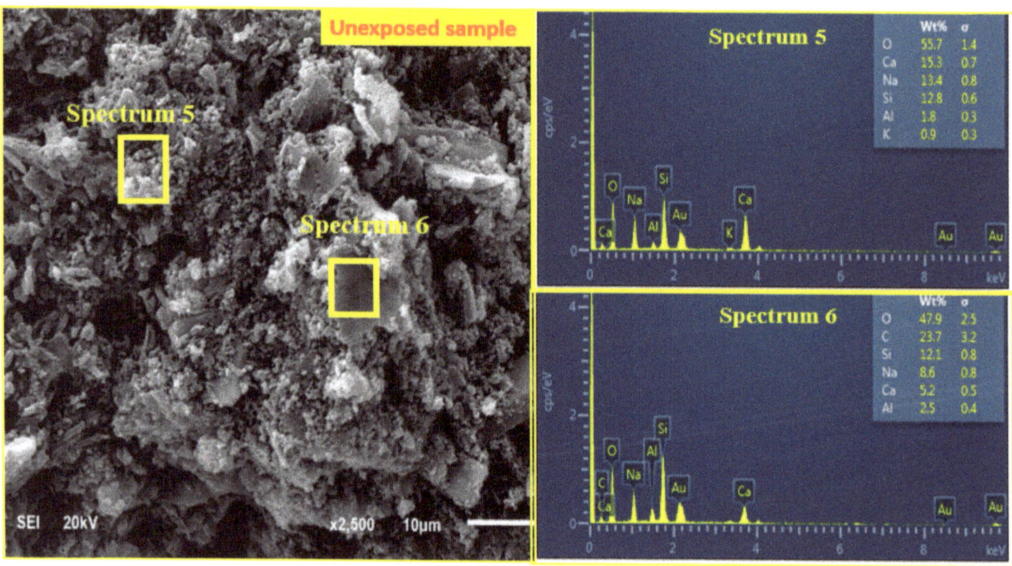

Figure 11. SEM and EDS of unexposed control sample.

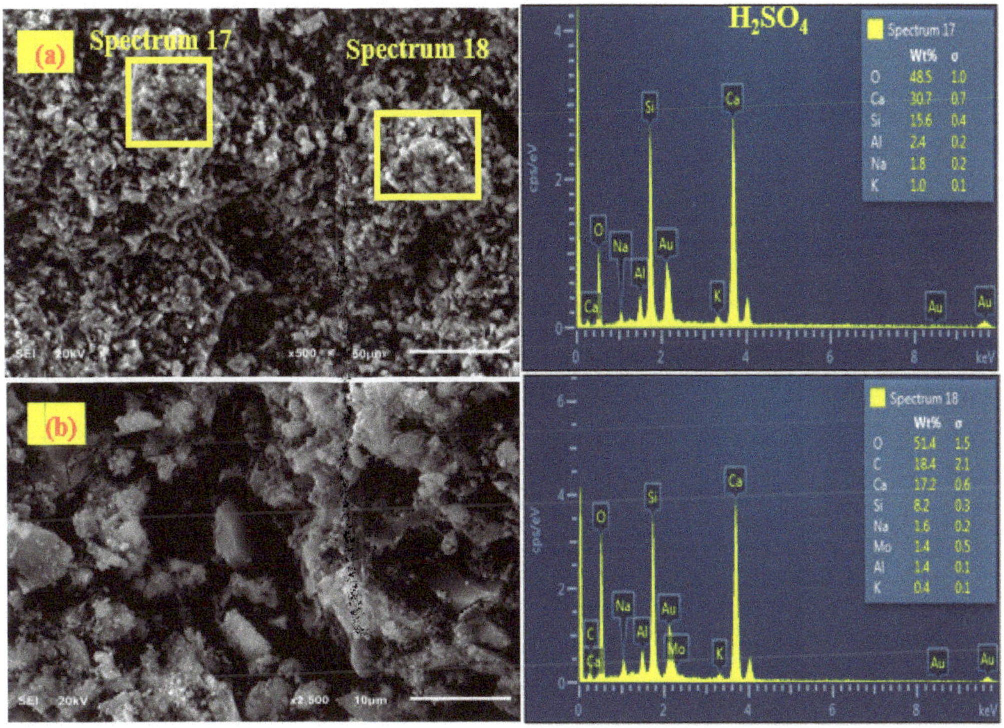

Figure 12. SEM and EDS of $AAN_{40}L_{60}$ exposed to 6% H_2SO_4: (**a**) low magnification; (**b**) high magnification.

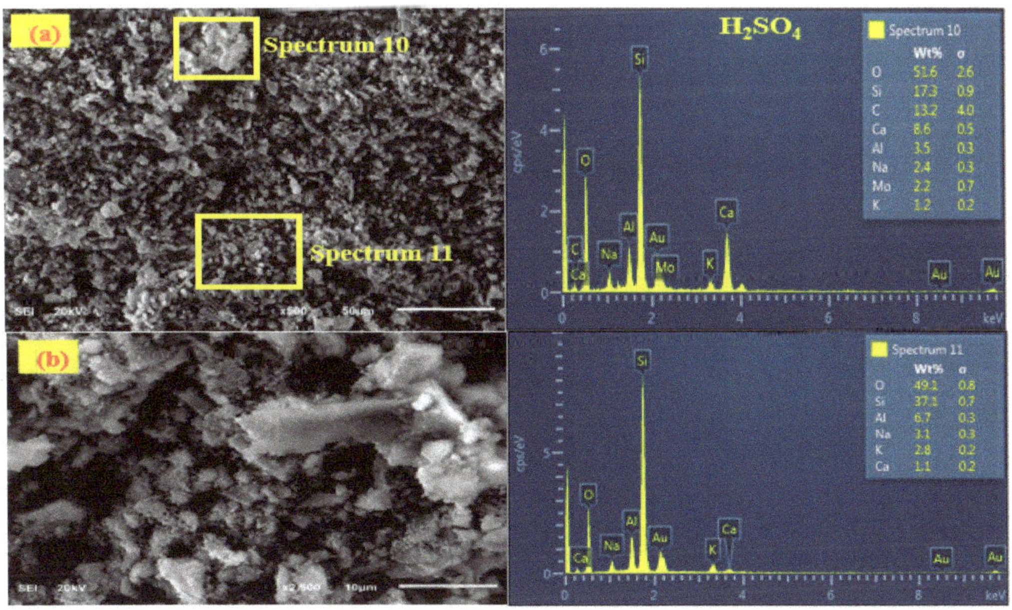

Figure 13. SEM and EDS of $AAN_{50}L_{50}$ exposed to 6% H_2SO_4: (**a**) low magnification; (**b**) high magnification.

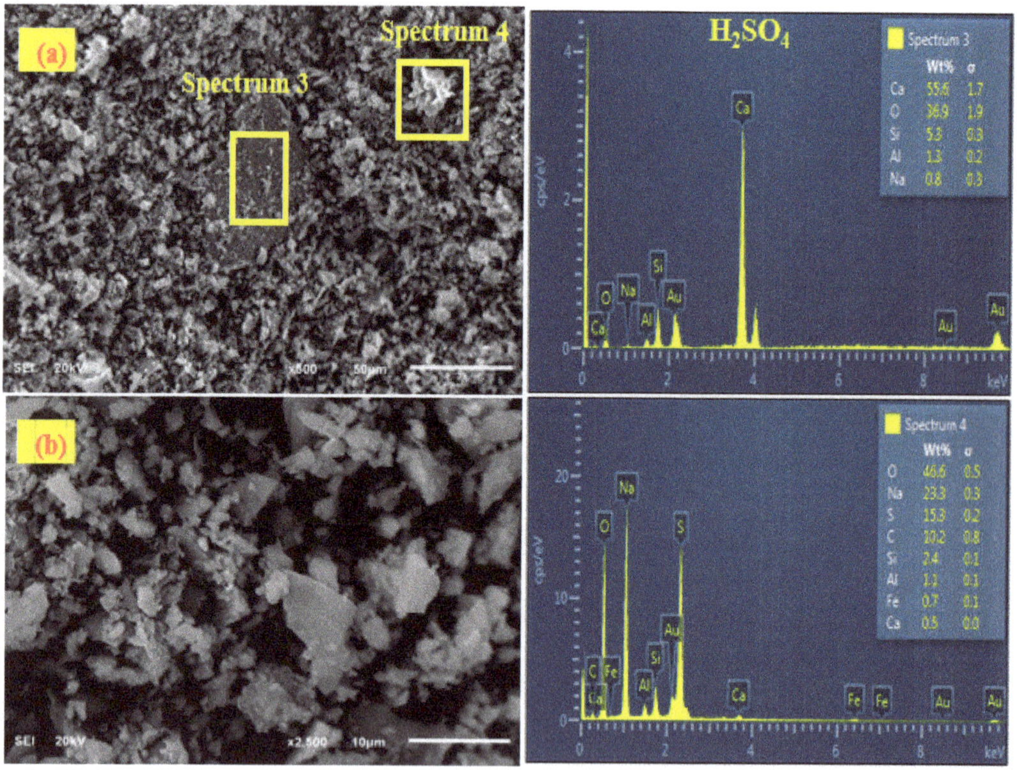

Figure 14. SEM and EDS of $AAN_{60}L_{40}$ exposed to 6% H_2SO_4: (**a**) low magnification; (**b**) high magnification.

4. Conclusions

The impacts of high-volume natural pozzolan on the sulfuric acidic resistance of alkali-activated mortar after 1 year of exposure to sulfuric acid are summarized as follows.

1. The mortar exhibited more expansion in the presence of a low volume of NP.
2. The maximum strengths of 20.8 MPa and 6.68 MPa were noted in mortar developed using a high volume of NP ($AAN_{60}L_{40}$) and a low volume of NP ($AAN_{40}L_{60}$), respectively.
3. The ultimate residual strengths were 16.8% and 1.4% for $AAN_{60}L_{40}$ and $AAN_{40}L_{60}$, respectively.
4. The mass gains were 75.6% and −10.64% for $AAN_{60}L_{40}$ and $AAN_{40}L_{60}$, respectively.
5. Samples synthesized using $AAN_{60}L_{40}$ (60% NP:40% LSP) exhibited no surface deterioration, while samples synthesized using $AAN_{40}L_{60}$ (40% NP:60% LSP) exhibited major surface cracks with minor multiple-edge delamination.
6. The high sulfuric acid resistance of $AAN_{60}L_{40}$ mortar is attributed to the presence of high values of Si/Al = 7 and Ca/Si = 10 present in C-A-S-H and N-A-S-H products, which resulted in the pore-filling effects within the microstructure.
7. The low strength recorded in $AAN_{40}L_{60}$ was due to the formation of gypsum in the binder product, whereas samples synthesized with a high volume of natural pozzolan ($AAN_{60}L_{40}$) showed more stability against acid attack.
8. $AAN_{60}L_{40}$ exhibited the highest resistance to sulfuric acid attack.

Author Contributions: Data curation, B.A.S., A.H.A. and M.I. (Mohammed Ismail); Formal analysis, K.A.A.A.-S. and M.M.H.A.-T.; Funding acquisition, K.A.A.A.-S.; Investigation, K.A.A.A.-S., A.A.A., M.I. (Mohammed Ibrahim) and M.I. (Mohammed Ismail); Methodology, M.A.M.A., A.H.A., M.M.H.A.-T. and M.I. (Mohammed Ismail); Project administration, A.A.A., M.O.Y. and A.H.A.; Supervision, S.M.I.S.; Validation, M.M.H.A.-T.; Writing—original draft, K.A.A.A.-S., A.A.A., B.A.S. and M.I. (Mohammed Ibrahim); Writing—review & editing, M.A.M.A., M.O.Y. and S.M.I.S. All authors have read and agreed to the published version of the manuscript.

Funding: The authors extend their appreciation to the Deputyship for Research & Innovation, Ministry of Education in Saudi Arabia, for funding this research work through project no. IFP-A-2022-2-1-06.

Informed Consent Statement: Not applicable.

Data Availability Statement: The raw data required to reproduce these findings are available in the in Sections 2.1 and 2.2 of this manuscript.

Acknowledgments: The authors appreciate the continuous support of the University of Hafr Al Batin.

Conflicts of Interest: The authors declare no conflict of interest.

References

1. Godinho, J.P.; de Medeiros, M.H.F. Biogenic sulfur attack in a reinforced concrete sewage treatment plant. Re-visited mechanism and rehabilitation proposal. *Eng. Fail. Anal.* **2021**, *124*, 105354. [CrossRef]
2. Madhuri, P.; Rao, B.K.; Chaitanya, A. Improved performance of concrete incorporated with natural zeolite powder as supplementary cementitious material. *Mater. Today Proc.* **2021**, *47*, 5369–5378. [CrossRef]
3. Mustapha, F.A.; Sulaiman, A.; Mohamed, R.; Umara, S. The effect of fly ash and silica fume on self-compacting high-performance concrete. *Mater. Today Proc.* **2020**, *39*, 965–969. [CrossRef]
4. Du, H.; Pang, S.D. Long-Term Influence of Nanosilica on the Microstructures, Strength, and Durability of High-Volume Fly Ash Mortar. *J. Mater. Civ. Eng.* **2021**, *33*, 04021185. [CrossRef]
5. Liu, S.; Zhu, M.; Ding, X.; Ren, Z.; Zhao, S.; Zhao, M.; Dang, J. High-Durability Concrete with Supplementary Cementitious Admixtures Used in Corrosive Environments. *Crystals* **2021**, *11*, 196. [CrossRef]
6. Kumar, M.; Sinha, A.K.; Kujur, J. Mechanical and durability studies on high-volume fly-ash concrete. *Struct. Concr.* **2020**, *22*, E1036–E1049. [CrossRef]
7. Huseien, G.F.; Joudah, Z.H.; Khalid, N.H.A.; Sam, A.R.M.; Tahir, M.M.; Lim, N.H.A.S.; Alyousef, R.; Mirza, J. Durability performance of modified concrete incorporating fly ash and effective microorganism. *Constr. Build. Mater.* **2021**, *267*, 120947. [CrossRef]
8. AlBiajawi, M.I.; Embong, R.; Muthusamy, K. An overview of the utilization and method for improving pozzolanic performance of agricultural and industrial wastes in concrete. *Mater. Today Proc.* **2021**, *48*, 778–783. [CrossRef]
9. Raheem, A.; Abdulwahab, R.; Kareem, M. Incorporation of metakaolin and nanosilica in blended cement mortar and concrete- A review. *J. Clean. Prod.* **2021**, *290*, 125852. [CrossRef]
10. Elavarasan, S.; Priya, A.; Ajai, N.; Akash, S.; Annie, T.; Bhuvana, G. Experimental study on partial replacement of cement by metakaolin and GGBS. *Mater. Today Proc.* **2020**, *37*, 3527–3530. [CrossRef]
11. Tambe, Y.; Nemade, P. Performance of Conventional Concrete Integrated with RHA and GGBS As a Cementitious Material. *Techno Soc.* **2020**, *2021*, 341–353. [CrossRef]
12. Sata, V.; Sathonsaowaphak, A.; Chindaprasirt, P. Resistance of lignite bottom ash geopolymer mortar to sulfate and sulfuric acid attack. *Cem. Concr. Compos.* **2012**, *34*, 700–708. [CrossRef]
13. Bakharev, T. Resistance of geopolymer materials to acid attack. *Cem. Concr. Res.* **2005**, *35*, 658–670. [CrossRef]
14. Tho-In, T.; Sata, V.; Chindaprasirt, P.; Jaturapitakkul, C. Pervious high-calcium fly ash geopolymer concrete. *Constr. Build. Mater.* **2012**, *30*, 366–371. [CrossRef]
15. Chen, K.; Wu, D.; Yi, M.; Cai, Q.; Zhang, Z. Mechanical and durability properties of metakaolin blended with slag geopolymer mortars used for pavement repair. *Constr. Build. Mater.* **2021**, *281*, 122566. [CrossRef]
16. Ren, J.; Zhang, L.; Nicolas, R.S. Degradation of Alkali-Activated Slag and Fly Ash Mortars under Different Aggressive Acid Conditions. *J. Mater. Civ. Eng.* **2021**, *33*, 04021140. [CrossRef]
17. Wang, A.; Zheng, Y.; Zhang, Z.; Liu, K.; Li, Y.; Shi, L.; Sun, D. The Durability of Alkali-Activated Materials in Comparison with Ordinary Portland Cements and Concretes: A Review. *Engineering* **2020**, *6*, 695–706. [CrossRef]
18. Salami, B.A.; Johari, M.A.M.; Ahmad, Z.A.; Maslehuddin, M. POFA-Engineered Alkali-activated Cementitious Composite Performance in Acid Environment. *J. Adv. Concr. Technol.* **2017**, *15*, 684–699. [CrossRef]
19. Idir, R.; Cyr, M.; Pavoine, A. Investigations on the durability of alkali-activated recycled glass. *Constr. Build. Mater.* **2019**, *236*, 117477. [CrossRef]
20. Joudah, Z.H.; Huseien, G.F.; Samadi, M.; Lim, N.H.A.S. Sustainability evaluation of alkali-activated mortars incorporating industrial wastes. *Mater. Today Proc.* **2021**, *46*, 1971–1977. [CrossRef]

21. Huseien, G.F.; Sam, A.R.M.; Shah, K.W.; Mirza, J.; Tahir, M.M. Evaluation of alkali-activated mortars containing high volume waste ceramic powder and fly ash replacing GBFS. *Constr. Build. Mater.* **2019**, *210*, 78–92. [CrossRef]
22. Jeon, I.K.; Kim, H.G.; Jakhrani, S.H.; Ryou, J.-S. Evaluation of the microstructure, mechanical, and durability properties of alkali-activated slag-based mortar with light-burnt dolomite powder. *J. Mater. Res. Technol.* **2021**, *13*, 2220–2228. [CrossRef]
23. Sasui, S.; Kim, G.; Nam, J.; van Riessen, A.; Hadzima-Nyarko, M. Effects of waste glass as a sand replacement on the strength and durability of fly ash/GGBS based alkali activated mortar. *Ceram. Int.* **2021**, *47*, 21175–21196. [CrossRef]
24. Yang, Y.; Zeng, H.; Chang, J.; Shi, J.; Liu, B. Waste glass powder and its effect on the fresh and mechanical properties of concrete: A state of the art review. *Adv. Concr. Constr.* **2020**, *10*, 417–429. [CrossRef]
25. Tome, S.; Nana, A.; Kaze, C.R.; Djobo, J.N.Y.; Alomayri, T.; Kamseu, E.; Etoh, M.-A.; Etame, J.; Kumar, S. Resistance of Alkali-Activated Blended Volcanic Ash-MSWI-FA Mortar in Sulphuric Acid and Artificial Seawater. *Silicon* **2021**, *14*, 2687–2694. [CrossRef]
26. Kaze, C.R.; Tome, S.; Lecomte-Nana, G.L.; Adesina, A.; Essaedi, H.; Das, S.K.; Alomayri, T.; Kamseu, E.; Melo, U.C. Development of alkali-activated composites from calcined iron-rich laterite soil. *Materialia* **2021**, *15*, 101032. [CrossRef]
27. Nasir, M.; Johari, M.A.M.; Maslehuddin, M.; Yusuf, M.O. Magnesium sulfate resistance of alkali/slag activated silico-manganese fume-based composites. *Constr. Build. Mater.* **2020**, *265*, 120851. [CrossRef]
28. Bailey, R.A.; Clark, H.M.; Ferris, J.P.; Krause, S.; Strong, R.L. *Chemistry and Environment*; Elsevier: Amsterdam, The Netherlands, 2002; pp. 443–482. Available online: https://www.elsevier.com/books/chemistry-of-the-environment/9780120734610 (accessed on 31 October 2022).
29. Adewumi, A.A.; Ariffin, M.A.M.; Yusuf, M.O.; Maslehuddin, M.; Ismail, M. Effect of sodium hydroxide concentration on strength and microstructure of alkali-activated natural pozzolan and limestone powder mortar. *Constr. Build. Mater.* **2020**, *271*, 121530. [CrossRef]
30. Silva, G.; Castañeda, D.; Kim, S.; Castañeda, A.; Bertolotti, B.; Ortega-San-Martin, L.; Nakamatsu, J.; Aguilar, R. Analysis of the production conditions of geopolymer matrices from natural pozzolana and fired clay brick wastes. *Constr. Build. Mater.* **2019**, *215*, 633–643. [CrossRef]
31. Firdous, R.; Stephan, D. Effect of silica modulus on the geopolymerization activity of natural pozzolans. *Constr. Build. Mater.* **2019**, *219*, 31–43. [CrossRef]
32. Mageed, A.A.; AbdelHaffez, S. Utilization of Limestone Dust in Brick Making. *J. Eng. Sci.* **2012**, *40*, 913–922.
33. Ghafoori, N.; Najimi, M.; Radke, B. Natural Pozzolan-based geopolymers for sustainable construction. *Environ. Earth Sci.* **2016**, *75*, 1110. [CrossRef]
34. Ibrahim, M.; Johari, M.A.M.; Maslehuddin, M.; Rahman, M.K.; Salami, B.A.; Mohamed, H.D. Influence of composition and concentration of alkaline activator on the properties of natural-pozzolan based green concrete. *Constr. Build. Mater.* **2019**, *201*, 186–195. [CrossRef]
35. Ibrahim, M.; Salami, B.A.; Algaifi, H.A.; Rahman, M.K.; Nasir, M.; Ewebajo, A.O. Assessment of acid resistance of natural pozzolan-based alkali-activated concrete: Experimental and optimization modelling. *Constr. Build. Mater.* **2021**, *304*, 124657. [CrossRef]
36. Ibrahim, M.; Johari, M.A.M.; Rahman, M.K.; Maslehuddin, M. Effect of alkaline activators and binder content on the properties of natural pozzolan-based alkali activated concrete. *Constr. Build. Mater.* **2017**, *147*, 648–660. [CrossRef]
37. Ibrahim, M.; Rahman, M.K.; Johari, M.A.M.; Maslehuddin, M. Effect of Incorporating Nano-silica on the Strength of Natural Pozzolan-Based Alkali-Activated Concrete. In *International Congress on Polymers in Concrete (ICPIC 2018)*; Springer International Publishing: Berlin/Heidelberg, Germany, 2018; pp. 703–709. [CrossRef]
38. Ghasemi, M.; Rasekh, H.; Berenjian, J.; AzariJafari, H. Dealing with workability loss challenge in SCC mixtures incorporating natural pozzolans: A study of natural zeolite and pumice. *Constr. Build. Mater.* **2019**, *222*, 424–436. [CrossRef]
39. Aragón, P.; Robayo-Salazar, R.A.; de Gutiérrez, R.M. Alkali-Activated Concrete Based on Natural Volcanic Pozzolan: Chemical Resistance to Sulfate Attack. *J. Mater. Civ. Eng.* **2020**, *32*, 04020106. [CrossRef]
40. Djobo, J.N.Y.; Elimbi, A.; Tchakouté, H.K.; Kumar, S. Mechanical properties and durability of volcanic ash based geopolymer mortars. *Constr. Build. Mater.* **2016**, *124*, 606–614. [CrossRef]
41. Bondar, D.; Lynsdale, C.J.; Milestone, N.B.; Hassani, N. Sulfate Resistance of Alkali Activated Pozzolans. *Int. J. Concr. Struct. Mater.* **2014**, *9*, 145–158. [CrossRef]
42. Oxygen and Chloride Permeability of Alkali-Activated Natural Pozzolan Concrete. *ACI Mater. J.* **2012**, *109*, 53–62. [CrossRef]
43. Aguirre-Guerrero, A.M.; Robayo-Salazar, R.A.; de Gutiérrez, R.M. Corrosion resistance of alkali-activated binary reinforced concrete based on natural volcanic pozzolan exposed to chlorides. *J. Build. Eng.* **2020**, *33*, 101593. [CrossRef]
44. Ibrahim, M.; Rahman, M.K.; Johari, M.A.M.; Nasir, M.; Oladapo, E.A. Chloride diffusion and chloride-induced corrosion of steel embedded in natural pozzolan-based alkali activated concrete. *Constr. Build. Mater.* **2020**, *262*, 120669. [CrossRef]
45. Ozata, S.; Akturk, B.; Yuzer, N. Utilization of waste Cappadocia earth as a natural pozzolan in alkali activation: A parametric study. *Constr. Build. Mater.* **2022**, *329*, 127192. [CrossRef]

sustainability

MDPI

Review

Mechanical and Durability Properties of Self-Compacted Concrete Incorporating Waste Crumb Rubber as Sand Replacement: A Review

Yarivan J. Zrar [1,*] and Khaleel H. Younis [2,3]

1 Department of Civil Engineering, Faculty of Engineering, Soran University, Soran 44008, Kurdistan Region, Iraq
2 Department of Surveying and Road Construction, Erbil Technology College, Erbil Polytechnic University, Erbil 44001, Kurdistan Region, Iraq
3 Civil Engineering Department, Tishk International University, Erbil 44001, Kurdistan Region, Iraq
* Correspondence: yjz450h@cive.soran.edu.iq

Abstract: The lack of disposal facilities for waste tires from various vehicles is a major environmental and economic problem. Crumb rubber (CR) generated from waste tires can be used to partially replace fine natural aggregates in self-compacted concrete (SCC), lowering sand usage and protecting raw material resources. The main objective of this study is to summarize the influence of CR as a partial replacement for sand on the behavior of SCC. For this aim, 42 papers were selected out of 89 that were relevant to the objective of this study. The mechanical properties, i.e., compressive strength, flexural strength, splitting tensile strength, modulus of elasticity, and bond strength, as well as the ultrasonic pulse velocity (UPV), were all reduced by the insertion of CR into SCC mixtures. With the addition of CR, fracture energy decreases, but the ductility of concrete in terms of characteristic length can be enhanced. Meanwhile, replacing sand with CR can also reduce the durability performance of SCC, such as sorptivity, free-drying shrinkage, rapid chloride permeability, and depth of chloride penetration, except for the electrical resistivity, depth of carbonation, and impact resistance, which exhibit a positive tendency. Based on the results of the reviewed articles, predicted reductions in the strength of the SCC incorporating CR were also recommended. Moreover, the results of the reviewed studies were employed to develop empirical models that demonstrate the relations between various mechanical properties.

Keywords: waste crumb rubber; rubberized self-compacted concrete; mechanical properties; durability properties; empirical model

Citation: Zrar, Y.J.; Younis, K.H. Mechanical and Durability Properties of Self-Compacted Concrete Incorporating Waste Crumb Rubber as Sand Replacement: A Review. *Sustainability* **2022**, *14*, 11301. https://doi.org/10.3390/su141811301

Academic Editor: Ahmed Salih Mohammed

Received: 25 June 2022
Accepted: 24 August 2022
Published: 8 September 2022

Publisher's Note: MDPI stays neutral with regard to jurisdictional claims in published maps and institutional affiliations.

1. Introduction

Concrete is one of the world's largest industries [1], with global use of approximately 25 gigatons per year [2]. Concrete has a very high negative impact on the environment, caused by the emission of CO_2 and exploitation of natural resources, due to its huge volume of use [3]. Moreover, the amount of excess waste tires from different kinds of vehicles continues to rise, and it is rapidly becoming one of the world's most critical environmental and ecological issues. Each year, nearly one billion used tires are discarded [4], and this number is expected to reach around 1.2 billion by 2030 [5]. Nearly 1 billion tires end their life cycle every year [4,5]. Because of the large increase in the number of cars on the road worldwide, the accumulation of massive amounts of scrap tires has now become an important waste management issue [6–8]. Reusing or recycling used tires effectively is critical [9–13]. Therefore, rubber waste utilization in the construction industry has advanced significantly in recent years, as it contributes to sustainability in two ways.

To begin with, it involves reusing materials that would otherwise pollute the environment and consume rare land resources. Second, it minimizes land and environmental degradation due to it requiring relatively less digging [14,15]. Additionally, conventional methods of tire disposal, such as disposal in landfills and combustion, may cause major ecological issues, either through site degradation or greenhouse gases. Burning the tires may appear to be the simplest and most cost-effective method. However, air pollution caused by toxic fumes and CO_2 emissions during the open burning procedure ended this illegal approach [16]. Waste-tire rubber is a material that is not biodegradable and has a relatively long life.

On the other hand, using natural aggregates in concrete has increased significantly. Therefore, to overcome this problem in the construction industry, building sustainable concrete needs to be considered. To achieve this, using waste tires in concrete production might be feasible [5,15–17].

Self-compacted concrete (SCC) is a superior type of concrete with excellent fluidity and segregation resistance and can be used for highly reinforced concrete elements without the aid of vibration [18–20]. SCC was established in Japan in the late 1980s and has lately been utilized in many countries for various purposes [20]. SCC is a novel category of high-strength concrete. SCC mixtures typically include supplementary cement-based materials such as fly ash and slag to develop rheological properties in fresh conditions, reduce costs, and minimize adverse environmental impact [21–23]. Compared to conventional vibrating concrete, SCC has some advantages because it does not require vibration, has low energy costs, and can decrease noise [24–26]. Despite recent developments, SCC is currently broadly utilized in the industry as a result of its benefits, including increased on-site productivity, improved quality of construction, and enhanced on-site working conditions [27]. Although, as indicated by Tuyan et al. [28], SCC costs higher than conventional concrete because of the higher binder content and chemical admixture required [28]. Therefore, to permit SCC to reach its full performance regarding industry engagement, it is essential to develop ways to make SCC a more environmentally friendly and eco-efficient composite type [28]. Moreover, in designing SCC mixes, 60–70% of the mix's total volumes are occupied by aggregates. Therefore, aggregates play a vital role in the fresh and hardened state behavior of the produced concrete [29]. Hence, using CR in SCC instead of a natural aggregate reduces SCC costs while reducing carbon dioxide emissions in concrete production [13,26].

In the literature, several investigations on self-compacted concrete (SCC) with different utilizations of recycled waste-tire aggregate as an alternative to natural gravel and sand in the manufacture of SCC were studied. Most research has focused on self-compacted rubberized concrete's mechanical properties [30]. Güneyisi et al. [31] studied the effects of substituting fine and coarse aggregates with tire rubber. Their study observed the largest decrease in compressive strength in mixtures where only fine particles were replaced by rubber. Ilker et al. [32] substituted waste-tire rubber with aggregates at rates of 60, 120, and 180 kg/m^3 by total weight for the different rubberized SCC mixes, and they observed a reduction in strength values as the rubber content increased [32]. Hilal et al. [33] discovered a similar trend when replacing fine and coarse aggregates with waste-tire rubber in concrete mixes containing 30% fly ash by weight [33]. This loss of strength is due to the low rigidity and hydrophobicity of crumb rubber relative to natural aggregates, which causes non-remarkable growth of the interfacial transition zone (ITZ). The utilization of CR similarly influences the properties of the bond between the aggregate and the cement paste, which later impacts the strength of concrete that includes CR [8,34–36]. When examining the mechanical characteristics of concrete using CR, most studies have found a reduction in mechanical strength [30]. On the other hand, only some studies have dealt with the durability properties of rubberized SCC.

From an ecological standpoint, incorporating rubber obtained from waste tires into concrete would reduce the number of waste tires disposed of and provide a resource of environmentally friendly concrete. From an engineering standpoint, incorporating scrap tires into concrete may produce a material with enhanced dynamic and durability properties, including ductility, carbonation resistance, etc. Mallek et al. [30] presented that increasing the rubber replacement in SCC causes a higher or similar carbonation depth to that of the control mix. On the other hand, they found that increasing the replacement level of rubber causes a reduction in carbonation depth after exposure to CO2 exceeds one day. On the other hand, they found that increasing the replacement level of rubber causes a reduction in carbonation depth after exposure to CO_2 exceeds one day [30]. A reduction in carbonation depth was also reported by Thomas et al. [37]. Gesoglu and Güneyisi [15] conducted rapid chloride penetration experiments on several rubberized SCC mixes with/without fly ash and discovered that when the rubber percentage increases, chloride ion penetration increases, particularly for the mixes without fly ash. Karahan et al. [38] observed a similar outcome. They designated four different categories of mixtures with 0%, 10%, 20%, and 30% replacement levels of sand with CR [38].

Table 1 illustrates the review papers published previously on rubberized SCC compared to the current review. Based on the above-mentioned table, there are a few review articles about the use of rubber in the production of SCC. However, it can be understood from Table 1 that there is a gap in studies about a comprehensive review of the effect of using CR as a partial sand replacement on the various properties of SCC. Another reason is that most experimental investigations work by substituting natural sand with CR due to rubberized SCC's better properties than the findings attained by replacing natural gravel with waste rubber [39]. To some extent, constructing models to estimate the performance of SCC comprising CR in mechanical and durability conditions has not been comprehensively studied yet. In addition, no effort has been made to review the investigations that have been released since 2018 about SCC that includes rubber. For this aim, in this study, 42 papers about the influence of CR as a sand replacement on the performance of SCC were reviewed for their mechanical and durability properties, as can be seen in Figure 1. The properties of SCC in its hardened state are distinct from normal concrete. Therefore, this research focuses on the mechanical properties of rubberized self-compacted concrete in terms of the compressive, splitting, flexural, elastic module, bond strength, and fracture parameters. The UPV test, used to examine the quality and homogeneity of concrete, was also reviewed. Durability properties were also assessed, i.e., sorptivity, electrical resistivity, free-drying shrinkage, rapid chloride permeability, depth of carbonation, depth of chloride penetration, and impact resistance. Based on the outcomes of the reviewed publications, predicted percentages of the SCC's strength, made including CR, were also suggested. Besides the properties mentioned above, empirical models between mechanical properties were also created to construct models for the developed type of concrete, named rubberized SCC.

Table 1. Comparison between published paper reviews on SCC using crumb rubber versus current review.

Ref	Year	Replacement Type	No. of Papers Reviewed	Range	Mechanical Properties	Durability Properties	Empirical Models
[40]	2010	Coarse and fine aggregate	4	2006–2009	Compressive, splitting, and flexural	Thermal resistance, noise reduction, air entrainment and shrinkage, impact resistance and ductility	–
[41]	2016	Coarse and fine aggregate	12	2006–2016	Compressive, splitting, flexural, and dynamic modulus of elasticity	Shrinkage	–
[39]	2018	Coarse and fine aggregate	27	2006–2018	Compressive, splitting, flexural, and elastic module	Water absorption, water sorptivity, impact resistance, ductility, brittleness, fracture energy, shrinkage, fatigue behavior, SEM	–
[42]	2020	Coarse and fine aggregate	14	2006–2018	Compressive strength, modulus of elasticity, flexural strength, ultrasonic pulse velocity	Stiffness, fracture energy, durability, deformability before failure, dynamic properties, strain capacity, fatigue strength	–
Current Review	Submitted review	Fine aggregate	42	2006–2021	Compressive, splitting, flexural, elastic module, and fracture energy, characteristic length, bond strength, and (UPV)	Sorptivity, electrical resistivity, free-drying shrinkage, rapid chloride permeability, depth of carbonation, depth of chloride penetration, and impact resistance	Compressive and splitting, compressive and flexural, compressive and modulus, compressive and UPV, and compressive and bond strength

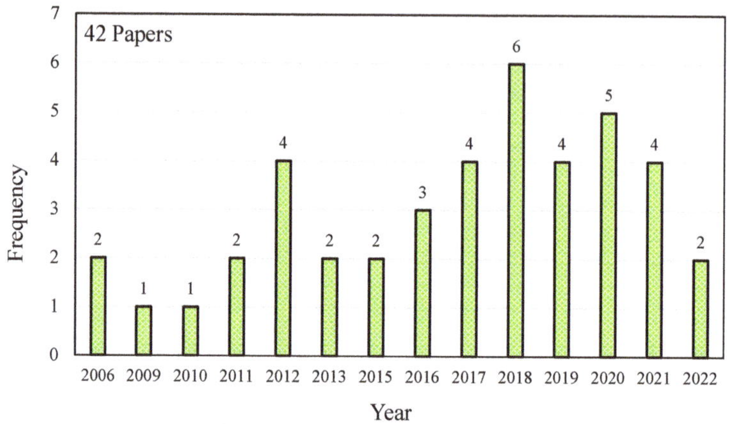

Figure 1. Reviewed papers about rubberized SCC published between 2006 and 2022. (2006) [43,44]; (2009) [32]; (2010) [45]; (2011) [15,46]; (2012) [4,38,47,48]; (2013) [49,50]; (2015) [51,52]; (2016) [16,31,53]; (2017) [18,33,54,55]; (2018) [56–61]; (2019) [19,62–64]; (2020) [65–69]; (2021) [7,30,70,71]; (2022) [72,73].

2. Methodology of Paper Selection

Using keywords such as rubberized self-compacted concrete, waste crumb rubber, and recycled crumb rubber, a search was undertaken in the databases of Scopus and Google Scholar to identify studies that were carried out until January 2022. The publications were restricted to rubberized self-compacted concrete and were published beginning in 2006. The titles of the articles were then evaluated to determine whether they were relevant to the objectives of the study or irrelevant. The relevant ones are those that focus on the use of crumb rubber as a fine aggregate replacement in the production of self-compacted concrete, and they were read in their full text in order to extract data. In total, 130 papers were found, and the duplicate papers (41 papers) were removed using Mendeley. Finally, only 42 papers were selected for the current review. The selected papers were about the influence of crumb rubber as a sand replacement on the performance of self-compacted concrete. Figure 2 shows a flow chart diagram of the process of selecting articles.

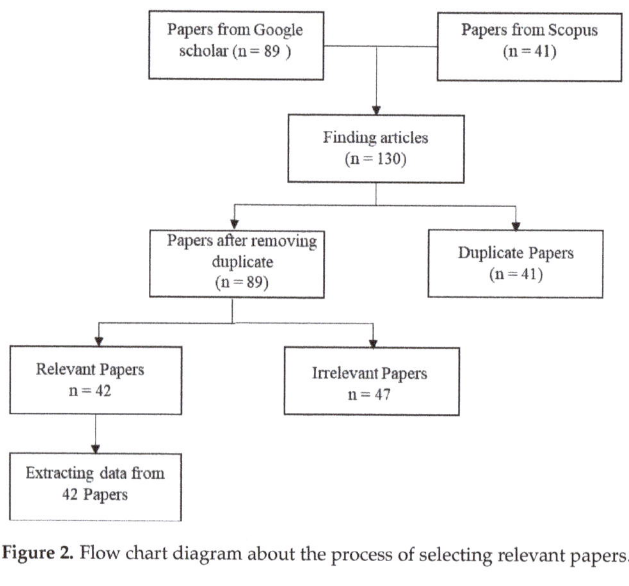

Figure 2. Flow chart diagram about the process of selecting relevant papers.

3. Manufacturing of CR Aggregates

Every year, one billion end-of-life tires are manufactured worldwide, of which 355 million end-of-life tires are produced in Europe [74]. Only approximately 5% of used tires are utilized for civil engineering purposes; the majority are landfilled. However, given the enormous demand for construction, about 32 billion tons annually [75], the utilization of recycled organic materials such as crumb rubber (CR) in Portland cement concrete can successfully alleviate environmental constraints [4]. Additionally, end-of-life tires are also used for a variety of projects, such as playground surfacing and sports fields [76].

Crumb rubber is manufactured from end-life tires in a variety of techniques. Ambient grinding and cryogenic processing are two of the most common techniques. Mechanical grinding at ambient temperature is frequently utilized in industry, where discarded tires are chopped into small pieces using "cracker mills" and "granulator" procedures. Ambient grinding is a multistep process that utilizes complete or pre-treated automobile or truck tires in the form of shreds or chips, as well as sidewalls or treads. Rubber, metals, and textiles are separated successively. The tires are shredded in a shredder. The chips are fed into a granulator, which shreds them into minute bits while simultaneously eliminating steel and fiber. Any remaining steel is magnetically removed, and the remaining fiber is sifted using a combination of shaking screens and wind sifters. Further grinding in secondary granulators and high-speed rotary mills can produce finer rubber particles [77].

Cryogenic processing involves freezing scrap tires in liquid nitrogen below the glass transition temperature and crushing them using automatic hammers [40,77]. In both methods, magnetic fields are used to separate the steel wires in the tires from the rubber particles, and vibrating sieves are used to separate the rubber particles from the wire mesh [40,78].

Moreover, the CR surface structure itself is highly influenced by its production process. A novel high-pressure water-jet-based technology (2–3 kbar) has been shown to produce CR with a rougher and larger surface area compared to traditional techniques such as mechanical shredding or cryogenic processing [79].

4. Utilization and Types of Crumb Rubber in SCC

In the literature, the utilization of CR as a partial substitution of natural aggregate in the production of SCC to enhance the sustainability of concrete has been investigated in some research. Table 2 highlights the types and utilization of CRs for sand substitution in the preparation of SCC composites. Figure 3 shows the different types of waste rubber aggregates that can be used in the production of concrete. Further, Table 3 illustrates the type of waste rubber used as a fine aggregate in the production of SCC in previous studies.

Table 2. Classification of scrap tire rubber.

Type	Size (mm)	Replacement Type	Refs.
Shredded or chipped tire rubber	>4.75 mm	Coarse aggregate partial)	[8,31]
Crumb rubber aggregate	4.75–0.425	Fine aggregate (partial)	[8,31]
Granular tire rubber	<0.425	Cement (partial)	[8,80]
Fiber rubber aggregate	Short fibers: 8.5–21.5 Strips: <8	Fiber/aggregate	[36,81]

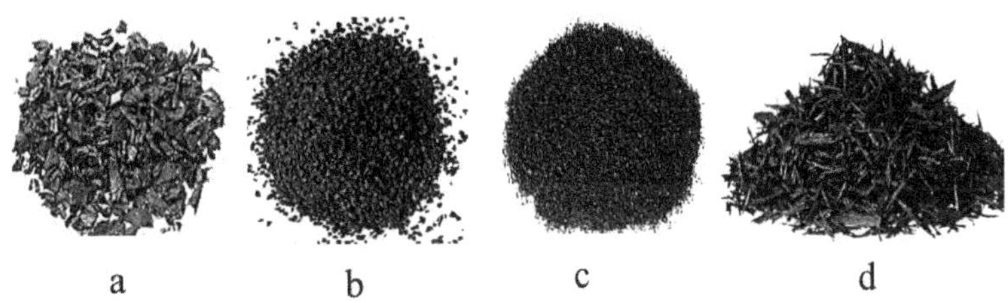

Figure 3. Different sorts of rubber aggregates: (**a**). chipped, (**b**). crumb, (**c**). granular, and (**d**). fiber.

Table 3. Types and use of waste crumb rubber as fine aggregate in the production of SCC composites.

Ref.	Composite Type	Binder Content (kg/m^3)	Type of Mineral Admixture	Type of CR	Replacement Amounts	Particle Size of CR	Properties of CR
[15]	SCC	550	Fly ash (0, 20, 40, and 60%.)	Crumb rubber	(0, 5, 15, and 25%)	0.13–4 mm	Specific gravity = 0.83
[50]	SCC	522	Fly ash (45%)	Shredded scrap rubber	(0, 15, and 20%)	4.75 mm	Specific gravity = 1.14
[30]	SCC	450	Inert calcareous (17.8%)	Waste-tire rubber	(0, 5, 10, and 15%)	0–4 mm	Specific gravity = 1.2
[45]	SCC	550	Fly ash (0, 20, 40, and 60%)	Crumb rubber	(0, 5, 15, and 25%)	4 mm	Specific gravity = 0.83
[43]	SCC	566	Fine filler (CaCO3) (34.6%)	Untreated tire waste	(0, 22.2, and 33.3 *v*/*v*%)	0.05–2 mm	Tire rubber density (0.9 g/cm^3)
[4]	SCC	5.92 kg/batch	-	Waste tire	(0, 10%)	1–4 mm	-
[70]	SCC	550	Fly ash (25.3%), silica fume (4.7%)	Shredded discarded tires	(0, 10, 20, and 30%)	1–5 mm	Apparent density = 1060 kg/m^3 bulk density = 433 kg/m^3
[47]	SCC	702	Limestone powder (41.2%)	Recycled crumb rubber	(0, 4, 8, and 12 wt.%)	3–5 mm	Elasticity module = 22 MPa, tensile strength = 28 MPa
[31]	SCC	520	Class F fly ash (30%)	Crumb rubber (CR No 5 and 18) and tire chips (TC)	(0, 5,10, 15, 20, and 25%)	1–4 mm	Density no.18 CR = 0.50 g/cm^3, no.5 CR = 0.67 g/cm^3, chips = 1.02 g/cm^3 elongated particles chips between 10 and 40 mm
[49]	SCC	600	Fly ash (50%), slag (50%)	Waste-tire rubber	(0, 5, 10, 15, and 20%)	Passing (0.3 mm and 0.6 mm) sieve	Specific gravity = 0.95
[48]	SCC	472	Pulverized fuel ash (23.1%)	Crumb rubber	(0, 5, 10, and 15 wt.%)	2–6 mm	Specific gravity = 1.12, apparent density = 489 kg/m^3, thermal conductivity = 0.11 W/mk, tensile resistance = 4.2–15 MPa, water absorption = 0.65
[51]	SCC	550	Fly ash (20%), slag (30%), metakaolin (20%)	Crumb rubber	(0, 5, 10, 15, 20, 30, and 40%)	4.75 mm	Specific gravity = 0.95
[52]	SCC	450	Cement kiln dust	Crumb rubber	(0, 10, 20, 30, and 40%)	2 mm	-
[46]	SCC	(522, 469, 428, 407)	Fly ash (21.5, 27.7, 37.6, 51.8%)	Worn-out tire	(0, 5, 10, 15 and 20%)	4.75 mm	Specific gravity = 1.14
[68]	SCC	498	Grade I fly ash (35%)	Waste-tire rubber	(0, 5, 10, 20, and 30%)	2–4 mm	Packing density = 710 kg/m^3, apparent density = 1600 kg/m^3
[7]	SCC	10.78 Kg/batch	Fly ash (55%), silica fume (5, 10%)	Crumb rubber from waste tires	(0, 15, and 30%)	600 µm to 2.36 mm	-
[66]	LWSCC	1.67 in mass	Silica fume (6%), expanded clay (12.6, 21.6%)	Waste-tire rubber	(0, 5, 10, and 15%)	0.15–9.5 mm	Density = 1.16 g/cm^3, saturated dry density = 0.4 g/cm^3, fineness modulus =3.49
[16]	SCC	500–550	Fly ash (20%), slag (30%), metakaolin (20%)	Crumb rubber	(0, 5, 10, 15, 20, 30, 40, and 50%)	5 mm	Specific gravity = 0.95
[18]	SCC	550–600	Metakaolin (20%), fly ash (30%)	Crumb rubber	(0, 5, 10, 15, 20, 25, 30, and 40%)	4.75 mm	Specific gravity = 0.95

Table 3. *Cont.*

Ref.	Composite Type	Binder Content (kg/m^3)	Type of Mineral Admixture	Type of CR	Replacement Amounts	Particle Size of CR	Properties of CR
[64]	SCC	530	Fly ash (15.1%), slag (20%)	Waste-tire rubber	(0, 10, 20, and 30%)	2–4 mm, 1–2 mm, and 0–0.3 mm	Specific gravity = 1.0, tensile strength = 8.0 MPa, ultimate tensile strain = 256%, initial elastic modulus = 3.4 MPa
[38]	SCC	500	Ground granulated blast furnace slag (25%)	Scrap tires	(0, 10, 20, and 30%)	0.15–4.75 mm	Specific gravity = 0.90
[19]	LWSCC	510	Fly ash (15.7%)	Rubber	(0, 10, 20, 30, 40, and 50%)	0.15–4.75 mm	Modulus of fineness 2.7, density 1.19 g/cm^3, and lose bulk density 365 kg/m^3
[71]	SCC	550	Fly ash (25.3%), silica fume (4.7%)	Tire rubber	(0, 10, 20, and 30%)	-	Ash content = 2.4, carbon black content = 25, density = 1060 kg/m^3, apparent density = 433kg/m^3, tensile strength = 11 MPa
[65]	SCC	600	-	Waste vehicle tires (WVT)	(0, 5, 10, 15, and 20%)	0–4 mm	Specific gravity = 1.050
[62]	SCC	450	Fly ash (40%), silica fume (7.5%), ground granulated blast furnace slag (22.5%)	Waste tires	(0, 10, 20, 30, and 40%)	2–5 mm	Specific gravity = 1.15
[53]	SCC	702	Limestone powder (41.2%)	Scrap tires of heavy vehicles	(0, 5, 10, and 15%)	4.75 mm	Specific gravity = 1.122 (g/cm^3), weight percentage of sulfur = 0.97%
[60]	SCC	550	Class-F fly ash (21.8%)	Crumb rubber	(0, 15, and 25%)	1.4–2.83 mm	-
[55]	SCC	495	Silica fume (10%)	Crumb rubber	(0, 10, 20, 30, and 40%)	4 mm	Specific gravity = 1.15, bulk density =489 kg/m^3
[59]	LWSCC	450	Fly ash (30%), silica fume (7.5%), slag (22.5%)	Recycled crumb rubber	20%	2–5 mm	Specific gravity = 1.15
[44]	SCC	470	Calcareous filler (25.55%)	End-of-life tires	(0, 20, 30, 40, and 50%)	0–4 mm	Specific gravity = 1.2
[57]	SCC	450	Fly ash (30%), silica fume (7.5%), slag (22.5%)	Scrap rubber	(0, 10, 20, 30, and 40%)	1–3 mm	The chemical composition = of 45% polymer, 40% carbon black, and 15% organic materials by weight
[58]	SCC	442	-	Waste tires	(5, 10, 15, 20, and 30%)	0–2 mm	Specific gravity = 0.77 and a water absorption coefficient = 0.24
[69]	SCC	520	Crushed dune sand (5–20%)	Pre-coating rubber	2.50%	0.5–4 mm	-
[67]	SCC	450	Fly ash (30%), silica fume (7.5%), slag (22.5%)	Waste tires	(0, 10, and 20%)	2–5 mm	Specific gravity= 1.15 kg.m^{-3}, the chemical composition = of 45% polymer, 40% carbon black, and 15% organic materials by weight
[63]	SCC	550	Fly ash (25.3%), silica fume (4.7%)	Crumb rubber	(0, 10, 20, and 20%)	1–5 mm	-

Table 3. *Cont.*

Ref.	Composite Type	Binder Content (kg/m³)	Type of Mineral Admixture	Type of CR	Replacement Amounts	Particle Size of CR	Properties of CR
[56]	SCC	550	Silica fume (10%)	Crumb rubber	(0, 5, 10, 15, and 20%)	<4.75 mm	Specific gravity = 0.95
[61]	SCC	550	Fly ash (30%) and metakaolin (20%)	Crumb rubber	(0, 5, 10, 15, 20, 25, and 30%)	<4.75 mm	Specific gravity = 0.95
[32]	SCC	530	Fly ash (28%)	Crumb rubber	(60, 120, and 180 kg/m³)	0–4 mm	-
[54]	SCC	550	Fly ash (30%) and metakaolin (20%)	Crumb rubber	(0, 5, 10, 15, 20, 25, and 30%)	<4.75 mm	Specific gravity = 0.95
[33]	SCC	520	Fly ash (30%)	Crumb rubber	(0, 5, 10, 15, 20, and 25%)	0–4 mm	The specific gravity of no. 18 CR and no. 5 CR is 0.50 and 0.67, respectively
[73]	SCC	454	Fly ash (40%) and calcium carbide waste (5, 10%)	Crumb rubber	(0, 10, and 20%)	0–4 mm	Specific gravity = 0.95
[72]	SCC	600	Fly ash (10, 25, C40%) and nano silica (2, 4%)	Crumb rubber	(0, 15, and 7.5%)	0–4 mm	Specific gravity = 0.95, fineness modulus = 0.92

5. Effect of Crumb Rubber on the Hardened Properties of SCC

Table 4 highlights the hardened properties of rubberized SCC that have been studied in the literature. The following table focuses on the studies that investigate the impact of CR (as a sand substitution) on the hardening of SCC. The properties that have been reviewed in this section were the compressive, splitting, elastic module, flexural, bond strength, fracture energy, characteristic length, and UPV of SCC made with or without rubber. The partial substitution of sand with CR particles ranged from 0% to 50%, although most studies utilized up to 30%. Introducing CR into SCC mixes causes a systematic decrease in hardened properties. The decline in mechanical characteristics of SCC caused by the CR inclusion is typically due to the lower strength of CR particles and their poor adhesion to the hardened cement paste matrix. [19,30,53]. This effect can be explained by air entrainment on the interfacial transition zone (ITZ) between rubber particles and cement paste and the lower modulus of elasticity of rubber aggregate than the natural aggregate modulus. However, it is feasible to produce rubberized SCC specimens that could be applicable for structural applications. On the other hand, the ductility behavior in characteristic length was enhanced by utilizing CR aggregates. Therefore, reviewing the hardened properties of the rubberized SCC is a topic of interest for future research in this field.

Table 4. Hardened properties of rubberized SCC concrete reported in the literature.

Ref.	Compressive Strength	Tensile Strength	Flexural Strength	Elasticity Modulus	Bond Strength	Fracture Energy	Characteristic Length	UPV
[15]	✓							
[50]	✓							
[30]	✓							
[45]	✓							
[43]	✓							
[4]	✓							✓
[47]	✓	✓	✓					
[31]	✓							
[49]	✓							✓
[48]	✓	✓	✓					✓
[51]	✓	✓	✓	✓				
[52]	✓	✓	✓					
[46]	✓	✓		✓				
[68]	✓			✓				
[7]	✓	✓	✓					
[66]	✓	✓						
[16]	✓	✓	✓	✓				
[18]	✓	✓	✓					
[64]	✓					✓		
[38]	✓	✓	✓		✓			
[19]	✓	✓	✓	✓				
[71]	✓							
[65]	✓	✓						
[62]	✓	✓		✓				
[53]	✓	✓	✓	✓				✓
[60]	✓	✓						✓
[59]	✓	✓		✓				
[44]	✓			✓				
[57]	✓	✓						
[58]	✓							
[69]	✓							
[67]	✓	✓						
[56]	✓	✓	✓					
[61]	✓	✓	✓	✓				
[32]	✓							
[54]	✓	✓	✓	✓				
[33]	✓	✓			✓	✓	✓	
[72]	✓	✓	✓					

5.1. Compressive Strength

When various forms of rubber aggregate are utilized to partially replace natural gravel and sand, irrespective of rubber aggregate size, replacement level, or aggregate type substituted, a reduction in the strength of SCC is predictable. Compressive strength is the main critical mechanical attribute in the concrete industry. Each new concrete mix must meet the structural minimal strength criteria. The 28-day compressive strength values published in the literature versus different replacement levels of CR are depicted in Figure 4. It is clear from the previous studies that rubberized SCC mixes were produced with compressive strengths ranging from 21 to 80 MPa. Therefore, based on the available compressive strength findings, rubberized SCC is suitable for structural applications [18,30,43,47,48,50,54,56,61]. Adding CR to SCC has a detrimental effect on its compressive strength. The SCC strength diminishes as the CR content increases, regardless of the CR particle type. The degree of strength loss is proportional to the amount of CR utilized [16,46,58,65–67]. Khalilpasha et al. [47] discovered that adding more rubber weakens the specimens when compressed. In addition, different studies detected that the compressive strength value at 28 days was measured at a rate of 56 MPa in the SCC mix made without CR. When 15% of the CR particles was replaced with natural fine aggregate, this value went down to 25 MPa [30,48].

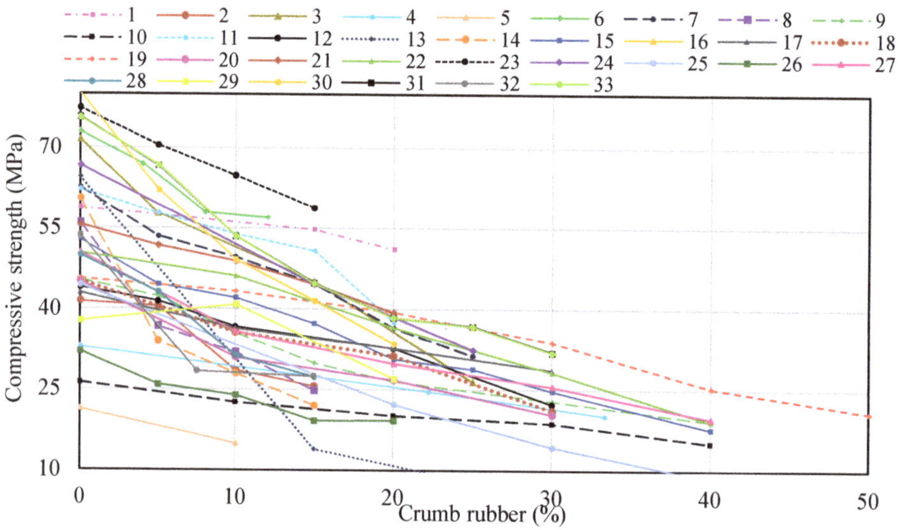

Figure 4. Variation of compressive strength of SCC made with different replacement levels of CR: 1 [50], 2 [30], 3 [45], 4 [43], 5 [4], 6 [47], 7 [31], 8 [48], 9 [51], 10 [52], 11 [46], 12 [68], 13 [7], 14 [66], 15 [16], 16 [18], 17 [64], 18 [38], 19 [19], 20 [71], 21 [65], 22 [62], 23 [53], 24 [60], 25 [44], 26 [49], 27 [57], 28 [58], 29 [67], 30 [56], 31 [61], 32 [72], 33 [54].

Similarly, Abdel Aleem and Hassan [56] reported that the reference compressive strength with no rubber was about 76 MPa and then dropped to 62 MPa, 49 MPa, 41 MPa, and 33 MPa as the CR content increased to 5%, 10%, 15%, and 20%, respectively. Bignozzi et al. [43] used CR at 0%, 22%, and 33% of the total volume of fine aggregates. They found that the compressive strength of SCC mixes decreased with an increasing CR particle percentage. Rubber particles have lower tensile strength and less adhesion to hardened cement paste, which causes a reduction in the compressive strength of the mixture [19,30,53]. Nevertheless, it was found in some studies that increasing compressive strength by more than 30% led to a drop in compressive strength of less than 20 MPa [4,7,44,49,52]. Therefore, it can be mainly concluded that the rubberized SCC could be used for structural applications with CR inclusion of up to 30%.

The variance in normalized residual compressive strength as a function of CR concentration is depicted in Figure 5. It was shown that increasing CR to 50% resulted in a 55% decrease in strength [19]. Likewise, some other studies investigated the use of 40% CR in the production of SCC, which could decrease the compressive strength by around 61% [51,57,62]. In the earlier studies, it was detected that raising CR to 30% decreased the normalized compressive strength value by 58% [18,54,61]. A similar phenomenon was found by some other researchers [38,64,68,71]. According to most research, growing the CR content up to 25% leads to a 50% loss in the compressive strength results [31,45,60]. To some extent, Ganesan et al. [50] obtained a compressive strength loss of around 13% when 20% CR granules were used as an alternative to natural fine aggregates. A study by Hesami et al. [53] stated that if 15% of the aggregate was composed of rubber, the compressive strength dropped by 29%. The compressive strength dropped by about 13% if 5% of the aggregate was made of rubber. Based on the summarized data, the percent reduction in compressive SCC for different CR replacement levels can be predicted using Table 5. According to the above-mentioned table, using CR at a 30% and 40% replacement level could decrease compressive strength in the range of 40–60% and 60–70%, respectively. However, with a 10% CR replacement level, the compressive strength decreases at a lower rate (10–25).

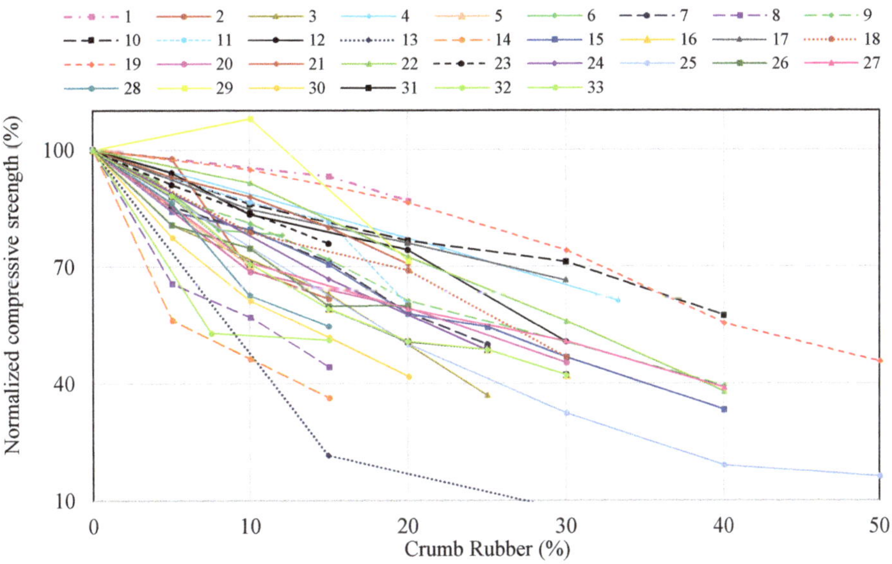

Figure 5. Normalized percentage of compressive strength: 1 [50], 2 [30], 3 [45], 4 [43], 5 [4], 6 [47], 7 [31], 8 [48], 9 [51], 10 [52], 11 [46], 12 [68], 13 [7], 14 [66], 15 [16], 16 [18], 17 [64], 18 [38], 19 [19], 20 [71], 21 [65], 22 [62], 23 [53], 24 [60], 25 [44], 26 [49], 27 [57], 28 [58], 29 [67], 30 [56], 31 [61], 32 [72], 33 [54].

Table 5. Predicted percentage reduction values in the compressive strength of SCC via CR content.

CR Content (%)	10	20	30	40
Compressive strength reduction (%)	10–30	25–45	30–60	40–70

5.2. Splitting Tensile Strength

The splitting tensile strength test was performed horizontally on cylinder-shaped samples between the compression testing machine's loading surfaces. The force was applied until the cylinder failed along its vertical diameter. Similar to compressive strength, the tensile strengths of SCC dropped as the CR concentration increased. Figure 6 summarizes previous experiments' splitting tensile strength results for different replacement amounts of CR. It is obvious that SCC's splitting strength diminishes as the CR content increases. When the CR replacement ratio rises from 0% to 12% at 28 days, splitting strength drops from 4 to 2.7 MPa [47]. Similar to the above findings, it was realized that when 15% of the sand volume is replaced with CR, the tensile strength drops from 5.7 to 4.9 MPa [53]. Cemalgil and Etli [65] used 0% to 20% of the CR as a part of the fine aggregate and found a compressive strength range of 5 to 7 MPa. In addition, Si et al. [60] determined that the splitting tensile strength outcomes eventually decline when the amount of CR reaches 25%.

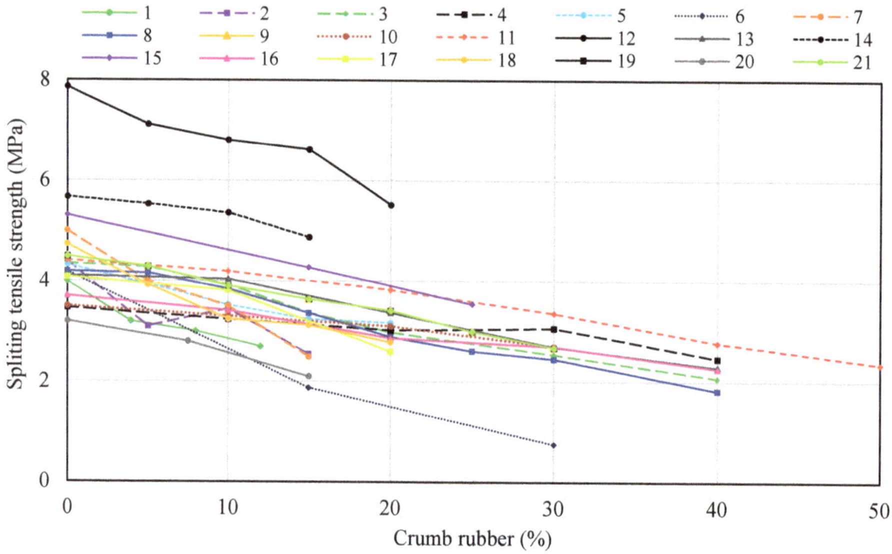

Figure 6. Variation of splitting tensile strength of SCC made with different replacement levels of CR: 1 [47], 2 [48], 3 [51], 4 [52], 5 [46], 6 [7], 7 [66], 8 [16], 9 [18], 10 [38], 11 [19], 12 [65], 13 [62], 14 [53], 15 [60], 16 [57], 17 [67], 18 [56], 19 [61], 20 [72], 21 [54].

Further, Alaloul et al. [7] discovered the same outcome when they employed 30% of CR. This reduction would be owing to the physical features of crumb rubber particles, specifically their high elastic deformation capacity before splitting. Unlike plain concrete, which splits abruptly, SCC with CR particles splits gradually [7,47,60].

Figure 7 depicts the variance in normalized residual split tensile strength as a function of CR content. Garros et al. [66] studied the performance of SCC with various CR replacement ratios. They found that by increasing CR to 50%, the residual tensile strength value was 53% [44]. Similarly, other studies found that using 40% CR in the manufacture of SCC resulted in a more than 45% loss in tensile strength [57,62]. It was also stated that the addition of 30% CR content led to a reduction of 40% in the tensile strength [18,54,61]. Karahan et al. [38] used the same proportion of CR and discovered that the residual in the splitting tensile strength results was approximately 77%. Valizadeh et al. [70] claimed that when 10% or 20% CR was used, the splitting strength of SCC mixtures was lowered by 6.5% and 36%, respectively, compared to the control mix prepared without CR. Comparable phenomena were also detected by Cemalgil and Etli [65] and AbdelAleem and Hassan [56]. Other research has shown that the tensile strength of SCC made with rubber decreases by more than 40% when 15% CR is added to the mixture [48,66]. From the plotted results shown in Figure 7, it can be concluded that using CR up to 40% might result in a loss in splitting values of about 50%. To predict the percent reduction in splitting tensile strength for various CR additions, Table 6 was developed based on the conducted literature data.

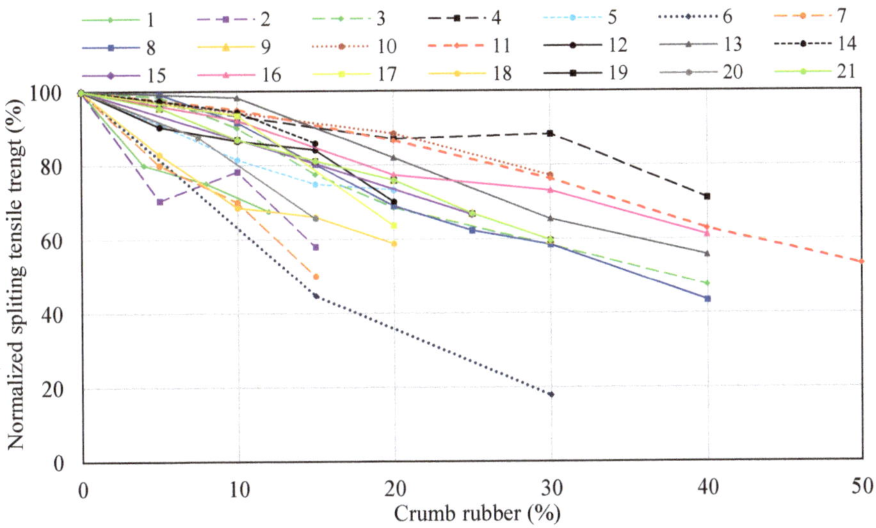

Figure 7. Normalized percentage of splitting tensile strength: 1 [47], 2 [48], 3 [51], 4 [52], 5 [46], 6 [7], 7 [66], 8 [16], 9 [18], 10 [38], 11 [19], 12 [65], 13 [62], 14 [53], 15 [60], 16 [57], 17 [67], 18 [56], 19 [61], 20 [72], 21 [54].

Table 6. Predicted percentage reduction values in the splitting tensile strength of SCC via CR content.

CR Content (%)	2	20	30	40
Splitting tensile strength reduction (%)	5–20	10–30	25–40	35–55

5.3. Modulus of Elasticity

The elastic module is the most critical characteristic of concrete since it describes the elastic properties of concrete and is mostly determined by the paste's qualities and the stiffness of the aggregates utilized. The elastic module of SCC is shown in Figure 8 as a function of the rubber content. The summarized investigations show that the CR replacement levels mainly used were 30% of the total sand volume. The substitution of CR particles for fine natural aggregates (sand) had a detrimental effect on the static elastic module. The static elastic module was reduced as the rubber replacement level increased. This concrete variation is due to air entrainment induced by CR aggregates and the rubber aggregate's relatively low elastic modulus compared to sand [48]. It was also declared that the lower stiffness of rubber particles than the natural aggregates caused that reduction. In other words, the aggregate type used greatly impacts measuring the modulus of elasticity of concrete [53,61,82,83]. For instance, the modulus of elasticity lowered from 44 GPa to 36 GPa as the CR addition increased from 0% to 15%, respectively [53]. Raj et al. [46] also testified that the elastic modulus increased as the water/powder decreased but decreased as the CR increased. The elastic modulus of SCC was found to be 33 GPa for the reference mix and then declined to 22 GPa for the mix made with 20% CR aggregate. Tian et al. [68] found that the elastic module value dropped from 34 GPa to 23 GPa as the CR percentage rose from 0% to 30%, respectively.

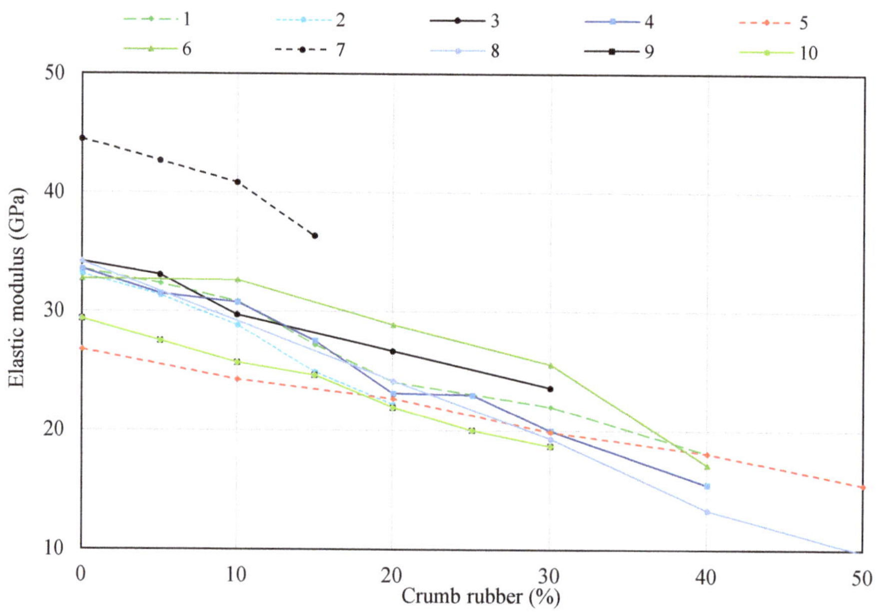

Figure 8. Variation of elastic modulus of SCC made with different replacement levels of CR: 1 [51], 2 [46], 3 [68], 4 [16], 5 [19], 6 [62], 7 [53], 8 [44], 9 [61], 10 [54].

In Figure 9, the normalized percentage of the elastic module versus CR addition is shown. As indicated in Figure 9, compared to the reference mix, modules of elasticity were reduced by 10%, 12%, 22%, and 48% for specimens with 10%, 20%, 30%, and 40% CR, respectively [62]. Ismail and Hassan [16] determined that the elastic module of rubberized SCC containing 40% CR aggregates was 46% less than that of the reference mix. Previous studies revealed that by substituting 30% CR for sand in SCC mixes, the elastic module could be reduced by approximately 36% [54,61]. Garros et al. [44] also reported the decline behavior of static modulus by adding CR aggregates. They claimed that the percent decrease in the elastic module was higher than 70% for the mix of 50% CR. Moreover, the percent decrease in the elastic module was 42% for the mix with 50% CR content [19]. Table 7 shows the predicted percentage reduction in the modulus of elasticity of SCC for different contents of CR, concerning the collected data in the literature. The reduction in the modulus of elasticity was 5–15%, 15–25%, 25–40%, and 35–55% for the 105, 20%, 30%, and 40% contents of CR.

Table 7. Predicts percentage reduction values in the modulus of elasticity of SCC via CR content.

CR Content (%)	10	20	30	40
Modules of elasticity reduction (%)	5–15	15–30	25–40	35–55

5.4. Flexural Strength

In Figure 10, which displays the outcomes published in the former investigations [7,16,18,19,38,47,48,51,52,54,56,61], the flexural strength of rubberized SCC follows the same pattern as that observed in the splitting test. From the above-mentioned figure, the highest replacement ratio of CR aggregates was 50%. Due to the weak connection between CR aggregates and the hardened cement matrix, SCC's flexural values were dramatically reduced as the CR concentration rose. This is like how the elastic modules' compressive and tensile strengths work. It was claimed that growing the amount of CR particles from 0 to 12% led to a significant decrease in flexural strength value from 5.5 to 4 MPa [47].

At 28 days, Najim and Hall [48] found that SCC with a CR replacement ratio of 15% by volume had a flexural strength of 5.5 MPa, which is lower than the mix made without rubber aggregates. Multiple studies demonstrated that growing the replacement level of CR from 0 to 30% reduced the flexural strength values from 6 to 4 MPa [18,38,54,61]. The reduction in flexural findings may be attributed to the fact that CR aggregates do not have as much elasticity as natural aggregates [56].

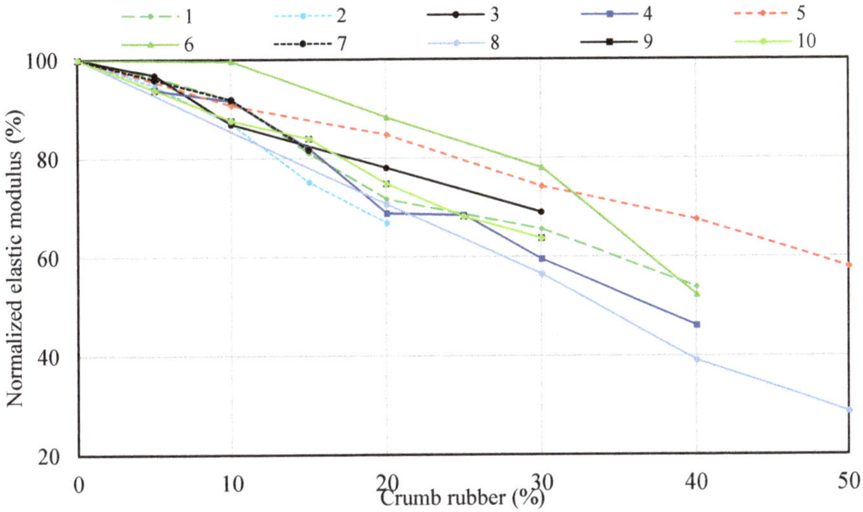

Figure 9. Normalized percentage of elastic modulus: 1 [51], 2 [46], 3 [68], 4 [16], 5 [19], 6 [62], 7 [53], 8 [44], 9 [61], 10 [54].

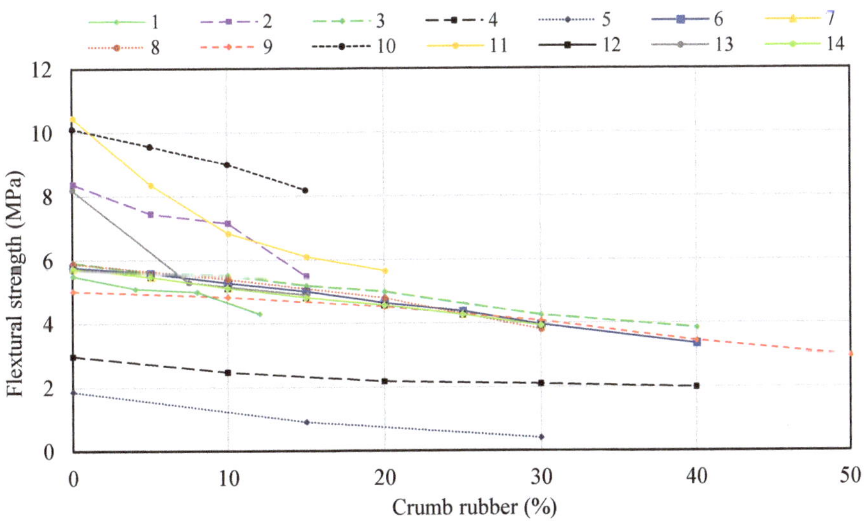

Figure 10. Variation of flexural strength of SCC made with different replacement levels of CRs: 1 [47], 2 [48], 3 [51], 4 [52], 5 [7], 6 [16], 7 [18], 8 [38], 9 [19], 10 [53], 11 [56], 12 [61], 13 [72], 14 [54].

As illustrated in Figure 11, the flexural strength of SCC reduces approximately linearly as the CR particle replacement ratio increases. In one study, for the CR additions of 10%, 20%, 30%, 40%, and 50%, the residual flexural strength in SCC specimens was 96%, 90%,

81%, 68%, and 59%, respectively [19]. In addition, the percent decrease in flexural strength was about 32% for the rubberized SCC made with 30% CR content [51], while for the equivalent CR replacement ratio, the reduction in compressive strength was 42% [53]. These findings are in line with those of other researchers [16,52]. According to AbdelAleem and Hassan [56], at 28 days, flexural test values decreased by 46% as the CR replacement ratio reached 20%. Likewise, to the above mechanical properties, the predicted percentage reduction in the flexural strength of SCC with various contents of CR is tabulated in Table 8 depending on the available literature studies. The reduction in the flexural strength was in the range of 30–40% as the CR content reached 40%.

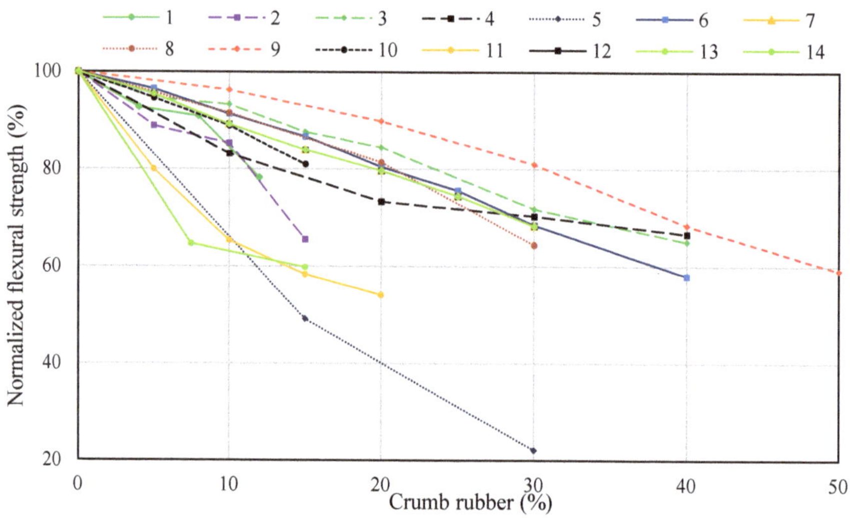

Figure 11. Normalized percentage of flexural strength: 1 [47], 2 [48], 3 [51], 4 [52], 5 [7], 6 [16], 7 [18], 8 [38], 9 [19], 10 [53], 11 [56], 12 [61], 13 [72], 14 [54].

Table 8. Predicted percentage reduction values in the flexural strength of SCC via CR content.

CR Content (%)	10	20	30	40
Flexural strength reduction (%)	5–15	15–25	20–35	30–40

5.5. Bond Strength

Figure 12 shows the bond strength values for the SCC mixtures as a function of CR. Karahan et al. [38] evaluated the bonding strength of a 15 mm diameter reinforcing bar to the SCC by performing a direct pullout test on the reinforcing bars embedded in the 100 × 200 mm cylinder specimens. Bond strength was calculated and evaluated in this investigation using the greatest pullout load sustained during the test. As shown in Figure 12, the rubberized SCC combinations had a lower bond strength capacity than the control mixture, most likely due to the cement paste's poor bonding to the CR aggregate. Although a steady decline in bond strength capacity was observed as CR content increased, the decreases in bond strength were only negligible. For example, raising the CR content from 10% to 30% did not affect the bond strength, which was reduced from 5.2 to 4.8 MPa [38].

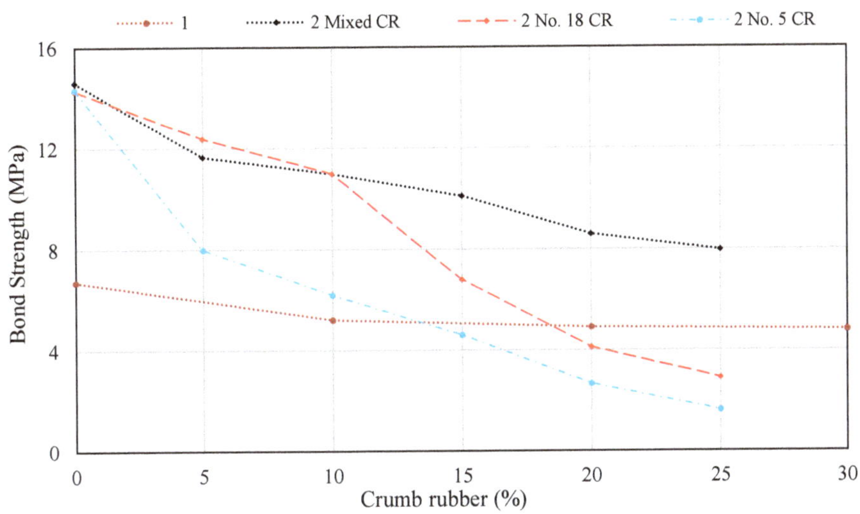

Figure 12. Variation of bond strength of SCC made with replacement levels of CR: 1 [38], 2 [33].

Hilal [33] also inspected the influence of CR on the bond strength behavior of rubberized SCC. A 150 × 150 × 150 mm specimen and a steel bar with a 16 mm diameter were employed for this purpose. In the current study, three different sizes of CR were utilized, number 18 CR, number 5 CR, and mixed CR. A testing machine with 600 KN capacity was outfitted with a specially designed test apparatus to carry out the loading. It was reported that the reference mix had maximum bond strength and that as the CR content increased, the bond strength declined progressively. However, the 25% CR mixture resulted in the lowest bond strength values on the 90th day, regardless of the CR type. This was because the CR particles had low adherence to the adjacent cement paste. The bonding strength of the mixes depicted in the above figure drops as the CR size and quantity rise. With the inclusion of CR replacement ratios of 5%, 10%, 15%, 20%, and 25%, there was a decrease in bond strength values by 20%, 24%, 30%, 40%, and 45%, respectively [33]. Consequently, using CR in the SCC could significantly reduce the bond strength property. Emiroglu et al. [84] also claimed that the reduction in bond strength is attributed to poor bonding qualities around rubber particles and cement matrices. In the rubberized concrete near the ITZ, there are numerous micro-cracks. In this instance, research suggests treating rubber to increase the adhesion between rubber aggregates and the hardened cement matrix.

5.6. Fracture Energy

Fracture energy can be calculated as the amount of energy needed to open a unit area of crack surface [85]. Compared to other mechanical parameters, SCC that includes CR exhibits better ductility than SCC made without rubber particles. Figure 13 depicts the fracture energy of SCC specimens made with or without CR particles. Li et al. [64] explored the impact of CR aggregates on the fracture energy of SCC at a curing time of 90 days. They found that raising the CR content from 0 to 30% improved the fracture energy from 109 to 130 N/m. Moustafa and ElGawady [86] studied the dynamic behavior of high-strength concrete made with rubber aggregate. They reported improved fracture energy outcomes with increased waste rubber particle replacement ratios.

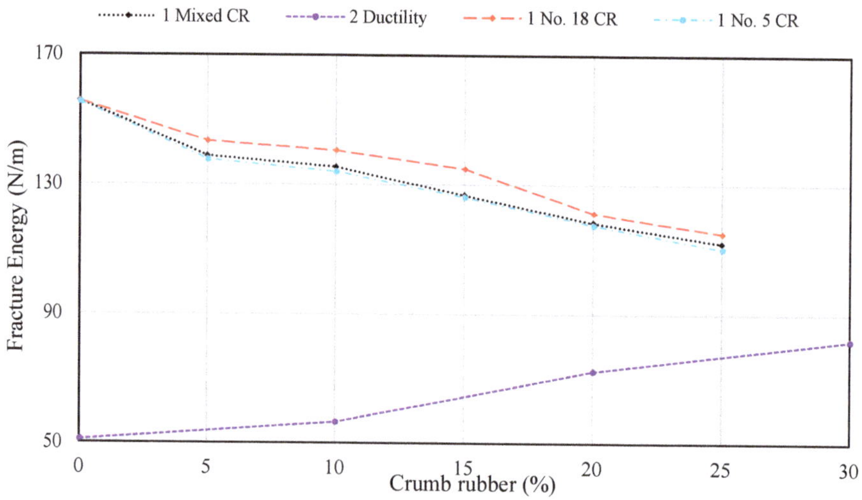

Figure 13. Variation of fracture energy of SCC made with different replacement levels of CRs: 1 [33], 2 [64].

On the other hand, Hilal [33] examined the influence of CR particles on the mechanical characteristics of SCC. It was noticed that the three different types of CR particles display the same influence on the fracture energy test values. Their study determined the fracture energy using 100 × 100 × 500 mm beam specimens. The size of CR particles ranged from 0 to 4 mm. As the mixed-type CR replacement level increased from 0 to 25%, the fracture energy began to decrease dramatically, from 156 N/m to 112 N/m, which meant that CR had a negative effect on the SCC's critical deflection. According to Gesoglu et al. [87], the fracture energy of porous concrete was reduced or increased depending on the category or size of waste rubber particles. With the addition of 20% CR and a 20% combination of tire chips and CR, the fracture energy was reduced by 25% and 74%, respectively, while the fracture energy improved by up to 42% when 10% tire chips or a 20% combination of tire chips and CR were added to the concrete mix. As a result of the preceding, it may be inferred that test values from rubberized concrete, SCC, and conventional concrete combinations are inconsistent. Because both increases and reductions in fracture energy have been documented, it is hard to say whether rubber incorporation enhances or lowers fracture energy. However, fracture energy can be improved with the right quantity of rubber particles, so more research into the fracture energy of rubberized SCC is prospective and should be pursued.

5.7. Characteristic Length

The characteristic length is an indicator of material brittleness. It is primarily determined by important concrete mechanical properties such as fracture energy, elastic module, and splitting strength. The concrete could be less brittle because its characteristic length is higher [88]. It was previously declared that the characteristic length of normal concrete is higher than that of SCC. In Figure 14, the influence of rubber aggregates on the characteristic length parameter of SCC was plotted. Hilal [33] used three types of fine rubber aggregates as a fractional substitution for sand. The three types of crumb rubber aggregates were number 18 CR (0.1–1.00 mm), number 5 CR (0.1–4.0 mm), and mixed CR (0.1–4 mm). An increasing tendency in the characteristic length values can be observed from the above figure as the replacement ratios of rubber aggregate growth, regardless of the rubber size. It was found that the addition of CR at a 25% replacement ratio obtained the maximum value of the characteristic lengths of 718 mm, 681 mm, and 670 mm for the number 18, number 5, and mixed CR types, respectively. Further, with increasing the CR replacement ratios to 5%, 10%, 15%, and 25%, the characteristic length grew by 20%, 17%, 24%, 32%, and 72%, respectively, with respect to the mixed CR type. This fracture characteristic of SCC made with scrap tire rubber requires more research due to a shortage of experimental findings. Other authors, on the other hand, examined the ductility of normal concrete made with scrap tire rubber substituting sand or gravel. According to Vadivel et al. [89], adding rubber aggregate content increased ductility. The impact of different types and sizes of rubber aggregates on concrete ductility was examined by Gesoglu et al. [87]. They found that using CR with a size of 1.00–4.00 mm increased ductility but using tire chips with a maximum size of 10 mm or a very fine CR range of 0.1–1.00 mm reduced ductility.

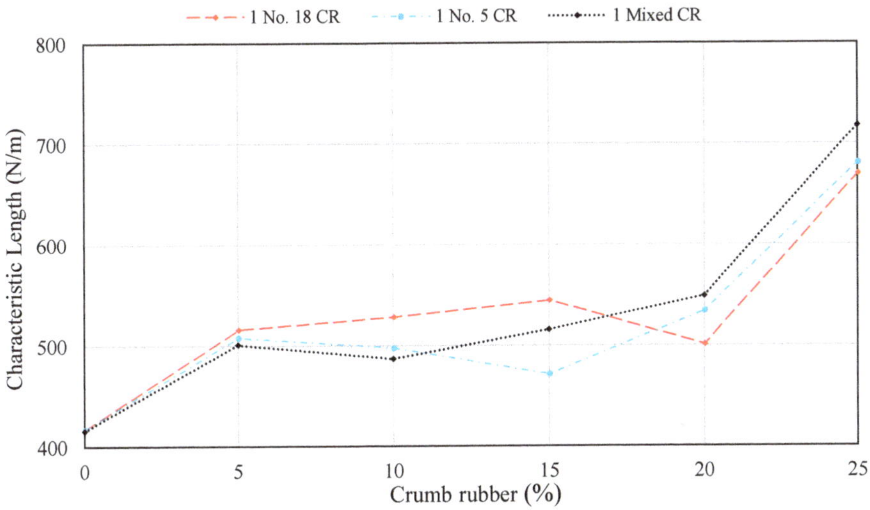

Figure 14. Variation of characteristic length of SCC made with different replacement levels of CR: 1 [33].

Further research is needed due to the limitation of test data on the ductility of rubberized SCC. Because of the improvement in ductility of rubberized concrete, the results presented in Figure 14 are encouraging. SCC ductility could be improved with the proper amount and the correct size of rubber particles.

5.8. Ultrasonic Pulse Velocity (UPV)

Ultrasonic pulse velocity (UPV) testing is used to measure the homogeneity of concrete and its voids. The results of the UPV test of rubberized SCC mixtures are shown in Figure 15. Rahman et al. [4] determined that as the percentage of CR in SCC grew from 0% to 10%, the UPV decreased from 3780 m/s to 3540 m/s at 28 days. Si et al. [60] demonstrated that the UPV of the rubberized SCC was 4260–4015 m/s. There were less UPV readings when more CR was added to the SCC mixture. They achieved the lowest UPV of 4015 m/s for the mix incorporating 25% CR. Yung et al. [49] reported the UPV at 28 days and achieved 4160 m/s for the control mix, but the inclusion of 20% CR could decrease the UPV to 3800 m/s. Hesami et al. [53] highlighted the reduction in the UPV values due to the increase in the CR addition. It was stated that this reduction in UPV values by introducing CR into the SCC mixes could be referred to as the formation of a pore system and the reduction in the elastic module of rubberized concrete. This variance is compatible with the well-established fact that UPV is affected by several variables, including air and CR content, which trap air on its surface. As a result, the UPV value drops as the air content in the rubberized concrete mixture grows due to the increase in the CR percent replacement level [90]. Moreover, the percent decrease in UPV outcomes as a function of CR is shown in

Figure 16. Oikonomou and Mavridou [4] measured the UPV test results for the rubberized SCC mixes with a percent decrease of 7%, 11%, and 17% for the CR addition of 5%, 10%, and 15%, respectively. Moreover, for the 15% CR content, a percent decrease in UPV values of 8% was obtained [48,53]. In the summarized UPV results, the highest CR replacement ratio was 20%, which caused a reduction of 6% [60].

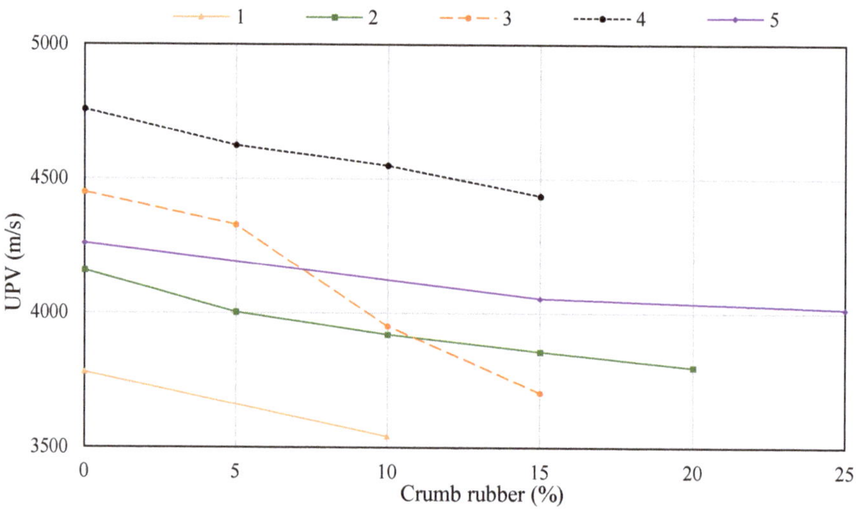

Figure 15. Variation of UPV of SCC made with different replacement levels of CR: 1 [4], 2 [49], 3 [48], 4 [53], 5 [60].

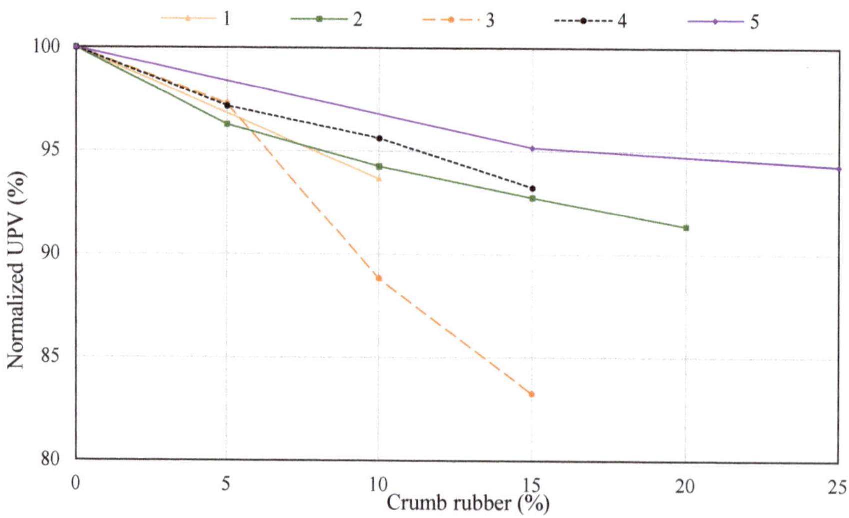

Figure 16. Normalized percentage of UPV: 1 [4], 2 [49], 3 [48], 4 [53], 5 [60].

6. Impact of CR on Durability Behavior of SCC

The following table (Table 9) summarizes the durability characteristics of the rubberized SCC as reported in the summarized literature. Major research that examined the impact of CR particles on the durability behavior of SCC used a CR addition of less than 30% of the total fine aggregate volume. The durability aspects of the rubberized SCC that have been investigated in the literature were the sorptivity, electrical resistivity, shrinkage, rapid chloride permeability, carbonation depth, chloride penetration depth, and impact resistance. Except for electrical resistivity, depth of carbonation, and impact resistance, which show a different trend, the utilization of CR in SCC mixes resulted in a decrease in the durability of concrete samples. Compared to mechanical properties, there is a dearth of research on the durability of rubberized SCC composites.

Table 9. Durability properties of rubberized SCC concrete reported in the literature.

Ref.	Sorptivity (mm/min 0.5)	Rapid Chloride Permeability (coulombs)	Depth of Chloride Penetration (mm)	Shrinkage (%)	Electrical Resistivity (kΩ-cm)	Depth of Carbonation (mm)	Impact Resistance
[15]	✓	✓					
[30]			✓		✓	✓	
[49]				✓	✓		
[52]							✓
[64]	✓	✓					
[38]		✓					
[60]				✓	✓		
[44]				✓			
[61]							✓
[54]							✓

6.1. Sorptivity

For the sorptivity test, silica gel was applied on each specimen's side surface, while a plastic sheet was placed over each top. In the pan, the samples were positioned with the test surface side submerged from 1 mm to 3 mm in water. After initial contact with water, each specimen's mass was recorded and weighed at various time intervals to determine the mass increase. Figure 17 shows the sorptivity height in mm of SCC made with four percent levels of CR. According to Li et al. [64], the sorptivity height decreased gradually with the inclusion of CR particles. They measured sorptivity that fell from 1.71 to 1.04 mm for the CR content ranging between 0% and 30%, respectively. Segre and Joekes [91] and Oikonomou and Mavridou [88] have said that there are similar trends. The reason for the above could be due to two factors: Firstly, the hydrophobic property of rubber results in an interaction angle greater than 90 degrees between CR particles and the cement matrix, reducing the capillary action of water to penetrate specimens. Secondly, CR particles can make capillary tubes longer and more curved, stopping water and chloride ions from entering specimens [92].

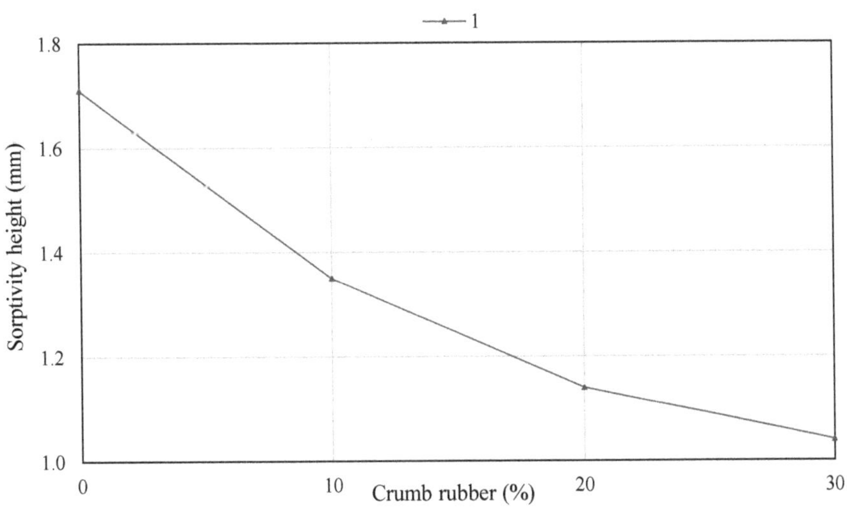

Figure 17. Variation of sorptivity height in mm of SCC with different replacement levels of CR: 1 [64].

On the other hand, Figure 18 shows the sorptivity coefficients of SCC versus CR and fly ash replacement levels for 28 and 90 days of curing. The fly ash replacement levels of 0%, 20%, 40%, and 60% as a substitution for cement were cured for 28 days and 90 days. Unlike the findings depicted in Figure 17, it can be noted from Figure 18 that the sorptivity increased significantly as the CR percentage increased, regardless of the fly ash content and curing time. After 28 days, the sorptivity of the reference mix was 0.078 mm/min 0.5, but it increased to 0.106 mm/min 0.5 when 25% CR was added. In addition, for the SCC mixes made without CR, the sorptivity coefficients evaluated after 90 days showed superior sorptivity values than those at 28 days of curing age [15]. However, the rubberized SCC mixes that have a different percent level of fly ash as a substitution for cement binder exhibit a reduction phenomenon in the sorptivity values more so than the rubberized concrete mixes made without the fly ash binder. For instance, the non-fly ash content of concrete mixes at 90 days of sorptivity was approximately 7% lower than that observed at 28 days.

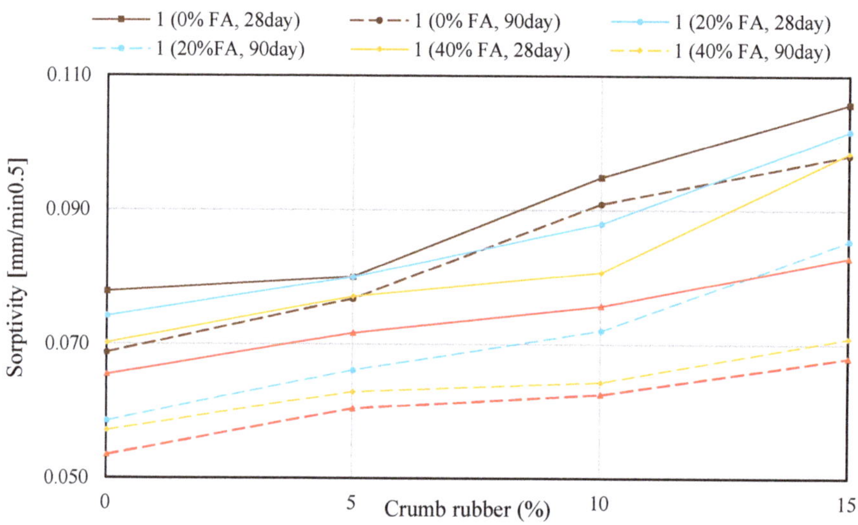

Figure 18. Variation of sorptivity in mm/min 0.5 of SCC made with different replacement levels of CR and fly ash at different curing days: 1 [15].

Meanwhile, this change was shown to have a bigger impact on the fly-ash-containing rubberized SCCs. This was because of the fly ash's long-term pozzolanic action [15]. Additionally, it was formally found that the type of aggregates, supplementary cementitious materials, and curing time all play a vital role in the water sorptivity values of a concrete specimen [15]. Therefore, to produce a rubberized concrete with CR particles, the utilization of fly ash as a substitution instead of cement and prolonged curing time could be an efficient way to advance its sorptivity property.

6.2. Chloride Ion Permeability

The impact of CR content on the chloride ion permeability values (in coulombs) of SCC mixes is shown in Figure 19. Karahan et al. [38] demonstrate the impact of CR particles on the resistance of SCC to chloride ion penetration. The electrical charge transmitted through concrete specimens was used to express the chloride ion permeability resistance in coulombs. It was observed that the lowest value of chloride ion penetration was recorded for the control mix that contained no CR (755 coulombs). In total, 860, 994, and 1196 coulombs of electricity flowed over the SCC mixes containing 10%, 20%, and 30% CR. As shown from this experiment, the rubberized SCC had a weaker restriction to chloride ion absorption than the reference mix. The percent increases in chloride ion permeability were 14%, 32%, and 58% for the mixes with 10%, 20%, and 30% CR content, respectively.

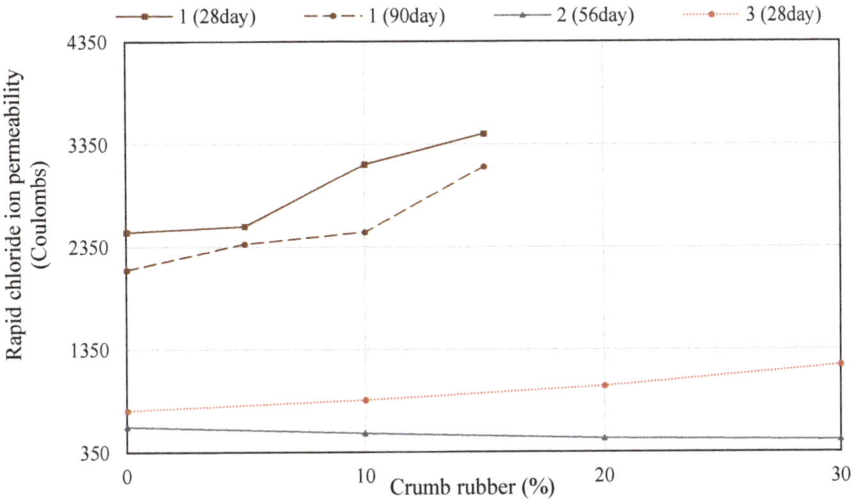

Figure 19. Variation of rapid chloride permeability in coulombs of SCC with different replacement levels of CR: 1 [15], 2 [64], 3 [38].

Similarly, Gesoglu and Guneyisi [15] revealed an increasing chloride ion permeability value after 28 days when 15% CR is employed in SCC production. On the contrary, Li et al. [64] demonstrated that as the CR replacement ratio grew from 0% to 30%, the chloride ion permeability dropped from 597 to 469 coulombs. Further, Figure 20 demonstrates the influence of CR on the chloride permeability depth in mm. From the above-mentioned figure, increases in CR content increase chloride penetration depth. For example, an SCC mixture containing 15% CR increased chloride ion penetration by 35%. This could be because CR particles tend to make concrete materials more porous, make internal packing less dense, and create micro-cracks in the ITZ [30].

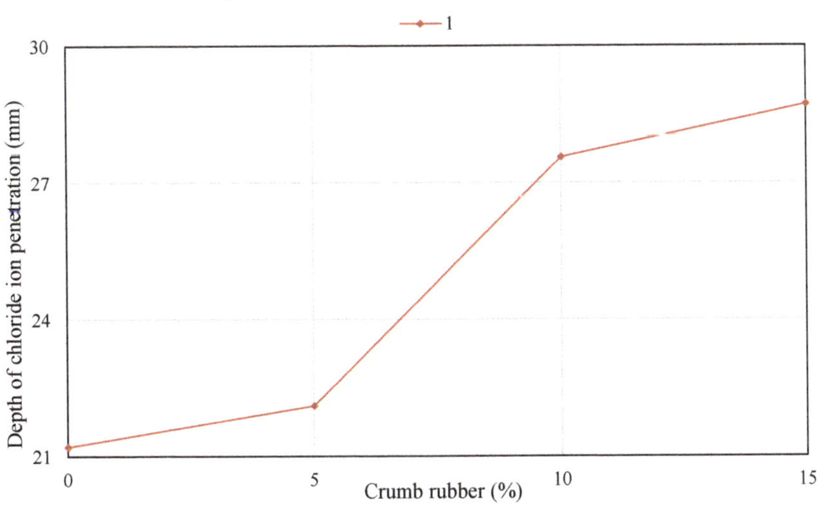

Figure 20. Variation of the depth of chloride penetration in mm of SCC with different replacement levels of CR: 1 [30].

Figure 21 illustrates the chloride ion permeability findings via CR and fly ash levels and the testing age. According to Figure 21, the rubberized SCC's chloride depth was between 1904 coulombs and 3460 coulombs and 476 coulombs and 3139 coulombs after 28 days and 90 days, correspondingly. The chloride ion penetration increased linearly with increasing CR content, particularly in concrete without fly ash. Increasing CR from 0% to 15% in the first batch of concrete (at 28 days and 0% fly ash), the chloride depth of SCC rose from 2491 to 3451 coulombs, respectively. While for the comparable mixes the chloride ion values decreased as the curing time prolonged to 90 days. As a result, the increased CR content increased the chloride ion penetration, but the results diminished with increasing curing time. Moreover, the addition of fly ash did not result in a significant decrease in the chloride ion permeability of the concrete at 28 days of curing. After 28 days, the chloride ion permeability of the concrete that had fly ash added to it dropped from 2491 coulombs to 2320, 2180, and 1904 coulombs when 20%, 40%, and 60% of the fly ash was added. However, it was found that introducing fly ash into rubberized SCC mixes greatly increased their resistance to chloride penetration when the curing time was prolonged to 90 days. Regardless of the CR replacement ratio, the SCC with 20%, 40%, and 60% fly ash additions had an average reduction in chloride ion permeability of 67%, 79%, and 78%, respectively. This meant that the concrete went from moderate to very low. This can be explained by the fact that the detrimental effect of CR on chloride penetration was significantly reduced with the addition of fly ash after 90 days. As Manmohan and Mehta [90] noted, this conclusion is related to the long-term interaction of fly ash, which improves the concrete's pore structure, reducing chloride ion infiltration. The improved pore structure reduces the permeability of chloride ions.

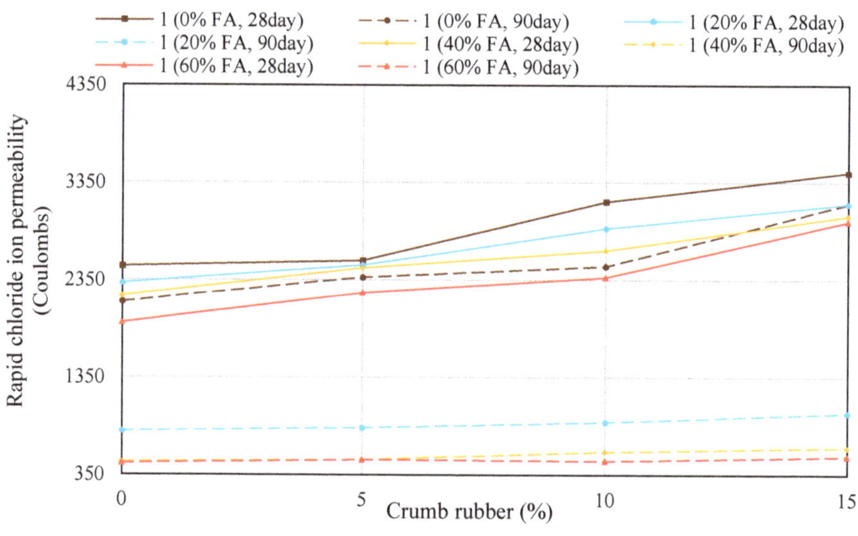

Figure 21. Variation of rapid chloride ion permeability of SCC made with different replacement levels of CR and fly ash at different curing days: 1 [15].

6.3. Shrinkage

Figure 22 depicts SCC's free-drying shrinkage values with different rubber replacement levels and curing times. It is clear from the test results shown in Figure 22 that the use of CR causes an increase in free-drying shrinkage measurement in comparison to the mix manufactured without CR. Yung et al. [49] investigated the shrinkage behavior of SCC made with different levels of CR particles. The CR replacement levels were 0%, 5%, 10%, 15%, and 20% as a substitution for natural fine aggregates by volume. The specimens were subjected to drying conditions for up to 60 days. Their investigation discovered that adding CR caused a higher shrinkage value than the mix made with no CR, regardless of the drying age. The shrinkage value of the 28-day rubberized SCC of the reference mix was about 0.0183%, and the overall reduction in shrinkage outcome of concrete with CR that had been passed through a # 30 sieve was 0.0294% when the CR addition reached 5% of the total fine aggregate volume. At 5% CR addition, the reduction in shrinkage was the smallest, nearly 35% higher than the reference mix. However, with increasing the CR addition to 20%, the shrinkage value was the maximum, and the average shrinkage value was 0.0336%, which was 90% greater than the reference mix.

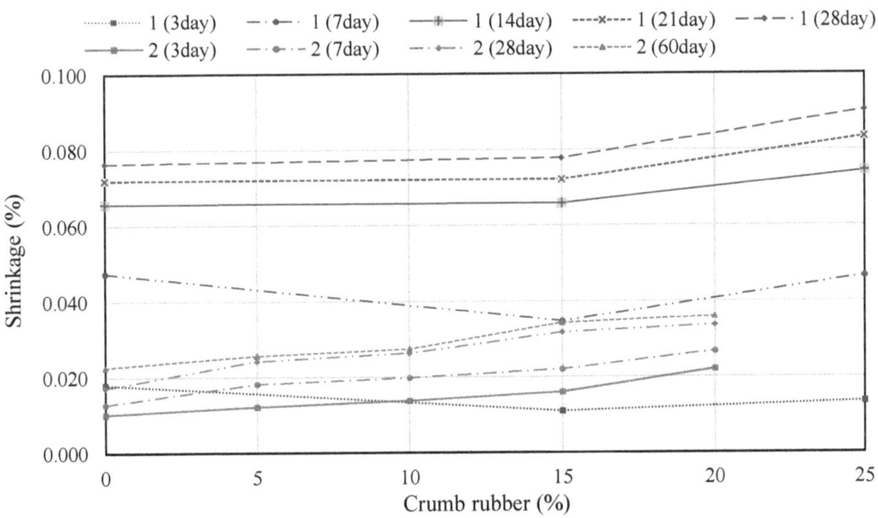

Figure 22. Variation of shrinkage of SCC with different replacement levels of CR and drying times: 1 [60], 2 [49].

Since a portion of the fine aggregate was replaced with rubber, which had a substantially lower elastic module than the other materials, its deformation capability was limited; consequently, its shrinkage was greater than that of ordinary concrete [93]. Si et al. [60] also examined the impact of rubber particles on the shrinkage performance of SCC. The shrinkage value of SCC mixes increased with the increased CR content, irrespective of the curing time. This could be because the elastic module of rubber aggregates is lesser than that of sand. Zaoiai et al. [94] conducted another study investigating the shrinking of SCC using rubber aggregate. They substituted a portion of the natural gravel and sand with rubber aggregate at various levels and recorded shrinkage reductions at 28, 90, 200, and 300 days when the rubber content was increased. After 300 days, the greatest degree of shrinkage was observed. The results indicated that after 300 days, shrinkage was reduced by 16%, 33%, 20%, and 5%, accordingly, when 2.5, 5, 10, and 20% rubber contents were added. As compared to the reference SCC, rubberized SCC undergoes greater displacement while the cement matrix starts to shrink.

Further, shrinkage could be decreased by treating the rubber aggregate in the rubberized SCC mix with NaOH [60]. Due to the length change over time, shrinkage can be a critical factor in the design of concrete elements [94]. For this aim, shrinkage values of normal concrete were also discussed. Bravo and de Brito [95] investigated the shrinkage properties of normal concrete by substituting tire rubber aggregate (a size range of 4–11.2 mm) for natural aggregates at varied replacement percentages of 5%, 10%, and 15%. They observed an increase in shrinkage as the volume of rubber aggregate in the concrete mixture increased. The rise in shrinkage after 90 days was roughly 45% more than that of the control mix, with a rubber replacement level of 15%. As seen from the above, the shrinkage of SCC and conventional concrete increases with the use of more waste-tire rubber in the mix, irrespective of the rubber size.

6.4. Electrical Resistivity

CR can significantly impact the electrical resistivity over time, as shown in Figure 23. The electrical resistivity of concrete is determined by the pore structure of the material and the ionic concentration in the pore water [96]. The high resistivity of the concrete may help limit the ionic current between the anodic and cathodic sites, reducing steel corrosion in steel-reinforced concrete [97]. Thus, the structure's durability could be enhanced by increasing the concrete mixture's electrical resistivity. Yung et al. [49] indicate that adding CR aggregate increased electrical resistance. They discovered that when the CR content is increased from 0% to 20%, the electric resistivity increases by 19%, 37%, 215, and 35% after 7, 28, 56, and 91 days, respectively. Similarly, Si et al. [60] stated that incorporating CR aggregate improved electrical resistance. They discovered that an increased CR concentration from 0 to 25% enhanced electrical resistivity by 9%, 22%, 28%, 37%, and 52% after 3, 7, 14, 21, and 28 days, respectively. Moreover, at day 28, the electrical resistivity of SCC containing 15% and 25% rose by 51.27% and 52.28%, respectively. Rubber is an insulating, waterproof, and flexible substance. It acts as a barrier to pore fluid transmission [98,99]. Thus, the rubber particles can be used to improve the resistance of concrete to electricity. Contrary to the above findings, Mallek et al. [28] demonstrate that electrical resistivity lowers when the CR concentration increases. They declared that rubber improves porosity and hence facilitates current flow. The relationship between electrical resistivity and porosity is inverse. The drop in electrical resistivity with increasing CR addition could be explained by rubber's lower resistivity compared to natural aggregates. In summary, if the right amount and design are considered, the use of CR in the production of SCC may be beneficial in terms of electrical resistivity. Increasing the electrical resistivity of concrete can enhance concrete's resistance to corrosion. According to a previous study, there is a clear tendency for the corrosion rate to decrease as the resistivity of concrete increases [97].

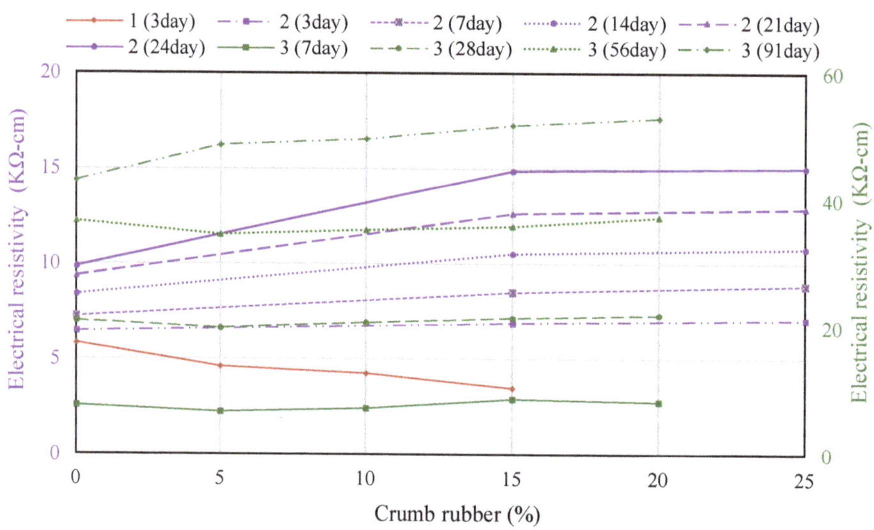

Figure 23. Variation of electrical resistivity of SCC with different replacement levels of CR and curing ages: 1 [30], 2 [60], 3 [49].

6.5. Carbonation

The utilization of CR changes the carbonation depth after different CO_2 exposure times (see Figure 24). Mallek et al. [30] showed that after 15 days of CO_2 exposure, the carbonation depth of rubberized SCC rose by approximately 1.7 %. On the other hand, the outcomes generally indicate that carbonation depth increases with increasing CO_2 exposure time. Besides that, it has also been found that the carbonation level drops as the CR content rises. A decrease of 3.0%, 5.6%, and 4.8% in carbonation depth can be seen after 30 to 45 days of CO_2 contact, when the CR concentration is increased from 0% to 15%, respectively. This finding may be due to CR's hydrophobic nature, which retards the carbonation movement. Rubber is characterized by its hydrophobicity, which means that it does not interact readily with water. This can minimize the amount of liquid CO_2 surrounding these CR particles, thus reducing carbonation depth. Additionally, CR aggregates can deceive hydrates that help the carbonation process by forming a physical obstacle that retards the carbonation phenomenon. Further, Luhar et al. [100] showed that when sand was substituted with rubber fiber, the carbonation depth decreased. They demonstrated that a rise in water permeability and absorption could lead to a rise in carbon depth. The maximum carbonation depth of 9.0 mm was measured for normal concrete exposed to CO_2 for 90 days with a 30% rubber content.

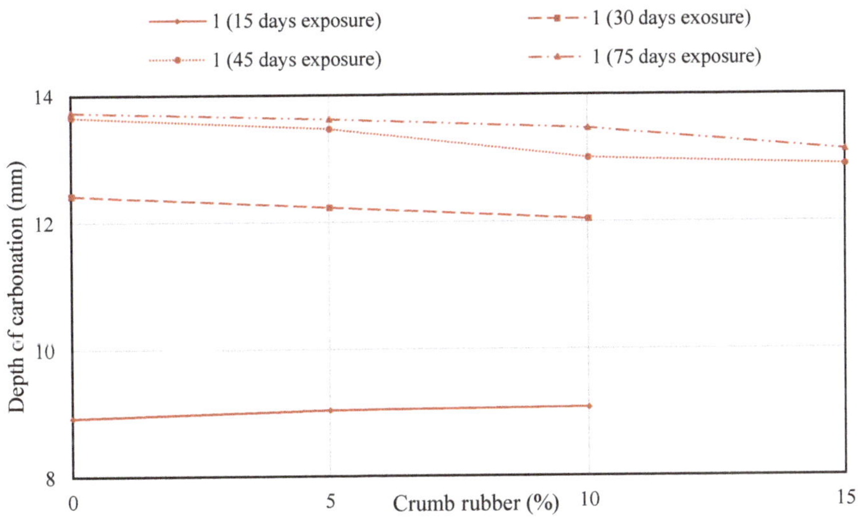

Figure 24. Variation of the depth of carbonation of SCC with replacement level of CR: 1 [30].

6.6. Impact Resistance

Figure 25 shows the effect of CR particles on the impact resistance of SCC as a function of the number of blows. AbdelAleem et al. [61] studied the influence of a CR aggregate on the impact resistance of SCC. It was observed that when CR was added to SCC mixtures, it made concrete more impact resistant. The impact resistance of concrete can be measured using two different approaches: impact resistance by drop-weight test and impact resistance by flexural loading tests (impact energy). The drop-weight test values indicate an increasing trend in blow numbers at ultimate failure by nearly 91% and in blow numbers required for the first crack by about 89% as the CR inclusion rises from 0% to 30%. The flexural loading test indicated an improvement in ultimate impact energy of up to 2.42 times as the CR percentage rises from 0% to 25%. Khalil et al. [52] studied the impact resistance of SCC with the presence of CR aggregates. When the proportion of CR varies between 0% and 30%, drop-weight values reveal an increase in ultimate failure and the first crack of 2.8 and 3 times, respectively. Ismail and Hassan [54] observed an enhancement in the impact resistance of SCC with CR content in both impact resistance tests (drop-weight and flexural loading). Flexural loading values indicated that the utilization of 30% CR enhanced ultimate failure by 116%.

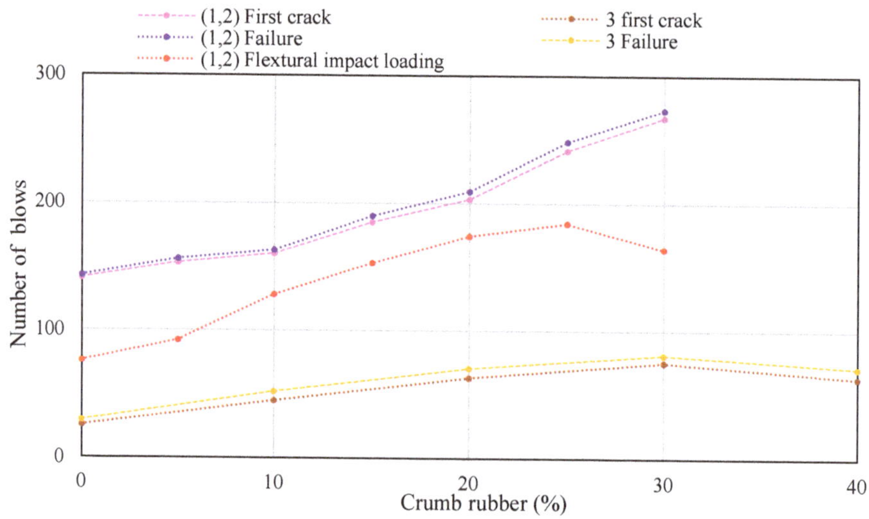

Figure 25. Variation of average blows of SCC with different replacement levels of CR: 1 [61], 2 [54], 3 [52].

Furthermore, when rising the CR addition from 0% to 30%, blow numbers for the first crack and the number of blows for the ultimate failure increased from 141 to 277 and 263 to 277 and 263, respectively. As expected, the addition of CR improves the energy absorption capacity of rubberized SCC in contrast to SCC, with an increase in the required blow numbers to induce the initial crack and failure. This behavior could be explained by the CR particles' ability to absorb energy due to their high deformability. In other words, CR particles act as springs, causing fracture propagation to be slowed [52,89,101]. This may be related to a weak ITZ between the CR and the matrix, which acts as a starting point for cracks, as well as the angularity of these components, which elongates the crack pathway, i.e., the crack deformation increases. Other researchers have previously established the above conclusion by investigating the impact resistance of rubberized normal concrete mixtures [89,102].

In summary, introducing CR particles in manufacturing SCC or normal concrete increased the material's impact resistance. However, the behavior of rubberized SCC and impact resistance loading remains mostly unclear. Additional experimental investigations are warranted in light of the potential application of rubberized SCC in concrete structures.

7. Microstructure

One of the most important factors that affects hardened self-compacted concrete (SCC) properties is the zone between the cement paste matrix and the rubber particles called the interfacial transition zone (ITZ). The mechanical and durability characteristics of concrete are directly influenced by the aggregates' adhesion to the cement matrix. This microstructural analysis could be made by scanning electron microscopy (SEM) to examine the ITZ between the rubber particles and the cement paste matrix. According to Guo et al. [103], one of the primary disadvantages of rubberized concrete is the poor bond between rubber particles and the cement paste matrix, which diminishes compressive strength. To solve this drawback, the surface of the rubber particles should be treated using surface modifiers (such as NaOH) or an admixture (such as silica fume) [104]. Bignozzi and Sandrolini [43] demonstrated good adhesion between tire rubber and the cement matrix after rubber particles had been pre-treated and coated with cement matrix. Zaouai et al. [69] revealed a poor bond between the cementitious matrix and rubber particles. This resulted in a highly porous region. This weak adhesion may contribute to the decrease in mechanical strength (see Figure 26a). Ziebelmann [105] demonstrated that rubber particles do not react chemically with cement paste and that the use of rubber particles cannot form chemical bonds. These findings explain the lack of adhesion observed in the SEM and justify the increased air content and large porosity reported in composites containing rubber particles. By modifying the poor adherence of the ITZ, it is feasible to obtain enhancement in the properties of SCC-containing rubber aggregates. The pre-coating of rubber aggregates with crushed dune sand in SCC improves the ITZ (see Figure 26b). This mineral admixture acts as a micro-filler to fill the ITZ between the surface of rubber particles and the cement paste matrix [69]. Najim and Hall [106] utilized several pre-treatment techniques for rubber particles, including NaOH treatment, cement paste coating, water washing, and mortar coating. Using SEM, microstructural analysis and porosity were investigated, as can be seen in Figure 27. In the specimens that were pre-coated with mortar, the compressive strength and splitting tensile strength increased by 37 and 19%, respectively, compared to un-treated rubber. This was due to a significant improvement in the ITZ. There is still a lack of studies about microstructure analysis of rubberized SCC; therefore, further studies are required.

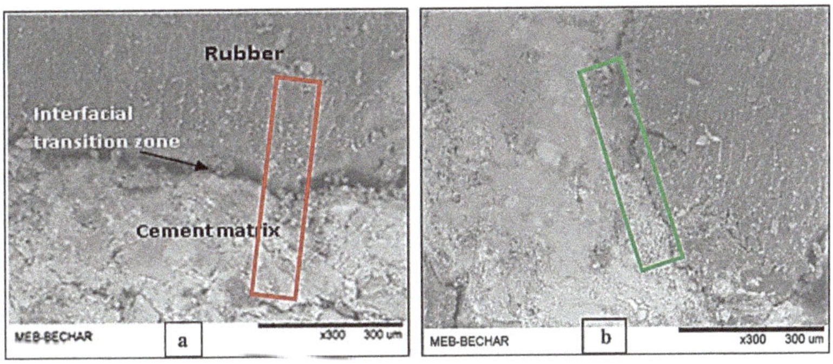

Figure 26. SEM image of ITZ between rubber aggregates and cement paste, (**a**) un-coated CR particles, and (**b**) pre-coated CR particles [69].

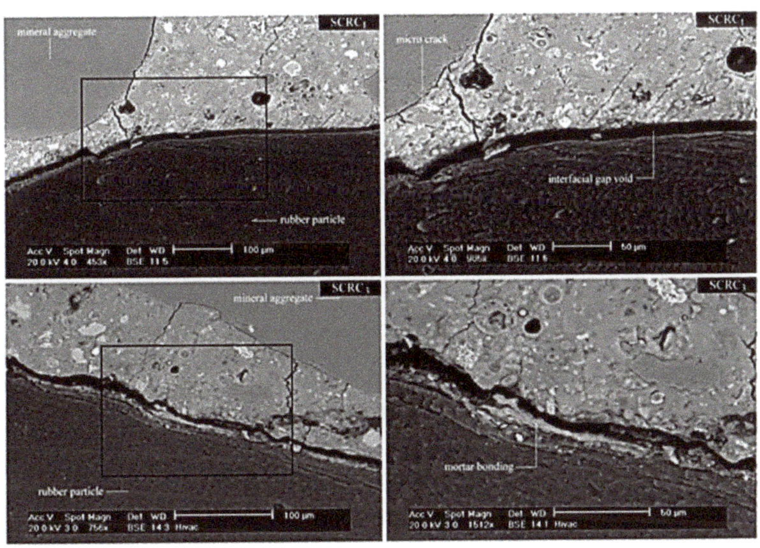

Figure 27. SEM image of ITZ between cement paste and rubber aggregate, without pre-treatment (SCRC1), and with pre-coating with mortar—(SCRC3) [106].

8. Empirical Relationships among Hardened Properties

The hardened properties of several SCCs with different replacement levels of CR were conducted in this section to generate empirical models between different hardened properties under the influence of CR particles (see Figures 28–32 and Equations (1)–(5)). In order to determine whether the current relations are strong, the coefficient of determination (R-square) of the rubberized SCC for the various hardened properties was also determined. The constructed models could help researchers estimate the hardened properties of this type of developed concrete.

8.1. Compressive vs. Splitting

The relation between the compressive and splitting strengths of SCC comprising CR particles is depicted in Figure 28. Experimental data from prior research studies demonstrate a power function between the compressive and splitting tensile strengths of rubberized SCC. Both parameters were in good agreement with a coefficient of determination (R-square) of 0.66. Equation (1) can be implemented to predict the splitting tensile values of rubberized SCC regarding compressive strength.

$$\text{Splitting tensile strength} = 0.22\text{compressive strength}^{0.72} \tag{1}$$

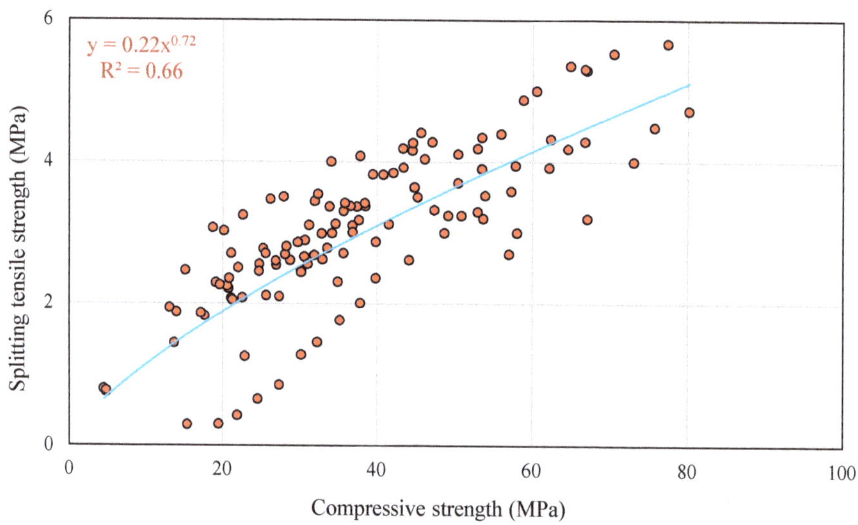

Figure 28. Compressive strength and splitting tensile strength relation of rubberized SCC.

8.2. Compressive vs. Flexural

From the conducted results in the summarized literature, a relation between compressive and flexural strength for rubberized SCC was drawn, as depicted in Figure 29. As a result, a power relation between the compressive and flexural strength of SCC mixes made with various replacement levels of CR aggregates was constructed with an R-square value of 0.58, which shows an exponential relation. Moreover, Equation (2) was proposed to predict the flexural strength of rubberized SCC in correspondence to its compressive strength.

$$\text{Flexural strength} = 0.51\text{compressive srength}^{0.62} \tag{2}$$

8.3. Compressive vs. Modulus of Elasticity

Compressive strength for rubberized SCC has been shown to correlate with its corresponding modulus of elasticity based on the reviewed literature, as shown in Figure 30. As a result, an R-squared value of 0.77 was calculated between the properties above rubberized SCC. According to Equation (3), the modulus of elasticity of SCC made with rubber can be predicted concerning its compressive strength.

$$\text{Modulus of elasticty} = 3.43\text{compressive strength}^{0.55} \quad (3)$$

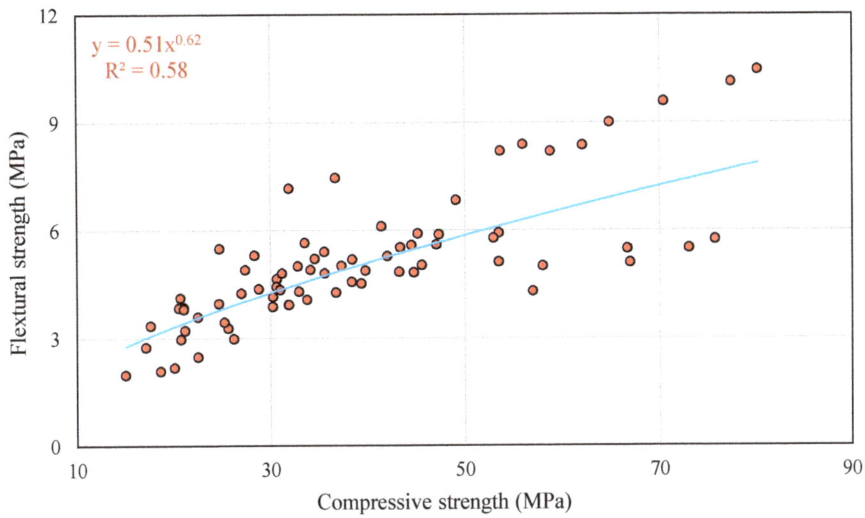

Figure 29. Compressive strength and flexural strength relation of rubberized SCC.

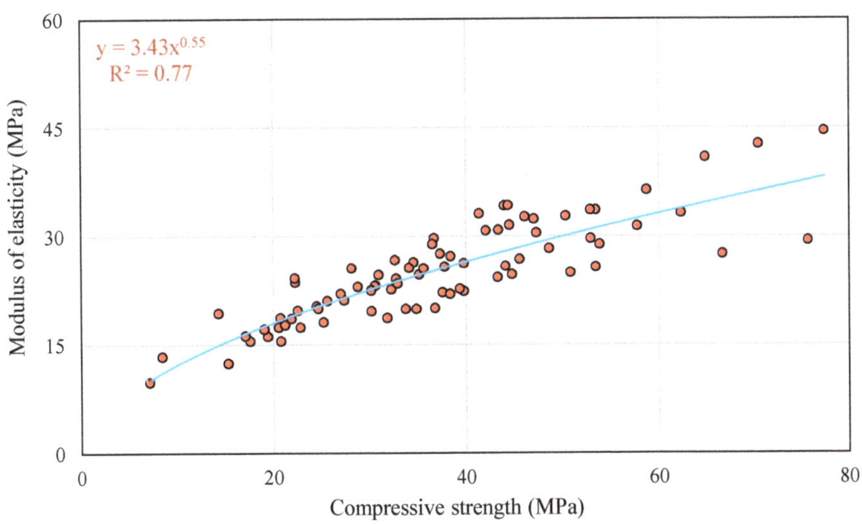

Figure 30. Compressive strength and modules of elasticity relation of rubberized SCC.

8.4. Compressive vs. UPV

Figure 31 displays the connection between UPV and compressive strength of the previous experimental results for SCC produced with CRs. As shown in the figure below, fitting a power curve can result in a very strong relationship between the two property equations, with an R-square of 0.88. Unlike the previous relations, the compressive strength of rubberized SCC can be predicted depending on the existing UPV results, as shown in Equation (4).

$$\text{compressive strength} = 0.01 \times \text{UPV}^{5.72} \tag{4}$$

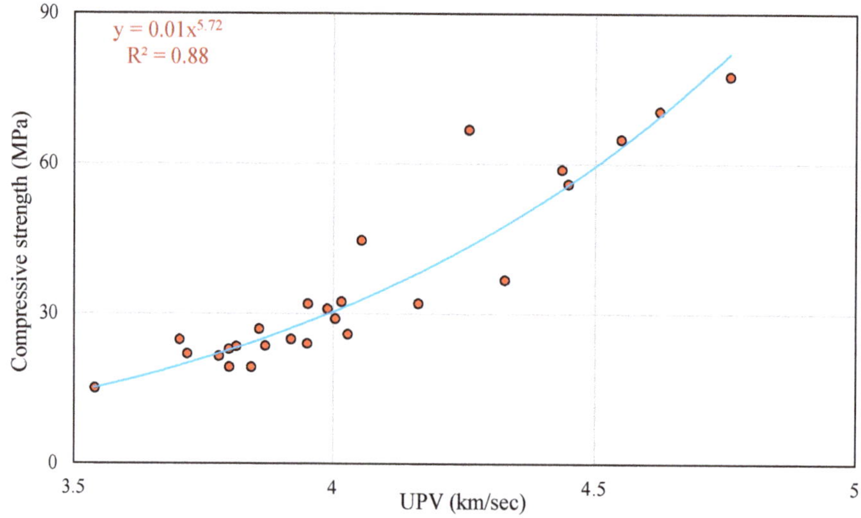

Figure 31. UPV and compressive strength relation of rubberized SCC.

8.5. Compressive Strength vs. Bond Strength

According to the earlier experimental studies, Figure 32 shows that the compressive and the bond strength of the rubberized SCC are well connected with a power function. The R-square value of 0.88 suggests that they have a very good relationship. From Equation (5), the bond strength of SCC made with CR can be forecasted if the compressive strength data exist.

$$\text{Bond strength} = 0.24 \times \text{compressive strength}^{0.91} \tag{5}$$

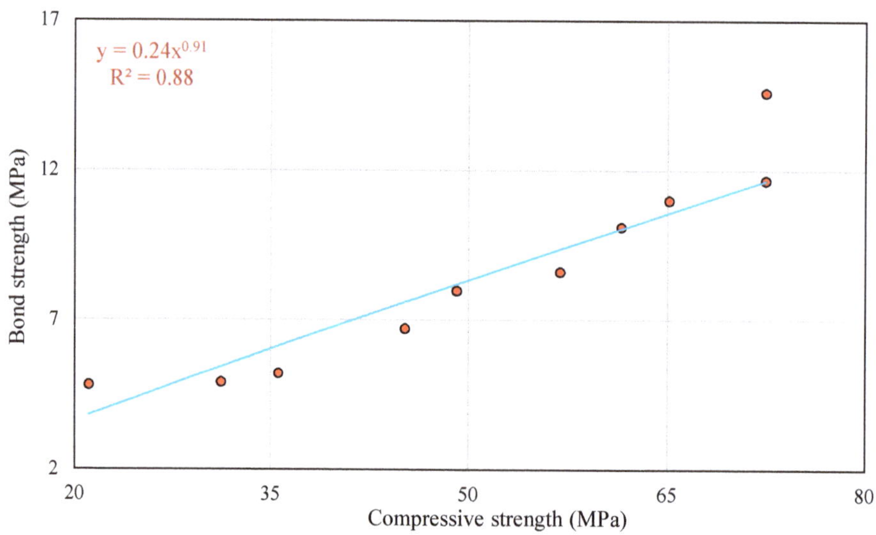

Figure 32. Compressive strength and bond strength relation of rubberized SCC.

9. Discussion

This study aims to better understand the mechanical and durability properties of self-compacted concrete (SCC) with different amounts of crumb rubber replacement (CR). According to the existing research, substituting fine aggregates with CR particles in forming SCC mixes does not exceed 50% of the total fine aggregate volume, whereas most studies employ 30% CR content. This paper presents a complete evaluation of the prior literature on this subject. The following findings can be taken from the comprehensive critical review of the literature data that were previously published:

1. The incorporation of CRs can reduce the mechanical properties of SCC, including compressive, splitting tensile, elastic module, and flexural parameters. The utilization of CR as a partial substitution for natural fine aggregate significantly reduces the strength of SCC, irrespective of the amount of rubber aggregate used. This impact can be related to the limited adhesion and bond strength between CR particles and the cement matrix, the poor stiffness of rubber aggregate in comparison to natural aggregates, and an increased amount of air entrapped within rubber CR particles and the cement matrix. However, by adding mineral admixture (such as fly ash), these qualities of rubberized SCC can be even better.
2. It seems that rubberized SCC with CR could be used in certain structural parts, but substitute levels must be sensibly calculated in order to keep a significant level of mechanical properties. This article presents test findings that support this hypothesis. According to the available literature data, predicted reductions in some of the mechanical properties of SCC were shown in Tables 5–8, and it can be realized that using CR up to 30% of total sand volume could be an appropriate replacement level.
3. For SCC that includes CR, very little is known about fracture parameters. The increased and decreased fracture energy of SCC have been reported in the literature due to the addition of CR aggregates. Additionally, using more CR volume fractions enhanced the ductility of SCC in terms of characteristic length. However, the ductility of rubberized SCC needs to be studied further.
4. Similar to the aforementioned mechanical properties, the inclusion of CR particles into the SCC mixes causes a reduction in bond strength values as well. Because of this, it is clear that the cement matrix and CR aggregates have weak bonding characteristics, causing a decrease in binding strength. Research suggests treating rubber to improve the bond between the rubber and the cement matrix to address this issue.
5. Incorporating CR into the SCC mixture decreased the ultrasonic pulse velocity (UPV) values, indicating that the CR impacted the pore structure of the SCC mixture. The pretreatment method for rubber aggregate eliminated the decline in ultrasonic pulse velocity (UPV).
6. A few studies are investigating the durability behavior of rubberized SCC, and a lack of information exists.
7. An earlier research study discovered that the sorptivity of SCC decreases with the increase in CR content as a partial alternative to sand. However, a different investigation claims the opposite. It was also reported that extending curing time and adding fly ash partially instead of cement can improve the sorptivity coefficient of rubberized concrete. Rubberized SCC's sorptivity quality has to be properly understood through further experimentation.
8. Regardless of the curing time and fly ash content, a significant increase in chloride ion permeability results was recorded as the percentage of CR increased. In theory, this may be related to the fact that CR particles tend to increase the porosity of concrete, making it less dense and creating micro-fissures in the ITZ. Prolonged curing time and using fly ash could improve the chloride ion permeability of rubberized SCC. With using fly ash as a partial substitution for cement, a steady decrease in chloride ion permeability was experimented with at 28 days of curing age, whilst a remarkable reduction in chloride permeability outcomes of rubberized SCC mixes was noticed for a long-term curing age (90 days). This result comes from the long-term action of fly

ash on concrete, which modifies the pore structure of the concrete, lowering chloride ion infiltration.

9. Adding rubber aggregates to SCC results in greater shrinkage. The behavior of normal rubberized concrete is similar to the rubberized SCC. Shrinkage values of rubberized SCC increase as CR% rises. This happened because rubber particles have a lower elastic modulus than natural fine aggregate.
10. When CR particles were added to SCC, the electric resistance increased. Only a few studies came to the inverse conclusion. Rubberized SCC needs to be studied further to comprehend its potential applications in concrete construction.
11. Based on the findings of accelerated carbonation tests, it can be inferred that the existence of CR enhances carbonation resistance. This was due to the hydrophobicity of CR particles that do not easily combine with water, leading to a reduction in the amount of CO_2 around the CR particles. Rubberized SCC, on the other hand, may be good for buildings that are exposed to harsh weather and need to be resistant to carbonation.
12. Enhancement in the impact resistance of rubberized SCC is expected with the utilization of CR particles. Including CR improves rubberized SCC's ability to absorb energy when compared to SCC. This is because CR can absorb energy because of its unique stiffness property.

10. Conclusions and Recommendations for Further Research

10.1. Conclusions

The main purpose of this recent review was to show and investigate a complete literature evaluation on the influence of CR as a partial substitution for sand (by volume) on the mechanical and durability properties of SCC, which led to important findings:

1. Self-compacted concrete can be made by utilizing crumb rubber (CR) as a part of sand to make it more sustainable and environmentally friendly.
2. From this review, it was clarified that most studies implemented CR as a partial alternative to fine aggregate due to the superior properties of rubberized SCC compared to the one that replaced coarse aggregate with waste-tire rubber.
3. Rubberized SCC's mechanical and durability qualities are substantially influenced by the morphology of CR particles and their replacement level.
4. Increased CR content considerably affected the different mechanical characteristics of rubberized SCC.
5. Reduction in SCC strength is expected using CR. This reduction could be something in the range of 30–40% for compressive strength, 20–35% for tensile strength, 15–35% for modulus of elasticity, and 15–30% for flexural strength, depending on the content of CR.
6. If properly designated, the ductility of SCC can be enhanced by adding CR as a partial alternative to sand.
7. The addition of CR can greatly enhance several of the durability attributes of rubberized SCC, including electrical resistivity, carbonation depth, and impact resistance. However, additional research into the various durability features of rubberized SCC is required.
8. Rubberized SCC could be utilized in several structural applications. However, to maintain appropriate mechanical and durability properties, the replacement level must be carefully designed. In addition, there is still a lack of studies on the performance of rubberized SCC.
9. The empirical models developed in this review article demonstrate that significant relations exist between the various mechanical parameters of rubberized SCC.
10. Microstructure analysis of the ITZ in rubberized SCC reveals that this weak zone between CR particles and the cement paste matrix can be improved by pretreating the CR particles using surface modifiers or admixtures.

10.2. Rcomendations for Further Research

Based on the research conducted in this study, the following recommendations need to be considered:

1. For future research, using CR in concrete production needs to be further investigated due to its potential use in structural concrete elements (walls, slabs, and columns) or even using rubber powder partially instead of sand or cement.
2. Incorporating CR into the SCC mixtures for structural components subjected to impact loading is strongly suggested. For this reason, it might be important to examine the durability properties of this material under different types of loading.
3. The ductility behavior of rubberized SCC is not completely explored and understood. Therefore, investigating fracture parameters of rubberized SCC could be a topic of interest.
4. There are a rare number of studies about the durability properties of rubberized SCC. Experimental studies on the various durability properties of rubberized SCC could answer the possibility of using this type of material in hazardous conditions.

Author Contributions: Y.J.Z.: writing—original draft, writing—review and editing, methodology, and formal analysis; K.H.Y.: writing—original draft, writing—review and editing, formal analysis, visualization, project administration, supervision, and resources. All authors have read and agreed to the published version of the manuscript.

Funding: This research received no external funding.

Institutional Review Board Statement: Not applicable.

Conflicts of Interest: The authors declare no conflict of interest.

References

1. European Ready Mixed Concrete Organization (ERMCO). *Ready-Mixed Concrete Industry Statistics Year 2015*; European Ready Mixed Concrete Organization (ERMCO): Bruxelles, Belgium, 2016.
2. Ivel, J.; Watson, R.; Abbassi, B.; Abu-Hamatteh, Z.S. Life cycle analysis of concrete and asphalt used in road pavements. *Environ. Eng. Res.* **2020**, *25*, 52–61. [CrossRef]
3. Coffetti, D.; Crotti, E.; Gazzaniga, G.; Carrara, M.; Pastore, T.; Coppola, L. Pathways towards sustainable concrete. *Cem. Concr. Res.* **2022**, *154*, 106718. [CrossRef]
4. Rahman, M.M.; Usman, M.; Al-Ghalib, A.A. Fundamental properties of rubber modified self-compacting concrete (RMSCC). *Constr. Build. Mater.* **2012**, *36*, 630–637. [CrossRef]
5. Liu, H.; Wang, X.; Jiao, Y.; Sha, T. Experimental investigation of the mechanical and durability properties of crumb rubber concrete. *Materials* **2016**, *9*, 172. [CrossRef] [PubMed]
6. Sofi, A. Effect of waste tyre rubber on mechanical and durability properties of concrete—A review. *Ain Shams Eng. J.* **2018**, *9*, 2691–2700. [CrossRef]
7. Alaloul, W.S.; Musarat, M.A.; Haruna, S.; Law, K.; Tayeh, B.A.; Rafiq, W.; Ayub, S. Mechanical properties of silica fume modified high-volume fly ash rubberized self-compacting concrete. *Sustainability* **2021**, *13*, 5571. [CrossRef]
8. Ganjian, E.; Khorami, M.; Maghsoudi, A.A. Scrap-tyre-rubber replacement for aggregate and filler in concrete. *Constr. Build. Mater.* **2009**, *23*, 1828–1836. [CrossRef]
9. Ling, T.-C. Prediction of density and compressive strength for rubberized concrete blocks. *Constr. Build. Mater.* **2011**, *25*, 4303–4306. [CrossRef]
10. Ling, T.-C.; Nor, H.M.; Hainin, M.R.; Chik, A.A. Laboratory performance of crumb rubber concrete block pavement. *Int. J. Pavement Eng.* **2009**, *10*, 361–374. [CrossRef]
11. Ling, T.-C.; Nor, H.M.; Hainin, M.R. Properties of crumb rubber concrete paving blocks with SBR latex. *Road Mater. Pavement Des.* **2009**, *10*, 213–222. [CrossRef]
12. Ling, T.-C.; Nor, H.M.; Hainin, M.R.; Lim, S.-K. Long-term strength of rubberised concrete paving blocks. *Proc. Inst. Civ. Eng. Mater.* **2010**, *163*, 19–26. [CrossRef]
13. Ling, T.-C. Effects of compaction method and rubber content on the properties of concrete paving blocks. *Constr. Build. Mater.* **2012**, *28*, 164–175. [CrossRef]
14. Terro, M.J. Properties of concrete made with recycled crushed glass at elevated temperatures. *Build. Environ.* **2006**, *41*, 633–639. [CrossRef]
15. Gesolu, M.; Güneyisi, E.; Gesoğlu, M.; Güneyisi, E. Permeability properties of self-compacting rubberized concretes. *Constr. Build. Mater.* **2011**, *25*, 3319–3326. [CrossRef]

16. Ismail, M.K.; Hassan, A.A. Use of metakaolin on enhancing the mechanical properties of self-consolidating concrete containing high percentages of crumb rubber. *J. Clean. Prod.* **2016**, *125*, 282–295. [CrossRef]
17. Younis, K.H.; Pilakoutas, K. Strength prediction model and methods for improving recycled aggregate concrete. *Constr. Build. Mater.* **2013**, *49*, 688–701. [CrossRef]
18. Ismail, M.K.; Hassan, A.A. Use of steel fibers to optimize self-consolidating concrete mixtures containing crumb rubber. *ACI Mater. J.* **2017**, *114*, 581–594. [CrossRef]
19. Lv, J.; Du, Q.; Zhou, T.; He, Z.; Li, K. Fresh and mechanical properties of self-compacting rubber lightweight aggregate concrete and corresponding mortar. *Adv. Mater. Sci. Eng.* **2019**, *2019*, 8372547. [CrossRef]
20. Akram, T.; Memon, S.A.; Obaid, H. Production of low cost self compacting concrete using bagasse ash. *Constr. Build. Mater.* **2009**, *23*, 703–712. [CrossRef]
21. Ahari, R.S.; Erdem, T.K.; Ramyar, K. Permeability properties of self-consolidating concrete containing various supplementary cementitious materials. *Constr. Build. Mater.* **2015**, *79*, 326–336. [CrossRef]
22. Melo, K.A.; Carneiro, A.M.P. Effect of metakaolin's finesses and content in self-consolidating concrete. *Constr. Build. Mater.* **2010**, *24*, 1529–1535. [CrossRef]
23. Boukendakdji, O.; Kenai, S.; Kadri, E.H.; Rouis, F. Effect of slag on the rheology of fresh self-compacted concrete. *Constr. Build. Mater.* **2009**, *23*, 2593–2598. [CrossRef]
24. Ozawa, K. High-Performance Concrete Based on the Durability Design of Concrete Structures. In Proceedings of the Second East Asia-Pacific Conference on Structural Engineering and Construction, Chlang Mai, Thailand, 11–13 January 1989.
25. Sonebi, M. Medium strength self-compacting concrete containing fly ash: Modelling using factorial experimental plans. *Cem. Concr. Res.* **2004**, *34*, 1199–1208. [CrossRef]
26. Kebaïli, O.; Mouret, M.; Arabi, N.; Cassagnabere, F. Adverse effect of the mass substitution of natural aggregates by air-dried recycled concrete aggregates on the self-compacting ability of concrete: Evidence and analysis through an example. *J. Clean. Prod.* **2015**, *87*, 752–761. [CrossRef]
27. Zhu, W.; Bartos, P.J.M. Permeation properties of self-compacting concrete. *Cem. Concr. Res.* **2003**, *33*, 921–926. [CrossRef]
28. Tuyan, M.; Mardani-Aghabaglou, A.; Ramyar, K. Freeze–thaw resistance, mechanical and transport properties of self-consolidating concrete incorporating coarse recycled concrete aggregate. *Mater. Des.* **2014**, *53*, 983–991. [CrossRef]
29. Okamura, H.; Ozawa, K.; Ouchi, M. Self-compacting concrete. *Struct. Concr.* **2000**, *1*, 3–17. [CrossRef]
30. Mallek, J.; Daoud, A.; Omikrine-Metalssi, O.; Loulizi, A. Performance of self-compacting rubberized concrete against carbonation and chloride penetration. *Struct. Concr.* **2021**, *22*, 2720–2735. [CrossRef]
31. Güneyisi, E.; Gesoglu, M.; Naji, N.; Ipek, S. Evaluation of the rheological behavior of fresh self-compacting rubberized concrete by using the herschel-bulkley and modified bingham models. *Arch. Civ. Mech. Eng.* **2016**, *16*, 9–19. [CrossRef]
32. Topçu, I.B.; Bilir, T. Experimental investigation of some fresh and hardened properties of rubberized self-compacting concrete. *Mater. Des.* **2009**, *30*, 3056–3065. [CrossRef]
33. Hilal, N.N. Hardened properties of self-compacting concrete with different crumb rubber size and content. *Int. J. Sustain. Built Environ.* **2017**, *6*, 191–206. [CrossRef]
34. Onuaguluchi, O.; Panesar, D.K. Hardened properties of concrete mixtures containing pre-coated crumb rubber and silica fume. *J. Clean. Prod.* **2014**, *82*, 125–131. [CrossRef]
35. Bateni, A.; Susnar, S.S.; Amirfazli, A.; Neumann, A.W. A high-accuracy polynomial fitting approach to determine contact angles. *Colloids Surfaces A Physicochem. Eng. Asp.* **2003**, *219*, 215–231. [CrossRef]
36. Emiroglu, M.; Kelestemur, M.H.; Yildiz, S. An investigation on ITZ microstructure of the concrete containing waste vehicle tire. In Proceedings of the 8th International Fracture Conference, Shanghai, China, 14–17 August 2007; Volume 7.
37. Thomas, B.S.; Gupta, R.C.; Mehra, P.; Kumar, S. Performance of high strength rubberized concrete in aggressive environment. *Constr. Build. Mater.* **2015**, *83*, 320–326. [CrossRef]
38. Karahan, O.; Özbay, E.; Hossain, K.M.A.; Lachemi, M.; Atiş, C.D. Fresh, mechanical, transport, and durability properties of self-consolidating rubberized concrete. *ACI Mater. J.* **2012**, *109*, 413–420. [CrossRef]
39. Bušić, R.; Miličević, I.; Šipoš, T.K.; Strukar, K. Recycled rubber as an aggregate replacement in self-compacting concrete-literature overview. *Materials* **2018**, *11*, 1729. [CrossRef]
40. Najim, K.B.; Hall, M.R. A review of the fresh/hardened properties and applications for plain (PRC) and self-compacting rubberised concrete (SCRC). *Constr. Build. Mater.* **2010**, *24*, 2043–2051. [CrossRef]
41. Ghosh, S.K.; Bera, D.K. Fundamental properties of self-compacting concrete utilizing waste rubber tires—A review. *Int. J. Res. Eng. Technol.* **2016**, *5*, 254–261. [CrossRef]
42. Jafari, M.; Mozhdehi, A.M.; Ganjali, A. Positive and negative influences of waste tires on self-compacting concrete: A summarized review. *J. Sci. Eng. Elites* **2020**, *5*, 40–62.
43. Bignozzi, M.C.; Sandrolini, F. Tyre rubber waste recycling in self-compacting concrete. *Cem. Concr. Res.* **2006**, *36*, 735–739. [CrossRef]
44. Garros, M.; Turatsinze, A.; Granju, J.-L. Effect of rubber aggregates from grinding of end-of-life tires on the properties of SCC. *Spec. Publ.* **2006**, *235*, 177–188.
45. Güneyisi, E. Fresh properties of self-compacting rubberized concrete incorporated with fly ash. *Mater. Struct.* **2010**, *43*, 1037–1048. [CrossRef]

46. Raj, B.; Ganesan, N.; Shashikala, A.P. Engineering properties of self-compacting rubberized concrete. *J. Reinf. Plast. Compos.* **2011**, *30*, 1923–1930. [CrossRef]
47. Khalilpasha, M.H.; Sadeghi-Nik, A.; Lotfi-Omran, O.; Kimiaeifard, K.; Amirpour-Molla, M. Sustainable Development Using Recyclable Rubber in Self-Compacting Concrete. In Proceedings of the Third International Conference on Construction in Developing Countries (Advancing Civil, Architectural and Construction Engineering & Management), Bangkok, Thailand, 4–6 July 2012; pp. 580–585.
48. Najim, K.B.; Hall, M.R. Mechanical and dynamic properties of self-compacting crumb rubber modified concrete. *Constr. Build. Mater.* **2012**, *27*, 521–530. [CrossRef]
49. Yung, W.H.; Yung, L.C.; Hua, L.H. A study of the durability properties of waste tire rubber applied to self-compacting concrete. *Constr. Build. Mater.* **2013**, *41*, 665–672. [CrossRef]
50. Ganesan, N.; Raj, J.B.; Shashikala, A.P. Flexural fatigue behavior of self compacting rubberized concrete. *Constr. Build. Mater.* **2013**, *44*, 7–14. [CrossRef]
51. Ismail, M.K.; De Grazia, M.T.; Hassan, A.A. Mechanical properties of self-consolidating rubberized concrete with different supplementary cementing materials. *J. Mater. Civ. Eng.* **2015**, *28*, 68–75. [CrossRef]
52. Khalil, E.; Abd-Elmohsen, M.; Anwar, A.M. Impact resistance of rubberized self-compacting concrete. *Water Sci.* **2015**, *29*, 45–53. [CrossRef]
53. Hesami, S.; Hikouei, I.S.; Emadi, S.A.A. Mechanical behavior of self-compacting concrete pavements incorporating recycled tire rubber crumb and reinforced with polypropylene fiber. *J. Clean. Prod.* **2016**, *133*, 228–234. [CrossRef]
54. Ismail, M.K.; Hassan, A.A. Impact resistance and mechanical properties of self-consolidating rubberized concrete reinforced with steel fibers. *J. Mater. Civ. Eng.* **2017**, *29*, 4016193. [CrossRef]
55. Younis, K.H.; Naji, H.S.; Najim, K.B. Rheological behavior of self-compacting concrete incorporating crumb rubber particles as fine aggregate. *Dev. Civ. Comput. Eng.* **2017**, *2017*, 62–74. [CrossRef]
56. AbdelAleem, B.H.; Hassan, A.A. Development of self-consolidating rubberized concrete incorporating silica fume. *Constr. Build. Mater.* **2018**, *161*, 389–397. [CrossRef]
57. Aslani, F.; Ma, G.; Wan, D.L.Y.; Le, V.X.T. Experimental investigation into rubber granules and their effects on the fresh and hardened properties of self-compacting concrete. *J. Clean. Prod.* **2018**, *172*, 1835–1847. [CrossRef]
58. Hamza, B.; Belkacem, M.; Said, K.; Walid, Y. Performance du béton autoplaçant à base de granulats en caoutchouc recyclés. *MATEC Web Conf.* **2018**, *149*, 1070. [CrossRef]
59. Aslani, F.; Kelin, J. Assessment and development of high-performance fibre-reinforced lightweight self-compacting concrete including recycled crumb rubber aggregates exposed to elevated temperatures. *J. Clean. Prod.* **2018**, *200*, 1009–1025. [CrossRef]
60. Si, R.; Wang, J.; Guo, S.; Dai, Q.; Han, S. Evaluation of laboratory performance of self-consolidating concrete with recycled tire rubber. *J. Clean. Prod.* **2018**, *180*, 823–831. [CrossRef]
61. AbdelAleem, B.H.; Ismail, M.K.; Hassan, A.A. The combined effect of crumb rubber and synthetic fibers on impact resistance of self-consolidating concrete. *Constr. Build. Mater.* **2018**, *162*, 816–829. [CrossRef]
62. Aslani, F.; Khan, M. Properties of high-performance self-compacting rubberized concrete exposed to high temperatures. *J. Mater. Civ. Eng.* **2019**, *31*, 4019040. [CrossRef]
63. Yang, G.; Chen, X.; Guo, S.; Xuan, W. Dynamic mechanical performance of self-compacting concrete containing crumb rubber under high strain rates. *KSCE J. Civ. Eng.* **2019**, *23*, 3669–3681. [CrossRef]
64. Li, N.; Long, G.; Ma, C.; Fu, Q.; Zeng, X.; Ma, K.; Xie, Y.; Luo, B. Properties of self-compacting concrete (SCC) with recycled tire rubber aggregate: A comprehensive study. *J. Clean. Prod.* **2019**, *236*, 117707. [CrossRef]
65. Cemalgil, S.; Etli, S. Effects of specimen size on the compressive strength of rubber modified self-compacting concrete. *Int. J. Pure Appl. Sci.* **2020**, *6*, 118–129. [CrossRef]
66. Angelin, A.F.; Cecche Lintz, R.C.; Osório, W.R.; Gachet, L.A. Evaluation of efficiency factor of a self-compacting lightweight concrete with rubber and expanded clay contents. *Constr. Build. Mater.* **2020**, *257*, 119573. [CrossRef]
67. Valizadeh, A.; Hamidi, F.; Aslani, F.; Shaikh, F.U.A. The Effect of Specimen Geometry on the Compressive and Tensile Strengths of Self-Compacting Rubberised Concrete Containing Waste Rubber Granules. *Structures* **2020**, *27*, 1646–1659. [CrossRef]
68. Tian, L.; Qiu, L.; Li, J.; Yang, Y. Experimental study of waste tire rubber, wood-plastic particles and shale ceramsite on the performance of self-compacting concrete. *J. Renew. Mater.* **2020**, *8*, 153–170. [CrossRef]
69. Zaouai, S.; Tafraoui, A.; Makani, A.; Benmerioul, F. Hardened and transfer properties of self-compacting concretes containing pre-coated rubber aggregates with crushed dune sand. *J. Rubber Res.* **2020**, *23*, 5–12. [CrossRef]
70. Liu, Z.; Chen, X.; Wu, P.; Cheng, X. Investigation on micro-structure of self-compacting concrete modified by recycled grinded tire rubber based on x-ray computed tomography technology. *J. Clean. Prod.* **2021**, *290*, 125838. [CrossRef]
71. Chen, C.; Chen, X.; Zhang, J. Experimental study on flexural fatigue behavior of self-compacting concrete with waste tire rubber. *Mech. Adv. Mater. Struct.* **2021**, *28*, 1691–1702. [CrossRef]
72. Rahim, N.I.; Mohammed, B.S.; Abdulkadir, I.; Dahim, M. Effect of crumb rubber, fly ash, and nanosilica on the surface methodology. *Materials* **2022**, *15*, 1501. [CrossRef]
73. Kelechi, S.E.; Adamu, M.; Mohammed, A.; Ibrahim, Y.E.; Obianyo, I.I. Durability performance of self-compacting concrete containing crumb rubber, fly ash and calcium carbide waste. *Materials* **2022**, *15*, 488. [CrossRef]

74. Qin, J.-L.; Qiao, W.-G.; Lin, D.-G.; Zhang, S.; Wang, J.-Y. Mechanical properties and numerical analyses of basalt fiber crumb rubber mortars in soft rock roadways. *Adv. Civ. Eng.* **2019**, *2019*, 5159094. [CrossRef]
75. Meddah, M.S. Recycled aggregates in concrete production: Engineering properties and environmental impact. *MATEC Web Conf.* **2017**, *101*, 5021. [CrossRef]
76. Fořt, J.; Kobetičová, K.; Böhm, M.; Podlesný, J.; Jelínková, V.; Vachtlová, M.; Bureš, F.; Černý, R. Environmental consequences of rubber crumb application: Soil and water pollution. *Polymers* **2022**, *14*, 1416. [CrossRef] [PubMed]
77. Jedidi, M.; Benjeddou, O. Crumb rubber effect on acoustic properties of self- consolidating concrete. *Int. J. Therm. Environ. Eng.* **2014**, *8*, 69–76. [CrossRef]
78. Sherwood, P.T.; TRL. The Use of Waste and Recycled Materials in Roads. In Proceedings of the Institution of Civil Engineers-Transport; Thomas Telford-ICE Virtual Library: London, UK, 1995. Volume 111. pp. 116–124.
79. Loderer, C.; Partl, M.N.; Poulikakos, L.D. Effect of crumb rubber production technology on performance of modified bitumen. *Constr. Build. Mater.* **2018**, *191*, 1159–1171. [CrossRef]
80. Hernandez-Olivares, F.; Barluenga, G.; Bollati, M.; Witoszek, B. Static and dynamic behaviour of recycled tyre rubber-filled concrete. *Cem. Concr. Res.* **2002**, *32*, 1587–1596. [CrossRef]
81. Huang, B.; Li, G.; Pang, S.-S.; Eggers, J. Investigation into waste tire rubber-filled concrete. *J. Mater. Civ. Eng.* **2004**, *16*, 187–194. [CrossRef]
82. Duarte, A.P.C.; Silva, B.A.; Silvestre, N.; De Brito, J.; Júlio, E. Mechanical characterization of rubberized concrete using an image-processing/XFEM coupled procedure. *Compos. Part B Eng.* **2015**, *78*, 214–226. [CrossRef]
83. Rahmani, E.; Dehestani, M.; Beygi, M.H.A.; Allahyari, H.; Nikbin, I.M. On the mechanical properties of concrete containing waste PET particles. *Constr. Build. Mater.* **2013**, *47*, 1302–1308. [CrossRef]
84. Emiroglu, M.; Yildiz, S.; Kelestemur, M.H. An investigation on its microstructure of the concrete containing waste vehicle tire. *Comput. Concr.* **2008**, *5*, 503–508. [CrossRef]
85. Mohammed, B.H.; Sherwani, A.F.H.; Faraj, R.H.; Qadir, H.H.; Younis, K.H. Mechanical properties and ductility behavior of ultra-high performance fiber reinforced concretes: Effect of low water-to-binder ratios and micro glass fibers. *Ain Shams Eng. J.* **2021**, *12*, 1557–1567. [CrossRef]
86. Moustafa, A.; ElGawady, M.A. Dynamic properties of high strength rubberized concrete. *ACI Spec. Publ* **2017**, *314*, 1–22.
87. Gesoğlu, M.; Güneyisi, E.; Khoshnaw, G.; Ipek, S. Investigating properties of pervious concretes containing waste tire rubbers. *Constr. Build. Mater.* **2014**, *63*, 206–213. [CrossRef]
88. Qadir, H.H.; Faraj, R.H.; Sherwani, A.F.H.; Mohammed, B.H.; Younis, K.H. Mechanical properties and fracture parameters of ultra high performance steel fiber reinforced concrete composites made with extremely low water per binder ratios. *SN Appl. Sci.* **2020**, *2*, 1594. [CrossRef]
89. Vadivel, T.S.; Thenmozhi, R.; Doddurani, M. Experimental behaviour of waste tyre rubber aggregate concrete under impact loading. *Iran. J. Sci. Technol. Trans. Civ. Eng.* **2014**, *38*, 251.
90. Mohammed, B.S.; Azmi, N.J.; Abdullahi, M. Evaluation of rubbercrete based on ultrasonic pulse velocity and rebound hammer tests. *Constr. Build. Mater.* **2011**, *25*, 1388–1397. [CrossRef]
91. Segre, N.; Joekes, I. Use of tire rubber particles as addition to cement paste. *Cem. Concr. Res.* **2000**, *30*, 1421–1425. [CrossRef]
92. Fu, Q.; Xie, Y.J.; Long, G. Study on capillary water absorption properties of rubberized self-compacting concrete. *J. Build. Mater.* **2015**, *18*, 17–23. [CrossRef]
93. Paul, J. Management of used or scrap tyres. *Encycl. Polym. Sci. Eng.* **1985**, *14*, 787–802.
94. Zaouai, S.; Makani, A.; Tafraoui, A.; Benmerioul, F. Optimization and mechanical characterization of self-compacting concrete incorporating rubber aggregates. *Mater.Sci. Eng.* **2016**, *17*, 817–829.
95. Bravo, M.; De Brito, J. Concrete made with used tyre aggregate: Durability-related performance. *J. Clean. Prod.* **2012**, *25*, 42–50. [CrossRef]
96. Elkey, W.; Sellevold, E.J. Electrical Resistivity of Concrete. *Concr. Int.* **1995**, *37*, 41–46.
97. Hornbostel, K.; Larsen, C.K.; Geiker, M.R. Relationship between concrete resistivity and corrosion rate—A literature review. *Cem. Concr. Compos.* **2013**, *39*, 60–72. [CrossRef]
98. Kaewunruen, S.; Meesit, R. Sensitivity of crumb rubber particle sizes on electrical resistance of rubberised concrete. *Cogent Eng.* **2016**, *3*, 1126937. [CrossRef]
99. Mohammed, B.S.; Hossain, K.M.A.; Swee, J.T.E.; Wong, G.; Abdullahi, M. Properties of crumb rubber hollow concrete block. *J. Clean. Prod.* **2012**, *23*, 57–67. [CrossRef]
100. Luhar, S.; Luhar, I.; Nicolaides, D.; Gupta, R. Durability performance evaluation of rubberized geopolymer concrete. *Sustainability* **2021**, *13*, 5969. [CrossRef]
101. Al-Tayeb, M.M.; Abu Bakar, B.H.; Akil, H.M.; Ismail, H. Performance of rubberized and hybrid rubberized concrete structures under static and impact load conditions. *Exp. Mech.* **2013**, *53*, 377–384. [CrossRef]
102. Pedro, D.; de Brito, J.; Veiga, R. Mortars made with fine granulate from shredded tires. *J. Mater. Civ. Eng.* **2013**, *25*, 519–529. [CrossRef]
103. Guo, S.; Dai, Q.; Si, R.; Sun, X.; Lu, C. Evaluation of properties and performance of rubber-modified concrete for recycling of waste scrap tire. *J. Clean. Prod.* **2017**, *148*, 681–689. [CrossRef]

104. Chen, Z.; Li, L.; Xiong, Z. Investigation on the interfacial behaviour between the rubber-cement matrix of the rubberized concrete. *J. Clean. Prod.* **2019**, *209*, 1354–1364. [CrossRef]
105. Onuaguluchi, O. Effects of surface pre-coating and silica fume on crumb rubber-cement matrix interface and cement mortar properties. *J. Clean. Prod.* **2015**, *104*, 339–345. [CrossRef]
106. Najim, K.B.; Hall, M.R. Crumb rubber aggregate coatings/pre-treatments and their effects on interfacial bonding, air entrapment and fracture toughness in self-compacting rubberised concrete (SCRC). *Mater. Struct.* **2013**, *46*, 2029–2043. [CrossRef]

MDPI
St. Alban-Anlage 66
4052 Basel
Switzerland
www.mdpi.com

Sustainability Editorial Office
E-mail: sustainability@mdpi.com
www.mdpi.com/journal/sustainability

www.ingramcontent.com/pod-product-compliance
Lightning Source LLC
Chambersburg PA
CBHW061618220926
51850CB00005B/541
* 9 7 8 3 0 3 6 5 9 4 0 3 3 *